Lecture Notes in Mathematics   2213

More information about this series at http://www.springer.com/series/304

Chaire Jean-Morlet

The CIRM Jean-Morlet Series is a collection of scientific publications centering on the themes developed by successive holders of the Jean Morlet Chair.

This chair has been hosted by the *Centre International de Rencontres Mathématiques* (CIRM, Luminy, France) since its creation in 2013. The Chair is named in honour of Jean Morlet (1931–2007). He was an engineer at the French oil company Elf (now Total) and, together with the physicist Alex Grossman, conducted pioneering work in wavelet analysis. This theory has since become a building block of modern mathematics. It was at CIRM that they met on several occasions, and the center then played host to some of the key conferences in this field.

Appointments to the *Jean-Morlet* Chair are made to worldclass researchers based outside France and who work in collaboration with local project leaders in order to conduct original and ambitious scientific programs.The Chair is supported financially by CIRM, Aix-Marseille Université and the City of Marseille.

A key feature of the Chair is that it does not focus solely on the research themes developed by Jean Morlet. The idea is to support the freedom of pioneers in mathematical sciences and to nurture the enthusiasm that comes from opening new avenues of research.

CIRM: a beacon for international cooperation

Situated at the heart of the *Parc des Calanques*, an area of outstanding natural beauty, CIRM is one of the largest conference centers dedicated to mathematical and related sciences in the world, with close to 3500 visitors per year. Jointly supervised by SMF (the French Mathematical Society) and CNRS (French National Center for Scientific Research), CIRM has been a hub for international research in mathematics since 1981. CIRM's *raison d'être* is to be a venue that fosters exchanges, pioneering research in mathematics in interaction with other sciences and the dissemination of knowledge to the younger scientific community

www.chairejeanmorlet.com
www.cirm-math.fr

Sébastien Ferenczi • Joanna Kułaga-Przymus •
Mariusz Lemańczyk

Editors

Ergodic Theory and Dynamical Systems in their Interactions with Arithmetics and Combinatorics

CIRM Jean-Morlet Chair, Fall 2016

Springer

Editors
Sébastien Ferenczi
CNRS UMR 7373
Institut de Mathématiques de Marseille
Marseille, France

Joanna Kułaga-Przymus
Faculty of Mathematics and Computer
Science
Nicolaus Copernicus University
Toruń, Poland

Mariusz Lemańczyk
Faculty of Mathematics and Computer
Science
Nicolaus Copernicus University
Toruń, Poland

ISSN 0075-8434   ISSN 1617-9692   (electronic)
Lecture Notes in Mathematics
ISBN 978-3-319-74907-5   ISBN 978-3-319-74908-2   (eBook)
https://doi.org/10.1007/978-3-319-74908-2

Library of Congress Control Number: 2018940898

Mathematics Subject Classification (2010): 37-XX, 11-XX, 5-XX, 51-XX

Printed on acid-free paper

This Springer imprint is published by the registered company Springer International Publishing AG part
of Springer Nature.
The registered company address is: Gewerbestrasse 11, 6330 Cham, Switzerland

Foreword

The interaction between number theory and ergodic theory can be traced to the birth of the latter with Birkhoff's pointwise ergodic theorem. The early applications were naturally concerned with typical behavior, for example in the metrical theory of diophantine approximation. Most number theoretic questions which can be connected to ergodic theory are concerned with the dynamics of specific orbits or systems constructed from an arithmetic or combinatorial input, and it is with the classification or determination of the basic properties of the possible systems that can arise that the interaction becomes powerful. Starting with Furstenberg's introduction of such concepts as unique ergodicity, disjointness of dynamical systems, and nonconventional ergodic averages, and thanks to advances by many ergodic/number theorists, there is by now a body of striking applications. Homogeneous dynamics takes place on parameter spaces of arithmetic objects, and as a consequence, rigidity theorems such as that of Ratner for unipotent orbits become powerful tools which underlie many of the most striking applications in homogeneous dynamics. There have also been major advances and arithmetic applications in various nonhomogeneous dynamical settings. For example it turns out that Vinogradov's bilinear method in the study of sums over primes for a sequence which is an observable in a dynamical system is intimately connected with the Birkhoff sums for joinings of the system with itself. An example exploiting this is the proof by Mauduit and Rivat of a conjecture of Gelfond about the distribution of the parity of the sum of the binary digits of prime numbers. As far as combinatorial/additive number theory, the path developed by Furstenberg in his proof of Szemeredi's theorem on arithmetic progressions in sets of positive density is at the center of this well-developed modern tool from ergodic theory.

The above are just a small sample (and biased to my taste and knowledge) of what is today a thriving interaction between ergodic theory and number theory. The well-timed 2016 fall semester activity at CIRM (Luminy) focused on this theme, with the aim of exposing these interactions and the theories that underlie the progress and the latest developments, as well advancing them. From my own experience and accounts by others, the minicourses and the workshops and seminars were a great success and there were a number of exciting new developments.

Fortunately many of the experts who are responsible for this success prepared and expanded their presentations for this volume. The result is an instructive and insightful account of the basic techniques from ergodic theory and number theory that have facilitated the recent developments. There are also excellent survey papers that bring the reader up to forefront of the latest developments and open problems in this fast moving area.

Princeton, NJ, USA   Peter Sarnak
October 3, 2017

Preface

This volume consists of minicourses notes, survey, research/survey, and research articles that have arisen as an outcome of workshops, research in pairs, and other scientific work held under the auspices of the Jean Morlet Chair at CIRM between August 1, 2016 and January 31, 2017. The semester had a substantial core support and funding by CIRM, Aix-Marseille University, and the city of Marseille. Additionally, it was supported by the LABEX Archimède, and the ANR grants of Christian Mauduit (Aix-Marseille University) and Joël Rivat (Aix-Marseille University).

The minicourses were those given in the framework of the doctoral school *Applications of Ergodic Theory in Number Theory* organized by Sébastien Ferenczi (Aix-Marseille University), Joanna Kułaga-Przymus (Nicolaus Copernicus University Toruń and Aix-Marseille University), Mariusz Lemańczyk (Nicolaus Copernicus University Toruń), and Serge Troubetzkoy (Aix-Marseille University). The main aim of this school was, on one hand, to provide participants with modern methods of ergodic theory and topological dynamics oriented toward applications in number theory and combinatorics, and, on the other hand, to present them with a broad spectrum of number theory problems that can be treated with the use of such tools. These tasks were realized in four minicourses by Vitaly Bergelson (Ohio State University), "Mutually enriching connections between ergodic theory and combinatorics," Manfred Einsiedler (ETH Zürich), "Equidistribution on homogeneous spaces, a bridge between dynamics and number theory," Carlos Matheus Silva Santos (CNRS - Université Paris 13), "The Lagrange and Markov spectra from the dynamical point of view," and Joël Rivat "Introduction to analytic number theory."

The main conference *Ergodic Theory and its Connections with Arithmetic and Combinatorics* was organized by Julien Cassaigne (Aix-Marseille University), Sébastien Ferenczi, Pascal Hubert (Aix-Marseille University), Joanna Kułaga-Przymus, Mariusz Lemańczyk with the scientific committee consisting of Artur Avila (University Paris Diderot and IMPA, Rio de Janeiro), Vitaly Bergelson, Mandred Einsiedler, Hillel Furstenberg (The Hebrew University of Jerusalem), Anatole Katok (Penn State University), Christian Mauduit, Imre Ruzsa (Alfred Rényi Institute Budapest), and Peter Sarnak (IAS Princeton). The conference was aimed at interactions between ergodic theory and dynamical systems and number theory. Its main subjects were disjointness in ergodic theory and randomness in number theory, ergodic theory and combinatorial number theory, and homogenous dynamics and its applications.

Important events of the semester were two smaller specialized workshops. The first one *Ergodic Theory and Möbius Disjointness* was organized by Sébastien Ferenczi, Joanna Kułaga-Przymus, Mariusz Lemańczyk, Christian Mauduit, and Joël Rivat. The meeting focused on the recent progress on Sarnak's conjecture on Möbius disjointness: methods, results, and the feedback in ergodic theory. The second one *Spectral Theory of Dynamical Systems and Related Topics* was organized by Alexander Bufetov (Aix-Marseille University), Sébastien Ferenczi, Joanna Kułaga-Przymus, Mariusz Lemańczyk, and Arnaldo Nogueira (Aix-Marseille University). The meeting was aimed at the recent progress in the spectral theory and joinings of dynamical systems, especially, in the recent spectacular progress toward the solutions of some open classical problems of ergodic theory: Rokhlin problem on mixing of all orders, stability of spectral properties under smooth changes for the parabolic systems, the Banach problem on the existence of dynamical systems with simple Lebesgue spectrum, and the problem of spectral multiplicity.

The scientific part of the semester was completed by two research in pairs: *Dynamical Properties of Systems Determined by Free Points in Lattices* and *On the Stability of Möbius Disjointness in Topological Models* and a special program of invitations with participation of Michael Baake (University of Bielefeld), Jean-Pierre Conze (University of Rennes 1), Alexandre Danilenko (Institute of Low Temperature, Kharkov), Christian Huck (University of Bielefeld), Joanna Kułaga-Przymus, El Houcein El Abdalaoui (University of Rouen), Mariusz Lemańczyk and Thierry de la Rue (University of Rouen).

The contents of this volume are as follows. It begins with Part I which is entirely the course.

- Joël Rivat, *Bases of Analytic Number Theory*. Among other aspects, the course contains a presentation of the main properties of the Riemann ζ function with a generous introduction to the theory of Dirichlet series. Large sieve method together with a beautiful application to Twin Prime conjecture and deep relations with the theory of multiplicative functions are dealt with. We find also a detailed presentation of Vinogradov's method of major and minor arcs, together with a deep analysis of sums of type I and II which are of great use in current research. The final chapter is devoted to the van der Corput method of computing and estimating trigonometric sums.

Part II of the volume consists of articles devoted to interactions between arithmetic and dynamics. They are all of research/survey/course type:

- M. Baake, *A Brief Guide to Reversing and Extended Symmetries of Dynamical Systems* is a survey which presents the basic notions and reviews facts concerning the reversing symmetry of dynamical systems, focusing on systems (subshifts) of algebraic and number-theoretic origin.

- M. Einsiedler, M. Luethi, *Kloosterman Sums, Disjointness, and Equidistribution* summarizes the aforementioned minicourse of M. Einsiedler. Various applications of Kloosterman sums are shown: equidistribution properties of sparse subsets of horocycle orbits in the modular case, disjointness results on the torus, mixing properties.
- S. Ferenczi, J. Kułaga-Przymus, M. Lemańczyk, *Sarnak's Conjecture: What's New?* is a survey presenting an exhaustive list of methods and results concerning the problem of Möbius disjointness. Some new results are also included.
- A. Gomilko, D. Kwietniak, M. Lemańczyk, *Sarnak's Conjecture Implies the Chowla Conjecture Along a Subsequence* proves this elementary but new result.
- C. Huck, *On the Logarithmic Probability That a Random Integral Ideal Is A-free* is an article which extends a theorem of Davenport and Erdös on sets of multiples with integers to the existence of logarithmic density for unions of integral ideals in number fields.
- C. Matheus, *The Lagrange and Markov Spectra from the Dynamical Point of View* summarizes the aforementioned minicourse of C. Matheus. The notes introduce the world of Lagrange and Markov spectra with a special focus on the proof of Moreira's theorem on the intricate structure of such spectra.
- O. Ramaré, *On the Missing Log Factor* is a "journey" around the Axer-Landau Equivalence Theorem of the Prime Number Theorem and properties of the Möbius and von Mangoldt functions.
- O. Ramaré, *Chowla's Conjecture: From the Liouville Function to the Möbius Function* is a note focusing on proofs of implications between various versions of the Chowla conjecture in which we use either Liouville or Möbius function.

Part III of the volume consists of three articles of survey or research/survey type from selected topics in dynamics:

- T. Adams, C. Silva, *Weak Mixing for Infinite Measure Invertible Transformations* surveys and studies mixing properties of transformations preserving infinite measure.
- E. Glasner, M. Megrelishvili, *More on Tame Dynamical Systems* surveys and amplifies old results in (topological) tame dynamical systems, proves some new results, and provides new examples of tame systems.
- K. Inoue, H. Nakada, *A Piecewise Rotation of the Circle, IPR Maps and Their Connection with Translation Surfaces* reviews a construction of translation surfaces in terms of a continuous version of the cutting-and-stacking systems and proves a new result of realization of Rauzy classes.

Marseille, France   Sébastien Ferenczi
Toruń, Poland   Joanna Kułaga-Przymus
Toruń, Poland   Mariusz Lemańczyk

Contents

Contributors

Terrence Adams U.S. Government, Ft. Meade, MD, USA

Michael Baake Faculty of Mathematics, Universität Bielefeld, Bielefeld, Germany

Manfred Einsiedler Departement Mathematik, ETH Zürich, Rämistrasse, Zürich, Switzerland

Sébastien Ferenczi Aix Marseille Université, CNRS, Centrale Marseille, Institut de Mathématiques de Marseille, I2M – UMR 7373, Marseille, France

Eli Glasner Department of Mathematics, Tel-Aviv University, Ramat Aviv, Israel

Alexander Gomilko Faculty of Mathematics and Computer Science, Nicolaus Copernicus University, Toruń, Poland

Christian Huck Fakultät für Mathematik, Universität Bielefeld, Bielefeld, Germany

Kae Inoue Faculty of Pharmacy, Keio University, Tokyo, Japan

Joanna Kułaga-Przymus Faculty of Mathematics and Computer Science, Nicolaus Copernicus University, Toruń, Poland

Aix Marseille Université, CNRS, Centrale Marseille, Institut de Mathématiques de Marseille, I2M – UMR 7373, Marseille, France

Dominik Kwietniak Faculty of Mathematics and Computer Science, Jagiellonian University in Kraków, Kraków, Poland

Institute of Mathematics, Federal University of Rio de Janeiro, Rio de Janeiro, Brazil

Mariusz Lemańczyk Faculty of Mathematics and Computer Science, Nicolaus Copernicus University, Toruń, Poland

Manuel Luethi Departement Mathematik, ETH Zürich, Zürich, Switzerland

Carlos Matheus Université Paris 13, Sorbonne Paris Cité, LAGA, CNRS (UMR 7539), Villetaneuse, France

Michael Megrelishvili Department of Mathematics, Bar-Ilan University, Ramat-Gan, Israel

Hitoshi Nakada Department of Mathematics, Keio University, Yokohama, Japan

Olivier Ramaré Aix Marseille Université, CNRS, Centrale Marseille, Institut de Mathématiques de Marseille, I2M – UMR 7373, Marseille, France

Joël Rivat Aix Marseille Université, CNRS, Centrale Marseille, Institut de Mathématiques de Marseille, I2M – UMR 7373, Marseille, France

Cesar E. Silva Department of Mathematics, Williams College, Williamstown, MA, USA

Part I
Bases of Analytic Number Theory

Joël Rivat

These lecture notes were written in French in 2000 with no plan to be published, and I used them several times to give lectures. Many thanks to Sébastien Ferenczi for the English translation. They should not be compared with reference books like Tenenbaum [6], Iwaniec and Kowalski [3] and Montgomery and Vaughan [4], but an invitation to read these books.

The zeta function part owes much to Davenport's book [1]. The chapter on the large sieve uses the complete works of Selberg [5]. Our upper bounds on exponential sums are adapted from Graham and Kolesnik [2], with an effort to make the constants explicit but without attempting at optimality; they were then used later by Tenenbaum [6]. We think that the constant factor 16 instead of $2\pi^2$ in the Bombieri-Iwaniec inequality (Theorem 6.38) is new.

September 29, 2017

References

1. H. Davenport, *Multiplicative Number Theory*. Graduate Texts in Mathematics, 3rd edn., vol. 74 (Springer, New York, 2000). Revised and with a preface by Hugh L. Montgomery
2. S. Graham, G. Kolesnik, *Van der Corput's Method of Exponential Sums*. London Mathematical Society Lecture Note Series, vol. 126 (Cambridge University Press, Cambridge, 1991)
3. H. Iwaniec, E. Kowalski, *Analytic Number Theory*. American Mathematical Society Colloquium Publications, vol. 53 (American Mathematical Society, Providence, RI, 2004)

J. Rivat (✉)
Aix Marseille Université, CNRS, Centrale Marseille, Institut de Mathématiques de Marseille, I2M – UMR 7373, Marseille, France
e-mail: joel.rivat@univ-amu.fr

4. H.L. Montgomery, R.C. Vaughan, *Multiplicative Number Theory. I. Classical Theory*. Cambridge Studies in Advanced Mathematics, vol. 97 (Cambridge University Press, Cambridge, 2007)
5. A. Selberg, *Collected Papers*, vol. II (Springer, Berlin, 1991). With a foreword by K. Chandrasekharan
6. G. Tenenbaum, *Introduction à la théorie analytique et probabiliste des nombres*, quatrième edn. (Belin, Paris, 2015)

Chapter 1
Prime Numbers

Joël Rivat

1.1 Historical Notes

We denote by $\pi(x)$ the number of prime numbers smaller than or equal to x, and decide that p will always denote a prime number.

Euclides (third–second century BC) was the first to prove that

$$\pi(x) \to +\infty, \ x \to +\infty,$$

by noticing that, if there were only finitely many prime numbers, then for n large enough all would be smaller than n, and by considering

$$N := n! + 1$$

we could build an integer N which is divisible by no d such that $2 \leqslant d \leqslant n$ (the remainder of the division is 1), so that N would have only prime factors larger than n. Contradiction.

Eratosthenes (second–first century BC) devised an excellent method to compile a full finite list of prime numbers. To delete the multiples of p up to x requires $\lfloor x/p \rfloor$ operations, hence building the table of prime numbers up to x does not exceed

$$x + x \sum_{p \leqslant x} \frac{1}{p} \sim x \log \log x, \ x \to +\infty,$$

J. Rivat (✉)
Aix Marseille Université, CNRS, Centrale Marseille, Institut de Mathématiques de Marseille,
I2M – UMR 7373, Marseille, France
e-mail: joel.rivat@univ-amu.fr

© Springer International Publishing AG, part of Springer Nature 2018
S. Ferenczi et al. (eds.), *Ergodic Theory and Dynamical Systems in their Interactions with Arithmetics and Combinatorics*, Lecture Notes in Mathematics 2213, https://doi.org/10.1007/978-3-319-74908-2_1

operations, to find about $x/\log x$ prime numbers, which is optimal up to a factor $O(\log x \log\log x)$.

Legendre (1752–1833) was the first to propose a reasonable conjecture about the size of $\pi(x)$:

$$\pi(x) \approx \frac{x}{\log x - 1.08\ldots},$$

but it was Gauß (1777–1855) who formulated the "good" conjecture:

$$\pi(x) \approx \mathrm{li}(x) := \int_2^x \frac{dt}{\log t}.$$

The first proven results on the behavior of $\pi(x)$ at infinity are the work of Tchebychev (1821–1894) who was first able to prove (in 1851) that

$$\liminf \frac{\pi(x)}{\mathrm{li}(x)} \leqslant 1 \leqslant \limsup \frac{\pi(x)}{\mathrm{li}(x)},$$

so that the limit, if it exists, is 1. The following year (1852), Tchebychev got the inequalities

$$(0.92\ldots)\frac{x}{\log x} < \pi(x) < (1.105\ldots)\frac{x}{\log x},$$

for x large enough.

Mertens (1840–1927) made a substantial progress in 1874 by showing that

$$\sum_{p \leqslant x} \frac{1}{p} = \log\log x + A + O((\log x)^{-1}).$$

Before stating modern results about the repartition of prime numbers, we need to introduce the von Mangoldt arithmetic function, defined by

$$\Lambda(n) = \begin{cases} \log p & \text{if } n \text{ is a power of } p \text{ prime,} \\ 0 & \text{otherwise.} \end{cases}$$

It constitutes a convenient alternative to the indicator function of prime numbers, and will be used frequently, as well in statements as in proofs. Its main quality is to satisfy the convolution formula

$$\sum_{d \mid n} \Lambda(d) = \log n,$$

which makes it much more pleasant to handle without introducing an important distorsion, as a simple summing by parts allows to show that $\pi(x) \sim \mathrm{li}(x)$, $x \to +\infty$ if and only if

$$\psi(x) := \sum_{n \leqslant x} \Lambda(n) \sim x, \ x \to +\infty.$$

Euler (1707–1783) had got that for all real $s > 1$,

$$\zeta(s) := \sum_{n=1}^{+\infty} \frac{1}{n^s} = \prod_p (1 - p^{-s})^{-1},$$

thus establishing a link between prime numbers and the most famous among the series of Dirichlet, but it was Riemann (1860), by considering this function for complex values of s, who opened a decisive way by formulating a few remarkable conjectures about the repartition of zeros of ζ. The famous Riemann hypothesis locates all non real zeros of ζ on the vertical straight line of first coordinate $\sigma = \frac{1}{2}$ (the "critical line"). Let us mention also the following explicit formula, a conjecture proved by von Mangoldt in 1895:

$$\psi_0(x) := \tfrac{1}{2}(\psi(x^+) + \psi(x^-)) = x - \lim_{T \to +\infty} \sum_{\substack{\rho \\ |\Im\rho| \leqslant T}} \frac{x^\rho}{\rho} - \frac{\zeta'(0)}{\zeta(0)} - \tfrac{1}{2} \log(1 - x^{-2}),$$

where the summation over ρ involves the non trivial zeros ρ of ζ with imaginary part satisfying $|\Im\rho| \leqslant T$.

We note that $\frac{\zeta'(0)}{\zeta(0)} = \log(2\pi)$.

Finally, Hadamard and de la Vallée Poussin showed independently in 1896 the famous prime numbers theorem:

$$\pi(x) \sim \mathrm{li}(x), \ x \to +\infty,$$

which was quickly made more precise under the form

$$\pi(x) = \mathrm{li}(x) + O(x \exp(-c\sqrt{\log x}))$$

for some $c > 0$.

This result did not know much improvement during the twentieth century, as the best error term, due to Vinogradov and Korobov (1958) after a very difficult proof, remains hardly better:

$$\pi(x) = \mathrm{li}(x) + O(x \exp(-c(\log x)^{3/5}/(\log\log x)^{1/5})),$$

while the Riemann hypothesis would imply

$$\pi(x) = \mathrm{li}(x) + O(x^{1/2} \log x).$$

1.2 The von Mangoldt Function

Definition 1.1 The arithmetic von Mangoldt function is defined by

$$\Lambda(n) = \begin{cases} \log p & \text{if } n \text{ is a power of } p \text{ (prime),} \\ 0 & \text{sinon.} \end{cases}$$

Proposition 1.2 *For every integer* $n \geqslant 1$,

$$\sum_{d \mid n} \Lambda(d) = \log n.$$

Proof It is easy to check this formula by writing $n = p_1^{a_1} \cdots p_k^{a_k}$, but it is more enlightening, for proving this formula and similar identities, to compare the coefficients of two equal Dirichlet series. By Euler's formula, for all real $s > 1$,

$$\zeta(s) := \prod_p (1 - p^{-s})^{-1},$$

thus on one side

$$\log \zeta(s) = -\sum_p \log(1 - p^{-s}) = \sum_p \sum_{k=1}^{+\infty} \frac{p^{-ks}}{k}$$

and by making a derivation under the sum,

$$-\frac{\zeta'(s)}{\zeta(s)} = \sum_p \sum_{k=1}^{+\infty} p^{-ks} \log p = \sum_{n=1}^{+\infty} \Lambda(n) n^{-s}$$

while on the other side

$$\left(\sum_{n=1}^{+\infty} \Lambda(n) n^{-s} \right) \left(\sum_{n=1}^{+\infty} n^{-s} \right) = -\frac{\zeta'(s)}{\zeta(s)} \zeta(s) = \sum_{n=1}^{+\infty} n^{-s} \log n$$

and the comparison of coefficients gives the result.

1.3   The Tchebychev Inequalities

Theorem 1.3 *For $x \geqslant 2$, we have the inequalities*

$$x \log 2 + O(\log x) \leqslant \psi(x) \leqslant x \log 4 + O((\log x)^2).$$

Proof Summing on n the convolution formula on Λ, we get

$$T(x) := \sum_{n \leqslant x} \log n = \sum_{n \leqslant x} \sum_{d \mid n} \Lambda(d) = \sum_{md \leqslant x} \Lambda(d) = \sum_{d \leqslant x} \Lambda(d) \lfloor x/d \rfloor$$

and as the log function is concave, we get the bounds

$$x \log x - x = \int_0^x \log t \, dt \leqslant \sum_{n \leqslant x} \log n \leqslant \int_1^{x+1} \log t \, dt = (x+1) \log(x+1) - x$$

which gives a weak form of the Stirling formula:

$$\sum_{n \leqslant x} \log n = x \log x - x + O(\log x).$$

We consider the equality

$$T(x) - 2T(x/2) = \sum_{n \leqslant x} \Lambda(n) \left(\left\lfloor \frac{x}{n} \right\rfloor - 2 \left\lfloor \frac{x}{2n} \right\rfloor \right).$$

The left term is equal to $x \log 2 + O(\log x)$. The function $u \mapsto \lfloor u \rfloor - 2 \lfloor u/2 \rfloor$ is 2-periodic and satisfies

$$\lfloor u \rfloor - 2 \lfloor u/2 \rfloor = \begin{cases} 0 \text{ if } 0 \leqslant u < 1, \\ 1 \text{ if } 1 \leqslant u < 2. \end{cases}$$

Thus we have on one side

$$x \log 2 + O(\log x) \leqslant \sum_{n \leqslant x} \Lambda(n) = \psi(x)$$

and on the other side

$$x \log 2 + O(\log x) \geqslant \sum_{x/2 < n \leqslant x} \Lambda(n) = \psi(x) - \psi(x/2).$$

Take $K = \lfloor \log x / \log 2 \rfloor$. We have

$$\psi(x/2^k) - \psi(x/2^{k+1}) \leqslant \frac{x}{2^k} \log 2 + O(\log x), \quad k = 0, \ldots, K.$$

Adding all these inequalities, we get

$$\psi(x) \leqslant x(\log 2)\frac{1 - 2^{-K-1}}{1 - 2^{-1}} + O((\log x)^2) \leqslant 2x \log 2 + O((\log x)^2).$$

Corollary 1.4 *For $x \geqslant 2$, we have the inequalities*

$$\frac{x}{\log x} \log 2 + O(1) \leqslant \pi(x) \leqslant \frac{x}{\log x} \log 4 + O\left(\frac{x}{(\log x)^2}\right).$$

Proof We get the lower bound by observing that

$$x \log 2 + O(\log x) \leqslant \sum_{n \leqslant x} \Lambda(n) = \sum_{p \leqslant x} \left\lfloor \frac{\log x}{\log p} \right\rfloor \log p \leqslant \pi(x) \log x.$$

Taking

$$\theta(x) = \sum_{p \leqslant x} \log p$$

we have

$$\pi(x) - \pi(\sqrt{x}) = \sum_{\sqrt{x} < p \leqslant x} \frac{\log p}{\log p} = \int_{\sqrt{x}}^{x} \frac{1}{\log t} d\theta(t)$$

$$= \frac{\theta(x)}{\log x} - \frac{\theta(\sqrt{x})}{\log \sqrt{x}} + \int_{\sqrt{x}}^{x} \frac{\theta(t)}{t(\log t)^2} dt.$$

Thus, by deleting the negative term and observing that $\theta(t) \leqslant \psi(t)$, we get the upper bound

$$\pi(x) - \pi(\sqrt{x}) \leqslant \frac{\psi(x)}{\log x} + \int_{\sqrt{x}}^{x} \frac{\psi(t)}{t(\log t)^2} dt.$$

We use

$$\psi(x) \leqslant x \log 4 + O((\log x)^2)$$

to bound the first term, while for the integral we need only $\psi(t) = O(t)$ and $\log t \geqslant \frac{1}{2} \log x$, which allows to bound it by $O(x(\log x)^{-2})$.

1.4 The Mertens Theorems

Theorem 1.5 *For $x \geqslant 2$,*

$$\sum_{p \leqslant x} \frac{\log p}{p} = \log x + O(1).$$

Proof We have already proved that

$$T(x) = \sum_{d \leqslant x} \Lambda(d) \left\lfloor \frac{x}{d} \right\rfloor = x \log x + O(x).$$

The contribution to this sum of prime values of d is

$$\sum_{p \leqslant x} \lfloor x/p \rfloor \log p = x \sum_{p \leqslant x} \frac{\log p}{p} + O(x).$$

The contribution of other values of d is bounded by

$$\sum_{p \leqslant \sqrt{x}} (\log p) \sum_{k=2}^{\infty} \frac{x}{p^k} = x \sum_{p \leqslant \sqrt{x}} \frac{\log p}{p^2} \cdot \frac{1}{1 - \frac{1}{p}} = O(x).$$

thus we get the result after dividing by x.

Theorem 1.6 *There exists a constant A such that for $x \geqslant 2$,*

$$\sum_{p \leqslant x} \frac{1}{p} = \log \log x + A + O((\log x)^{-1}).$$

Proof Put

$$S(t) = \sum_{p \leqslant t} \frac{\log p}{p}.$$

We have

$$\sum_{p \leqslant x} \frac{1}{p} = \sum_{p \leqslant x} \frac{1}{\log p} \frac{\log p}{p} = \int_{2-}^{x^+} \frac{1}{\log t} \, dS(t)$$

$$= \frac{S(x)}{\log x} - \frac{S(2^-)}{\log 2} + \int_{2}^{x} S(t) \frac{dt}{t(\log t)^2}$$

We observe that $S(2^-) = 0$ and from Theorem 1.5 we have

$$\frac{S(x)}{\log x} = 1 + O((\log x)^{-1}).$$

The last integral is

$$\int_2^x \frac{dt}{t \log t} + \int_2^\infty (S(t) - \log t) \frac{dt}{t(\log t)^2} - \int_x^\infty (S(t) - \log t) \frac{dt}{t(\log t)^2}.$$

The integral on the right is bounded by

$$O\left(\int_x^\infty \frac{dt}{t(\log t)^2}\right) = O\left((\log x)^{-1}\right),$$

which implies also the convergence of the middle integral, and we have

$$\int_2^x \frac{dt}{t \log t} = \log \log x - \log \log 2.$$

By putting all the terms together we get the claimed result, with

$$A = 1 - \log \log 2 + \int_2^\infty (S(t) - \log t) \frac{dt}{t(\log t)^2}.$$

Theorem 1.7 (The Mertens Formula) *For $x \geqslant 2$,*

$$\prod_{p \leqslant x} (1 - 1/p)^{-1} = e^\gamma \log x + O(1),$$

where γ denotes the Euler constant.

Proof See for example Tenenbaum [1, p. 17].

Reference

1. G. Tenenbaum, *Introduction à la théorie analytique et probabiliste des nombres*, quatrième edn. (Belin, Paris, 2015)

Chapter 2
Arithmetic Functions

Joël Rivat

Definition 2.1 We call an *arithmetic function* any application from $\mathbb{N}^*$ to $\mathbb{C}$.

Definition 2.2 We call the *Dirichlet convolution* of two arithmetic functions f and g the arithmetic function, denoted by $f * g$, defined by

$$f * g(n) = \sum_{d \mid n} f(d) g(n/d).$$

Proposition 2.3 *The convolution is associative and commutative. It has a neutral element, the function δ, defined by*

$$\delta(n) = \begin{cases} 1 \ \textit{if } n = 1, \\ 0 \ \textit{if } n > 1. \end{cases}$$

Furthermore, if f_1, f_2, g_1, g_2 are arithmetic functions and $\alpha, \beta \in \mathbb{C}$, we have

$$(\alpha f_1 + f_2) * (\beta g_1 + g_2) = (\alpha\beta)(f_1 * g_1) + (\alpha)(f_1 * g_2) + (\beta)(f_2 * g_1) + (f_2 * g_2).$$

Proof If $f_1, \ldots, f_k$ are arithmetic functions we get that for whatever parentheses and order of factors,

$$f_1 * \cdots * f_k(n) = \sum_{d_1 \cdots d_k = n} f_1(d_1) \cdots f_k(d_k).$$

J. Rivat (✉)
Aix Marseille Université, CNRS, Centrale Marseille, Institut de Mathématiques de Marseille, I2M – UMR 7373, Marseille, France
e-mail: joel.rivat@univ-amu.fr

© Springer International Publishing AG, part of Springer Nature 2018
S. Ferenczi et al. (eds.), *Ergodic Theory and Dynamical Systems in their Interactions with Arithmetics and Combinatorics*, Lecture Notes in Mathematics 2213, https://doi.org/10.1007/978-3-319-74908-2_2

and for the neutral element

$$f * \delta(n) = \sum_{d \mid n} f(d)\, \delta(n/d) = f(n).$$

The linearity properties are trivial.

Definition 2.4 An arithmetic function f is said to be *invertible* if there exists an arithmetic function g such that $f * g = \delta$.

Proposition 2.5 *An arithmetic function f is invertible if and only if $f(1) \neq 0$. In that case, the inverse of f is unique.*

Proof If f is invertible, there exists g such that $f * g = \delta$. In particular, we have $f(1)g(1) = 1$ hence $f(1) \neq 0$. Conversely, if $f(1) \neq 0$, we put $g(1) = 1/f(1)$. The relation

$$f(1)\, g(n) + \sum_{d \mid n,\ d < n} f(n/d)\, g(d) = 0, \quad (n > 1),$$

allows to build g inductively, which proves both existence and unicity.

Definition 2.6 An arithmetic function f is *multiplicative* if $f(1) = 1$ and

$$(m, n) = 1 \Rightarrow f(mn) = f(m)f(n).$$

Proposition 2.7 *An arithmetic function f is multiplicative if and only if*

$$f(1) = 1, \ \forall n \geqslant 2, \ f(n) = \prod_{p^v \| n} f(p^v),$$

where the product is taken on the powers of prime numbers p^v which divide n exactly, namely such that $p^v \mid n$ and $p^{v+1} \nmid n$.

Proposition 2.8 *If f and g are multiplicative, then $f * g$ is multiplicative.*

Proof Let m and n be coprime. The divisors of mn are all of the form uv where $u \mid m$ and $v \mid n$. Thus we have

$$f * g(mn) = \sum_{d \mid mn} f(d)\, g(mn/d) = \sum_{u \mid m} \sum_{v \mid n} f(uv)\, g(mn/uv).$$

We have $(u, v) = 1$ and $(m/u, n/v) = 1$, hence by multiplicativity

$$f * g(mn) = \sum_{u \mid m} f(u)\, g(m/u) \sum_{v \mid n} f(v)\, g(n/v) = f * g(m) f * g(n),$$

which shows that $f * g$ is multiplicative.

Example 2.9 The function *number of divisors*, denoted by d or τ, is multiplicative.

Proof We have

$$\tau(n) = \sum_{d \mid n} 1 = \mathbb{1} * \mathbb{1}(n),$$

and $\mathbb{1}$ is multiplicative.

Example 2.10 The function *sum of divisors*, denoted by σ, is multiplicative.

Proof By introducing the identity function $\mathrm{Id}(n) = n$, we have

$$\sigma(n) = \sum_{d \mid n} d = \sum_{d \mid n} \mathrm{Id}(d)\ \mathbb{1}(n/d) = \mathbb{1} * \mathrm{Id}(n),$$

and Id and $\mathbb{1}$ are multiplicative.

Proposition 2.11 *If f is multiplicative, then f is invertible and its inverse is multiplicative.*

Proof If f is multiplicative, then $f(1) = 1 \neq 0$ hence f is invertible. To show that the inverse g of f is multiplicative, we prove by induction on N that

$$\forall m, n \geqslant 1, (mn \leqslant N \text{ and } (m, n) = 1) \Rightarrow g(mn) = g(m)g(n).$$

As $g(1) = 1$, this property is satisfied for $N = 1$. Suppose it is true for $N - 1$, and let $(m, n) = 1$, $mn = N > 1$. We can write

$$g(mn) = - \sum_{\substack{u \mid m, v \mid n \\ uv < mn}} f(mn/uv)\, g(uv),$$

and as $(u, v) = 1$ and $(m/u, n/v) = 1$, we get by multiplicativity

$$g(mn) = - \sum_{u \mid m} \sum_{v \mid n} f(m/u) f(n/v)\, g(u)\, g(v) + g(m)\, g(n).$$

The double sum is $\delta(m)\delta(n) = 0$ (as $mn > 1$), which achieves the induction.

Example 2.12 The inverse of $\mathbb{1}$ for convolution, denoted by μ, is a multiplicative function, called the *Möbius function*:

$$\mathbb{1} * \mu = \delta, \quad \text{i.e.,} \quad \sum_{d \mid n} \mu(d) = \begin{cases} 1 \text{ if } n = 1, \\ 0 \text{ if } n > 1. \end{cases}$$

Proposition 2.13 *We have $\mu(1) = 1$ and for $n > 1$,*

$$\mu(n) = \begin{cases} (-1)^k & \text{if } n \text{ is a product of } k \text{ distinct prime numbers,} \\ 0 & \text{otherwise.} \end{cases}$$

Proof The Möbius function is the inverse of a multiplicative function, hence μ is multiplicative. Thus we have $\mu(1) = 1$ and we need only to determine $\mu(p^\nu)$ for $\nu \geqslant 1$. By applying $\mathbb{1} * \mu = \delta$, we get

$$\mu(1) + \mu(p) = 0$$

hence $\mu(p) = -1$, and for all $\nu \geqslant 2$,

$$\mu(1) + \mu(p) + \mu(p^2) + \cdots + \mu(p^\nu) = 0$$

which allows to show by induction that $\mu(p^\nu) = 0$ for $\nu \geqslant 2$.

Example 2.14 The Euler function $\varphi(n)$ is multiplicative.

Proof We have $\text{Id} = \mathbb{1} * \varphi$ hence $\varphi = \mu * \text{Id}$.

Proposition 2.15 *If f is multiplicative and $\lim_{p^\nu \to +\infty} f(p^\nu) = 0$, then*

$$\lim_{n \to +\infty} f(n) = 0.$$

Proof By deciding that an empty product is equal to 1, we can write for all $Q > 0$,

$$|f(n)| = \left(\prod_{\substack{p^\nu \| n, \, p^\nu \leqslant Q \\ |f(p^\nu)| \leqslant 1}} |f(p^\nu)| \right) \left(\prod_{\substack{p^\nu \| n, \, p^\nu \leqslant Q \\ |f(p^\nu)| > 1}} |f(p^\nu)| \right) \left(\prod_{\substack{p^\nu \| n, \, p^\nu > Q}} |f(p^\nu)| \right).$$

The first product is $\leqslant 1$, the second is bounded by the product

$$A := \prod_{\substack{p^\nu \\ |f(p^\nu)| > 1}} |f(p^\nu)| \geqslant 1,$$

which is a finite product as $\lim_{p^\nu \to +\infty} f(p^\nu) = 0$.

For a fixed $0 < \varepsilon < 1$, we can choose Q large enough so that

$$p^\nu > Q \Rightarrow |f(p^\nu)| \leqslant \varepsilon.$$

With these hypotheses, the third product, if nonempty, is $\leqslant \varepsilon$. But the integers n such that

$$p^\nu \| n \Rightarrow p^\nu \leqslant Q$$

divide and thus are smaller than the product

$$B := \prod_{p^{\nu} \leqslant Q} p^{\nu}.$$

By choosing n large enough, the third product is not empty, and is thus $\leqslant \varepsilon$, which shows that $|f(n)| \leqslant A\varepsilon$ and finishes the proof.

Corollary 2.16 *For all $\varepsilon > 0$, we have*

$$\tau(n) = O_{\varepsilon}(n^{\varepsilon}).$$

Proof Set $f(n) = \tau(n)/n^{\varepsilon}$. We have

$$0 \leqslant f(p^{\nu}) = \frac{1 + \nu}{p^{\nu\varepsilon}} \leqslant \frac{1 + (\log p^{\nu})/(\log 2)}{p^{\nu\varepsilon}} \to 0 \qquad (p^{\nu} \to +\infty),$$

hence $f(n) \to 0$ by the previous proposition.

Theorem 2.17 (Hyperbola Principle) *Let f and g be two arithmetic functions whose summing functions are respectively*

$$F(x) = \sum_{n \leqslant x} f(n), \quad G(x) = \sum_{n \leqslant x} g(n).$$

For $1 \leqslant y \leqslant x$, we have

$$\sum_{n \leqslant x} f * g(n) = \sum_{n \leqslant y} F(x/n)\, g(n) + \sum_{n \leqslant x/y} f(n)\, G(x/n) - F(x/y)G(y).$$

Proof The left side can be written as

$$\sum_{uv \leqslant x} f(u)g(v) = \sum_{\substack{uv \leqslant x \\ v \leqslant y}} f(u)g(v) + \sum_{\substack{uv \leqslant x \\ v > y}} f(u)g(v)$$

$$= \sum_{v \leqslant y} F(x/v)g(v) + \sum_{u \leqslant x/y} f(u)\, (G(x/u) - G(y)),$$

hence the result by expanding the last term.

Corollary 2.18 (Dirichlet) *For $x \geqslant 2$, we have*

$$\sum_{n \leqslant x} \tau(n) = x \log x + (2\gamma - 1)x + O(\sqrt{x}),$$

where γ is the Euler constant.

Proof We apply the previous theorem with $f = g = \mathbb{1}$ and $y = \sqrt{x}$. Then

$$F(t) = G(t) = \lfloor t \rfloor,$$

and

$$\sum_{n \leqslant x} \tau(n) = 2 \sum_{n \leqslant \sqrt{x}} \left\lfloor \frac{x}{n} \right\rfloor - \lfloor \sqrt{x} \rfloor^2 = 2 \sum_{n \leqslant \sqrt{x}} \frac{x}{n} - x + O(\sqrt{x}).$$

The conclusion comes from the classic identity

$$\sum_{n \leqslant t} \frac{1}{n} = \log t + \gamma + O\left(\frac{1}{t}\right),$$

applied with $t = \sqrt{x}$.

Chapter 3
Dirichlet Series

Joël Rivat

Definition 3.1 We call *Dirichlet series* any function of a complex variable

$$F(s) := \sum_{n \geqslant 1} \frac{a_n}{n^s},$$

where $a_n \in \mathbb{C}$, defined wherever it converges.

Notation 3.2 *We write $s = \sigma + i\tau$ with $(\sigma, \tau) \in \mathbb{R}^2$.*

Proposition 3.3 *Given two arithmetic functions f and g, formally we have*

$$\sum_{n \geqslant 1} \frac{f * g(n)}{n^s} = \left(\sum_{n \geqslant 1} \frac{f(n)}{n^s} \right) \left(\sum_{n \geqslant 1} \frac{g(n)}{n^s} \right).$$

Theorem 3.4 (Eulerian Product) *Let f be a multiplicative function. We have*

$$\sum_{n \geqslant 1} \left| \frac{f(n)}{n^s} \right| < +\infty \quad \text{if and only if} \quad \sum_{p} \sum_{v \geqslant 1} \left| \frac{f(p^v)}{p^{vs}} \right| < +\infty.$$

J. Rivat (✉)
Aix Marseille Université, CNRS, Centrale Marseille, Institut de Mathématiques de Marseille,
I2M – UMR 7373, Marseille, France
e-mail: joel.rivat@univ-amu.fr

© Springer International Publishing AG, part of Springer Nature 2018   17
S. Ferenczi et al. (eds.), *Ergodic Theory and Dynamical Systems in their
Interactions with Arithmetics and Combinatorics*, Lecture Notes
in Mathematics 2213, https://doi.org/10.1007/978-3-319-74908-2_3

When these series converge, we get the Eulerian product *expansion*

$$\sum_{n \geqslant 1} \frac{f(n)}{n^s} = \prod_p \left(1 + \sum_{v \geqslant 1} \frac{f(p^v)}{p^{vs}}\right).$$

Proof The absolute convergence of the Dirichlet series

$$F(s) := \sum_{n \geqslant 1} f(n) n^{-s}$$

implies the absolute convergence of the double series. Conversely suppose that

$$\sum_p \sum_{v \geqslant 1} \left|\frac{f(p^v)}{p^{vs}}\right| < +\infty,$$

and consider the infinite product

$$M := \prod_p \left(1 + \sum_{v \geqslant 1} \left|\frac{f(p^v)}{p^{vs}}\right|\right) = \exp\left(\sum_p \log\left(1 + \sum_{v \geqslant 1} \left|\frac{f(p^v)}{p^{vs}}\right|\right)\right).$$

As for $u \geqslant 0$, we have $\log(1 + u) \leqslant u$, we get the bound

$$M \leqslant \exp\left(\sum_p \sum_{v \geqslant 1} \left|\frac{f(p^v)}{p^{vs}}\right|\right) < +\infty.$$

Denote by $P^+(n)$ the largest prime factor of n. For $x \geqslant 1$, we have the bounds

$$\sum_{n \leqslant x} |f(n)n^{-s}| \leqslant \sum_{P^+(n) \leqslant x} |f(n)n^{-s}| = \prod_{p \leqslant x}\left(1 + \sum_{v \geqslant 1} \left|\frac{f(p^v)}{p^{vs}}\right|\right) \leqslant M,$$

which establishes the absolute convergence of $F(s)$.

To prove the Eulerian product formula, it is enough to observe that

$$\left|\sum_{n \geqslant 1} f(n)n^{-s} - \prod_{p \leqslant x}\left(1 + \sum_{v \geqslant 1} \frac{f(p^v)}{p^{vs}}\right)\right| = \left|\sum_{P^+(n) > x} f(n)n^{-s}\right| \leqslant \sum_{n > x} |f(n)n^{-s}|$$

and to make x go to $+\infty$.

Corollary 3.5 *For all $s \in \mathbb{C}$ with $\Re(s) > 1$,*

$$\zeta(s) = \sum_{n=1}^{\infty} \frac{1}{n^s} = \prod_p (1 - 1/p^s)^{-1}.$$

Proof Apply the previous theorem with $f = \mathbb{1}$.

Theorem 3.6 *Let $F(s) := \sum_{n \geq 1} a_n n^{-s}$ be a Dirichlet series.*

1. *If the series converges at $s_0 := \sigma_0 + i\tau_0$, then it converges in the half-plane $\sigma > \sigma_0$, and the convergence is uniform in every angular sector*

$$S(\vartheta) := \{s \in \mathbb{C} : |\arg(s - s_0)| < \vartheta\}, \text{ with } \vartheta < \pi/2.$$

2. *If the series converges absolutely for $s = s_0$, then it converges absolutely and uniformly for $\sigma \geq \sigma_0$.*
3. *The function $F(s)$ is holomorphic in every open convergence domain, and in such a domain we have*

$$F^{(k)}(s) = (-1)^k \sum_{n \geq 1} \frac{a_n (\log n)^k}{n^s}.$$

Proof

1. By setting $a'_n = a_n n^{-s_0}$ and $s' = s - s_0$ if necessary, we can assume that $s_0 = 0$, or equivalently that $\sum a_n$ converges.

   Thus for a fixed $\varepsilon > 0$, there exists $n_0 > 0$ such that for all integers $M, N \geq n_0$,

$$\left| \sum_{M < n \leq N} a_n \right| \leq \varepsilon.$$

   Integrating by parts, we get

$$\sum_{M < n \leq N} a_n n^{-s} = N^{-s} \sum_{M < n \leq N} a_n + s \int_M^N \left(\sum_{M < n \leq t} a_n \right) \frac{dt}{t^{s+1}},$$

   which gives

$$\left| \sum_{M < n \leq N} a_n n^{-s} \right| \leq N^{-\sigma} \varepsilon + |s| \varepsilon \int_M^\infty \frac{dt}{t^{\sigma+1}} = \varepsilon \left(N^{-\sigma} + \frac{|s|}{\sigma} M^{-\sigma} \right)$$

$$\leq \varepsilon \left(1 + \frac{|s|}{\sigma} \right).$$

   If $s \in S(\vartheta)$, then (with $s_0 = 0$) we get $\sigma = |s| \cos(\arg(s)) \geq |s| \cos \vartheta$, hence

$$\left| \sum_{M < n \leq N} a_n n^{-s} \right| \leq \varepsilon \left(1 + \frac{1}{\cos \vartheta} \right),$$

   which establishes the uniform convergence of $\sum a_n n^{-s}$.

If the series converges at s_0, then for all s such that $\sigma > \sigma_0$, there exists $\vartheta < \pi/2$ such that $s \in S(\vartheta)$, hence the series converges at s.

2. This comes from the initial bound

$$\left| \sum_{M < n \leqslant N} a_n n^{-s} \right| \leqslant \sum_{M < n \leqslant N} |a_n|\, n^{-\sigma}.$$

3. For all compact K included in an open convergence domain, we can find s_0 and $0 < \vartheta < \pi/2$ such that $K \subset S(\vartheta)$. On K, $F(s)$ is thus a uniform limit of the partial sums $\sum_{n=1}^{N} a_n n^{-s}$, hence from a theorem of Weierstrass, F is a holomorphic function and its derivatives are the limits of the derivatives of the partial sums.

The first two points of this theorem show that the convergence of a Dirichlet series (as well as its absolute convergence) takes place in a half-plane.

Definition 3.7 Let $F(s) := \sum_{n \geqslant 1} a_n n^{-s}$ be a Dirichlet series.

1. We call convergence abscissa, and denote by σ_c:

$$\sigma_c := \inf \left\{ \sigma \in \mathbb{R} : \sum_{n \geqslant 1} a_n n^{-\sigma} \text{ converges} \right\}.$$

2. We call absolute convergence abscissa, and denote by σ_a:

$$\sigma_a := \inf \left\{ \sigma \in \mathbb{R} : \sum_{n \geqslant 1} |a_n|\, n^{-\sigma} \text{ converges} \right\}$$

We decide that σ_c and σ_a can possibly be $\pm\infty$.

Proposition 3.8 *Let $F(s) := \sum_{n \geqslant 1} a_n n^{-s}$ be a Dirichlet series, of convergence abscissa σ_c and absolute convergence abscissa σ_a. Then*

$$\sigma_c \leqslant \sigma_a \leqslant \sigma_c + 1.$$

Proof Clearly, $\sigma_c \leqslant \sigma_a$.

Let $\varepsilon > 0$. The convergence of the series $\sum_{n \geqslant 1} a_n n^{-\sigma_c - \varepsilon}$ implies that $a_n n^{-\sigma_c - \varepsilon}$ is bounded, hence the series

$$\sum_{n \geqslant 1} a_n n^{-\sigma_c - \varepsilon} n^{-1 - \varepsilon}$$

converges absolutely, which proves that $\sigma_a \leqslant \sigma_c + 1 + 2\varepsilon$, hence the result by going to the limit.

Example 3.9 The Dirichlet series

$$F(s) := \sum_{n \geqslant 1} (-1)^n n^{-s}$$

satisfies $\sigma_c = 0$ (alternate series criterion) and $\sigma_a = 1$.

Proposition 3.10 *Let* $F(s) := \sum_{n \geqslant 1} a_n n^{-s}$ *be a Dirichlet series equal to 0 for* σ *large enough. Then* $a_n = 0$ *for all* $n \geqslant 1$.

Proof Let m be the smallest integer such that $a_m \neq 0$. We have for σ large enough

$$0 = F(s) = \sum_{n \geqslant m} a_n n^{-s} = a_m m^{-s} (1 + G(s)),$$

hence $G(s) = -1$ for σ large enough, where

$$G(s) := \frac{m^s}{a_m} \sum_{n > m} \frac{a_n}{n^s}.$$

For $\sigma \geqslant \sigma_c + 2$, the convergence is absolute and uniform, hence

$$|G(s)| \leqslant \frac{m^\sigma}{|a_m|} \sum_{n > m} \frac{|a_n|}{n^{\sigma_c + 2}} \frac{1}{n^{\sigma - \sigma_c - 2}}$$

$$\leqslant |a_m|^{-1} \frac{m^\sigma}{(m+1)^{\sigma - \sigma_c - 2}} \sum_{n > m} \frac{|a_n|}{n^{\sigma_c + 2}} = c_m \left(\frac{m}{m+1} \right)^\sigma,$$

thus $G(s)$ tends to 0 when σ goes to $+\infty$, which contradicts $G(s) = -1$ for σ large enough. $\qquad \blacksquare$

The following theorem is the analogue of Hadamard's theorem, which for power series gives the convergence radius from the size of coefficients.

Theorem 3.11 *Let* $F(s) := \sum_{n \geqslant 1} a_n n^{-s}$ *be a Dirichlet series. Set*

$$A(x) := \sum_{n \leqslant x} a_n, \quad \kappa := \limsup_{x \to +\infty} \frac{\log |A(x)|}{\log x}.$$

1. *If* $\kappa \neq 0$, *then* $\sigma_c = \kappa$.
2. *If* $\kappa = 0$ *and if* $A(x)$ *has no finite limit when* $x \to +\infty$, *then* $\sigma_c = 0$.
3. *If* $\kappa = 0$ *and* $A(x) \to \alpha \in \mathbb{R}$ *when* $x \to +\infty$, *then*

$$\sigma_c = \limsup_{x \to +\infty} \frac{\log |A(x) - \alpha|}{\log x} \leqslant 0.$$

Proof Let us first show that $\sigma_c \leqslant \kappa$. Suppose $\sigma > \kappa$ and let $\varepsilon > 0$ be such that $\kappa + \varepsilon < \sigma$. We have thus $|A(x)| = O_\varepsilon(x^{\kappa+\varepsilon})$. Summing by parts,

$$\sum_{M<n\leqslant N} a_n n^{-s} = A(N)N^{-s} - A(M)M^{-s} + s \int_M^N A(t)\frac{dt}{t^{s+1}},$$

so that

$$\left| \sum_{M<n\leqslant N} a_n n^{-s} \right| = O_\varepsilon(N^{\kappa+\varepsilon-\sigma}) + O_\varepsilon(M^{\kappa+\varepsilon-\sigma}) + O_\varepsilon\left(|s| \int_M^\infty t^{\kappa+\varepsilon-\sigma-1} dt \right)$$

which tends to 0 when M goes to $+\infty$.

We can write for $0 < y \leqslant x$,

$$\sum_{y<n\leqslant x} \frac{a_n}{n^s} n^s = \left(\sum_{n\leqslant x} \frac{a_n}{n^s} \right) x^s - \left(\sum_{n\leqslant y} \frac{a_n}{n^s} \right) y^s - \int_y^x \left(\sum_{n\leqslant t} \frac{a_n}{n^s} \right) st^{s-1} dt,$$

hence for $\sigma > \sigma_c$, as $t \mapsto \sum_{n\leqslant t} \frac{a_n}{n^s}$ is bounded, we have

$$|A(x) - A(y)| = \left| \sum_{y<n\leqslant x} \frac{a_n}{n^s} n^s \right| = O(x^\sigma) + O(y^\sigma).$$

Suppose $\kappa > 0$. By taking $y = 1^-$, we get $A(x) = O(x^\sigma) + O(1)$, hence $\kappa \leqslant \sigma$. As σ can be taken arbitrarily close to σ_c, in the limit we get $\kappa \leqslant \sigma_c$ and finally $\kappa = \sigma_c$.

Suppose $\kappa < 0$. Then $\sigma_c < 0$ and $A(y) \to 0$ when $y \to +\infty$. For $\sigma_c < \sigma < 0$, we get by letting $y \to +\infty$, that $A(x) = O(x^\sigma)$ thus $\kappa \leqslant \sigma$ and again $\kappa = \sigma_c$.

If $\kappa = 0$ and $A(x)$ has no limit when $x \to +\infty$, then $\sigma_c \geqslant 0 = \kappa$ and thus $\sigma_c = 0$. If $\kappa = 0$ and $A(x) \to \alpha$ when $x \to +\infty$, set

$$\varrho_0 = \inf\{\varrho \leqslant 0 : A(x) = \alpha + o(x^\varrho)\}.$$

Summing by parts, we get

$$\sum_{M<n\leqslant N} a_n n^{-s} = A(N)N^{-s} - A(M)M^{-s} + s \int_M^N A(t)\frac{dt}{t^{s+1}}$$

and when we replace $A(u)$ by $\alpha + O(u^\varrho)$, the terms in α disappear and we get

$$\left| \sum_{M<n\leqslant N} a_n n^{-s} \right| = O(N^{\varrho-\sigma}) + O(M^{\varrho-\sigma})$$

which shows that $F(s)$ converges as soon as $\sigma > \varrho_0$, hence $\sigma_c \leqslant \varrho_0$. If we had $\sigma_c < \sigma < \varrho_0 \leqslant 0$, then by letting $y \to +\infty$, we would get $A(x) - \alpha = O(x^\sigma)$, and this would contradict the definition of ϱ_0.

Chapter 4
Euler's Gamma Function

Joël Rivat

For proofs of the following results, see for example Chapters 12 and 13 of [2].

Definition 4.1 For all $z \in \mathbb{C}$ with $\mathfrak{R}(z) > 0$, we define

$$\Gamma(z) = \int_0^\infty e^{-t} t^{z-1} dt.$$

Theorem 4.2 (Weierstrass Product) *For all $z \in \mathbb{C}$,*

$$\frac{1}{\Gamma(z)} = z \, e^{\gamma z} \prod_{n=1}^\infty \left(1 + \frac{z}{n}\right) e^{-z/n},$$

where $\gamma = 0.5772156649\ldots$ is the Euler constant.

Thus the function Γ is meromorphic on $\mathbb{C}$, has no zero, and admits simple poles at $z = 0, -1, -2, \ldots$.

Proposition 4.3 *For $z \in \mathbb{C} \setminus (-\mathbb{N})$, we have*

$$\Gamma(z + 1) = z \, \Gamma(z).$$

Corollary 4.4 *For $n \in \mathbb{N}$, we have $\Gamma(n + 1) = n!$.*

J. Rivat (✉)
Aix Marseille Université, CNRS, Centrale Marseille, Institut de Mathématiques de Marseille, I2M – UMR 7373, Marseille, France
e-mail: joel.rivat@univ-amu.fr

© Springer International Publishing AG, part of Springer Nature 2018
S. Ferenczi et al. (eds.), *Ergodic Theory and Dynamical Systems in their Interactions with Arithmetics and Combinatorics*, Lecture Notes in Mathematics 2213, https://doi.org/10.1007/978-3-319-74908-2_4

Proposition 4.5 *For $z \in \mathbb{C} \setminus \mathbb{Z}$,*

$$\Gamma(z)\, \Gamma(1-z) = \frac{\pi}{\sin \pi z}.$$

Corollary 4.6 *We have $\Gamma(\frac{1}{2}) = \sqrt{\pi}$.*

Proposition 4.7 (Legendre Duplication Formula) *For $z \in \mathbb{C} \setminus (-\frac{1}{2}\mathbb{N})$,*

$$\Gamma(z)\, \Gamma(z+\tfrac{1}{2}) = \pi^{1/2}\, 2^{1-2z}\, \Gamma(2z).$$

Corollary 4.8 *For $z \in \mathbb{C} \setminus (2\mathbb{Z})$, we have*

$$\Gamma(\tfrac{z}{2})/\Gamma(\tfrac{1-z}{2}) = \pi^{-1/2}\, 2^{1-z}\, \cos(\tfrac{\pi z}{2})\, \Gamma(z).$$

Theorem 4.9 (Stirling Formula) *For fixed $0 \leqslant \vartheta < \pi$ and $|arg(z)| < \vartheta$, we have*

$$\log \Gamma(z) = (z - \tfrac{1}{2})\log z - z - \tfrac{1}{2}\log 2\pi + O(|z|^{-1}), \quad |z| \to +\infty.$$

Proposition 4.10 *For fixed $0 \leqslant \vartheta < \pi$ and $|arg(z)| < \vartheta$, we have*

$$\frac{\Gamma'(z)}{\Gamma(z)} = \log z + O(|z|^{-1}), \quad |z| \to +\infty.$$

Proof See [1, footnote, p.57].

References

1. A.E. Ingham, *The Distribution of Prime Numbers*. Cambridge Mathematical Library (Cambridge University Press, Cambridge, 1990); Reprint of the 1932 original, With a foreword by R. C. Vaughan
2. E.T. Whittaker, G.N. Watson, *A Course of Modern Analysis*. Cambridge Mathematical Library (Cambridge University Press, Cambridge, 1996); An introduction to the general theory of infinite processes and of analytic functions; with an account of the principal transcendental functions, Reprint of the fourth (1927) edition

Chapter 5
Riemann's Zeta Function

Joël Rivat

5.1 The Functional Equation of Zeta

The zeta function is defined for $\Re(s) > 1$ by

$$\zeta(s) = \sum_{n=1}^{+\infty} \frac{1}{n^s}.$$

Theorem 5.1 *The zeta function can be extended analytically to the whole complex plane, into a meromorphic function, with one pole, which is a simple pole, at the point* 1, *with residue* 1.

We can write both

$$\zeta(s) = \frac{s}{s-1} - s \int_1^{+\infty} t^{-s-1} \{t\} \, dt = \frac{1}{s-1} + s \int_1^{+\infty} t^{-s-1} (1 - \{t\}) \, dt.$$

J. Rivat (✉)
Aix Marseille Université, CNRS, Centrale Marseille, Institut de Mathématiques de Marseille, I2M – UMR 7373, Marseille, France
e-mail: joel.rivat@univ-amu.fr

© Springer International Publishing AG, part of Springer Nature 2018
S. Ferenczi et al. (eds.), *Ergodic Theory and Dynamical Systems in their Interactions with Arithmetics and Combinatorics*, Lecture Notes in Mathematics 2213, https://doi.org/10.1007/978-3-319-74908-2_5

Proof We show here the extension to $\Re(s) > 0$, the extension to all $\mathbb{C}$ will be done in the next theorem. We write for $\Re(s) > 1$,

$$\sum_{n=1}^{+\infty} \frac{1}{n^s} = \int_{1-}^{+\infty} t^{-s} d[t]$$

$$= \int_{1-}^{+\infty} t^{-s} dt - \int_{1-}^{+\infty} t^{-s} d\{t\}$$

$$= \frac{1}{s-1} + 1 - s \int_{1}^{+\infty} t^{-s-1} \{t\} dt$$

and the integral on the right side converges absolutely for $\Re(s) > 0$, hence the right side is a meromorphic function on the half-plane $\Re(s) > 0$, which admits one pole, on $s = 1$, and this pole is simple, with residue 1.

Corollary 5.2 *For $s > 0$, $s \neq 1$, we have*

$$\frac{1}{s-1} < \zeta(s) < \frac{s}{s-1}.$$

Theorem 5.3 *The function $\pi^{-s/2} \Gamma(s/2) \zeta(s)$ can be extended into a meromorphic function on $\mathbb{C} \setminus \{0, 1\}$, and we have the functional equation:*

$$\pi^{-s/2} \Gamma(s/2) \zeta(s) = \pi^{-(1-s)/2} \Gamma((1-s)/2) \zeta(1-s).$$

Remark 5.4 The functional equation expresses a symmetry with respect to $s = \frac{1}{2}$ which allows us to deduce the properties for $\sigma < 0$ from those for $\sigma > 1$.

Proof We give one of Riemann's original proofs, cf. Titchmarsh [4] For $\sigma > 0$, we have

$$\Gamma(s/2) = \int_0^\infty e^{-t} t^{\frac{s}{2}-1} dt,$$

hence by putting $t = n^2 \pi x$, we get

$$\pi^{-s/2} \Gamma(s/2) n^{-s} = \int_0^\infty e^{-n^2 \pi x} x^{\frac{s}{2}-1} dx,$$

and by summing on n, for a real $s > 1$, we get (everything being positive)

$$\pi^{-s/2} \Gamma(s/2) \zeta(s) = \int_0^\infty \left(\sum_{n \geq 1} e^{-n^2 \pi x} \right) x^{\frac{s}{2}-1} dx,$$

We write

$$\omega(x) = \sum_{n \geq 1} e^{-n^2 \pi x}, \quad \theta(x) = \sum_{n \in \mathbb{Z}} e^{-n^2 \pi x}.$$

The function θ is a particular case of Jacobi's functions ϑ, and satisfies the functional equation

$$\theta(1/x) = x^{1/2}\theta(x), \quad x > 0.$$

which we shall prove below. As

$$2\omega(x) = \theta(x) - 1,$$

we deduce that

$$\omega(1/x) = -\tfrac{1}{2} + \tfrac{1}{2}x^{1/2} + x^{1/2}\,\omega(x).$$

We write

$$\pi^{-s/2}\Gamma(s/2)\zeta(s) = \int_0^1 \omega(x)x^{\frac{s}{2}-1}dx + \int_1^\infty \omega(x)x^{\frac{s}{2}-1}dx$$

$$= \int_1^\infty \omega(1/x)x^{-\frac{s}{2}-1}dx + \int_1^\infty \omega(x)x^{\frac{s}{2}-1}dx$$

and

$$\int_1^\infty \omega(1/x)x^{-\frac{s}{2}-1}dx = \int_1^\infty \left(-\tfrac{1}{2} + \tfrac{1}{2}x^{1/2} + x^{1/2}\,\omega(x)\right)x^{-\frac{s}{2}-1}dx$$

$$= -\frac{1}{s} + \frac{1}{s-1} + \int_1^\infty \omega(x)x^{-\frac{s}{2}-\frac{1}{2}}dx$$

which gives for all real $s > 1$,

$$\pi^{-s/2}\Gamma(s/2)\zeta(s) = \frac{1}{s(s-1)} + \int_1^\infty \omega(x)\left(x^{\frac{s}{2}-1} + x^{\frac{1-s}{2}-1}\right)dx.$$

For $x \geq 1$,

$$\omega(x) = e^{-\pi x}\sum_{n \geq 1} e^{-(n^2-1)\pi x} \leq e^{-\pi x}\sum_{n \geq 1} e^{-(n^2-1)\pi}$$

hence

$$\omega(x) = O(e^{-\pi x}), \quad x \to +\infty,$$

the integral on the right side converges for all $s \in \mathbb{C}$, uniformly on every compact subset of $\mathbb{C}$. Thus this integral is an entire function (holomorphic on $\mathbb{C}$), and the right side gives an analytic extension of zeta to the whole complex plane. Moreover it is invariant when we replace s by $1 - s$, which proves the functional equation.

Corollary 5.5 *The function*

$$\xi(s) := \tfrac{1}{2}s(s - 1)\pi^{-s/2}\Gamma(s/2)\zeta(s)$$

is holomorphic on $\mathbb{C}$.

Proof We have neutralized the term $\frac{1}{s(s-1)}$.

Corollary 5.6 *The only pole of zeta is $s = 1$, and it is a simple pole with residue 1.*

Proof From the Weierstrass product formula applied to the function Γ,

$$\left(\frac{s}{2}\right)\Gamma\left(\frac{s}{2}\right) = e^{-\gamma s/2}\prod_{n=1}^{\infty}\left(1 + \frac{s}{2n}\right)^{-1}e^{s/2n},$$

hence the function $s \mapsto (s/2)\Gamma(s/2)$ has no zero.

Corollary 5.7 $\zeta(0) = -\tfrac{1}{2}$.

Proof As $(s/2)\Gamma(s/2) \to 1$ when $s \to 0$, in the formula

$$\pi^{-s/2}(s/2)\Gamma(s/2)\zeta(s) = \frac{1}{2(s - 1)} + \frac{s}{2}\int_{1}^{\infty}\omega(x)\left(x^{\frac{s}{2}-1} + x^{-\frac{s}{2}-1}\right)dx$$

the left side tends to $\zeta(0)$ when $s \to 0$, and the right side tends to $-\tfrac{1}{2}$.

Corollary 5.8 *The zeros of zeta with $\sigma < 0$ are the negative even integers ($s = -2, -4, -6, \ldots$). We call them the* trivial zeros *of zeta.*

Proof These are the poles of $\Gamma\left(\frac{s}{2}\right) = \frac{2}{s}e^{-\gamma s/2}\prod_{n=1}^{\infty}\left(1 + \frac{s}{2n}\right)^{-1}e^{s/2n}$.

Definition 5.9 The subset of the plane defined by $0 < \sigma < 1$ is called the *critical stripe*.

What is still to prove is the claimed property of θ:

Lemma 5.10 *The function $\theta(a) = \sum_{n\in\mathbb{Z}}e^{-\pi an^2}$ satisfies the functional equation*

$$\theta(1/a) = a^{1/2}\theta(a), \quad a > 0.$$

Proof It is a special case of Poisson's summation formula

$$\sum_{n \in \mathbb{Z}} f(n) = \lim_{N \to +\infty} \sum_{n=-N}^{N} \widehat{f}(n), \quad \text{with} \quad \widehat{f}(t) = \int_{\mathbb{R}} f(x) \, e^{-2i\pi xt} dx,$$

which holds (see Zygmund [5, p. 68]) as soon as f is integrable on $\mathbb{R}$, with bounded variation on $\mathbb{R}$ and normalized:

$$\forall x \in \mathbb{R}, \ f(x) = \tfrac{1}{2}(f(x^+) + f(x^-)).$$

Here we have Gaussian functions, for $a > 0$,

$$f(x) = e^{-\pi a x^2}, \quad \widehat{f}(t) = a^{-1/2} e^{-\pi t^2/a},$$

which satisfy the conditions of Poisson's formula, hence

$$\sum_{n \in \mathbb{Z}} e^{-\pi a n^2} = a^{-1/2} \sum_{n \in \mathbb{Z}} e^{-\pi n^2/a},$$

which is precisely the expected formula.

5.2 Entire Functions of Finite Order

Definition 5.11 An entire function $f(z)$ is said to be of *finite order* if there exists a real number α such that

$$f(z) = O(e^{|z|^\alpha}), \quad x \to +\infty.$$

The lower bound of these numbers α is called the *order* of $f(z)$.

Proposition 5.12 (Jensen Formula) *Let $f(z)$ be a holomorphic function on $|z| < R'$ such that $f(0) \neq 0$. If $z_1, \ldots, z_k$ are the zeros of $f(z)$ for $|z| \leqslant R < R'$, then*

$$\frac{1}{2\pi} \int_0^{2\pi} \log \left| f(Re^{i\theta}) \right| d\theta - \log |f(0)| = \log \frac{R^k}{|z_1| \cdots |z_k|} = \int_0^R \frac{n(r)}{r} dr,$$

where $n(r)$ is the number of zeros of $f(z)$ for $|z| \leqslant r$.

Remark 5.13 This formula links the modulus of a holomorphic function with the modulus of its zeros.

Proof See Titchmarsh [3, 3.61].

Theorem 5.14 *An entire function $f(z)$ of order 1 with $f(0) \neq 0$ must be of the form*

$$f(z) = e^{A+Bz} \prod_{n=1}^{\infty} (1 - z/z_n)\, e^{z/z_n},$$

where the z_n are the zeros of $f(z)$, repeated if they are multiple, and for every $\varepsilon > 0$,

$$\sum_{n \geq 1} |z_n|^{-1-\varepsilon} < +\infty.$$

If also $\sum_{n \geq 1} |z_n|^{-1} < +\infty$, then there exists a constant $C > 0$ such that

$$|f(z)| < e^{C|z|}.$$

Proof See Davenport [1, pp. 74–78].

5.3 Infinite Product for $\xi(s)$

We recall the definition of the entire function

$$\xi(s) = \tfrac{1}{2} s(s-1) \pi^{-s/2} \Gamma(s/2) \zeta(s).$$

Proposition 5.15 *There exists a constant $C > 0$ such that*

$$|\xi(s)| < \exp(C\,|s|\log|s|).$$

Corollary 5.16 *The function $\xi(s)$ is of order at most 1.*

Proof As $\xi(s) = \xi(1-s)$, it is enough to bound $|\xi(s)|$ for $\sigma \geq \tfrac{1}{2}$. We have

$$\left| \tfrac{1}{2} s(s-1) \pi^{-s/2} \right| < \exp(c_1 |s|).$$

and using the Stirling formula for $-\pi/2 < \arg(s) < \pi/2$,

$$|\Gamma(s/2)| < \exp(c_2 |s| \log|s|).$$

We have still to bound $|\zeta(s)|$, and for that we recall that for $\sigma > 0$,

$$\zeta(s) = \frac{s}{s-1} - s \int_1^{\infty} \{t\} t^{-s-1} dt.$$

The integral is bounded for $\sigma \geqslant \frac{1}{2}$, hence

$$|\zeta(s)| < c_3 \, |s| \,, \quad \text{for } \sigma \geqslant \tfrac{1}{2}, \ |s| \geqslant 2,$$

which completes the proof.

Remark 5.17 When $s \to +\infty$ (s real), we have $\log \Gamma(s) \sim s \log s$ and $\zeta(s) \to 1$, thus the bound we got for $\xi(s)$ is, except for the constants, optimal. In particular, $\xi(s)$ is not bounded by $\exp(C \, |s|)$.

Theorem 5.18 *The function* $\xi(s)$ *admits infinitely many zeros, denoted by* $\rho_1, \rho_2, \ldots$, *such that*

$$\sum_{n \geqslant 1} |\rho_n|^{-1} = +\infty; \quad \forall \varepsilon > 0, \ \sum_{n \geqslant 1} |\rho_n|^{-1-\varepsilon} < +\infty,$$

and the expansion into an infinite product

$$\xi(s) = e^{A+Bs} \prod_{\rho} (1 - s/\rho) \, e^{s/\rho}$$

with

$$A = -\log 2 \approx -0.69314718\ldots$$
$$B = -\tfrac{1}{2}\gamma - 1 + \tfrac{1}{2}\log 4\pi \approx -0.023095708966121\ldots.$$

Proof The results on the series come immediately from the fact that $\xi(s)$ is of order 1 and is not bounded by $\exp(C \, |s|)$. In particular, $\sum |\rho|^{-1} = +\infty$ implies the existence of infinitely many zeros. We have yet to compute A and B, which will be done later.

Corollary 5.19 *When* $\xi(s) \neq 0$, *we have*

$$\frac{\xi'(s)}{\xi(s)} = B + \sum_{\rho} \left(\frac{1}{s-\rho} + \frac{1}{\rho} \right).$$

Proof This is the logarithmic derivative of the infinite product.

Corollary 5.20 *When* $s \neq 1$ *and* $\zeta(s) \neq 0$, *we have*

$$\frac{\zeta'(s)}{\zeta(s)} = B - \frac{1}{s-1} + \tfrac{1}{2}\log\pi - \frac{\Gamma'(s/2+1)}{2\,\Gamma(s/2+1)} + \sum_{\rho} \left(\frac{1}{s-\rho} + \frac{1}{\rho} \right).$$

Proof We write

$$\xi(s) = \tfrac{1}{2}s(s-1)\pi^{-s/2}\Gamma(s/2)\zeta(s) = (s-1)\pi^{-s/2}\Gamma(s/2+1)\zeta(s),$$

thus the logarithmic derivative of ξ equals

$$\frac{\xi'(s)}{\xi(s)} = \frac{1}{s-1} - \tfrac{1}{2}\log\pi + \frac{\Gamma'(s/2+1)}{2\,\Gamma(s/2+1)} + \frac{\zeta'(s)}{\zeta(s)},$$

hence the result by comparing both expressions of ξ'/ξ.

Remark 5.21 This formula exhibits the pole $\zeta(s)$ at $s=1$, and the nontrivial zeros at $s=\rho$. The trivial zeros of $\zeta(s)$ appear in the expression in Γ, as by computing the logarithmic derivative of Γ from the Weierstrass product we get

$$-\frac{\Gamma'(s/2+1)}{2\,\Gamma(s/2+1)} = \tfrac{1}{2}\gamma + \sum_{n=1}^{\infty}\left(\frac{1}{s+2n} - \frac{1}{2n}\right).$$

Lemma 5.22 *We have $A = -\log 2$ and $B = -\tfrac{1}{2}\gamma - 1 + \tfrac{1}{2}\log 4\pi$.*

Proof As

$$\xi(1) = \tfrac{1}{2}\pi^{-1/2}\Gamma(1/2)\lim_{s\to 1}(s-1)\zeta(s) = \tfrac{1}{2},$$

we get $\xi(0) = \tfrac{1}{2}$, thus $e^A = \tfrac{1}{2}$ and $A = -\log 2$.

To get B, we have

$$B = \frac{\xi'(0)}{\xi(0)} = -\frac{\xi'(1)}{\xi(1)}$$

by using the logarithmic derivative of the infinite product and $\xi(s) = \xi(1-s)$. The other formula for the logarithmic derivative of ξ gives

$$\frac{\xi'(s)}{\xi(s)} = \frac{\zeta'(s)}{\zeta(s)} + \frac{1}{s-1} - \tfrac{1}{2}\log\pi + \frac{\Gamma'(s/2+1)}{2\,\Gamma(s/2+1)},$$

and we can make $s \to 1$.

As

$$-\frac{\Gamma'(1/2+1)}{2\,\Gamma(1/2+1)} = \tfrac{1}{2}\gamma + \sum_{n=1}^{\infty}\left(\frac{1}{1+2n} - \frac{1}{2n}\right) = \tfrac{1}{2}\gamma - 1 + \log 2,$$

we get

$$B = \tfrac{1}{2}\gamma - 1 + \tfrac{1}{2}\log 4\pi - \lim_{s\to 1}\left(\frac{\zeta'(s)}{\zeta(s)} + \frac{1}{s-1}\right).$$

Recall that

$$\zeta(s) = \frac{s}{s-1} - sI(s), \quad I(s) = \int_1^\infty (x - \lfloor x\rfloor)x^{-s-1}dx,$$

hence $(s-1)\zeta(s) = s(1 - (s-1)I(s))$, thus

$$\frac{\zeta'(s)}{\zeta(s)} + \frac{1}{s-1} = \frac{1}{s} + \frac{-I(s) - (s-1)I'(s)}{1 - (s-1)I(s)} \to 1 - I(1), \quad \text{when } s \to 1.$$

Finally

$$\int_1^N (x - \lfloor x\rfloor)x^{-2}dx = \log N - \sum_{n=1}^{N-1}\int_n^{n+1} nx^{-2}dx$$

$$= \log N - \sum_{n=1}^{N-1}\frac{1}{n+1} = 1 + \log N - \sum_{n=1}^N \frac{1}{n},$$

thus $I(1) = 1 - \gamma$ and it yields $B = -\tfrac{1}{2}\gamma - 1 + \tfrac{1}{2}\log 4\pi$.

Proposition 5.23 *Writing $\rho = \beta + i\gamma$, we get*

$$B = -\sum_\rho \frac{1}{\rho} = -\sum_{\gamma>0}\frac{2\beta}{\beta^2 + \gamma^2},$$

where the sum is taken simultaneously on ρ and $\overline{\rho}$.

Proof The convergence holds because $0 \leqslant \dfrac{1}{\rho} + \dfrac{1}{\overline{\rho}} = \dfrac{2\beta}{\beta^2 + \gamma^2} \leqslant \dfrac{2}{|\rho|^2}$, and the series $\sum |\rho|^{-2}$ converges by Theorem 5.18. We have

$$\frac{\xi'(1-s)}{\xi(1-s)} = B + \sum_\rho\left(\frac{1}{1-s-\rho} + \frac{1}{\rho}\right) = -\frac{\xi'(s)}{\xi(s)} = -B - \sum_\rho\left(\frac{1}{s-\rho} + \frac{1}{\rho}\right).$$

while if ρ is a zero, then $1 - \rho$ is a zero also, thus the terms $1 - s - \rho$ and $s - \rho$ compensate.

Corollary 5.24 *Every zero $\rho = \beta + i\gamma$ satisfies $|\gamma| > 6.5611$.*

Proof Possibly by changing ρ into $1 - \rho = 1 - \beta - i\gamma$, we can assume $\frac{1}{2} \leqslant \beta \leqslant 1$. We have

$$\frac{2\beta}{\beta^2 + \gamma^2} \leqslant -B,$$

thus

$$|\gamma| \geqslant \sqrt{-\frac{2\beta}{B} - \beta^2} \geqslant \sqrt{-\frac{1}{B} - \frac{1}{4}} > 6.5611.$$

5.4   A Zero Free Region for Zeta

Remark 5.25 The meromorphic function $\zeta'(s)/\zeta(s)$ admits as only poles for $\sigma > 0$ the zeros of $\zeta(s)$ and $s = 1$.

Proposition 5.26 (Mertens) *For $\sigma > 1$ and $\tau \in \mathbb{R}$,*

$$3\left(-\frac{\zeta'(\sigma)}{\zeta(\sigma)}\right) + 4\left(-\mathfrak{R}\frac{\zeta'(\sigma + i\tau)}{\zeta(\sigma + i\tau)}\right) + \left(-\mathfrak{R}\frac{\zeta'(\sigma + 2i\tau)}{\zeta(\sigma + 2i\tau)}\right) \geqslant 0.$$

Proof For $k \in \{0, 1, 2\}$ we can write

$$-\mathfrak{R}\frac{\zeta'(\sigma + ik\tau)}{\zeta(\sigma + ik\tau)} = \mathfrak{R}\sum_{n=1}^{\infty} \Lambda(n)n^{-\sigma - ik\tau} = \sum_{n=1}^{\infty} \Lambda(n)n^{-\sigma}\cos(k\tau \log n).$$

It is enough to prove that

$$\sum_{n=1}^{\infty} \Lambda(n)n^{-\sigma}\left(3 + 4\cos(k\tau \log n) + \cos(2\tau \log n)\right) \geqslant 0,$$

which follows from the inequality $3 + 4\cos\theta + \cos 2\theta = 2(1 + \cos\theta)^2 \geqslant 0$.

Remark 5.27 The behaviour of $-\zeta'(\sigma)/\zeta(\sigma)$ is well known when $\sigma \to 1^+$. Indeed, as $s = 1$ is a simple pole of ζ, we have

$$-\frac{\zeta'(\sigma)}{\zeta(\sigma)} < \frac{1}{\sigma - 1} + O(1),$$

for $1 < \sigma \leqslant 2$. It is thus clear that the other two terms will influence each other in the case when ζ would admit a zero slightly to the left of $1 + i\tau$ or $1 + 2i\tau$.

Theorem 5.28 *The function zeta has no zero in the region*

$$\sigma \geqslant 1 - \frac{1}{35\log(|\tau| + 2) + O(1)}.$$

Proof We recall the formula

$$-\frac{\zeta'(s)}{\zeta(s)} = \frac{1}{s-1} - B - \tfrac{1}{2}\log\pi + \frac{\Gamma'(s/2+1)}{2\,\Gamma(s/2+1)} - \sum_\rho\left(\frac{1}{s-\rho} + \frac{1}{\rho}\right)$$

and also

$$\frac{\Gamma'(s)}{\Gamma(s)} = \log s + O(|s|^{-1}), \quad |s| \to +\infty, \ |\arg s| \leqslant \theta < \pi.$$

We have for $1 \leqslant \sigma \leqslant 2$ and $\tau \geqslant 2$,

$$\Re\log(s/2+1) = \log\sqrt{(\sigma/2+1)^2 + \tau^2/4} = \log\tau + O(1),$$

thus, for $1 \leqslant \sigma \leqslant 2$ and $\tau \geqslant 2$,

$$-\Re\frac{\zeta'(s)}{\zeta(s)} < \tfrac{1}{2}\log\tau - \sum_\rho\Re\left(\frac{1}{s-\rho} + \frac{1}{\rho}\right) + O(1).$$

We remark that

$$\Re\frac{1}{s-\rho} = \frac{\sigma-\beta}{|s-\rho|^2}, \quad \text{and} \quad \Re\frac{1}{\rho} = \frac{\beta}{|\rho|^2},$$

so that

$$\sum_\rho\Re\left(\frac{1}{s-\rho} + \frac{1}{\rho}\right) \geqslant 0.$$

It results, by applying what precedes to $s = \sigma + 2i\tau$, and by omitting the sum on ρ, that

$$-\Re\frac{\zeta'(\sigma + 2i\tau)}{\zeta(\sigma + 2i\tau)} < \tfrac{1}{2}\log\tau + O(1),$$

which coincides with the second coordinate γ of a zero $\beta + i\gamma$, and in the sum on ρ, we only keep the term $1/(s - \rho)$ corresponding to that zero. We get

$$-\Re\frac{\zeta'(\sigma + i\tau)}{\zeta(\sigma + i\tau)} < \tfrac{1}{2}\log\tau - \frac{1}{\sigma - \beta} + O(1).$$

Combining these estimates with Proposition 5.26 we get

$$\frac{3}{\sigma - 1} + \frac{5}{2} \log \tau - \frac{4}{\sigma - \beta} + O(1) \geqslant 0.$$

We take $\sigma = 1 + \delta / \log \tau$, with $\delta > 0$. Then

$$\beta < 1 + \frac{\delta}{\log \tau} - \frac{4}{(3/\delta + 5/2) \log \tau + O(1)},$$

and by taking for example $\delta = 1/5$, we get the result for $\tau \geqslant 2$. For $|\tau| \leqslant 2$, we already know there is no zero, hence the result holds for every τ.

Remark 5.29 The zero free region was extended by Littlewood (1922) to

$$\sigma \geqslant 1 - C \frac{\log \log \tau}{\log \tau},$$

then, independently, by Vinogradov and Korobov (1958), to

$$\sigma \geqslant 1 - C(\log \tau)^{-2/3} (\log \log \tau)^{-1/3}.$$

5.5 The Number of Zeros of Zeta

Theorem 5.30 *If $f(z)$ is holomorphic inside and in the neighborhood of a closed path C, nonzero on C, then the number N of zeros of $f(z)$, counted with their multiplicity, inside C, is given by*

$$N = \frac{1}{2i\pi} \int_C \frac{f'(z)}{f(z)} dz = \frac{1}{2\pi} \Im \int_C \frac{f'(z)}{f(z)} dz = \Delta_C \arg f(z),$$

where Δ_C means generically the variation along the path C.

Proof See e.g. [3, 3.41].

Definition 5.31 For $T > 0$, we denote by $N(T)$ the number of zeros (counted with their multiplicity) of $\zeta(\sigma + i\tau)$ in the rectangle $0 < \sigma < 1, 0 < \tau < T$.

Proposition 5.32 *If $T > 0$ is not the second coordinate of a zero of zeta then*

$$2\pi N(T) = \Delta_{\mathcal{R}} \arg \xi(s),$$

where $\mathcal{R}$ is the rectangle with vertices $2, 2+iT, -1+iT, -1$, spanned in the positive sense.

Proof The zeros of $\zeta(s)$ and $\xi(s)$ coincide in this region, and ξ is entire.

Proposition 5.33 *If $T > 0$ is not the second coordinate of a zero of zeta then*

$$\pi N(T) = \Delta_{\mathcal{L}} \arg \xi(s),$$

where $\mathcal{L}$ is the broken line joining 2 with $2 + iT$ then $\frac{1}{2} + iT$.

Proof When s spans the base of the rectangle, $\arg \xi(s)$ does not change as in this place $\xi(s)$ is real and nonzero.

As

$$\xi(\sigma + i\tau) = \xi(1 - \sigma - i\tau) = \overline{\xi(1 - \sigma + i\tau)},$$

the variation of $\arg \xi(s)$ when s goes from $\frac{1}{2} + iT$ to $-1 + iT$ then -1 equals the variation of $\arg \xi(s)$ along $\mathcal{L}$.

Proposition 5.34 *If $T > 0$ is not the second coordinate of a zero of zeta then*

$$N(T) = \frac{T}{2\pi} \log \frac{T}{2\pi} - \frac{T}{2\pi} + \frac{7}{8} + S(T) + O(T^{-1}),$$

with $S(T)$ defined by

$$\pi S(T) = \Delta_{\mathcal{L}} \arg \zeta(s) = \arg \zeta(\tfrac{1}{2} + iT).$$

Proof We can write $\xi(s)$ under the form

$$\xi(s) = (s - 1)\pi^{-s/2} \Gamma(s/2 + 1)\zeta(s).$$

We have

$$\Delta_{\mathcal{L}} \arg(s - 1) = \arg(iT - \tfrac{1}{2}) = \pi/2 + O(T^{-1}),$$

$$\Delta_{\mathcal{L}} \arg \pi^{-s/2} = \Delta_{\mathcal{L}}(-\tfrac{1}{2}\tau \log \pi) = -\tfrac{1}{2}T \log \pi.$$

By Stirling's formula,

$$\Delta_{\mathcal{L}} \arg \Gamma(s/2 + 1)$$
$$= \Im \log \Gamma(\tfrac{1}{2}iT + 5/4)$$
$$= \Im \left((\tfrac{1}{2}iT + 3/4) \log(\tfrac{1}{2}iT + 5/4) - \tfrac{1}{2}iT - 5/4 + \tfrac{1}{2} \log 2\pi + O(T^{-1}) \right)$$
$$= \tfrac{1}{2}T \log \tfrac{1}{2}T - \tfrac{1}{2}T + \frac{3\pi}{8} + O(T^{-1})$$

and the result follows.

Remark 5.35 To get an asymptotic formula for $N(T)$, it suffices to estimate $S(T)$. Littlewood showed that

$$\int_0^T S(\tau)d\tau = O(\log T),$$

thus hopefully at least $S(T) = o(1)$, when $T \to +\infty$, but this result was never proved. If it was true, the term $7/8$ would be significant. The best known result is

$$S(T) = O(\log T),$$

which we are going to prove.

Lemma 5.36 *If $\rho = \beta + i\gamma$ runs over all the nontrivial zeros of zeta, then for large $T > 0$,*

$$\sum_\rho \frac{1}{1 + (T - \gamma)^2} = O(\log T).$$

Proof We recall the inequality

$$-\Re \frac{\zeta'(s)}{\zeta(s)} < \tfrac{1}{2} \log \tau - \sum_\rho \Re \left(\frac{1}{s - \rho} + \frac{1}{\rho} \right) + O(1),$$

which holds for $1 \leqslant \sigma \leqslant 2$ and $\tau \geqslant 2$.

Taking $s = 2 + iT$, we remark that $\left| \zeta'(s)/\zeta(s) \right|$ is bounded independently of T, thus we get

$$\sum_\rho \Re \left(\frac{1}{s - \rho} + \frac{1}{\rho} \right) < \tfrac{1}{2} \log T + O(1).$$

Observing that

$$\Re \frac{1}{\rho} = \frac{\beta}{|\rho|^2} \geqslant 0,$$

$$\Re \frac{1}{s - \rho} = \frac{2 - \beta}{(2 - \beta)^2 + (T - \gamma)^2} \geqslant \frac{1}{4 + (T - \gamma)^2}$$

permits to bound from below the left hand sum and the result follows.

Corollary 5.37 *For all $T \geqslant 1$,*

$$N(T + 1) - N(T - 1) = O(\log T).$$

Proof We have the inequalities

$$N(T+1) - N(T-1) \leqslant \sum_{\substack{\rho \\ |T-\gamma|\leqslant 1}} 1 \leqslant \sum_{\rho} \frac{2}{1 + (T-\gamma)^2} = O(\log T).$$

Corollary 5.38 *For all* $T \geqslant 1$,

$$\sum_{\substack{\rho \\ |T-\gamma|\geqslant 1}} \frac{1}{(T-\gamma)^2} = O(\log T).$$

Proof We have the inequalities

$$\sum_{\substack{\rho \\ |T-\gamma|\geqslant 1}} \frac{1}{(T-\gamma)^2} \leqslant \sum_{\rho} \frac{2}{1 + (T-\gamma)^2} = O(\log T).$$

Corollary 5.39 *For all* $-1 \leqslant \sigma \leqslant 2$, *and all large enough* τ *which does not coincide with the second coordinate of a zero of zeta,*

$$\frac{\zeta'(s)}{\zeta(s)} = \sum_{\substack{\rho \\ |\tau-\gamma|<1}} \frac{1}{s-\rho} + O(\log \tau).$$

Proof We recall the formula

$$\frac{\zeta'(s)}{\zeta(s)} = B - \frac{1}{s-1} + \frac{1}{2}\log \pi - \frac{\Gamma'(s/2+1)}{2\,\Gamma(s/2+1)} + \sum_{\rho} \left(\frac{1}{s-\rho} + \frac{1}{\rho} \right)$$

which we apply for $s = \sigma + i\tau$ and $2 + i\tau$. We remark that for all large enough τ,

$$\frac{1}{\sigma + i\tau - 1} = O(1), \quad \frac{1}{2 + i\tau - 1} = O(1),$$

and as $\Gamma'(z)/\Gamma(z) = \log z + O(|z|^{-1})$, we have

$$\frac{\Gamma'(s/2+1)}{2\,\Gamma(s/2+1)} - \frac{\Gamma'((2+i\tau)/2+1)}{2\,\Gamma((2+i\tau)/2+1)} = O(1),$$

thus by subtracting we get

$$\frac{\zeta'(s)}{\zeta(s)} = O(1) + \sum_{\rho} \left(\frac{1}{s-\rho} - \frac{1}{2+i\tau-\rho} \right).$$

As $\Im(s - \rho) = \Im(2 + i\tau - \rho) = \tau - \gamma$, we get

$$\left| \frac{1}{s - \rho} - \frac{1}{2 + i\tau - \rho} \right| = \frac{2 - \sigma}{|(s - \rho)(2 + i\tau - \rho)|} \leqslant \frac{3}{|\gamma - \tau|^2},$$

hence

$$\sum_{\substack{\rho \\ |\gamma - \tau| \geqslant 1}} \left| \frac{1}{s - \rho} - \frac{1}{2 + i\tau - \rho} \right| = O(\log \tau).$$

We have still to deal with terms such as $|\gamma - \tau| < 1$. Their number is at most $O(\log \tau)$, and, as $|2 + i\tau - \rho| \geqslant 1$, we get the claimed result.

Lemma 5.40 *We have*

$$S(T) = O(\log T).$$

Proof We can write

$$\pi S(T) = \int_2^{2+iT} \Im \frac{\zeta'(s)}{\zeta(s)} ds - \int_{\frac{1}{2}+iT}^{2+iT} \Im \frac{\zeta'(s)}{\zeta(s)} ds = O(1) - \int_{\frac{1}{2}+iT}^{2+iT} \Im \frac{\zeta'(s)}{\zeta(s)} ds$$

because

$$\int_2^{2+iT} \Im \frac{\zeta'(s)}{\zeta(s)} ds = -\sum_2^\infty \frac{\Lambda(n)}{n^2} \int_0^T \sin(-\tau \log n) d\tau$$

$$= \sum_2^\infty \frac{\Lambda(n)}{n^2} \frac{1 - \cos(T \log n)}{\log n} = O(1).$$

Now,

$$\left| \int_{\frac{1}{2}+iT}^{2+iT} \Im \frac{1}{s - \rho} ds \right| = |\Delta \arg(s - \rho)| \leqslant \pi.$$

As the number of terms to be summed (i.e., the ρ such that $|\gamma - T| < 1$) is $O(\log T)$, we deduce the result.

Theorem 5.41 *We have*

$$N(T) = \frac{T}{2\pi} \log \frac{T}{2\pi} - \frac{T}{2\pi} + O(\log T).$$

Proof Immediate from what precedes.

Corollary 5.42 *If we enumerate the second coordinates $\gamma > 0$ of the zeros of zeta in increasing order by $\gamma_1, \gamma_2, \ldots,$ then*

$$\gamma_n \sim 2\pi n / \log n, \quad \text{when } n \to +\infty.$$

Remark 5.43 Littlewood proved in 1924 that

$$\gamma_{n+1} - \gamma_n \to 0, \quad \text{when } n \to +\infty.$$

5.6   An Explicit Formula for $\psi(x)$

Theorem 5.44 *For $x \geqslant 2$,*

$$\psi_0(x) := \tfrac{1}{2}(\psi(x^+) + \psi(x^-)) = x - \sum_\rho \frac{x^\rho}{\rho} - \frac{\zeta'(0)}{\zeta(0)} - \tfrac{1}{2}\log(1 - x^{-2}),$$

where, in the sum on nontrivial zeros ρ of zeta, ρ and $\overline{\rho}$ must be taken simultaneously. Note that $\zeta'(0)/\zeta(0) = \log(2\pi)$.

Our first objective is to establish a link between, on one side

$$\psi(x) = \sum_{n \leqslant x} \Lambda(n) = \sum_{p^m \leqslant x} \log p,$$

and on the other side

$$-\frac{\zeta'(s)}{\zeta(s)} = \sum_{n \geqslant 1} \frac{\Lambda(n)}{n^s}.$$

In this perspective, we shall use the famous Perron formula:

$$\frac{1}{2i\pi} \int_{c-i\infty}^{c+i\infty} y^s \frac{ds}{s} = \delta(y) := \begin{cases} 0 & \text{if } 0 < y < 1, \\ \tfrac{1}{2} & \text{if } y = 1, \\ 1 & \text{if } y > 1, \end{cases}$$

holding for $c > 0$, in a quantitative form which we shall prove.

Taking $y = x/n$ and $c = \sigma > 1$, we have

$$\psi_0(x) = \sum_{n \geq 1} \Lambda(n)\, \delta(x/n)$$

$$= \frac{1}{2i\pi} \int_{c-i\infty}^{c+i\infty} \left(\sum_{n \geq 1} \frac{\Lambda(n)}{n^s} \right) x^s \frac{ds}{s}$$

$$= \frac{1}{2i\pi} \int_{c-i\infty}^{c+i\infty} \left(-\frac{\zeta'(s)}{\zeta(s)} \right) x^s \frac{ds}{s}.$$

If we could move the vertical integration line infinitely to the left, then we could express $\psi_0(x)$ as the sum of residues of the function

$$\left(-\frac{\zeta'(s)}{\zeta(s)} \right) \frac{x^s}{s}$$

in its poles. The pole of zeta at $s = 1$ contributes with x, the pole at $s = 0$ contributes with $-\zeta'(0)/\zeta(0)$, and each zero of zeta, whether trivial or not, contributes with x^ρ/ρ.

Lemma 5.45 *For $y > 0$, $c > 0$, $T > 0$, put*

$$I(y, T) = \frac{1}{2i\pi} \int_{c-iT}^{c+iT} \frac{y^s}{s}\, ds.$$

Then

$$|I(y, T) - \delta(y)| < \begin{cases} y^c \min(\pi^{-1} T^{-1} |\log y|^{-1}, 1) & \text{if } y \neq 1, \\ c\,\pi^{-1} T^{-1} & \text{if } y = 1. \end{cases}$$

Proof Suppose first that $0 < y < 1$. The function y^s/s tends to 0 when $\sigma \to +\infty$, uniformly in τ. Hence, integrating on a rectangle with right side going to infinity, we have

$$I(y, T) = -\frac{1}{2i\pi} \int_{c+iT}^{+\infty+iT} \frac{y^s}{s}\, ds + \frac{1}{2i\pi} \int_{c-iT}^{+\infty-iT} \frac{y^s}{s}\, ds.$$

But

$$\left| \frac{1}{2i\pi} \int_{c+iT}^{+\infty+iT} \frac{y^s}{s}\, ds \right| < \frac{1}{2\pi T} \int_{c}^{+\infty} y^\sigma\, d\sigma = \frac{y^c}{2\pi T |\log y|},$$

and similarly for the other integral. This gives the first inequality.

For the second one, we replace the vertical integration line by an arc on the right with centre O. Its radius is $R = (c^2 + T^2)^{1/2}$. On this arc, we have $|y^s| \leqslant y^c$ and $|s| = R$, hence

$$|I(y, T)| \leqslant \frac{1}{2\pi} \pi R \frac{y^c}{R} < y^c.$$

The proof for $y > 1$ is similar by using a rectangle on the left:

$$\left| \frac{1}{2i\pi} \int_{-\infty+iT}^{c+iT} \frac{y^s}{s} ds \right| < \frac{1}{2\pi T} \int_{-\infty}^{c} y^\sigma d\sigma = \frac{y^c}{2\pi T |\log y|},$$

where we get the strict inequality by dealing separately with the $|\sigma| \leqslant \varepsilon$, for which we cannot do better than $|s| \geqslant T$. Elsewhere, we have $|s| \geqslant (T^2 + \varepsilon^2)^{1/2}$.

Then the integration closed path includes the pole in $s = 0$, with residue $1 = \delta(y)$.

In the same way we get the second bound with an arc on the left, and

$$|I(y, T) - 1| < \frac{1}{2\pi} 2\pi R \frac{y^c}{R} < y^c.$$

We have still to deal with $y = 1$. We do it by a direct computation: for $s = c + it$,

$$I(1, T) = \frac{1}{2\pi} \int_0^T \frac{2c}{c^2 + t^2} dt = \frac{1}{\pi} \int_0^{T/c} \frac{du}{1 + u^2} = \frac{1}{2} - \frac{1}{\pi} \int_{T/c}^{\infty} \frac{du}{1 + u^2}$$

and the last integral is $< c/T$.

Lemma 5.46 *For $c > 1$, set*

$$J(x, T) = \frac{1}{2i\pi} \int_{c-iT}^{c+iT} \frac{-\zeta'(s)}{\zeta(s)} \frac{x^s}{s} ds.$$

Then

$$|\psi_0(x) - J(x, T)| < \sum_{\substack{n \geqslant 1 \\ n \neq x}} \Lambda(n)(x/n)^c \min(1, T^{-1} |\log x/n|^{-1}) + c T^{-1} \Lambda(x).$$

where the term with $\Lambda(x)$ is present only when $x = p^m$.

Proof It comes immediately from what precedes.

Notation 5.47 *We introduce Vinogradov's notation*

$$A \ll B \text{ if and only if } A = O(B).$$

Lemma 5.48 *We have*

$$|\psi_0(x) - J(x, T)| \ll \frac{x(\log x)^2}{T} + \min\left(1, \frac{x}{T\langle x \rangle}\right) \log x,$$

where $\langle x \rangle$ denotes the distance from x to the nearest power of a prime number which is different from x.

Proof We choose $c = 1 + (\log x)^{-1}$. Then we observe that $x^c = ex \ll x$. We begin to bound the sum we got in the last lemma by considering the n such that $n \leqslant 3x/4$ or $n \geqslant 5x/4$. Then we have $|\log x/n| \gg 1$, thus the contribution of these terms is at most

$$\ll \frac{x^c}{T} \sum_{n \geqslant 1} \Lambda(n) n^{-c} = \frac{x^c}{T} \left(-\frac{\zeta'(c)}{\zeta(c)}\right) \ll \frac{x \log x}{T}.$$

Consider now the n such that $3x/4 < n < x$. Let x_1 be the largest power of a prime number which is strictly smaller than x. We can assume $3x/4 < x_1 < x$, as otherwise the relevant terms vanish. For $n = x_1$, we have

$$\log \frac{x}{n} = -\log\left(1 - \frac{x - x_1}{x}\right) \geqslant \frac{x - x_1}{x},$$

thus the contribution of this term is

$$\ll \Lambda(x_1) \min\left(1, \frac{x}{T(x - x_1)}\right) \ll \min\left(1, \frac{x}{T(x - x_1)}\right) \log x.$$

For the other terms, we write $n = x_1 - v$ with $0 < v < x/4$ (it is clear, by definition of x_1, that $\Lambda(n) = 0$ for $x_1 < n < x$), hence

$$\log \frac{x}{n} \geqslant \log \frac{x_1}{n} = -\log\left(1 - \frac{v}{x_1}\right) \geqslant \frac{v}{x_1}.$$

Thus the contribution of these terms is

$$\ll \sum_{0 < v < x/4} \Lambda(x_1 - v) \frac{x_1}{Tv} \ll \frac{x(\log x)^2}{T}.$$

The terms $x < n < 5x/4$ are dealt with similarly, except that we replace x_1 by x_2, the smallest power of a prime number which is strictly larger than x.

This completes the proof of the lemma, by bounding $x - x_1$ and $x_2 - x$ form below by $\langle x \rangle$.

Lemma 5.49 *For all $T' > 0$, we can choose $T = T' + O(1)$ such that*

$$|\gamma - T| \gg (\log T)^{-1},$$

for all the zeros $\rho = \beta + i\gamma$ of zeta.

Proof We have shown that

$$N(T' + 1) - N(T' - 1) \ll \log T,$$

so that among the second coordinates of the relevant zeros, there must be a gap of length $\gg (\log T)^{-1}$.

Lemma 5.50 *Let U be a large enough odd integer, and let $T > 0$ such that $|\gamma - T| \gg (\log T)^{-1}$ for every zero $\rho = \beta + i\gamma$ of zeta. We call $\mathcal{R}(T, U)$ the rectangle with vertices*

$$c - iT, \ c + iT, \ -U - iT, \ -U + iT.$$

Then

$$\frac{1}{2i\pi} \int_{\mathcal{R}(T,U)} \left(-\frac{\zeta'(s)}{\zeta(s)} \right) \frac{x^s}{s} ds = x - \sum_{|\gamma|<T} \frac{x^\rho}{\rho} - \frac{\zeta'(0)}{\zeta(0)} - \sum_{0<2m<U} \frac{x^{-2m}}{-2m}.$$

Proof The left vertical side of the closed path goes through the middle point between two trivial zeros of zeta. Hence we can apply the residues theorem.

Remark 5.51 The integral we want to estimate, $J(x, T)$, is the integral on the right vertical side of the closed path. To deal with the other three sides of the rectangle, we need upper bounds for ζ'/ζ.

Lemma 5.52 *For $s = \sigma + iT$, with $|\gamma - T| \gg (\log T)^{-1}$ for every zero $\rho = \beta + i\gamma$, and $-1 \leqslant \sigma \leqslant 2$, we have*

$$\frac{\zeta'(s)}{\zeta(s)} = O((\log T)^2)$$

Proof We recall that under these hypotheses,

$$\frac{\zeta'(s)}{\zeta(s)} = \sum_{|\gamma-T|<1} \frac{1}{s - \rho} + O(\log T),$$

and here, each term of the right-hand sum is $\ll \log T$, and also the number of terms is $\ll \log T$.

Lemma 5.53 *For $s \in \mathbb{C}$, with $\Re(s) \leqslant -1$ and $|s + 2m| \geqslant \frac{1}{2}$ for all $m \in \mathbb{N}$, we have*

$$\left| \zeta'(s)/\zeta(s) \right| \ll \log(2\,|s|).$$

Proof We work with $1 - s$ instead of s (hence $\sigma \geqslant 2$), and we use the asymmetric form of the functional equation:

$$\zeta(1 - s) = 2^{1-s} \pi^{-s} \cos(\pi s/2) \Gamma(s) \zeta(s),$$

which we got through the complements formula and the Legendre duplication formula (for the function Gamma).

The logarithmic derivative gives

$$\frac{\zeta'(1 - s)}{\zeta(1 - s)} = -\log 2\pi - \frac{\pi}{2} \tan\left(\frac{\pi s}{2}\right) + \frac{\Gamma'(s)}{\Gamma(s)} + \frac{\zeta'(s)}{\zeta(s)}.$$

When $|s - (2m + 1)| \geqslant \frac{1}{2}$, the term $\tan(\pi s/2)$ is bounded, hence the condition

$$|(1 - s) + 2m| \geqslant \frac{1}{2}.$$

The term in Γ is $\ll \log |s|$, hence $\ll \log 2\,|1 - s|$ for $\sigma \geqslant 2$.

The last term is bounded, hence the result.

Theorem 5.54 *For $x \geqslant 2$,*

$$\psi_0(x) = x - \sum_{|\rho| < T} \frac{x^\rho}{\rho} - \frac{\zeta'(0)}{\zeta(0)} - \frac{1}{2} \log(1 - x^{-2}) + R(x, T),$$

with

$$R(x, T) \ll \frac{x \log^2(xT)}{T} + \min\left(1, \frac{x}{T\langle x \rangle}\right) \log x.$$

Proof We recall the choice $c = 1 + (\log x)^{-1}$. We have

$$\int_{-1+iT}^{c+iT} \frac{\zeta'(s)}{\zeta(s)} \frac{x^s}{s}\,ds \ll \frac{\log^2 T}{T} \int_{-1}^{c} x^\sigma\,d\sigma \ll \frac{x \log^2 T}{T \log x}$$

and

$$\int_{-U+iT}^{-1+iT} \frac{\zeta'(s)}{\zeta(s)} \frac{x^s}{s}\,ds \ll \frac{\log 2T}{T} \int_{-U}^{-1} x^\sigma\,d\sigma \ll \frac{\log T}{Tx \log x}$$

thus the contribution from the horizontal sides of the rectangle $\mathcal{R}(T, U)$ is

$$\ll \frac{x \log^2 T}{T \log x}$$

The contribution from the left vertical side is

$$\int_{-U-iT}^{-U+iT} \frac{\zeta'(s)}{\zeta(s)} \frac{x^s}{s} ds \ll \frac{\log 2U}{U} \int_{-T}^{T} x^{-U} dt \ll \frac{T \log U}{U x^U}$$

which tends to 0 when $U \to +\infty$.

Gathering the terms, and remarking that

$$\sum_{m=1}^{\infty} \frac{x^{-2m}}{-2m} = \log(1 - x^2),$$

we get the claimed result, with the restriction that $|\gamma - T| \gg (\log T)^{-1}$. This restriction can be lifted, as by modifying T by $O(1)$, we change the sum of ρ by at most $O(\log T)$ terms, and each of them is $O(x/T)$, then the total sum is changed at most by $O(x(\log T)/T)$, and this quantity is absorbed by the error term.

Remark 5.55 Obviously we have

$$R(x, T) \to 0, \quad \text{when } T \to +\infty.$$

Furthermore, if x is an integer, then $\langle x \rangle \geqslant 1$, and then

$$R(x, T) \ll \frac{x \log^2 xT}{T}.$$

Remark 5.56 Ingham [2, p. 81] showed that a slight modification of the estimate for $R(x, T)$ allows to prove that the formula holds for $1 < x < 2$.

5.7   The Prime Numbers Theorem

Lemma 5.57 *We have*

$$\sum_{|\gamma| < T} \left| \frac{x^\rho}{\rho} \right| \ll x \log^2 T \exp\left(-c_1 \frac{\log x}{\log T} \right).$$

Proof We have already studied the region without zeros of zeta. If $\rho = \beta + i\gamma$ is a nontrivial zero with $|\gamma| < T$, and T is large enough, then

$$\beta < 1 - \frac{c_1}{\log T},$$

where $c_1 > 0$ is an absolute constant. We deduce that

$$|x^\rho| = x^\beta < x\exp\left(-c_1\frac{\log x}{\log T}\right).$$

Furthermore, we have $|\rho| \geqslant \gamma$, for $\gamma > 0$, thus we have to estimate

$$\sum_{0<\gamma<T} \frac{1}{\gamma}.$$

Also, denoting by $N(t)$ the number of zeros in the critical strip with second coordinate between 0 and t, we have

$$\sum_{0<\gamma<T} \frac{1}{\gamma} = \int_0^T t^{-1}dN(t) = \frac{N(T)}{T} + \int_0^T t^{-2}N(t)dt,$$

and as $N(t) \ll t\log t$ for t large enough, we get

$$\sum_{0<\gamma<T} \frac{1}{\gamma} \ll \log^2 T$$

which ends the proof.

Theorem 5.58 *For $x \geqslant 2$,*

$$\psi(x) := \sum_{n\leqslant x} \Lambda(n) = x + O(x\exp(-c_2\sqrt{\log x})).$$

Proof Without loss of generality, we can choose x integer, which allows us to use the simple upper bound for $R(x, T)$. Thus we get

$$|\psi(x) - x| \ll \frac{x\log^2 xT}{T} + x\log^2 T\exp\left(-c_1\frac{\log x}{\log T}\right),$$

for x large enough. We choose T such that

$$\log^2 T = \log x, \text{ i.e., } T^{-1} = \exp(-\sqrt{\log x}).$$

We get

$$|\psi(x) - x| \ll x(\log x)^2 \exp(-\sqrt{\log x}) + x(\log x) \exp(-c_1\sqrt{\log x})$$
$$\ll x \exp(-c_2\sqrt{\log x})$$

by choosing $0 < c_2 < \min(1, c_1)$.

Corollary 5.59 *For $x \geqslant 2$,*

$$\theta(x) := \sum_{p \leqslant x} \log p = x + O(x \exp(-c_2\sqrt{\log x})).$$

Proof We have

$$0 \leqslant \psi(x) - \theta(x) = \sum_{\substack{p^m \leqslant x \\ m \geqslant 2}} \log p \leqslant \log x \sum_{m \geqslant 2} \pi(x^{1/m}) = O(\sqrt{x}).$$

Corollary 5.60 *For $x \geqslant 2$,*

$$\pi(x) = \mathrm{li}(x) + O(x \exp(-c_3\sqrt{\log x})).$$

Proof We have

$$\pi(x) = \sum_{p \leqslant x} \frac{\log p}{\log p} = \int_{2^-}^{x} \frac{d\theta(t)}{\log t}$$

$$= \int_{2^-}^{x} \frac{dt}{\log t} + \int_{2^-}^{x} \frac{d(\theta(t) - t)}{\log t}$$

$$= \mathrm{li}(x) + \frac{\theta(x) - x}{\log x} + O(1) + \int_{2^-}^{x} \frac{\theta(t) - t}{t \log^2 t} dt,$$

and we can cut this last integral into two parts. For $t < x^{1/4}$, the contribution is $O(x^{1/4})$, and for $x^{1/4} < t < x$, we have $(\log t)^{1/2} > \frac{1}{2}(\log x)^{1/2}$, hence the result.

Theorem 5.61 *If all the zeros of zeta satisfy $\beta \leqslant \Theta$, where $\frac{1}{2} \leqslant \Theta < 1$, then*

$$\psi(x) = x + O(x^{\Theta} \log^2 x), \qquad \pi(x) = \mathrm{li}(x) + O(x^{\Theta} \log x).$$

Remark 5.62 The Riemann hypothesis corresponds to $\Theta = \frac{1}{2}$.

Proof Same proof, with $|x^\rho| \leqslant x^\Theta$. The explicit formula gives

$$|\psi(x) - x| \ll \frac{x \log^2 xT}{T} + x^\Theta \log^2 T.$$

and we choose $T = x^{1-\Theta}$.

Theorem 5.63 *If there exists $\frac{1}{2} \leqslant \alpha < 1$ such that $\psi(x) = x + O(x^\alpha)$, then all the zeros of zeta satisfy $\beta \leqslant \alpha$.*

Proof We have

$$-\frac{\zeta'(s)}{\zeta(s)} = \sum_{n=1}^{\infty} \frac{\Lambda(n)}{n^s} = \int_{1^-}^{\infty} \frac{d\psi(x)}{x^s} = \int_{1^-}^{\infty} \frac{dx}{x^s} + \int_{1^-}^{\infty} \frac{d(\psi(x) - x)}{x^s}$$

$$= \frac{1}{s-1} + s \int_{1}^{\infty} (\psi(x) - x)x^{-s-1} dx,$$

and if $\psi(x) = x + O(x^\alpha)$, then the integral represents a holomorphic function on $\sigma > \alpha$, hence $\zeta(s)$ cannot have any zero in this half-plane.

References

1. H. Davenport, *Multiplicative Number Theory*. Graduate Texts in Mathematics, vol. 74, 3rd edn. (Springer, New York, 2000); Revised and with a preface by Hugh L. Montgomery.
2. A.E. Ingham, *The Distribution of Prime Numbers*. Cambridge Mathematical Library (Cambridge University Press, Cambridge, 1990); Reprint of the 1932 original, With a foreword by R. C. Vaughan
3. E.C. Titchmarsh, *The Theory of Functions* (Oxford University Press, Oxford, 1932)
4. E.C. Titchmarsh, *The Theory of the Riemann Zeta-Function, Revised by D.R. Heath-Brown*. Oxford Science Publications, 2nd edn. (Oxford University Press, Oxford, 1986)
5. A. Zygmund, *Trigonometric Series. Vol. I, II*. Cambridge Mathematical Library. 3rd edn. (Cambridge University Press, Cambridge, 2002); With a foreword by Robert A. Fefferman

Chapter 6
The Large Sieve

Joël Rivat

6.1 Analytic Form of the Large Sieve

Notation 6.1 *For $x \in \mathbb{R}$, we denote by $\|x\|$ the distance from x to $\mathbb{Z}$:*

$$\|x\| = \min_{n \in \mathbb{Z}} |x - n| \, .$$

Definition 6.2 Let $0 < \delta < 1$. We say that the real numbers $x_1, \ldots, x_R$ are δ-well spaced if

$$\min_{1 \leqslant r < s \leqslant R} \|x_r - x_s\| \geqslant \delta > 0.$$

Remark 6.3 We have $R \leqslant \delta^{-1}$.

Proof By considering the fractional parts we may assume that the x_r are in $[0, 1[$, then re-order them in the increasing order, and then we have

$$x_1 + (R - 1)\delta \leqslant x_R \leqslant x_1 + 1 - \delta,$$

hence $R\delta \leqslant 1$.

J. Rivat (✉)
Aix Marseille Université, CNRS, Centrale Marseille, Institut de Mathématiques de Marseille, I2M – UMR 7357, Marseille, France
e-mail: joel.rivat@univ-amu.fr

© Springer International Publishing AG, part of Springer Nature 2018
S. Ferenczi et al. (eds.), *Ergodic Theory and Dynamical Systems in their Interactions with Arithmetics and Combinatorics*, Lecture Notes in Mathematics 2213, https://doi.org/10.1007/978-3-319-74908-2_6

Theorem 6.4 (Analytic Form of the Large Sieve) *If $x_1, \ldots, x_R$ are δ-well spaced real numbers $(0 < \delta < 1)$ and $a_1, \ldots, a_N \in \mathbb{C}$ then*

$$\sum_{r=1}^{R} \left| \sum_{n=1}^{N} a_n \, e(nx_r) \right|^2 \leq \left(N - 1 + \frac{1}{\delta} \right) \sum_{n=1}^{N} |a_n|^2 .$$

Remark 6.5 We can translate the indices n by observing that $e((n + t)x_r) = e(tx_r) \, e(nx_r)$ and $|e(tx_r)| = 1$.

Remark 6.6 This inequality is optimal: for an integer $R \geq 1$, set

$$x_r := \frac{r}{R}, \text{ for } 1 \leq r \leq R, \quad a_n := \begin{cases} 1 \text{ if } R \mid n, \\ 0 \text{ if } R \nmid n. \end{cases}$$

then for each integer $N \equiv 1 \bmod R$, we have

$$\sum_{r=1}^{R} \left| \sum_{n=0}^{N-1} a_n \, e(nx_r) \right|^2 = \sum_{r=1}^{R} \left| \sum_{\substack{0 \leq n \leq N-1 \\ R \mid n}} 1 \right|^2 = R \left(1 + \frac{N-1}{R} \right)^2,$$

$$\left(N - 1 + \frac{1}{\delta} \right) \sum_{n=0}^{N-1} |a_n|^2 = (N - 1 + R) \left(1 + \frac{N-1}{R} \right) = R \left(1 + \frac{N-1}{R} \right)^2.$$

Notation 6.7 *For each integrable function $f : \mathbb{R} \to \mathbb{C}$, we denote by $\widehat{f}$ its Fourier transform, defined by*

$$\widehat{f}(x) = \int_{\mathbb{R}} f(t) \, e(-xt) \, dt.$$

Lemma 6.8 *Let $F_\delta : \mathbb{R} \to \mathbb{R}$ be a continuous integrable function with bounded variation such that*

$$(i) \ F_\delta \geq \mathbb{1}_{[1,N]}, \quad (ii) \ \widehat{F}_\delta(u) = 0 \text{ for } |u| \geq \delta.$$

Then

$$\sum_{r=1}^{R} \left| \sum_{n=1}^{N} a_n \, e(nx_r) \right|^2 \leq \widehat{F}_\delta(0) \sum_{n=1}^{N} |a_n|^2 .$$

Proof (see Selberg [8, p. 220]) For convenience take $a_n = 0$ for $n \in \mathbb{Z} \setminus [1, N]$ and

$$S(\alpha) = \sum_{n=1}^{N} a_n \, e(n\alpha) = \sum_{n} a_n \, e(n\alpha).$$

By expanding the left-hand term of the inequality

$$\sum_{n \in \mathbb{Z}} \left| \frac{a_n}{\sqrt{F_\delta(n)}} \widehat{F_\delta}(0) - \sum_{r=1}^{R} \sqrt{F_\delta(n)} S(x_r) \, \mathrm{e}(-nx_r) \right|^2 \geqslant 0,$$

we get

$$\sum_{n} \frac{|a_n|^2}{F_\delta(n)} \left(\widehat{F_\delta}(0) \right)^2 - 2 \, \widehat{F_\delta}(0) \, \mathfrak{R} \left(\sum_{r=1}^{R} S(x_r) \sum_{n} \overline{a_n} \, \mathrm{e}(-nx_r) \right)$$

$$+ \sum_{r=1}^{R} \sum_{s=1}^{R} S(x_r) \overline{S(x_s)} \sum_{n} F_\delta(n) \, \mathrm{e}(-n(x_r - x_s)) \geqslant 0.$$

By Poisson summing formula, we have

$$\sum_{n} F_\delta(n) \, \mathrm{e}(-n(x_r - x_s)) = \sum_{m \in \mathbb{Z}} \int_{\mathbb{R}} F_\delta(t) \, \mathrm{e}(-t(x_r - x_s)) \, \mathrm{e}(-tm) dt$$

$$= \sum_{m \in \mathbb{Z}} \widehat{F_\delta}(m + x_r - x_s)$$

$$= \begin{cases} \widehat{F_\delta}(0) & \text{if } r = s, \\ 0 & \text{otherwise.} \end{cases}$$

The previous inequality can be written under the form

$$\sum_{n} \frac{|a_n|^2}{F_\delta(n)} \left(\widehat{F_\delta}(0) \right)^2 - 2\widehat{F_\delta}(0) \sum_{r=1}^{R} |S(x_r)|^2 + \widehat{F_\delta}(0) \sum_{r=1}^{R} |S(x_r)|^2 \geqslant 0,$$

which gives

$$\sum_{r=1}^{R} |S(x_r)|^2 \leqslant \widehat{F_\delta}(0) \sum_{n} \frac{|a_n|^2}{F_\delta(n)} \leqslant \widehat{F_\delta}(0) \sum_{n} |a_n|^2,$$

as expected.

In order to prove Theorem 6.4 it remains to prove that we can find a function F_δ such that

(i) $F_\delta \geqslant \mathbb{1}_{[1,N]}$,   (ii) $\widehat{F_\delta}(u) = 0$ for $|u| \geqslant \delta$,   (iii) $\widehat{F_\delta}(0) = N - 1 + \delta^{-1}$.

6.2   The Beurling–Selberg Function

Proposition 6.9 *The complex variable functions $H(z)$ and $K(z)$ defined by*

$$H(z) := \left(\frac{\sin \pi z}{\pi}\right)^2 \left(\sum_{n=-\infty}^{\infty} \frac{\operatorname{sgn}(n)}{(z-n)^2} + \frac{2}{z}\right); \quad K(z) := \left(\frac{\sin \pi z}{\pi z}\right)^2,$$

are entire, and of exponential type 2π (i.e., bounded from above by $O(e^{2\pi|\Im z|})$).

Proof The function K is obviously entire, and on every compact set H is a uniform limit of a sequence of holomorphic functions, hence H is entire. The exponential type comes from the Euler formula

$$\sin(x + iy) = \frac{e^{ix-y} - e^{-ix+y}}{2i}.$$

The function H is odd and $H(n) = \operatorname{sgn} n$ for all $n \in \mathbb{Z}$. We shall study the difference $\operatorname{sgn} x - H(x)$ for $x \in \mathbb{R}$. As H is odd, it suffices to consider $x > 0$, for which we have $\operatorname{sgn} x - H(x) = 1 - H(x)$. We can compute this quantity:

Lemma 6.10 *For all $z \in \mathbb{C}$,*

$$1 - H(z) = \left(\frac{\sin \pi z}{\pi}\right)^2 \sum_{n=0}^{\infty} \frac{1}{(z+n)^2(z+n+1)^2}. \tag{6.1}$$

Proof By Euler formula

$$\sum_{n=-\infty}^{\infty} \frac{1}{(z-n)^2} = \left(\frac{\pi}{\sin \pi z}\right)^2$$

we get for $z \in \mathbb{C}$,

$$1 - H(z) = \left(\frac{\sin \pi z}{\pi}\right)^2 \left(\sum_{n=-\infty}^{\infty} \frac{1 - \operatorname{sgn}(n)}{(z-n)^2} - \frac{2}{z}\right)$$

$$= \left(\frac{\sin \pi z}{\pi}\right)^2 \left(2\sum_{n=0}^{\infty} \frac{1}{(z+n)^2} - \frac{2}{z} - \frac{1}{z^2}\right).$$

Also, for $z \in \mathbb{C} \setminus \mathbb{Z}$,

$$\frac{2}{z} = 2\sum_{n=0}^{\infty} \left(\frac{1}{z+n} - \frac{1}{z+n+1}\right) = 2\sum_{n=0}^{\infty} \frac{1}{(z+n)(z+n+1)},$$

and

$$\frac{1}{z^2} = \sum_{n=0}^{\infty} \left(\frac{1}{(z+n)^2} - \frac{1}{(z+n+1)^2} \right),$$

thus, for all $z \in \mathbb{C}$ (by analytic continuation),

$$1 - H(z) = \left(\frac{\sin \pi z}{\pi} \right)^2 \sum_{n=0}^{\infty} \left(\frac{1}{(z+n)^2} + \frac{1}{(z+n+1)^2} - \frac{2}{(z+n)(z+n+1)} \right)$$

$$= \left(\frac{\sin \pi z}{\pi} \right)^2 \sum_{n=0}^{\infty} \left(\frac{1}{z+n} - \frac{1}{z+n+1} \right)^2.$$

and equality (6.1) follows.

Theorem 6.11 *For all $x \in \mathbb{R}$,*

$$|H(x)| \leqslant 1, \quad |\text{sgn}(x) - H(x)| \leqslant K(x).$$

Proof For $x > 0$, from (6.1), we get on one side, by summing positive terms, the inequality $1 - H(x) \geqslant 0$, and on the other side as $1 \leqslant (x+n+1)^2 - (x+n)^2$ when $x > 0$ and $n \geqslant 0$, we can write

$$1 - H(x) \leqslant \left(\frac{\sin \pi x}{\pi} \right)^2 \sum_{n=0}^{\infty} \frac{(x+n+1)^2 - (x+n)^2}{(x+n)^2(x+n+1)^2} = \left(\frac{\sin \pi x}{\pi} \right)^2 \frac{1}{x^2} = K(x),$$

hence $0 \leqslant 1 - H(x) \leqslant K(x) \leqslant 1$. As both functions H and sgn are odd, and the function K is even, we get the claimed inequalities.

The equality (6.1) shows that the previous result does not give the order of magnitude of $1 - H(x)$ when $x \to +\infty$. We can improve this result:

Proposition 6.12 *For all $x \in \mathbb{R}$,*

$$|\text{sgn}(x) - H(x)| \leqslant K(x) \min \left(1, \frac{1}{3|x|} \right).$$

Proof For $x > 0$,

$$\frac{1}{x^3} = \sum_{n=0}^{\infty} \left(\frac{1}{(x+n)^3} - \frac{1}{(x+n+1)^3} \right) = \sum_{n=0}^{\infty} \frac{3(x+n)^2 + 3(x+n) + 1}{(x+n)^3(x+n+1)^3},$$

hence

$$\frac{1}{x^3} \geqslant \sum_{n=0}^{\infty} \frac{3(x+n)^2 + 3(x+n)}{(x+n)^3(x+n+1)^3} = 3 \sum_{n=0}^{\infty} \frac{1}{(x+n)^2(x+n+1)^2},$$

thus

$$1 - H(x) \leqslant \left(\frac{\sin \pi x}{\pi}\right)^2 \frac{1}{3x^3} = \frac{K(x)}{3x}.$$

As both functions H and sgn are odd, and the function K is even, we get the claimed inequality.

Corollary 6.13 *The (Beurling–Selberg) function*

$$B(z) := H(z) + K(z) = \left(\frac{\sin \pi z}{\pi}\right)^2 \left(\sum_{n=0}^{\infty} \frac{1}{(z-n)^2} - \sum_{n=1}^{\infty} \frac{1}{(z+n)^2} + \frac{2}{z}\right),$$

is entire, of exponential type 2π, and satisfies

$$\forall x \in \mathbb{R}, \ 0 \leqslant B(x) - \text{sgn}(x) \leqslant 2K(x), \quad \int_{\mathbb{R}} (B(t) - \text{sgn}(t))dt = 1.$$

Proof We just have to compute the last integral, and as $H(x) - \text{sgn}(x)$ is odd and integrable (its absolute value is below $K(x)$), we have

$$\int_{\mathbb{R}} (B(t) - \text{sgn}(t))dt = \int_{\mathbb{R}} (H(t) - \text{sgn}(t))dt + \int_{\mathbb{R}} K(t)dt = \int_{\mathbb{R}} K(t)dt = 1.$$

Corollary 6.14 *Let $a \leqslant b$ and $\delta > 0$. The function F_δ defined by*

$$F_\delta(x) := \tfrac{1}{2}B(\delta(x-a)) + \tfrac{1}{2}B(\delta(b-x)) \tag{6.2}$$

is entire, of exponential type $2\pi\delta$, in $L^1(\mathbb{R}) \cap L^2(\mathbb{R})$, and satisfies

(i) $F_\delta \geqslant \mathbb{1}_{[a,b]}$,   (ii) $\widehat{F_\delta}(u) = 0$ for $|u| \geqslant \delta$,   (iii) $\widehat{F_\delta}(0) = b - a + \delta^{-1}$.

Proof We have for all real number x different from a and b,

$$F_\delta(x) \geqslant \tfrac{1}{2}\,\text{sgn}(\delta(x-a)) + \tfrac{1}{2}\,\text{sgn}(\delta(b-x)) = \begin{cases} 0 \text{ if } x < a \text{ ou } x > b, \\ 1 \text{ if } a < x < b. \end{cases}$$

and by continuity of F_δ in a and b, we get $F_\delta \geqslant \mathbb{1}_{[a,b]}$.

As $F_\delta \geqslant 0$, we show that $F_\delta \in L^1(\mathbb{R})$ by computing $\widehat{F_\delta}(0)$:

$$\widehat{F_\delta}(0) = \int_{\mathbb{R}} F_\delta(x)dx$$

$$= \frac{1}{2} \int_{\mathbb{R}} (B - \mathrm{sgn})(\delta(x - a))dx + \frac{1}{2} \int_{\mathbb{R}} (B - \mathrm{sgn})(\delta(b - x))dx$$

$$+ \int_{\mathbb{R}} \left(\frac{1}{2} \mathrm{sgn}(\delta(x - a)) + \frac{1}{2} \mathrm{sgn}(\delta(b - x)) \right) dx$$

$$= \frac{1}{2\delta} + \frac{1}{2\delta} + b - a.$$

We have $F_\delta(z) = O(e^{2\pi\delta|\Im z|})$, thus in particular, F_δ is bounded on $\mathbb{R}$, and as $F_\delta \in L^1(\mathbb{R})$, we deduce that $F_\delta \in L^2(\mathbb{R})$:

$$\int_{\mathbb{R}} F_\delta^2(t)dt \leqslant \left(\sup_{\mathbb{R}} F_\delta \right) \int_{\mathbb{R}} F_\delta(t)dt.$$

The fact that $\widehat{F_\delta}$ has its support in $[-\delta, \delta]$ comes from the following theorem (see for example Rudin [7, Theorem 19.3]):

Theorem 6.15 (Paley and Wiener) *For every entire function f of exponential type $2\pi A$, such that $f \in L^2(\mathbb{R})$, there exists $\varphi \in L^2(\mathbb{R})$ such that*

$$\forall z \in \mathbb{C}, \ f(z) = \int_{-A}^{A} \varphi(t) \, e(tz)dt.$$

It is indeed possible to compute explicitly $\widehat{F_\delta}$ and thus avoid using the theorem of Paley and Wiener.

6.3 The Sieve

How does the famous Eratosthenes sieve work? Let $c_1, \ldots, c_N \in \mathbb{C}$ (it quite usual to take $c_n = 1$) be ordered from 1 to N in an indexed table. For each prime number $p \leqslant \sqrt{N}$, we "sieve" by p, namely in the table we replace c_n by 0 for $n > p$ such that $n \equiv 0 \bmod p$. Let $a_1, \ldots, a_N$ be the sequence we get. It satisfies

$$a_n \neq 0 \Rightarrow (n = 1) \text{ or } (n \text{ prime}).$$

A natural generalization of this process consists, for each prime number p, in eliminating several classes modulo p. Let $c_{M+1}, \ldots, c_{M+N}$ be complex numbers and $\mathcal{P}$ a set of prime numbers. For each prime number p, we choose to delete

$w(p)$ classes modulo p, say $0 \leqslant r(p, 1) < r(p, 2) < \cdots < r(p, w(p)) < p$. Let $a_{M+1}, \ldots, a_{M+N}$ be the resulting sequence. It satisfies

$$a_n \neq 0 \Rightarrow \forall p \in \mathcal{P}, \; \forall i \in \{1, \ldots, w(p)\}, \; n \not\equiv r(p, i) \bmod p.$$

This "constructive" approach is the right one to implement "concretely" a sieve, on a computer for example, but in theoretical problems we do rather ask the question directly from the resulting sequence, and this is the point of view we shall adopt in the whole sequel.

Given a sequence of complex numbers $a_{M+1}, \ldots, a_{M+N}$, we set

$$Z := \sum_{n=M+1}^{M+N} a_n, \quad Z(q, r) := \sum_{\substack{M+1 \leqslant n \leqslant M+N \\ n \equiv r \bmod q}} a_n.$$

and for all prime p we define $w(p)$ as the number of classes r modulo p such that $a_n = 0$ for all $n \equiv r \bmod p$. Excluding the case where the sequence (a_n) is identically zero, we have $0 \leqslant w(p) < p$ for all p. The principle of the sieve is to deduce information on the size of Z from the knowledge of $w(p)$ for p prime.

6.4 The Arithmetic Form of the Large Sieve

To detect the integers n congruent to r modulo q, we use

$$\sum_{a=1}^{q} e\left(\frac{a(n-r)}{q}\right) = \begin{cases} q \text{ if } n \equiv r \bmod q, \\ 0 \text{ otherwise.} \end{cases}$$

which leads us to introduce

$$S(\alpha) := \sum_{n=M+1}^{M+N} a_n \, e(n\alpha).$$

In particular

$$S\left(\frac{a}{q}\right) = \sum_{n=M+1}^{M+N} a_n \, e\left(\frac{an}{q}\right) = \sum_{r=1}^{q} e\left(\frac{ar}{q}\right) Z(q, r).$$

We have

$$\sum_{a=1}^{q}\left|S\left(\frac{a}{q}\right)\right|^2 = \sum_{a=1}^{q}\left|\sum_{r=1}^{q}e\left(\frac{ar}{q}\right)Z(q,r)\right|^2$$

$$= \sum_{r=1}^{q}\sum_{r'=1}^{q}Z(q,r)\overline{Z(q,r')}\sum_{a=1}^{q}e\left(\frac{a(r-r')}{q}\right)$$

$$= q\sum_{r=1}^{q}|Z(q,r)|^2.$$

As

$$Z = S(0) = \sum_{r=1}^{q}Z(q,r),$$

we may imagine that with a good distribution, the mean value of $Z(q,r)$ is close to Z/q. We can estimate on average the variance of this approximation:

Proposition 6.16 *We have*

$$q\sum_{r=1}^{q}\left|Z(q,r) - \frac{Z}{q}\right|^2 = \sum_{a=1}^{q-1}\left|S\left(\frac{a}{q}\right)\right|^2.$$

Proof We have

$$q\sum_{r=1}^{q}\left|Z(q,r) - \frac{Z}{q}\right|^2 = q\sum_{r=1}^{q}\left(|Z(q,r)|^2 - \frac{2}{q}\Re(Z(q,r)\overline{Z}) + \frac{|Z|^2}{q^2}\right)$$

$$= q\sum_{r=1}^{q}|Z(q,r)|^2 - 2|Z|^2 + |Z|^2$$

$$= \sum_{a=1}^{q}\left|S\left(\frac{a}{q}\right)\right|^2 - |S(1)|^2$$

$$= \sum_{a=1}^{q-1}\left|S\left(\frac{a}{q}\right)\right|^2.$$

Proposition 6.17 *For all real number $Q > 1$, we have*

$$\sum_{\substack{q\leqslant Q}}\sum_{\substack{1\leqslant a\leqslant q \\ (a,q)=1}}|S(a/q)|^2 \leqslant (N-1+Q^2)\sum_{n=M+1}^{M+N}|a_n|^2.$$

Proof We apply the analytic form of the large sieve by replacing x_r with a/q, $(a, q) = 1, 1 \leqslant a \leqslant q, q \leqslant Q$. We get for $r \neq s$,

$$\|x_r - x_s\| = \left\| \frac{a}{q} - \frac{a'}{q'} \right\| = \left\| \frac{aq' - a'q}{qq'} \right\| \geqslant \frac{1}{Q^2},$$

hence the x_r are Q^{-2}-well spaced.

We remark that the summation conditions on a are different in the last two propositions. However, they coincide when q is prime. By bounding q from below by the sum restricted to prime numbers, we get the following result, which is very useful for applications:

Theorem 6.18 *We have*

$$\sum_{p \leqslant Q} p \sum_{r=1}^{p} \left| Z(p, r) - \frac{Z}{p} \right|^2 \leqslant (N - 1 + Q^2) \sum_{n=M+1}^{M+N} |a_n|^2.$$

Corollary 6.19 *Let $\mathcal{A}$ be a set of integers in the interval $[M+1, M+N]$. Let $\mathcal{P}$ be a set of prime numbers $p \leqslant Q$. We assume that there exists a real number $0 < \tau < 1$ such that, for all $p \in \mathcal{P}$, $\#\{a \in \mathcal{A} : a \equiv r \bmod p\} = 0$ for at least τp classes r modulo p. Then*

$$|\mathcal{A}| \leqslant \frac{N - 1 + Q^2}{\tau |\mathcal{P}|}.$$

Remark 6.20 This result illustrates the denomination *large sieve*: we delete a positive proportion (τ) of classes modulo p.

Proof We take $a_n = 1$ if $n \in \mathcal{A}$ and $a_n = 0$ otherwise. Thus we have, for $p \in \mathcal{P}$,

$$\sum_{r=1}^{p} \left| Z(p, r) - \frac{Z}{p} \right|^2 \geqslant \tau p \left(\frac{Z}{p} \right)^2,$$

and, applying the theorem,

$$\sum_{p \in \mathcal{P}} \tau Z^2 \leqslant \sum_{p \leqslant Q} p \sum_{r=1}^{p} \left| Z(p, r) - \frac{Z}{p} \right|^2$$

$$\leqslant (N - 1 + Q^2) \sum_{n=M+1}^{M+N} |a_n|^2 = (N - 1 + Q^2)Z,$$

which gives the result.

Example 6.21 Let $\mathcal{A}$ be the set of squares in the interval $[1, N]$. We take $Q = \sqrt{N}$ and $\mathcal{P}$ the set of odd prime numbers $p \leq Q$. Then $Z(p, r) = 0$ when r is not a quadratic residue modulo p, hence $Z(p, r) = 0$ for at least $\frac{1}{2}(p - 1)$ values of r. Thus we can take $\tau = \frac{1}{3}$ and we get

$$|\mathcal{A}| \ll \frac{N}{Q/\log Q} \asymp \sqrt{N} \log N$$

which is not far from the truth $|\mathcal{A}| \sim \sqrt{N}$.

Proposition 6.22 *For every prime number $p \geq 1$,*

$$|S(0)|^2 \frac{w(p)}{p - w(p)} \leq \sum_{a=1}^{p-1} \left| S\left(\frac{a}{p}\right) \right|^2 .$$

Proof We write

$$Z = \sum_{r=1}^{p} Z(p, r),$$

and observing that $Z(p, r)$ is zero for $w(p)$ classes modulo p, we get by Cauchy-Schwarz

$$|Z|^2 \leq (p - w(p)) \sum_{r=1}^{p} |Z(p, r)|^2 .$$

But we have seen that

$$p \sum_{r=1}^{p} |Z(p, r)|^2 = \sum_{a=1}^{p} \left| S\left(\frac{a}{p}\right) \right|^2 = |Z|^2 + \sum_{a=1}^{p-1} \left| S\left(\frac{a}{p}\right) \right|^2 ,$$

hence

$$p\, |Z|^2 \leq (p - w(p))p \sum_{r=1}^{p} |Z(p, r)|^2$$

$$\leq (p - w(p))\, |Z|^2 + (p - w(p)) \sum_{a=1}^{p-1} \left| S\left(\frac{a}{p}\right) \right|^2 .$$

In order to generalize this result to nonprime integers q, we introduce a multiplicative function ℓ and its summing function L by setting

$$\ell(q) = \mu^2(q) \prod_{p \mid q} \frac{w(p)}{p - w(p)}, \quad L(Q) := \sum_{q \leqslant Q} \ell(q).$$

Theorem 6.23 *For every integer $q \geqslant 1$,*

$$\left| \sum_{n=M+1}^{M+N} a_n \right|^2 \mu^2(q) \prod_{p \mid q} \frac{w(p)}{p - w(p)} \leqslant \sum_{\substack{1 \leqslant a \leqslant q \\ (a,q)=1}} \left| S\left(\frac{a}{q}\right) \right|^2.$$

Proof If q is not square-free, then $\mu^2(q) = 0$ and the result is trivial. Let us show by induction on $q \geqslant 1$ square-free that for each sequence $a_{M+1}, \ldots, a_{M+N}$, we have

$$\left| \sum_{n=M+1}^{M+N} a_n \right|^2 \ell(q) \leqslant \sum_{\substack{1 \leqslant a \leqslant q \\ (a,q)=1}} \left| \sum_{n=M+1}^{M+N} a_n \, e\left(\frac{an}{q}\right) \right|^2.$$

The property is trivially true for $q = 1$, and by the previous proposition for $q = 2$, $q = 3$, and all prime q.

Thus we take a nonprime $q \geqslant 4$, square-free and we suppose the property is true for all $q' < q$. Let p be the smallest prime factor (for example) and $q' = q/p$. We take a sequence $a_{M+1}, \ldots, a_{M+N}$ and we write

$$S(\alpha) = \sum_{n=M+1}^{M+N} a_n \, e(n\alpha).$$

Applying the induction hypothesis, on one side with p and the sequence $(a_n \, e(n\beta))$, we get

$$\forall \beta \in \mathbb{R}, \ |S(\beta)|^2 \ell(p) \leqslant \sum_{1 \leqslant b < p} \left| S\left(\frac{b}{p} + \beta\right) \right|^2.$$

and on the other side with q' and the sequence (a_n), we get

$$|S(0)|^2 \ell(q') \leqslant \sum_{\substack{1 \leqslant a' \leqslant q' \\ (a',q')=1}} \left| S\left(\frac{a'}{q'}\right) \right|^2.$$

We deduce that

$$|S(0)|^2 \, \ell(p)\ell(q') \leqslant \sum_{\substack{1 \leqslant a' \leqslant q' \\ (a',q')=1}} \left| S\left(\frac{a'}{q'}\right) \right|^2 \ell(p) \leqslant \sum_{\substack{1 \leqslant a' \leqslant q' \\ (a',q')=1}} \sum_{1 \leqslant b < p} \left| S\left(\frac{b}{p} + \frac{a'}{q'}\right) \right|^2 .$$

As q is square-free we have $(p, q') = 1$. The map which to (a', b) associates a such that $0 \leqslant a < pq'$ and $a \equiv bq' + a'p \bmod pq'$ is a bijection because by the Chinese Remainder Theorem, the system

$$\begin{cases} a \equiv bq' \bmod p \\ a \equiv a'p \bmod q' \end{cases}$$

admits a unique solution a with $0 \leqslant a < pq'$ (see [5, Theorem 59] for a different argument). When $(b, p) = 1$ and $(a', q') = 1$, we have $(a, q) = 1$. Thus

$$S\left(\frac{b}{p} + \frac{a'}{q'}\right) = S\left(\frac{bq' + a'p}{q}\right) = S\left(\frac{a}{q}\right),$$

hence finally

$$|S(0)|^2 \, \ell(q) \leqslant \sum_{\substack{1 \leqslant a \leqslant q \\ (a,q)=1}} \left| S\left(\frac{a}{q}\right) \right|^2 ,$$

which completes the induction.

Theorem 6.24 *We have*
$$\left| \sum_{n=M+1}^{M+N} a_n \right|^2 \leqslant \frac{N - 1 + Q^2}{L(Q)} \sum_{n=M+1}^{M+N} |a_n|^2 .$$

Proof It suffices to sum on q in the previous theorem, and to apply the large sieve to the right-hand side.

6.5   Twin Prime Numbers

Theorem 6.25 *The number $J(x)$ of prime numbers $p \leqslant x$ such that $p + 2$ is also prime, satisfies*

$$J(x) \leqslant 8 \frac{Cx}{\log^2 x} + O\left(\frac{x \log \log x}{\log^3 x}\right),$$

where

$$C := 2 \prod_{p \geqslant 3} \left(1 - \frac{1}{(p-1)^2} \right).$$

Remark 6.26 This bound is asymptotically eight times larger than the value conjectured by Hardy and Littlewood (1922) on the basis of an analytic approach in conformity with a probabilistic heuristic (see Hardy and Wright [5, pp. 371–373 and the notes on p. 374]), which is also corroborated by computer experiments.

Proof We apply the arithmetic large sieve with $M = 0$, $N = \lfloor x \rfloor$, $Q = \sqrt{x/(\log x)}$ and

$$a_n = \begin{cases} 1 \text{ if } P^-(n(n+2)) > Q, \\ 0 \text{ otherwise.} \end{cases}$$

Thus, counting for each prime number $p \leqslant Q$ the number of sieved classes modulo p, we get

$$w(2) = 1, \quad \forall p \geqslant 3, \ w(p) = 2,$$

hence

$$\ell(2) = 1, \quad \forall p \geqslant 3, \ \ell(p) = \frac{2}{p-2}.$$

We have $|a_n|^2 = a_n$, hence

$$J(x) - J(\sqrt{x}) \leqslant \sum_{1 \leqslant n \leqslant x} a_n \leqslant \frac{x}{L(Q)} \left(1 + \frac{Q^2}{x} \right),$$

where

$$L(Q) = \sum_{q \leqslant Q} \ell(q), \quad \ell(q) = \mu^2(q) \prod_{p \mid q} \ell(p).$$

To finish the proof, it suffices then to determine the behaviour of $L(Q)$ at infinity, which will require several steps of manipulating arithmetic functions.

Proposition 6.27 *For all integer $n \geqslant 1$,*

$$\mu^2(n) = \sum_{d^2 \mid n} \mu(d).$$

Proof Each integer $n \geqslant 1$ is decomposed in a unique way under the form

$$n = qm^2, \quad \mu^2(q) = 1,$$

where q is the product of prime numbers which appear in the decomposition of n into prime factors with an odd exponent. We can write

$$\mu^2(n) = \delta(m) = \sum_{d \mid m} \mu(d),$$

but the condition $d \mid m$ is equivalent to $d^2 \mid n$, hence the result.

Proposition 6.28 *For all $x \geqslant 1$,*

$$\sum_{n \leqslant x} \mu^2(n) = \frac{6}{\pi^2} x + O(\sqrt{x}).$$

Proof We have

$$\sum_{n \leqslant x} \mu^2(n) = \sum_{n \leqslant x} \sum_{d^2 \mid n} \mu(d) = \sum_{d \leqslant \sqrt{x}} \mu(d) \left\lfloor \frac{x}{d^2} \right\rfloor = \frac{x}{\zeta(2)} + O\left(x \sum_{d > \sqrt{x}} d^{-2} + \sqrt{x} \right),$$

hence the result.

Proposition 6.29 *We have the convolution relations*

$$2^\omega = \mathbb{1} * \mu^2, \quad \delta = (-1)^\Omega * \mu^2,$$

where $\omega(n)$ (resp. $\Omega(n)$) denotes the number of prime factors of n counted with (resp. without) their multiplicity. We notice that $(-1)^\Omega$ is the Liouville function.

Proof As all these arithmetic functions are multiplicative, it suffices to check these relations for the p^ν, $\nu \geqslant 1$. We have

$$\mathbb{1} * \mu^2(p^\nu) = 1 \cdot \mu^2(1) + 1 \cdot \mu^2(p) = 2 = 2^{\omega(p^\nu)}$$

and

$$(-1)^\Omega * \mu^2(p^\nu) = (-1)^{\Omega(p^\nu)} \mu^2(1) + (-1)^{\Omega(p^{\nu-1})} \mu^2(p)$$
$$= (-1)^\nu + (-1)^{\nu-1} = 0 = \delta(p^\nu),$$

hence the result.

Lemma 6.30 *We have*

$$\sum_{n \leqslant x} 2^{\omega(n)} = \frac{6}{\pi^2} x \log x + O(x).$$

Proof We write

$$\sum_{n \leqslant x} 2^{\omega(n)} = \sum_{n \leqslant x} \mathbb{1} * \mu^2(n) = \sum_{m \leqslant x} \sum_{d \leqslant x/m} \mu^2(d) = \sum_{m \leqslant x} \left(\frac{6}{\pi^2} \frac{x}{m} + O(\sqrt{x/m}) \right),$$

and the result follows from the relations

$$\sum_{m \leqslant x} \frac{1}{m} = \log x + O(1), \quad \sum_{m \leqslant x} \frac{1}{m^{1/2}} = O(x^{1/2}).$$

Lemma 6.31 *We have for $x \geqslant 1$,*

$$\sum_{n \leqslant x} \frac{2^{\omega(n)}}{n} = \frac{3}{\pi^2} \log^2 x + O(\log x).$$

Proof We have

$$\sum_{n \leqslant x} \frac{2^{\omega(n)}}{n} = \frac{1}{x} \sum_{n \leqslant x} 2^{\omega(n)} + \int_{1^-}^{x} \left(\sum_{n \leqslant t} 2^{\omega(n)} \right) \frac{dt}{t^2}$$

$$= O(\log x) + \frac{6}{\pi^2} \int_{1}^{x} \frac{\log t}{t} dt + O\left(\int_{1}^{x} \frac{dt}{t} \right)$$

$$= \frac{3}{\pi^2} \log^2 x + O(\log x).$$

Lemma 6.32 *We have for $0 < \sigma < 1$ and $x \geqslant 1$,*

$$\sum_{n \leqslant x} \frac{2^{\omega(n)}}{n^\sigma} = O\left(\frac{x^{1-\sigma}}{1-\sigma} \log x \right).$$

Proof We have

$$\sum_{n \leqslant x} \frac{2^{\omega(n)}}{n^\sigma} = \frac{1}{x^\sigma} \sum_{n \leqslant x} 2^{\omega(n)} + \sigma \int_{1^-}^{x} \left(\sum_{n \leqslant t} 2^{\omega(n)} \right) \frac{dt}{t^{\sigma+1}}$$

$$\ll x^{1-\sigma} \log x + \sigma \int_{1}^{x} \frac{\log t}{t^\sigma} dt$$

$$\ll x^{1-\sigma} \log x + (\log x)\, \sigma \int_1^x \frac{dt}{t^\sigma}$$

$$\ll \frac{x^{1-\sigma}}{1-\sigma} \log x.$$

We come back now to the study of $L(Q)$. We can write

$$\ell(q) = \mu^2(q) \frac{2^{\omega(q)}}{q} \prod_{\substack{p\mid q \\ p\geq 3}} \frac{p}{p-2},$$

which gives the idea to introduce $f(q) = q\,\ell(q)$, and h such that

$$f = 2^\omega * h = \mathbb{1} * \mu^2 * h, \quad h = f * \mu * (-1)^\Omega.$$

This allows to write for $\nu \geqslant 1$,

$$h(p^\nu) = \sum_{i+j+k=\nu} f(p^i)\mu(p^j)(-1)^{\Omega(p^k)}$$

$$= \sum_{i=0}^{1} \sum_{j=0}^{\min(1,\nu-i)} f(p^i)\mu(p^j)(-1)^{\nu-i-j}.$$

As $f(2) = 2$ and $f(p) = 2p/(p-2)$ for $p \geqslant 3$, we get thus,

$$h(2) = 0,$$
$$h(2^\nu) = 2(-1)^{\nu-1} \quad (\nu \geqslant 2),$$
$$h(p) = \frac{4}{p-2},$$
$$h(p^\nu) = 2(-1)^{\nu-1}\frac{p+2}{p-2} \quad (\nu \geqslant 2).$$

For all $\sigma > \frac{1}{2}$, we have

$$\sum_p \sum_{\nu \geqslant 1} \frac{|h(p^\nu)|}{p^{\nu\sigma}} \ll \sum_{\nu \geqslant 1} \frac{1}{2^{\nu\sigma}} + \sum_p \frac{1}{p^{1+\sigma}} + \sum_p \sum_{\nu \geqslant 2} \frac{1}{p^{\nu\sigma}} < +\infty,$$

hence the Dirichlet series $\sum_d h(d)d^{-s}$ converges absolutely for $\sigma > \frac{1}{2}$.

We can write

$$\sum_{q \leqslant Q} \ell(q) = \sum_{q \leqslant Q} \frac{f(q)}{q} = \sum_{dm \leqslant Q} \frac{2^{\omega(m)}}{m} \frac{h(d)}{d},$$

hence

$$\sum_{q \leqslant Q} \ell(q) = \sum_{m \leqslant Q} \frac{2^{\omega(m)}}{m} \sum_{d \leqslant Q/m} \frac{h(d)}{d}$$

$$= \sum_{m \leqslant Q} \frac{2^{\omega(m)}}{m} \sum_{d=1}^{\infty} \frac{h(d)}{d} - \sum_{m \leqslant Q} \frac{2^{\omega(m)}}{m} \sum_{d > Q/m} \frac{h(d)}{d}.$$

For all $\frac{1}{2} < \sigma < 1$,

$$\left| \sum_{d > Q/m} \frac{h(d)}{d} \right| \leqslant \sum_{d > Q/m} \frac{|h(d)|}{d^{\sigma} d^{1-\sigma}} \leqslant (Q/m)^{\sigma-1} \sum_{d=1}^{\infty} \frac{|h(d)|}{d^{\sigma}},$$

hence

$$\sum_{q \leqslant Q} \ell(q) = \sum_{m \leqslant Q} \frac{2^{\omega(m)}}{m} \sum_{d=1}^{\infty} \frac{h(d)}{d} + O_{\sigma} \left(Q^{\sigma-1} \sum_{m \leqslant Q} \frac{2^{\omega(m)}}{m^{\sigma}} \right)$$

$$= \frac{3}{\pi^2} (\log Q)^2 \sum_{d=1}^{\infty} \frac{h(d)}{d} + O_{\sigma}(\log Q),$$

and we can take for example $\sigma = 2/3$ (this choice has influence only on the implicit constant).

Thus we get

$$J(x) \leqslant \frac{2Cx}{\log^2 Q} + O \left(\frac{x}{\log^3 x} \right)$$

with

$$C = \left(\frac{6}{\pi^2} \sum_{d=1}^{\infty} \frac{h(d)}{d} \right)^{-1} = \zeta(2) \left(\sum_{d=1}^{\infty} \frac{h(d)}{d} \right)^{-1}$$

$$= \prod_p \left(1 - \frac{1}{p^2} \right)^{-1} \prod_p \left(1 + \sum_{\nu \geqslant 1} \frac{h(p^{\nu})}{p^{\nu}} \right)^{-1}.$$

The contribution of $p = 2$ in this product is

$$\frac{4}{3}\left(1 + \sum_{\nu \geq 2} \frac{2(-1)^{\nu-1}}{2^\nu}\right)^{-1} = \frac{4}{3}\left(\frac{1}{1 - (-\frac{1}{2})}\right)^{-1} = 2.$$

For $p \geq 3$ we have

$$1 + \sum_{\nu \geq 1} \frac{h(p^\nu)}{p^\nu} = 1 + \frac{4}{p(p-2)} + 2\frac{p+2}{p-2} \sum_{\nu \geq 2} \frac{(-1)^{\nu-1}}{p^\nu}$$

$$= 1 + \frac{4}{p(p-2)} - 2\frac{p+2}{p-2}\frac{1/p^2}{1-(-1/p)}$$

$$= \frac{p^3 - 2p^2 + p^2 - 2p + 4p + 4 - 2p - 4}{p(p-2)(p+1)}$$

$$= \frac{p^2(p-1)}{p(p-2)(p+1)},$$

hence

$$C = 2\prod_{p \geq 3}\left(1 - \frac{1}{p^2}\right)^{-1}\left(1 + \sum_{\nu \geq 1} \frac{h(p^\nu)}{p^\nu}\right)^{-1}$$

$$= 2\prod_{p \geq 3} \frac{p^2}{p^2 - 1} \cdot \frac{p(p-2)(p+1)}{p^2(p-1)}$$

$$= 2\prod_{p \geq 3} \frac{p(p-2)}{(p-1)^2}$$

$$= 2\prod_{p \geq 3}\left(1 - \frac{1}{(p-1)^2}\right).$$

Finally,

$$\log Q = \log\sqrt{\frac{x}{\log x}} = \tfrac{1}{2}(\log x - \log\log x),$$

hence

$$\log^2 Q = \tfrac{1}{4}\log^2 x\left(1 + O\left(\frac{\log\log x}{\log x}\right)\right),$$

which yields

$$J(x) \leqslant \frac{8Cx}{\log^2 x} + O\left(\frac{x \log \log x}{\log^3 x}\right).$$

6.6 Sums of Characters

For an introduction to Dirichlet characters and primitive characters we refer to Davenport [3] and Montgomery and Vaughan [6].

Theorem 6.33 *Let $a_{M+1}, \ldots, a_{M+N}$ be complex numbers and χ a Dirichlet character modulo q. We write*

$$T(\chi) = \sum_{n=M+1}^{M+N} a_n \chi(n).$$

Then for all $Q \geqslant 1$,

$$\sum_{q \leqslant Q} \frac{q}{\varphi(q)} \sum_{\chi^*} |T(\chi^*)|^2 \leqslant (N - 1 + Q^2) \sum_{n=M+1}^{M+N} |a_n|^2,$$

where χ^* denotes a primitive character modulo q (i.e., a character for which the smallest period is q).

Proof For a character χ modulo q, the Gauß sums are defined by

$$\tau(a, \chi) = \sum_{n=1}^{q} \chi(n)\, e(an/q), \quad \tau(\chi) = \tau(1, \chi).$$

They satisfy, when χ is a primitive character modulo q,

$$\tau(a, \chi) = \overline{\chi}(a)\tau(\chi), \quad |\tau(\chi)| = \sqrt{q},$$

hence for χ primitive modulo q, we have for all n,

$$\chi(n) = \frac{1}{\tau(\overline{\chi})} \sum_{a=1}^{q} \overline{\chi}(a)\, e(an/q).$$

Thus we have, still for χ primitive modulo q,

$$T(\chi) = \frac{1}{\tau(\overline{\chi})} \sum_{a=1}^{q} \overline{\chi}(a) \sum_{n=M+1}^{M+N} a_n\, e(an/q) = \frac{1}{\tau(\overline{\chi})} \sum_{a=1}^{q} \overline{\chi}(a) S(a/q).$$

Furthermore, summing on the primitive characters modulo q,

$$\sum_{\chi^*} |T(\chi^*)|^2 = \sum_{\chi^*} \frac{1}{|\tau(\overline{\chi^*})|^2} \left| \sum_{a=1}^{q} \overline{\chi^*}(a)S(a/q) \right|^2 = \frac{1}{q} \sum_{\chi^*} \left| \sum_{a=1}^{q} \overline{\chi^*}(a)S(a/q) \right|^2.$$

We get an upper bound when we extend the right-hand sum to all the characters modulo q. Thus

$$\sum_{\chi^*} |T(\chi^*)|^2 \leqslant \frac{1}{q} \sum_{\chi} \left| \sum_{a=1}^{q} \overline{\chi}(a)S(a/q) \right|^2$$

$$\leqslant \frac{1}{q} \sum_{a=1}^{q} \sum_{b=1}^{q} S(a/q)\overline{S(b/q)} \sum_{\chi} \overline{\chi}(a)\chi(b),$$

but, from the orthogonality relations on the characters,

$$\sum_{\chi} \overline{\chi}(a)\chi(b) = \begin{cases} \varphi(q) & \text{if } (a, q) = 1 \text{ and } a \equiv b \bmod q, \\ 0 & \text{otherwise.} \end{cases}$$

It follows that

$$\sum_{\chi^*} |T(\chi^*)|^2 \leqslant \frac{\varphi(q)}{q} \sum_{\substack{a=1 \\ (a,q)=1}}^{q} |S(a/q)|^2.$$

Applying the large sieve, we get

$$\sum_{q \leqslant Q} \frac{q}{\varphi(q)} \sum_{\chi^*} |T(\chi^*)|^2 \leqslant \sum_{q \leqslant Q} \sum_{\substack{a=1 \\ (a,q)=1}}^{q} |S(a/q)|^2 \leqslant (N - 1 + Q^2) \sum_{n=M+1}^{M+N} |a_n|^2,$$

which completes the proof.

6.7   Complements

A defect of the large sieve is that it does not take into account the irregularities in the distribution of the x_r, which are very natural. Thus when the x_r are the a/q, $1 \leqslant a \leqslant q$, $(a, q) = 1$, we have for adjacent a/q and a'/q',

$$\left\| \frac{a}{q} - \frac{a'}{q'} \right\| = \frac{1}{qq'},$$

thus the distance from a/q to its nearest neighbour is very variable.

Theorem 6.34 (Montgomery and Vaughan) *If* $x_1, \ldots, x_R$ *are distinct real numbers modulo* 1 *and* $\delta_r = \min_{s \neq r} \|x_s - x_r\|$, *we have*

$$\sum_{r=1}^{R} |S(x_r)|^2 \left(N + \tfrac{3}{2}\delta_r^{-1}\right)^{-1} \leq \sum_{n=M+1}^{M+N} |a_n|^2.$$

When for each prime number, we sieve almost all the classes modulo p, we can use the following result:

Theorem 6.35 (Gallagher) *Let* $\mathcal{A}$ *be a set of integers in the interval* $[M + 1, M + N]$. *For each prime number* p, *we denote by* $\nu(p)$ *the number of residual classes modulo* p *which contain an element of* $\mathcal{A}$. *Then for a finite set* $\mathcal{P}$ *of prime numbers we have*

$$|\mathcal{A}| \leq \frac{\sum_{p \in \mathcal{P}} \log p - \log N}{\sum_{p \in \mathcal{P}} \frac{\log p}{\nu(p)} - \log N}$$

provided the denominator is strictly positive.

6.8 Double Large Sieve

Lemma 6.36 *Let* $\mathbf{Y}$ *be a finite set of points* $\mathbf{y} = (y_1, \ldots, y_K) \in \mathbb{R}^K$ *and* $c(\mathbf{y})$ *be complex numbers. For all real numbers* $\delta_1, \ldots, \delta_K > 0$ *and* $T_1, \ldots, T_K > 0$, *we have*

$$\int_{-T_1}^{T_1} \cdots \int_{-T_K}^{T_K} \left| \sum_{\mathbf{y} \in \mathbf{Y}} c(\mathbf{y}) \, \mathrm{e}(\mathbf{y} \cdot \mathbf{t}) \right|^2 dt_1 \cdots dt_K \leq \prod_{k=1}^{K} (2T_k + \delta_k^{-1}) \sum_{\substack{(\mathbf{y}, \mathbf{y}') \in \mathbf{Y}^2 \\ |y_k - y_k'| < \delta_k \\ (k=1,\ldots,K)}} |c(\mathbf{y}) c(\mathbf{y}')| .$$

Proof cf. Bombieri-Iwaniec [2, Lemma 2.3] (see also [4, Lemma 7.4]).

For each $(\delta_k, T_k) \in \mathbb{R}^2$ with $\delta_k > 0$ and $T_k > 0$, the Beurling-Selberg function, thanks to (6.2) permits to construct F_{δ_k} such that $F_{\delta_k}(t) \geq 1$ for $|t| \leq T_k$, $F_{\delta_k}(t) \geq 0$ for $t \in \mathbb{R}$, $\widehat{F_{\delta_k}}(u) = 0$ for $|u| \geq \delta_k$, $\left|\widehat{F_{\delta_k}}(u)\right| \leq \widehat{F_{\delta_k}}(0) = 2T_k + \delta_k^{-1}$ for $u \in \mathbb{R}$.

For $k = 1, \ldots, K$, majorizing 1 by $F_{\delta_k}(t)$ and then extending the integral, the left hand side of the inequality is bounded above by

$$\int_{-\infty}^{+\infty} \cdots \int_{-\infty}^{+\infty} F_{\delta_1}(t) \cdots F_{\delta_K}(t) \left| \sum_{\mathbf{y} \in \mathbf{Y}} c(\mathbf{y}) \, \mathrm{e}(\mathbf{y} \cdot \mathbf{t}) \right|^2 dt_1 \cdots dt_K,$$

and, expanding the square, this leads to

$$\sum_{(\mathbf{y},\mathbf{y}')\in\mathbf{Y}^2} \overline{c(\mathbf{y})}\, c(\mathbf{y}')\, \widehat{F_{\delta_1}}(y_1 - y_1')\cdots \widehat{F_{\delta_K}}(y_K - y_K').$$

By the properties of F_{δ_k}, the non zero terms of this sum satisfy $\left|y_k - y_k'\right| < \delta_k$ for $k = 1,\ldots,K$, and since $\left|\widehat{F_{\delta_k}}\right| \leqslant \widehat{F_{\delta_k}}(0) = 2T_k + \delta_k^{-1}$, the complete sum is majorized by the right hand side expression in the Lemma.

Lemma 6.37 *Let* $a \leqslant b$ *and* $\delta > 0$. *There exists a function* $f_\delta \in L^2(\mathbb{R})$ *such that*

$$|f_\delta|^2 \geqslant \mathbb{1}_{[a,b]}; \quad \int_{\mathbb{R}} |f_\delta(t)|^2\, dt = b - a + \delta^{-1}$$

with a Fourier Transform $\widehat{f_\delta}$ *supported in* $[-\delta/2, \delta/2]$:

$$f_\delta(t) = \int_{-\delta/2}^{\delta/2} \widehat{f_\delta}(u)\, e(tu)du.$$

Proof This result is mentioned by Vaaler [9, p.185] who refers to Boas [1, pp. 124–126 (in particular the notes p. 132)].

It follows from a Theorem of Fejér, stating that for the function F_δ defined by (6.2), which is of exponential type $2\pi\delta$ and $\geqslant 0$ on the real axis, there exists an entire function $f_\delta \in L^2(\mathbb{R})$ of exponential type $\pi\delta$ such that $F_\delta(t) = |f_\delta(t)|^2$ for all $t \in \mathbb{R}$. We have,

$$|f_\delta|^2 = F_\delta \geqslant \mathbb{1}_{[a,b]}; \quad \int_{\mathbb{R}} |f_\delta(t)|^2\, dt = \widehat{F_\delta}(0) = b - a + \delta^{-1}.$$

By a Theorem of Paley and Wiener (Theorem 6.15) the Fourier Transform $\widehat{f_\delta}$ is supported in $[-\delta/2, \delta/2]$ and we get the claimed Fourier inversion formula.

We are now ready to present a variant of an inequality of Bombieri and Iwaniec [2, Lemma 2.4] (see also [4, Lemma 7.5]), for which we obtain a slightly better constant (16 instead of $2\pi^2$).

Theorem 6.38 *Let* $\mathbf{X}$ *and* $\mathbf{Y}$ *be two finite sets of points of* $\mathbb{R}^K$, *and* $a(\mathbf{x})$, $b(\mathbf{y})$ *be complex numbers. Let* $X_1,\ldots,X_K > 0$ *and* $Y_1,\ldots,Y_K > 0$. *We write*

$$\mathcal{B} = \sum_{\substack{\mathbf{x}\in\mathbf{X} \\ |x_k|\leqslant X_k \\ (k=1,\ldots,K)}} \sum_{\substack{\mathbf{y}\in\mathbf{Y} \\ |y_k|\leqslant Y_k \\ (k=1,\ldots,K)}} a(\mathbf{x})\, b(\mathbf{y})\, e(\mathbf{x}\cdot\mathbf{y}),$$

$$\mathcal{B}_1 = \sum_{\substack{(\mathbf{x},\mathbf{x}')\in\mathbf{X}^2 \\ |x_k-x'_k|<(2Y_k)^{-1} \\ (k=1,\ldots,K)}} |a(\mathbf{x})\,a(\mathbf{x}')|\,; \qquad \mathcal{B}_2 = \sum_{\substack{(\mathbf{y},\mathbf{y}')\in\mathbf{Y}^2 \\ |y_k-y'_k|<(2X_k)^{-1} \\ (k=1,\ldots,K)}} |b(\mathbf{y})\,b(\mathbf{y}')|\,.$$

Then

$$|\mathcal{B}|^2 \leqslant 16^K \left(\prod_{k=1}^{K}(X_kY_k + \tfrac{1}{8})\right)\mathcal{B}_1\mathcal{B}_2.$$

Proof It is sufficient to show the result when all points $\mathbf{x}$ satisfy $|x_k| \leqslant X_k$ and all points $\mathbf{y}$ satisfy $|y_k| \leqslant Y_k$ (for $k = 1,\ldots,K$), as the potential other points may appear only in the right hand side of the inequality. We now assume that this hypothesis holds. By Lemma 6.37, we construct functions $f_1,\ldots,f_K \in L^2(\mathbb{R})$ such that for $k = 1,\ldots,K$,

$$|f_k|^2 \geqslant \mathbb{1}_{[-Y_k,Y_k]}; \quad \int_{\mathbb{R}} |f_k(t)|^2\,dt = 2Y_k + \delta_k^{-1}; \quad f_k(t) = \int_{-\delta_k/2}^{\delta_k/2} \widehat{f_k}(u)\,e(tu)du.$$

For $\mathbf{y} \in \mathbf{Y}$ such that $|y_k| \leqslant Y_k$ $(k = 1,\ldots,K)$, let

$$c(\mathbf{y}) = \frac{b(\mathbf{y})}{f_1(y_1)\cdots f_K(y_K)},$$

and observe that

$$|c(\mathbf{y})| \leqslant |b(\mathbf{y})|\,.$$

We can write

$$\mathcal{B} = \sum_{\mathbf{x}\in\mathbf{X}}\sum_{\mathbf{y}\in\mathbf{Y}} a(\mathbf{x})c(\mathbf{y})\,e(\mathbf{x}\cdot\mathbf{y})\int_{\mathbb{R}}\cdots\int_{\mathbb{R}}\widehat{f_1}(u_1)\cdots\widehat{f_K}(u_K)\,e(\mathbf{u}\cdot\mathbf{y})du_1\cdots du_K.$$

Let $\mathbf{v} = \mathbf{x}+\mathbf{u}$, i.e. $v_k = x_k + u_k$ for $k = 1,\ldots,K$. We obtain

$$\mathcal{B} = \int_{\mathbb{R}}\cdots\int_{\mathbb{R}}\sum_{\mathbf{x}\in\mathbf{X}} a(\mathbf{x})\widehat{f_1}(v_1-x_1)\cdots\widehat{f_K}(v_K-x_K)\sum_{\mathbf{y}\in\mathbf{Y}} c(\mathbf{y})\,e(\mathbf{v}\cdot\mathbf{y})dv_1\cdots dv_K.$$

Since $|x_k| \leqslant X_k$, we have $\widehat{f_k}(v_k-x_k) = 0$ for $|v_k| > V_k$ with

$$V_k = X_k + \tfrac{1}{2}\delta_k,$$

and we can restrict the interval of integration over v_k to $[-V_k, V_k]$. By the Cauchy-Schwarz inequality, we obtain

$$|\mathcal{B}|^2 \leqslant \mathcal{B}'_1 \mathcal{B}'_2$$

with

$$\mathcal{B}'_1 = \int_{-V_1}^{V_1} \cdots \int_{-V_K}^{V_K} \left| \sum_{\mathbf{x} \in \mathbf{X}} a(\mathbf{x}) \widehat{f_1}(v_1 - x_1) \cdots \widehat{f_K}(v_K - x_K) \right|^2 dv_1 \cdots dv_K,$$

and

$$\mathcal{B}'_2 = \int_{-V_1}^{V_1} \cdots \int_{-V_K}^{V_K} \left| \sum_{\mathbf{y} \in \mathbf{Y}} c(\mathbf{y}) \, e(\mathbf{v} \cdot \mathbf{y}) \right|^2 dv_1 \cdots dv_K,$$

Expanding the square we obtain

$$\mathcal{B}'_1 = \sum_{(\mathbf{x}, \mathbf{x}') \in \mathbf{X}^2} \overline{a(\mathbf{x})} a(\mathbf{x}') \prod_{k=1}^{K} \int_{\mathbb{R}} \overline{\widehat{f_k}(v_k - x_k)} \widehat{f_k}(v_k - x'_k) dv_k.$$

For a given k, if $\left| x_k - x'_k \right| \geqslant \delta_k$, by the triangle inequality we have for all $v_k \in \mathbb{R}$:

$$\max(|v_k - x_k|, |v_k - x_k|) \geqslant \delta_k/2,$$

with equality if and only if $\delta_k = \left| x_k - x'_k \right|$ and $v_k = \frac{1}{2}(x_k + x'_k)$. Since the support of $\widehat{f_k}$ is included in $[-\delta_k/2, \delta_k/2]$, the function $v_k \mapsto \overline{\widehat{f_k}(v_k - x_k)} \widehat{f_k}(v_k - x'_k)$ is equal to zero almost everywhere and we get

$$\int_{\mathbb{R}} \overline{\widehat{f_k}(v_k - x_k)} \widehat{f_k}(v_k - x'_k) dv_k = 0.$$

When $\left| x_k - x'_k \right| < \delta_k$, we majorize this integral by the Cauchy Schwarz inequality and we obtain

$$\left(\int_{\mathbb{R}} \left| \widehat{f_k}(v_k - x_k) \right|^2 dv_k \right)^{1/2} \left(\int_{\mathbb{R}} \left| \widehat{f_k}(v_k - x'_k) \right|^2 dv_k \right)^{1/2} = \int_{\mathbb{R}} \left| \widehat{f_k}(t) \right|^2 dt$$

which is equal to $2Y_k + \delta_k^{-1}$, by Plancherel's formula and the definition of f_k. Thus,

$$\mathcal{B}_1' \leqslant \prod_{k=1}^{K}(2Y_k + \delta_k^{-1}) \sum_{\substack{(\mathbf{x},\mathbf{x}')\in\mathbf{X}^2 \\ |x_k-x_k'|<\delta_k \\ (k=1,\ldots,K)}} \left|a(\mathbf{x})a(\mathbf{x}')\right|.$$

Applying Lemma 6.36, we have for all $\delta_1', \ldots, \delta_K' > 0$

$$\mathcal{B}_2' \leqslant \prod_{k=1}^{K}(2V_k + \delta_k'^{-1}) \sum_{\substack{(\mathbf{y},\mathbf{y}')\in\mathbf{Y}^2 \\ |y_k-y_k'|<\delta_k' \\ (k=1,\ldots,K)}} \left|c(\mathbf{y})c(\mathbf{y}')\right|.$$

Choosing $\delta_k = (2Y_k)^{-1}$ et $\delta_k' = (2X_k)^{-1}$, we get

$$(2Y_k + \delta_k^{-1})(2V_k + \delta_k'^{-1}) = 4Y_k\left(4X_k + (2Y_k)^{-1}\right) = 16\left(X_kY_k + \tfrac{1}{8}\right),$$

hence

$$|\mathcal{B}|^2 \leqslant 16^K\left(\prod_{k=1}^{K}(X_kY_k + \tfrac{1}{8})\right)\mathcal{B}_1\mathcal{B}_2,$$

which is the expected inequality.

References

1. R.P. Boas Jr., *Entire Functions* (Academic, New York, 1954)
2. E. Bombieri, H. Iwaniec, On the order of $\zeta(\tfrac{1}{2} + it)$. Ann. Sc. Norm. Sup. Pisa **13**(3), 449–472 (1986)
3. H. Davenport, *Multiplicative Number Theory*. Graduate Texts in Mathematics, 3rd edn., vol. 74 (Springer, New York, 2000). Revised and with a preface by Hugh L. Montgomery
4. S. Graham, G. Kolesnik, *van der Corput's Method of Exponential Sums*. London Mathematical Society Lecture Note Series, vol. 126 (Cambridge University Press, Cambridge, 1991)
5. G.H. Hardy, E.M. Wright, *An Introduction to the Theory of Numbers*. Oxford Science Publications, 5th edn. (Oxford University Press, Oxford, 1979)
6. H.L. Montgomery, R.C. Vaughan, *Multiplicative Number Theory. I. Classical Theory*. Cambridge Studies in Advanced Mathematics, vol. 97 (Cambridge University Press, Cambridge, 2007)
7. W. Rudin, *Real and Complex Analysis*, 3rd edn. (McGraw-Hill Book Co., New York, 1987)
8. A. Selberg, *Collected Papers*, vol. II (Springer, Berlin, 1991). With a foreword by K. Chandrasekharan
9. J. Vaaler, Some extremal functions in Fourier analysis. Bull. Am. Math. Soc. **12**(2), 183–216 (1985)

Chapter 7
The Theorem of Vinogradov

Joël Rivat

7.1 The Circle Method

Hardy and Littlewood have shown in 1922 that assuming the Riemann hypothesis, we could deduce that every large enough odd integer is the sum of three prime numbers.

To detect the condition $n_1 + n_2 + n_3 = N$, we use the equality

$$\forall n \in \mathbb{Z}, \quad \int_0^1 e(n\alpha)d\alpha = \begin{cases} 1 \text{ if } n = 0, \\ 0 \text{ if } n \neq 0, \end{cases}$$

where we recall the notation $e(u) = \exp(2i\pi u)$.

Let us examine first the simplest case without prime numbers. We have

$$\sum_{n_1+n_2+n_3=N} 1 = \sum_{k=2}^{N-1} \sum_{n_1+n_2=k} 1 = \sum_{k=2}^{N-1}(k-1) = \tfrac{1}{2}(N-1)(N-2)$$

and can write

$$\sum_{n_1+n_2+n_3=N} 1 = \int_0^1 T(\alpha)^3 \, e(-N\alpha)d\alpha,$$

J. Rivat (✉)

Aix Marseille Université, CNRS, Centrale Marseille, Institut de Mathématiques de Marseille, I2M – UMR 7373, Marseille, France

e-mail: joel.rivat@univ-amu.fr

© Springer International Publishing AG, part of Springer Nature 2018
S. Ferenczi et al. (eds.), *Ergodic Theory and Dynamical Systems in their Interactions with Arithmetics and Combinatorics*, Lecture Notes in Mathematics 2213, https://doi.org/10.1007/978-3-319-74908-2_7

where $T(\alpha)$ is the exponential sum defined by

$$T(\alpha) = \sum_{n=1}^{N} e(n\alpha).$$

Thus we can write the identity

$$\int_0^1 T(\alpha)^3\, e(-N\alpha)d\alpha = \tfrac{1}{2}N^2 + O(N),$$

to which we shall reduce the problem in the general case.

Instead of working with the number of representations of N as a sum of three prime numbers, we introduce a weighted version:

$$r(N) = \sum_{n_1+n_2+n_3=N} \Lambda(n_1)\,\Lambda(n_2)\,\Lambda(n_3),$$

which is easier to handle.

This allows to write

$$r(N) = \int_0^1 S(\alpha)^3\, e(-N\alpha)d\alpha,$$

where $S(\alpha)$ is the exponential sum defined by

$$S(\alpha) = \sum_{n \leqslant N} \Lambda(n)\, e(n\alpha).$$

The circle method consists in dealing separately with the numbers α close to a rational with a small denominator and those close to a rational with a large denominator. Indeed, $S(\alpha)$ should be small in the second case and large in the first case, as appears from observing the graph of $|S(\alpha)|$ (Fig. 7.1).

We shall prove

Theorem 7.1 (Vinogradov) *For all fixed $A > 0$,*

$$r(N) = \tfrac{1}{2}\mathfrak{S}(N)\, N^2 + O(N^2(\log N)^{-A}),$$

with

$$\mathfrak{S}(N) = \left(\prod_{p\,|\,N}\left(1 - \frac{1}{(p-1)^2}\right)\right)\left(\prod_{p\,\nmid\,N}\left(1 + \frac{1}{(p-1)^3}\right)\right)$$

Fig. 7.1 Graph of $|S(\alpha)|$

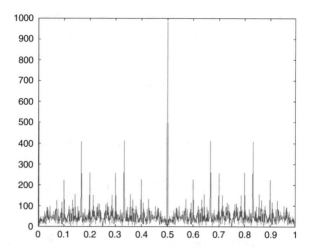

Remark 7.2 When N is even, this theorem has no interest. Indeed, on one side $\mathfrak{S}(N) = 0$, and on the other side n_1, n_2 or n_3 must be even, say n_1, and to get a contribution, $\Lambda(n_1) \neq 0$ implies $n_1 = 2^\ell$, hence $r(N) \ll N(\log N)^4$.

But when N is odd, we have $\mathfrak{S}(N) \approx 1$, thus $r(N) \gg N^2$. The contribution to $r(N)$ of representations containing at least one true power (≥ 2) of prime numbers does not exceed

$$\sum_{\substack{p_1^{k_1} \leqslant N \, p_2^{k_2} \leqslant N \\ k_1 \geqslant 1 \quad k_2 \geqslant 2}} (\log N)^3 \ll \left(\pi(N) + \pi(\sqrt{N}) + \cdots \right) \left(\pi(\sqrt{N}) + \cdots \right) (\log N)^3 \ll N^{\frac{3}{2}} \log N,$$

hence every large enough odd number can be written as the sum of three prime numbers in $\gg N^2 (\log N)^{-3}$ different ways.

Definition 7.3 Let $P = (\log N)^B$ and $Q = N(\log N)^{-B}$, where B is a constant which will be chosen later according to A. For $q \leqslant P$, $1 \leqslant a \leqslant q$, $(a, q) = 1$, we call *major arc* and denote by $\mathfrak{M}(a, q)$ the interval $|\alpha - a/q| \leqslant 1/Q$.

We denote by $\mathfrak{M}$ the union of major arcs, and by $\mathfrak{m}$ (the minor arcs) its complement in $[0, 1]$.

Remark 7.4 We work with real numbers modulo 1, hence $\mathfrak{M}(1, 1)$ can be seen as $|\alpha| \leqslant 1/Q$.

Proposition 7.5 *Two distinct major arcs are disjoint.*

Proof If $a/q \neq a'/q'$, then

$$\left| \frac{a}{q} - \frac{a'}{q'} \right| \geqslant \frac{1}{qq'} \geqslant \frac{1}{P^2} > \frac{2}{Q},$$

hence $\mathfrak{M}(a, q)$ and $\mathfrak{M}(a', q')$ are disjoint.

The proof of the Vinogradov theorem is quite long. It will consist on one side in proving that major arcs provide the main term

$$\int_{\mathfrak{M}} S(\alpha)^3 \, e(-N\alpha)d\alpha = \tfrac{1}{2}\mathfrak{S}(N) \, N^2 + O(N^2(\log N)^{-B+1}),$$

and on the other side in proving that minor arcs are negligible

$$\int_{\mathfrak{m}} S(\alpha)^3 \, e(-N\alpha)d\alpha \ll N^2(\log N)^{-B/2+5},$$

and then we shall get the final result by choosing $B = 2A + 10$.

7.2   Major Arcs

We suppose $\alpha = \frac{a}{q} + \beta$, with $1 \leqslant a \leqslant q \leqslant N$, $(a, q) = 1$ and $|\beta| \leqslant Q^{-1}$.

Lemma 7.6 *We have*

$$\sum_{\substack{n \leqslant N \\ (n,q)>1}} \Lambda(n) \leqslant \omega(q) \log N \leqslant \frac{\log N \log q}{\log 2}.$$

Proof The von Mangoldt function is zero except on powers of prime numbers, hence

$$\sum_{\substack{n \leqslant N \\ (n,q)>1}} \Lambda(n) \leqslant \sum_{\substack{p^\alpha \leqslant N \\ p \mid q}} \Lambda(p^\alpha) = \sum_{p \mid q} \log p \left\lfloor \frac{\log N}{\log p} \right\rfloor \leqslant \omega(q) \log N \leqslant \frac{\log N \log q}{\log 2},$$

the last inequality because $2^{\omega(q)} \leqslant q = p_1^{\alpha_1} \cdots p_{\omega(q)}^{\alpha_{\omega(q)}}$.

Corollary 7.7 *We have*

$$\sum_{n \leqslant N} \Lambda(n) \, e(n\alpha) = \sum_{\substack{r=1 \\ (r,q)=1}}^{q} e\left(\frac{ra}{q}\right) \sum_{\substack{n \leqslant N \\ n \equiv r \bmod q}} \Lambda(n) \, e(n\beta) + O((\log N)^2).$$

Theorem 7.8 (Siegel-Walfisz) *For each constant $B > 0$, and uniformly for*

$$x \geqslant 3, \ 1 \leqslant q \leqslant (\log x)^B, \ (r, q) = 1,$$

we have

$$\psi(x; r, q) = \frac{x}{\varphi(q)} + O(x \exp(-c\sqrt{\log x}))$$

$$\theta(x; r, q) = \frac{x}{\varphi(q)} + O(x \exp(-c\sqrt{\log x}))$$

$$\pi(x; r, q) = \frac{\mathrm{li}(x)}{\varphi(q)} + O(x \exp(-c\sqrt{\log x})),$$

where $c > 0$ is an absolute constant.

Corollary 7.9 *We have*

$$\sum_{\substack{n \leqslant N \\ n \equiv r \bmod q}} \Lambda(n) \, e(n\beta) = \frac{1}{\varphi(q)} T(\beta) + O(N \exp(-c_1 \sqrt{\log N})).$$

Proof We can write

$$\sum_{\substack{n \leqslant N \\ n \equiv r \bmod q}} \Lambda(n) \, e(n\beta)$$

$$= \int_{t=0^+}^{N^+} e(t\beta) d\psi(t; r, q)$$

$$= \int_{t=0^+}^{N^+} e(t\beta) d\frac{\lfloor t \rfloor}{\varphi(q)} + \int_{t=0^+}^{N^+} e(t\beta) d\left(\psi(t; r, q) - \frac{\lfloor t \rfloor}{\varphi(q)}\right)$$

$$= \frac{1}{\varphi(q)} \sum_{n=1}^{N} e(n\beta) + e(N\beta)\left(\psi(N; r, q) - \frac{\lfloor N \rfloor}{\varphi(q)}\right)$$

$$- \int_0^N 2i\pi\beta \, e(t\beta) \left(\psi(t; r, q) - \frac{\lfloor t \rfloor}{\varphi(q)}\right) dt.$$

The first error term is admissible by the theorem of Siegel-Walfisz. As for the integral, it is bounded by

$$\int_{\sqrt{N}}^{N} Q^{-1} t \exp(-c\sqrt{\log t}) dt + O(Q^{-1}N),$$

and as $\log t \gg \log N$ for $t > \sqrt{N}$, we get an upper bound

$$\ll Q^{-1}N^2 \exp(-c_1\sqrt{\log N}) + Q^{-1}N,$$

which is admissible because $Q = N(\log N)^{-B}$.

Corollary 7.10 *We have*

$$\sum_{n \leqslant N} \Lambda(n) \, e(n\alpha) = \frac{c_q(a)}{\varphi(q)} T(\beta) + O(N \exp(-c_2\sqrt{\log N})),$$

where $c_q(n)$ is the Ramanujan sum defined by

$$c_q(n) := \sum_{\substack{r=1 \\ (r,q)=1}}^{q} e\left(\frac{rn}{q}\right).$$

Proof The sum of error terms is

$$\ll qN \exp(-c_1\sqrt{\log N}),$$

which is admissible because $q \leqslant P = (\log N)^B$.

Proposition 7.11 *The Ramanujan sum is a multiplicative function in q and equals*

$$c_q(n) = \frac{\mu(q/(n,q))\varphi(q)}{\varphi(q/(n,q))}.$$

Proof In this proof n is fixed. We have

$$\sum_{r=1}^{q} e\left(\frac{rn}{q}\right) = \sum_{d \mid q} \sum_{\substack{r=1 \\ (r,q)=d}}^{q} e\left(\frac{rn}{q}\right) = \sum_{d \mid q} \sum_{\substack{r'=1 \\ (r',q/d)=1}}^{q/d} e\left(\frac{r'n}{q/d}\right) = \sum_{d \mid q} c_{q/d}(n),$$

thus, setting

$$f(m) := c_m(n), \quad F(m) := \sum_{r=1}^{m} e\left(\frac{rn}{m}\right) = \begin{cases} m \text{ if } m \mid n, \\ 0 \text{ otherwise,} \end{cases}$$

we get

$$F(q) = f * \mathbb{1}\,(q),$$

hence

$$c_q(n) = f(q) = F * \mu(q) = \sum_{d \mid q} F(d)\,\mu(q/d) = \sum_{\substack{d \mid q \\ d \mid n}} d\,\mu(q/d).$$

We can now deduce that $q \mapsto c_q(n)$ is multiplicative. Indeed if $(q_1, q_2) = 1$, and $d \mid q_1 q_2$, then d is written in a unique way as $d = d_1 d_2$ with $d_1 \mid q_1$, $d_2 \mid q_2$, and $(d_1, d_2) = 1$. We remark that $d \mid n$ is equivalent to $d_1 \mid n$ and $d_2 \mid n$, and that $(q_1/d_1, q_2/d_2) = 1$.

We get

$$c_{q_1 q_2}(n) = \sum_{\substack{d \mid q_1 q_2 \\ d \mid n}} d\,\mu(q_1 q_2/d) = \sum_{\substack{d_1 \mid q_1 \\ d_1 \mid n}} d_1\,\mu(q_1/d_1) \sum_{\substack{d_2 \mid q_2 \\ d_2 \mid n}} d_2\,\mu(q_2/d_2)$$

There remains for prime p and $\alpha \geqslant 1$ to check the equality

$$c_{p^\alpha}(n) = \frac{\mu(p^\alpha/(n, p^\alpha))\varphi(p^\alpha)}{\varphi(p^\alpha/(n, p^\alpha))}.$$

If $\alpha \leqslant v_p(n)$, we have $(n, p^\alpha) = p^\alpha$, and

$$c_{p^\alpha}(n) = \sum_{v=0}^{\alpha} p^v \mu(p^{\alpha-v}) = p^\alpha - p^{\alpha-1} = \varphi(p^\alpha).$$

If $\alpha \geqslant v_p(n) + 1$, we have $(n, p^\alpha) = p^{v_p(n)}$, and

$$c_{p^\alpha}(n) = \sum_{v=0}^{v_p(n)} p^v \mu(p^{\alpha-v}),$$

hence $c_{p^\alpha}(n) = 0$ if $\alpha > v_p(n) + 1$. If $\alpha = v_p(n) + 1$, we have $(p^\alpha, n) = p^{\alpha-1}$, and

$$c_{p^\alpha}(n) = -p^{\alpha-1} = \frac{\mu(p)\varphi(p^\alpha)}{\varphi(p)}.$$

which completes the proof.

Corollary 7.12 *We have*

$$S(\alpha) = \sum_{n \leqslant N} \Lambda(n) \, e(n\alpha) = \frac{\mu(q)}{\varphi(q)} T(\beta) + O(N \exp(-c_2\sqrt{\log N})).$$

We have trivially $|T(\beta)| \leqslant N$, thus, putting $S(\alpha)$ to the cube, we get,

$$S(\alpha)^3 = \frac{\mu(q)}{\varphi(q)^3} T(\beta)^3 + O(N^3 \exp(-c_2\sqrt{\log N})).$$

The total contribution of $\mathfrak{M}(a, q)$ is

$$\int_{\mathfrak{M}(a,q)} S(\alpha)^3 \, e(-N\alpha) d\alpha = \frac{\mu(q)}{\varphi(q)^3} \, e\left(\frac{-Na}{q}\right) \int_{-1/Q}^{1/Q} T(\beta)^3 \, e(-N\beta) d\beta$$
$$+ O(N^2 \exp(-c_3\sqrt{\log N})).$$

and, summing on all major arcs, we get

$$\int_{\mathfrak{M}} S(\alpha)^3 \, e(-N\alpha) d\alpha = \sum_{q \leqslant P} \frac{\mu(q)}{\varphi(q)^3} c_q(N) \int_{-1/Q}^{1/Q} T(\beta)^3 \, e(-N\beta) d\beta$$
$$+ O(N^2 \exp(-c_4\sqrt{\log N})).$$

Lemma 7.13 *We have*

$$\int_{-1/Q}^{1/Q} T(\beta)^3 \, e(-N\beta) d\beta = \tfrac{1}{2}N^2 + O(N^2 (\log N)^{-2B}).$$

Proof We had already

$$\int_0^1 T(\beta)^3 \, e(-N\beta) d\beta = \tfrac{1}{2}N^2 + O(N).$$

Furthermore

$$|T(\beta)| \leqslant \min\left(N, \frac{1}{\sin \pi |\beta|}\right),$$

thus

$$\int_{1/Q}^{1-1/Q} |T(\beta)|^3 \, d\beta \ll \int_{1/Q}^{1/2} \frac{d\beta}{\beta^3} \ll Q^2 = N^2 (\log N)^{-2B}$$

gives the result.

Lemma 7.14 *We have*

$$\mathfrak{S}(N) = \sum_{q=1}^{\infty} \frac{\mu(q)}{\varphi(q)^3} c_q(N).$$

Proof We have

$$\sum_{q=1}^{\infty} \frac{\mu(q)}{\varphi(q)^3} c_q(N) = \prod_{p} \left(1 - \frac{c_p(N)}{(p-1)^3} \right)$$

$$= \prod_{p \mid N} \left(1 - \frac{1}{(p-1)^2} \right) \prod_{p \nmid N} \left(1 + \frac{1}{(p-1)^3} \right)$$

$$= \mathfrak{S}(N).$$

Theorem 7.15 *We have*

$$\int_{\mathfrak{M}} S(\alpha)^3 \, e(-N\alpha) d\alpha = \tfrac{1}{2} \mathfrak{S}(N) \, N^2 + O(N^2 (\log N)^{-B+1}).$$

Proof As $\left| c_q(N) \right| \leqslant \varphi(q)$, we have

$$\sum_{q > P} \left| \frac{\mu(q)}{\varphi(q)^3} c_q(N) \right| \leqslant \sum_{q > P} \frac{1}{\varphi(q)^2}.$$

For all $0 < \delta < 1$, since $f(n) = \frac{n^{1-\delta}}{\varphi(n)}$ is multiplicative and

$$f(p^m) = \frac{p^{m(1-\delta)}}{p^m(1 - 1/p)} = \frac{p^{-m\delta}}{1 - 1/p} \to 0, \ p^m \to +\infty,$$

it follows that

$$f(n) = \frac{n^{1-\delta}}{\varphi(n)} \to 0, \ n \to +\infty,$$

Thus we get

$$\sum_{q > P} \frac{\mu(q)}{\varphi(q)^3} c_q(N) \ll \sum_{q > P} \frac{1}{q^{2(1-\delta)}} \ll (\log N)^{-B+1},$$

taking $\delta = \frac{1}{2B}$.

7.3 Minor Arcs

We must show that the contribution of minor arcs is $o(N^2)$.

Lemma 7.16 *We have*

$$\left| \int_{\mathfrak{m}} S(\alpha)^3 \, e(-N\alpha) d\alpha \right| \leqslant N(\log N) \sup_{\mathfrak{m}} |S(\alpha)| .$$

Proof We have

$$\left| \int_{\mathfrak{m}} S(\alpha)^3 \, e(-N\alpha) d\alpha \right| \leqslant \sup_{\mathfrak{m}} |S(\alpha)| \int_{\mathfrak{m}} |S(\alpha)|^2 \, d\alpha \leqslant \sup_{\mathfrak{m}} |S(\alpha)| \int_0^1 |S(\alpha)|^2 \, d\alpha$$

and the last integral, by the Parseval identity, becomes

$$\sum_{m \leqslant N} \Lambda(m) \sum_{n \leqslant N} \Lambda(n) \int_0^1 e((m-n)\alpha) d\alpha = \sum_{n \leqslant N} \Lambda(n)^2 \ll N \log N.$$

Remark 7.17 Thanks to the Parseval identity, we could bound the sum of N^2 terms by $N \log N$, which is excellent.

It remains now to bound $|S(\alpha)|$ on $\mathfrak{m}$.

Lemma 7.18 (Dirichlet) *For all real numbers α and $Q \geqslant 1$, there exists a rational number a/q such that*

$$\left| \alpha - \frac{a}{q} \right| \leqslant \frac{1}{qQ}, \quad 1 \leqslant q \leqslant Q, \ (a, q) = 1.$$

Proof Set $N = \lfloor Q \rfloor$. The $N+1$ numbers $\{n\alpha\}$, $n = 0, \ldots, N$ are in the interval $[0, 1[$, hence there exist two among them, say $\{m\alpha\}$ and $\{n\alpha\}$, with $m < n$, such that

$$|\{n\alpha\} - \{m\alpha\}| \leqslant \frac{1}{N+1},$$

that is

$$|(n-m)\alpha - \lfloor n\alpha \rfloor + \lfloor m\alpha \rfloor| \leqslant \frac{1}{N+1}.$$

It suffices to take $q = n - m$ and $a = \lfloor n\alpha \rfloor - \lfloor m\alpha \rfloor$. If $(a, q) = d > 1$, we replace a with $a' = a/d$ and q with $q' = q/d$.

Proposition 7.19 *If $\alpha \in \mathfrak{m}$, there exists a rational number a/q such that*

$$\left| \alpha - \frac{a}{q} \right| \leqslant \frac{1}{qQ}, \quad P = (\log N)^B < q \leqslant Q = \frac{N}{(\log N)^B}, \quad (a, q) = 1.$$

Proof Applying the Dirichlet lemma to α and Q, it suffices to prove that we do have $P < q$. Indeed, if $1 \leqslant q \leqslant P$, then as

$$\left| \alpha - \frac{a}{q} \right| \leqslant \frac{1}{qQ} \leqslant \frac{1}{Q},$$

we would have $\alpha \in \mathfrak{M}$, and thus $\alpha \notin \mathfrak{m}$.

Corollary 7.20 *If $\alpha \in \mathfrak{m}$, there exists a rational number a/q such that*

$$\left| \alpha - \frac{a}{q} \right| \leqslant \frac{1}{N}, \quad P < q \leqslant Q, \quad (a, q) = 1.$$

Proposition 7.21 *If $\alpha \in \mathfrak{m}$, there exists a rational number a/q such that $P < q \leqslant Q$, $(a, q) = 1$, and*

$$|S(\alpha)| \ll \max_{1 \leqslant x \leqslant N} |S(x; a, q)|, \quad S(x; a, q) := \sum_{n \leqslant x} \Lambda(n) \, \mathrm{e} \left(\frac{na}{q} \right).$$

Proof We can write $\alpha = a/q + \beta$ with $|\beta| \leqslant 1/N$, hence

$$S(\alpha) = \sum_{n \leqslant N} \Lambda(n) \, \mathrm{e} \left(\frac{na}{q} + n\beta \right)$$

$$= \mathrm{e}(N\beta) S(N; a, q) - 2i\pi\beta \int_1^N \mathrm{e}(x\beta) S(x; a, q) dx$$

$$\ll \max_{1 \leqslant x \leqslant N} |S(x; a, q)|.$$

To achieve the bound for the contribution of minor arcs, it suffices to prove that for $1 \leqslant x \leqslant N$, $P < q \leqslant Q$, $(a, q) = 1$, we have

$$S(x; a, q) \ll N(\log N)^{-B/2+4}.$$

7.4 The Vaughan Identity

Suppose we want to bound a sum

$$S := \sum_{n \leqslant x} \Lambda(n) f(n),$$

where f is a function which we can control relatively well. The principle of the method we shall use consists in transforming the sum S into a finite number of sums of the form

$$\sum_{m} \sum_{n} a_m f(mn) \qquad \text{(type I)}$$

$$\sum_{m} \sum_{n} a_m b_n f(mn) \qquad \text{(type II)}$$

with $a_m, b_n \in \mathbb{C}$, and the sizes of m and n are well determined. The sums of type I involve a summing on the variable n which is "smooth" hence "easy". For the sums of type II, which are more difficult, we can apply the Cauchy-Schwarz inequality to involve a smooth variable. Of course this operation has a price (we lose something in Cauchy-Schwarz).

The original method of Vinogradov was quite laborious, but Vaughan (1977) proposed a much simpler approach. We recall

$$\zeta(s) = \sum_{n=1}^{\infty} \frac{1}{n^s}, \quad \frac{1}{\zeta(s)} = \sum_{n=1}^{\infty} \frac{\mu(n)}{n^s}, \quad \frac{\zeta'(s)}{\zeta(s)} = -\sum_{n=1}^{\infty} \frac{\Lambda(n)}{n^s}.$$

To control the size of summed variables, we introduce the auxiliary functions

$$F_U(s) := \sum_{n=1}^{\infty} \frac{\Lambda_U(n)}{n^s} = \sum_{n \leqslant U} \frac{\Lambda(n)}{n^s}, \quad M_V(s) := \sum_{n=1}^{\infty} \frac{\mu_V(n)}{n^s} = \sum_{n \leqslant V} \frac{\mu(n)}{n^s}.$$

Then

$$1 - \zeta(s) M_V(s) = 1 - \sum_{n=1}^{\infty} \frac{\mathbb{1} * \mu_V(n)}{n^s} = -\sum_{n > V} \frac{\mathbb{1} * \mu_V(n)}{n^s}$$

and

$$-\frac{\zeta'(s)}{\zeta(s)} - F_U(s) = \sum_{n > U} \frac{\Lambda(n)}{n^s},$$

thus by multiplication we get

$$\left(-\frac{\zeta'(s)}{\zeta(s)} - F_U(s)\right)(1 - \zeta(s)M_V(s)) = \sum_{n=1}^{\infty} \frac{v(n)}{n^s},$$

where

$$v(n) = - \sum_{\substack{n_1 n_2 = n \\ n_1 > U,\, n_2 > V}} \Lambda(n_1)\, \mathbb{1} * \mu_V(n_2),$$

which gives a good control of the size of the divisors of n we used in v.

Proposition 7.22 (Vaughan Identity) *We have*

$$-\frac{\zeta'(s)}{\zeta(s)} = F_U(s) - \zeta(s)F_U(s)M_V(s) - \zeta'(s)M_V(s)$$

$$+ \left(-\frac{\zeta'(s)}{\zeta(s)} - F_U(s)\right)(1 - \zeta(s)M_V(s)).$$

Proof Obvious!

Corollary 7.23 *We have*

$$\Lambda(n) = \Lambda_U(n) - \sum_{\substack{n_1 n_2 n_3 = n \\ n_2 \leqslant U,\, n_3 \leqslant V}} \Lambda(n_2)\mu(n_3) + \sum_{\substack{n_1 n_2 = n \\ n_2 \leqslant V}} \log(n_1)\mu(n_2)$$

$$- \sum_{\substack{n_1 n_2 = n \\ n_1 > U,\, n_2 > V}} \Lambda(n_1)\, \mathbb{1} * \mu_V(n_2).$$

Corollary 7.24 *We have* $S = S_1 + S_2 + S_3 + S_4$*, with:*

$$S_1 = \sum_{n \leqslant U} \Lambda(n)f(n),$$

$$S_2 = - \sum_{m \leqslant UV} \Lambda_U * \mu_V(m) \sum_{n \leqslant x/m} f(mn),$$

$$S_3 = \sum_{m \leqslant V} \mu(m) \sum_{n \leqslant x/m} \log(n)f(mn),$$

$$S_4 = - \sum_{U < m \leqslant x/V} \Lambda(m) \sum_{V < n \leqslant x/m} \mathbb{1} * \mu_V(n)f(mn).$$

The sum S_1 is bounded trivially:

$$|S_1| \leqslant \sum_{n \leqslant U} \Lambda(n) \, |f(n)| \ll U \max_{n \leqslant U} |f(n)| \, .$$

The sum S_2 is of type I. We have for $m \leqslant UV$

$$|\Lambda_U * \mu_V(m)| \leqslant \Lambda * \mathbb{1}(m) = \log m \leqslant \log(UV),$$

hence

$$|S_2| \leqslant \log(UV) \sum_{m \leqslant UV} \left| \sum_{n \leqslant x/m} f(mn) \right| \, .$$

The sum S_3 is also of type I. Indeed

$$S_3 = \sum_{m \leqslant V} \mu(m) \sum_{n \leqslant x/m} f(mn) \int_1^n \frac{dt}{t} = \sum_{m \leqslant V} \mu(m) \int_1^x \sum_{t \leqslant n \leqslant x/m} f(mn) \frac{dt}{t},$$

hence

$$|S_3| \leqslant \log(x) \sum_{m \leqslant V} \sup_t \left| \sum_{t \leqslant n \leqslant x/m} f(mn) \right| \, .$$

The sum S_4 is of type II.

7.5   Sums of Type I

We want to bound

$$S_I = \sum_{m \leqslant M} \sup_{1 \leqslant t \leqslant x/m} \left| \sum_{t \leqslant n \leqslant x/m} e(mna/q) \right| \, ,$$

where the inner sum is geometric, hence

$$S_I \leqslant \sum_{m \leqslant M} \min \left(\frac{x}{m}, \frac{1}{|\sin(\pi ma/q)|} \right) \, .$$

When $ma \equiv 0 \bmod q$ (i.e. $m = kq$ because $(a, q) = 1$), we are obliged to use x/m, otherwise we take the sine term. We get

$$S_I \leqslant \sum_{k=1}^{\lfloor M/q \rfloor} \frac{x}{kq} + \left\lceil \frac{M}{q} \right\rceil \sum_{m=1}^{q-1} \frac{1}{|\sin(\pi m a/q)|}.$$

When m spans all the nonzero residual classes modulo q, so does am as $(a, q) = 1$. Furthermore we check easily that $t \mapsto 1/\sin t$ is convex on $]0, \pi[$, hence

$$\sum_{m=1}^{q-1} \frac{1}{|\sin(\pi m a/q)|} = \sum_{r=1}^{q-1} \frac{1}{\sin(\pi r/q)} \leqslant \int_{1/2}^{q-1/2} \frac{dt}{\sin(\pi t/q)} = \left[\frac{q}{\pi} \log \tan \frac{\pi t}{2q} \right]_{1/2}^{q-1/2}$$

and as $\tan(\frac{\pi}{2} - \theta) = \cot \theta$, we get for $q \geqslant 2$,

$$\sum_{m=1}^{q-1} \frac{1}{|\sin(\pi m a/q)|} \leqslant \frac{2q}{\pi} \log \cot \frac{\pi}{4q} \leqslant \frac{2q}{\pi} \log \frac{4q}{\pi}$$

and, for $3 \leqslant q \leqslant M$ (otherwise the term does not exist), we have

$$\sum_{k=1}^{\lfloor M/q \rfloor} \frac{1}{k} \leqslant 1 + \int_1^{M/q} \frac{dt}{t} \leqslant \log(3M/q) \leqslant \log M,$$

thus the bound for sums of type I:

$$S_I \leqslant \frac{x}{q} \log M + \frac{2}{\pi} (M + q) \log \frac{4q}{\pi}.$$

In particular, the sums S_2 and S_3 coming from the Vaughan identity are bounded for $f(n) = e(n\alpha)$ by

$$\frac{x}{q} \log(UV) + \frac{2}{\pi} (UV + q) \log \frac{4q}{\pi}.$$

7.6 Sums of Type II

We want to bound

$$S_{II} = \sum_{M < m \leqslant 2M} a_m \sum_{V < n \leqslant x/m} b_n \, e(mna/q).$$

By Cauchy-Schwarz, we get

$$|S_{II}|^2 \leqslant \left(\sum_{M<m\leqslant 2M} |a_m|^2 \right) \left(\sum_{M<m\leqslant 2M} \left| \sum_{V<n\leqslant x/m} b_n\, e(mna/q) \right|^2 \right).$$

The right-hand sum is bounded by

$$\sum_{V<n_1\leqslant x/M} \sum_{V<n_2\leqslant x/M} |b_{n_1}|\, |b_{n_2}| \left| \sum_{M<m\leqslant \min(2M,x/n_1,x/n_2)} e(m(n_1-n_2)a/q) \right|,$$

and as $|b_{n_1} b_{n_2}| \leqslant \frac{1}{2} |b_{n_1}|^2 + \frac{1}{2} |b_{n_2}|^2$, we get (using the symmetry between n_1 and n_2)

$$|S_{II}|^2 \leqslant \left(\sum_{M<m\leqslant 2M} |a_m|^2 \right) \sum_{V<n_1\leqslant x/M} |b_{n_1}|^2$$
$$\sum_{V<n_2\leqslant x/M} \min\left(M, \frac{1}{|\sin \pi (n_1-n_2)a/q|} \right).$$

When $n_2 \equiv n_1 \bmod q$, we bound by M, and otherwise by the sine term. We get, summing on n_2:

$$|S_{II}|^2 \leqslant \left(\sum_{M<m\leqslant 2M} |a_m|^2 \right) \sum_{V<n_1\leqslant x/M} |b_{n_1}|^2 \left\lceil \frac{x}{qM} \right\rceil \left(M + \sum_{r=1}^{q-1} \frac{1}{\sin \pi r/q} \right),$$

hence

$$|S_{II}|^2 \leqslant x^2 \left(\frac{1}{M} \sum_{M<m\leqslant 2M} |a_m|^2 \right) \left(\frac{M}{x} \sum_{V<n\leqslant x/M} |b_n|^2 \right)$$
$$\left(\frac{1}{q} + \frac{M}{x} + \left(\frac{1}{M} + \frac{q}{x} \right) \frac{2}{\pi} \log \frac{4q}{\pi} \right).$$

To bound the sum S_4 coming from the Vaughan identity, we decompose the sum in m, $U < m \leqslant x/V$ into $O(\log x)$ sums of the form S_{II} with $U < M \leqslant x/V$, thus

$$S_4 \ll x \left(\frac{1}{M} \sum_{M<m\leqslant 2M} \Lambda(m)^2 \right)^{1/2} \left(\frac{M}{x} \sum_{V<n\leqslant x/M} |\mathbb{1} * \mu_V(n)|^2 \right)^{1/2}$$

$$\left(\frac{1}{q} + \frac{1}{V} + \frac{1}{U} + \frac{q}{x}\right)^{1/2} (\log N)^{3/2}.$$

We have

$$\sum_{M < m \leqslant 2M} \Lambda(m)^2 \leqslant \log 2M \sum_{M < m \leqslant 2M} \Lambda(n) \ll M \log M$$

and

$$\sum_{V < n \leqslant x/M} |\mathbb{1} * \mu_V(n)|^2 \leqslant \sum_{n \leqslant x/M} \tau(n)^2,$$

where $\tau(n)$ denotes the number of divisors of n.

Lemma 7.25 *For all $x \geqslant 1$, we have*

$$\sum_{n \leqslant x} \tau(n)^2 \ll x(\log 2x)^3.$$

Proof As τ is multiplicative, we define a multiplicative function h by setting $h = \mu * \tau^2$. We have $h(p^a) = 2a + 1$ and

$$\tau(n)^2 = \sum_{d \mid n} h(d),$$

hence

$$\sum_{n \leqslant x} \tau(n)^2 = \sum_{d \leqslant x} h(d) \left\lfloor \frac{x}{d} \right\rfloor \leqslant x \sum_{d \leqslant x} \frac{h(d)}{d}$$

$$\leqslant x \prod_{p \leqslant x} \left(1 + \frac{h(p)}{p} + \frac{h(p^2)}{p^2} + \cdots\right)$$

$$\leqslant x \prod_{p \leqslant x} \left(1 - \frac{1}{p}\right)^{-3}$$

$$\ll x(\log 2x)^3.$$

Finally

$$S_4 \ll x(\log N)^{7/2} \left(\frac{1}{q} + \frac{1}{V} + \frac{1}{U} + \frac{q}{x}\right)^{1/2}.$$

7.7 Completing the Bound on Minor Arcs

From all what precedes, we have

$$S(x; a, q) \ll x(\log N)^{7/2} \left(\frac{1}{q^{1/2}} + \frac{1}{V^{1/2}} + \frac{1}{U^{1/2}} + \frac{q^{1/2}}{x^{1/2}} + \frac{1}{q} + \frac{UV + q}{x} \right)$$

and taking $U = V = x^{2/5}$, we get

$$S(x; a, q) \ll N(\log N)^{7/2} \left(\frac{1}{q^{1/2}} + \frac{1}{x^{1/5}} + \frac{q^{1/2}}{N^{1/2}} \right)$$

and for $P \leqslant q \leqslant Q$, we have

$$S(x; a, q) \ll N(\log N)^{-B/2 + 7/2}$$

which completes the proof of the Vinogradov theorem.

7.8 Combinatorial Identities

The key of the bound on minor arcs is a combinatorial identity, for example the Vaughan identity. This identity has been generalized by Heath-Brown:

$$\frac{\zeta'}{\zeta}(1 - \zeta M_X)^\ell = \frac{\zeta'}{\zeta} + \sum_{r=1}^{\ell} \binom{\ell}{r}(-1)^r \zeta' \zeta^{r-1} M_X^r$$

with

$$\zeta(s) = \sum_{n=1}^{\infty} \frac{1}{n^s}, \quad \frac{\zeta'}{\zeta}(s) = -\sum_{n=1}^{\infty} \frac{\Lambda(n)}{n^s}, \quad M_X(s) = \sum_{n \leqslant X} \frac{\mu(n)}{n^s} = \sum_{n=1}^{\infty} \frac{\mu_X(n)}{n^s}$$

Using elementary properties of Dirichlet series we have

$$1 - \zeta M_X = \sum_{n > X} \frac{a_X(n)}{n^s}, \quad (1 - \zeta M_X)^\ell = \sum_{n > X^\ell} \frac{a_X(n, \ell)}{n^s}$$

thus identifying the coefficients of the Dirichlet series in the previous identity, we get for $n \leqslant X^\ell$

$$\Lambda(n) = -\sum_{r=1}^{\ell} \binom{\ell}{r} (-1)^r \sum_{n_1 \cdots n_r n_{r+1} \cdots n_{2r}=n} \mu_X(n_1) \cdots \mu_X(n_r) \log n_{2r}.$$

To use this identity, we group a number of variables into a single one (which generates complicated coefficients but we can bound them by divisor functions). This is a very versatile tool which has a lot of applications.

We can use also the same idea to deal with

$$\sum_n \mu(n) f(n),$$

replacing ζ' with 1 in all what precedes.

Chapter 8
The van der Corput Method

Joël Rivat

8.1 Uniformly Distributed Sequences Modulo 1

Definition 8.1 A sequence of real numbers (u_n) is said to be *uniformly distributed modulo* 1 if for all interval I of length $|I| < 1$, we have

$$\lim_{N \to +\infty} \frac{1}{N} \#\{n : \ 1 \leqslant n \leqslant N, \ u_n \in I \bmod 1\} = |I|.$$

Notation 8.2 *We write* $\mathrm{e}(x) = \exp(2i\pi x)$.

Theorem 8.3 (Weyl Criterion) *A sequence of real numbers (u_n) is uniformly distributed modulo 1 if and only if*

$$\forall h \in \mathbb{N}^*, \quad \lim_{N \to +\infty} \frac{1}{N} \sum_{n=1}^{N} \mathrm{e}(h u_n) = 0.$$

Proof For compactness we will prefer to denote here by $\overline{u_n}$ the fractional part of u_n.

Suppose (u_n) is uniformly distributed modulo 1. For all $0 \leqslant a \leqslant b \leqslant 1$, we have

$$\frac{1}{N} \sum_{n=1}^{N} \mathbb{1}_{[a,b]}(\overline{u_n}) \to b - a = \int_0^1 \mathbb{1}_{[a,b]}(t) \, dt,$$

J. Rivat (✉)
Aix Marseille Université, CNRS, Centrale Marseille, Institut de Mathématiques de Marseille, I2M – UMR 7373, Marseille, France
e-mail: joel.rivat@univ-amu.fr

© Springer International Publishing AG, part of Springer Nature 2018
S. Ferenczi et al. (eds.), *Ergodic Theory and Dynamical Systems in their Interactions with Arithmetics and Combinatorics*, Lecture Notes in Mathematics 2213, https://doi.org/10.1007/978-3-319-74908-2_8

hence every step function f satisfies

$$\frac{1}{N} \sum_{n=1}^{N} f(\overline{u_n}) \to \int_0^1 f(t)\, dt.$$

Let f be Riemann integrable and $\varepsilon > 0$. There exist g and h step functions such that

$$g \leqslant f \leqslant h \quad \text{and} \quad \int_0^1 (h - g)(t)\, dt \leqslant \varepsilon.$$

Then for N large enough

$$\int_0^1 g(t)\, dt - \varepsilon \leqslant \frac{1}{N} \sum_{n=1}^{N} g(\overline{u_n}) \leqslant \frac{1}{N} \sum_{n=1}^{N} f(\overline{u_n}) \leqslant \frac{1}{N} \sum_{n=1}^{N} h(\overline{u_n}) \leqslant \int_0^1 h(t)\, dt + \varepsilon$$

and as $\int_0^1 (h - f)(t)\, dt \leqslant \varepsilon$ and $\int_0^1 (f - g)(t)\, dt \leqslant \varepsilon$, we get

$$\left| \frac{1}{N} \sum_{n=1}^{N} f(\overline{u_n}) - \int_0^1 f(t)\, dt \right| \leqslant 2\varepsilon.$$

Hence every Riemann integrable function f satisfies

$$\frac{1}{N} \sum_{n=1}^{N} f(\overline{u_n}) \to \int_0^1 f(t)\, dt.$$

In particular the function $f(x) = e(hx)$ is continuous hence Riemann integrable, thus

$$\frac{1}{N} \sum_{n=1}^{N} e(h\overline{u_n}) = \frac{1}{N} \sum_{n=1}^{N} e(hu_n) \to 0 = \int_0^1 e(ht)\, dt.$$

Conversely suppose that for all $h \in \mathbb{Z}^*$ (or $\mathbb{N}^*$),

$$\frac{1}{N} \sum_{n=1}^{N} e(hu_n) \to 0.$$

Then for each trigonometric polynomial $\varphi(x) = \sum_{h=-H}^{H} a_h\, e(hx)$, we have

$$\frac{1}{N} \sum_{n=1}^{N} \varphi(\overline{u_n}) = \sum_{h=-H}^{H} a_h \frac{1}{N} \sum_{n=1}^{N} e(hu_n) \to a_0 = \int_0^1 \varphi(t)\, dt, \quad N \to +\infty.$$

For each continuous 1-periodic function f and all $\varepsilon > 0$, by the Stone-Weierstrass theorem there exists a trigonometric polynomial φ such that $\|\varphi - f\|_\infty \leqslant \varepsilon$. Then

$$\left| \frac{1}{N} \sum_{n=1}^{N} f(\overline{u_n}) - \int_0^1 f(t)\, dt \right|$$

$$\leqslant \left| \frac{1}{N} \sum_{n=1}^{N} (f - \varphi)(\overline{u_n}) \right| + \left| \frac{1}{N} \sum_{n=1}^{N} \varphi(\overline{u_n}) - \int_0^1 \varphi(t)\, dt \right| + \left| \int_0^1 (\varphi - f)(t)\, dt \right|$$

$$\leqslant 3\varepsilon,$$

for N large enough.

If $0 \leqslant a \leqslant b \leqslant 1$ and $\varepsilon > 0$, there exist continuous 1-periodic ψ_1 and ψ_2 such that

$$\psi_1 \leqslant 1\!\!1_{[a,b]} \leqslant \psi_2 \quad \text{and} \quad \int_0^1 (\psi_2 - \psi_1)(t)\, dt \leqslant \varepsilon,$$

and applying what precedes we get finally

$$\frac{1}{N} \sum_{n=1}^{N} 1\!\!1_{[a,b]}(\overline{u_n}) \to b - a = \int_0^1 1\!\!1_{[a,b]}(t)\, dt$$

and (u_n) is uniformly distributed modulo 1.

Remark 8.4 The Weyl criterion gives no information on the "quality" of the uniform distribution.

Definition 8.5 We call *discrepancy* of real numbers $u_1, \ldots, u_N$ the quantity

$$D_N(u_1, \ldots, u_N) = \sup_I \left| \frac{1}{N} \sum_{n=1}^{N} 1\!\!1_I(\overline{u_n}) - |I| \right|,$$

where the sup is taken on all intervals I of length $|I| < 1$.

Theorem 8.6 (Erdős-Turan) *For $N > 0$, $H > 0$ integers and $u_1, \ldots, u_N$ real numbers we have*

$$D_N(u_1, \ldots, u_N) \leqslant \frac{1}{H+1} + \sum_{h=1}^{H} \frac{1}{h} \left| \frac{1}{N} \sum_{n=1}^{N} e(h u_n) \right|.$$

We shall only sketch the proof of the Erdős-Turán theorem under this strong form (with good constants). We write for $\alpha, \beta, u \in [0, 1[$,

$$\mathbb{1}_{[\alpha,\beta[}(u) = \lfloor u - \alpha \rfloor - \lfloor u - \beta \rfloor,$$

and we introduce the function $\psi(u) = u - \lfloor u \rfloor - \frac{1}{2}$, which allows to write

$$\mathbb{1}_{[\alpha,\beta[}(u) = \beta - \alpha - \psi(u - \alpha) + \psi(u - \beta).$$

The first Bernoulli function ψ can be approximated very closely by trigonometric polynomials, using the following result

Lemma 8.7 (Vaaler (1985)) *For $H \in \mathbb{N}$, $h \in \mathbb{Z}$, $1 \leqslant |h| \leqslant H$, let*

$$0 < b_H(h) := \pi \frac{|h|}{H+1} \left(1 - \frac{|h|}{H+1}\right) \cot\left(\pi \frac{|h|}{H+1}\right) + \frac{|h|}{H+1} < 1.$$

Then the trigonometric polynomial

$$\psi_H(x) = -\frac{1}{2i\pi} \sum_{1 \leqslant |h| \leqslant H} \frac{b_H(h)}{h} e(hx)$$

satisfies for every real number x,

$$|\psi(x) - \psi_H(x)| \leqslant \frac{1}{2H+2} \sum_{|h| \leqslant H} \left(1 - \frac{|h|}{H+1}\right) e(hx) = \frac{\sin^2 \pi (H+1)x}{2(H+1)^2 \sin^2 \pi x}.$$

Proof For $x \notin \mathbb{Z}$ this is inequality (7.14) from Vaaler (1985)—see also Theorem A.6 from Graham-Kolesnik [1]. For $x \in \mathbb{Z}$, both sides are equal to $\frac{1}{2}$, and thus the result is still true.

Replacing $\psi(u - \alpha)$ and $\psi(u - \beta)$ respectively by $\psi_H(u - \alpha)$ and $\psi_H(u - \beta)$, and managing the error we create, we get the result without any difficulty. In particular, the term $1/(H+1)$ comes from $h = 0$ in the error term.

Remark 8.8 After applying the Weyl criterion or Erdős-Turán inequality, we need to know how to bound sums

$$\sum_{n=1}^{N} e(hu_n).$$

Theorem 8.9 *For $\alpha \in \mathbb{R} \setminus \mathbb{Q}$, the sequence (αn) is uniformly distributed modulo 1.*

Proof The sum of exponentials is a geometric sum:

$$\left| \sum_{n=1}^{N} e(h\alpha n) \right| = \left| e(h\alpha) \frac{1 - e(h\alpha N)}{1 - e(h\alpha)} \right| \leqslant \frac{1}{|\sin(\pi h\alpha)|},$$

because for $h \in \mathbb{N}^*$, we have $h\alpha \notin \mathbb{Z}$. Thus,

$$\frac{1}{N} \left| \sum_{n=1}^{N} e(h\alpha n) \right| \leqslant \frac{1}{N |\sin(\pi h\alpha)|} \to 0, \ N \to +\infty,$$

and (αn) is uniformly distributed modulo 1 by the Weyl criterion.

8.2 Upper Bounds on Exponential Sums

Lemma 8.10 (Kusmin-Landau) *Let $0 < \theta \leqslant 1/2$. For all $x_1, \ldots, x_N \in \mathbb{R}$ such that*

$$0 < \theta \leqslant x_2 - x_1 \leqslant \cdots \leqslant x_N - x_{N-1} \leqslant 1 - \theta,$$

we have

$$\left| \sum_{n=1}^{N} e(x_n) \right| \leqslant \cot \frac{\pi \theta}{2}.$$

Remark 8.11 It is useful to notice that $\cot \frac{\pi \theta}{2} \leqslant \frac{2}{\pi \theta}$.

Proof To see $y_n = x_{n+1} - x_n$, we write

$$\sum_{n=1}^{N} e(x_n) = \sum_{n=1}^{N-1} (e(x_n) - e(x_{n+1})) c_n + e(x_N)$$

with

$$c_n = \frac{e(x_n)}{e(x_n) - e(x_{n+1})} = \frac{1}{1 - e(y_n)} = \frac{e(-y_n/2)}{-2i \sin \pi y_n} = \frac{1}{2}(1 + i \cot \pi y_n)$$

Then we have

$$|c_n| = |1 - c_n| \leqslant \frac{1}{2| \sin \pi y_n|} \leqslant \frac{1}{2 \sin \pi \theta}.$$

An Abel summation allows to write:

$$\sum_{n=1}^{N} e(x_n) = \sum_{1<n<N} e(x_n)(c_n - c_{n-1}) + e(x_1)c_1 + e(x_N)(1 - c_{N-1}).$$

In consequence:

$$\left| \sum_{n=1}^{N} e(x_n) \right| \leqslant \frac{1}{2} \sum_{1<n<N} |\cot \pi y_{n-1} - \cot \pi y_n| + |c_1| + |1 - c_{N-1}|.$$

The sequence $\cot \pi y_n$ is nonincreasing, thus we can delete the absolute values in the sum on the right hand side. Then, two by two, terms cancel one another and we get:

$$\left| \sum_{n=1}^{N} e(x_n) \right| \leqslant \frac{1}{2} (\cot \pi y_1 - \cot \pi y_{N-1}) + |c_1| + |1 - c_{N-1}|$$

$$\leqslant \frac{1}{2} (\cot \pi \theta + \cot \pi \theta) + \frac{1}{2 \sin \pi \theta} + \frac{1}{2 \sin \pi \theta} = \frac{\cos \pi \theta + 1}{\sin \pi \theta}$$

$$\leqslant \frac{2 \cos^2 (\pi \theta / 2)}{2 \sin(\pi \theta / 2) \cos(\pi \theta / 2)} = \cot \frac{\pi \theta}{2},$$

which yields the claimed inequality. The inequality in the remark is elementary:

$$\tan \frac{\pi \theta}{2} = \int_0^{\pi \theta / 2} (1 + \tan^2 t) \, dt \geqslant \int_0^{\pi \theta / 2} dt = \frac{\pi \theta}{2},$$

hence $\cot \frac{\pi \theta}{2} \leqslant \frac{2}{\pi \theta}$.

Notation 8.12 *For a real number x, $\|x\|$ denotes the distance from x to the nearest integer.*

Theorem 8.13 (Kusmin-Landau) *Let I be a bounded interval of $\mathbb{R}$ and $f : I \to \mathbb{R}$ a continuously differentiable function on I, f' monotonous and $\|f'\| \geqslant \lambda_1 > 0$ on I. Then*

$$\left| \sum_{n \in I} e(f(n)) \right| \leqslant \cot \frac{\pi \lambda_1}{2} \leqslant \frac{2}{\pi \lambda_1}.$$

Proof By translation, we can suppose $I = [1, N]$, and, possibly by changing f into $-f$, we can suppose that f' is nondecreasing.

As $\|f'\| \geqslant \lambda_1 > 0$ on I, the continuity of f' implies the existence of an integer k such that

$$\forall x \in I, \quad k + \lambda_1 \leqslant f'(x) \leqslant k + 1 - \lambda_1.$$

As $e(f(n)) = e(f(n) - kn)$, we can suppose that $\lambda_1 \leqslant f' \leqslant 1 - \lambda_1$ on I.

By the finite increments theorem, for every integer n with $1 \leqslant n \leqslant N - 1$, there exists $\theta_n \in]n, n + 1[$ such that $f(n+1) - f(n) = f'(\theta_n)$.

As f' is nondecreasing,

$$0 < \lambda_1 \leqslant f'(\theta_1) \leqslant \cdots \leqslant f'(\theta_{N-1}) \leqslant 1 - \lambda_1,$$

hence

$$0 < \lambda_1 \leqslant f(2) - f(1) \leqslant \cdots \leqslant f(N) - f(N-1) \leqslant 1 - \lambda_1,$$

and we can conclude by Lemma 8.10.

Remark 8.14 The Kusmin-Landau inequality is very precise, but unfortunately the condition $\|f'\| \geqslant \lambda_1 > 0$ is seldom fulfilled in practice. The following result allows to dispense with it.

Theorem 8.15 (van der Corput) *Let I be an interval of $\mathbb{R}$ containing N integers $(N \geqslant 0)$. Let $f : I \to \mathbb{R}$ be a twice continuously differentiable function on I. Furthermore we suppose there exists a real number $\lambda_2 > 0$ and a real number $\alpha \geqslant 1$ such that*

$$\forall x \in I, \quad \lambda_2 \leqslant |f''(x)| \leqslant \alpha \lambda_2.$$

Then we have the upper bound:

$$\left| \sum_{n \in I} e(f(n)) \right| \leqslant 3\,\alpha N \lambda_2^{1/2} + 6\,\lambda_2^{-1/2}.$$

Proof We can suppose $\lambda_2 < 1/9$ because if $\lambda_2 \geqslant 1/9$, we have the trivial bounds:

$$\left| \sum_{n \in I} e(f(n)) \right| \leqslant N \leqslant 3\alpha N \lambda_2^{1/2}.$$

Let $0 < \delta < 1/2$ which we shall choose later. The variation of f' between the first and last integer of I does not exceed $\alpha \lambda_2 N$, hence the interval I may be cut into $\leqslant \alpha N \lambda_2 + 2$ intervals on which $\|f'\| \geqslant \delta$ and $\leqslant \alpha N \lambda_2 + 2$ intervals on which $\|f'\| < \delta$.

For each interval on which $||f'|| \geqslant \delta$ we apply the Kusmin-Landau inequality (Theorem 8.13) and we get a contribution $\leqslant 2/(\pi\delta)$.

For each interval on which $||f'|| < \delta$, we bound trivially by the number of terms. The length of such an interval is $< 2\delta/\lambda_2$, and thus it contains at most $2\delta/\lambda_2 + 1$ integer points. Hence we have the bound:

$$\left|\sum_{n\in I} e(f(n))\right| \leqslant (\alpha N\lambda_2 + 2)\left(\frac{2}{\pi\delta} + \frac{2\delta}{\lambda_2} + 1\right).$$

We optimize the choice of δ by taking $\delta = (\lambda_2/\pi)^{1/2}$. As $\lambda_2 < 1/9$, we do have $0 < \delta < 1/2$ and also $1 \leqslant \frac{1}{3\sqrt{\lambda_2}}$, hence

$$\frac{2}{\pi\delta} + \frac{2\delta}{\lambda_2} + 1 \leqslant \frac{2}{\sqrt{\pi\lambda_2}} + \frac{2}{\sqrt{\pi\lambda_2}} + \frac{1}{3\sqrt{\lambda_2}} \leqslant \frac{3}{\sqrt{\lambda_2}},$$

and the theorem follows immediately.

Remark 8.16 If we have a more precise knowledge of the function f, in particular if we know its first derivative, we can improve the previous result. Thus when $f(n) = An^{-\sigma}$, or more generally when there exist $\lambda_1 > 0$, $\lambda_2 > 0$, $\alpha \geqslant 1$, with $\lambda_1 \asymp N\lambda_2$, and such that

$$\lambda_1 \leqslant |f'(x)| \leqslant \alpha\lambda_1, \quad \lambda_2 \leqslant |f''(x)| \leqslant \alpha\lambda_2, \quad (x \in I),$$

we have

$$\left|\sum_{n\in I} e(f(n))\right| \ll \alpha N\lambda_2^{1/2} + \lambda_1^{-1}.$$

Indeed, when $\alpha\lambda_1 < \frac{1}{2}$, we apply Kusmin-Landau which yields an upper bound in λ_1^{-1}, and in the opposite case we have $\alpha N\lambda_2 \gg 1$, hence $\alpha N\lambda_2^{1/2} \gg \lambda_2^{-1/2}$, and we apply van der Corput's lemma which yields an upper bound in $\alpha N\lambda_2^{1/2}$.

Application: the Voronoi Theorem
We denote by $\tau(n)$ the number of divisors of n. Dirichlet has shown that

$$\sum_{n\leqslant x} \tau(n) = x\log x + (2\gamma - 1)x + O(\sqrt{x}),$$

where γ is the Euler constant, defined by

$$\gamma = \lim_{N\to+\infty}\left(\sum_{n\leqslant N}\frac{1}{n} - \log N\right).$$

We shall prove the following result

Theorem 8.17 (Voronoi) *For every real number $x \geqslant 2$,*

$$\Delta(x) := \sum_{n \leqslant x} \tau(n) - x \log x - (2\gamma - 1)x = O(x^{1/3} \log x).$$

To establish this result, we need a fine knowledge of the harmonic series.

Lemma 8.18 *For every real number $y \geqslant 1$,*

$$\sum_{n \leqslant y} \frac{1}{n} = \log y + \gamma - \frac{\psi(y)}{y} + O\left(\frac{1}{y^2}\right),$$

where $\psi(u) = u - \lfloor u \rfloor - \frac{1}{2}$.

Corollary 8.19 *For all integer $N \geqslant 1$,*

$$\sum_{n \leqslant N} \frac{1}{n} = \log N + \gamma + \frac{1}{2N} + O\left(\frac{1}{N^2}\right).$$

Proof We write

$$\sum_{n \leqslant y} \frac{1}{n} = \int_{1-}^{y^+} \frac{d \lfloor u \rfloor}{u} = \int_{1-}^{y^+} \frac{d(u - \psi(u) - \frac{1}{2})}{u}$$

$$= \log y - \int_{1-}^{y^+} \frac{d\psi(u)}{u}$$

$$= \log y - \left[\frac{\psi(u)}{u}\right]_{1-}^{y^+} - \int_{1}^{y} \psi(u) \frac{du}{u^2}$$

$$= \log y - \frac{\psi(y)}{y} + \frac{\psi(1^-)}{1} - \int_{1}^{\infty} \psi(u) \frac{du}{u^2} + \int_{y}^{\infty} \psi(u) \frac{du}{u^2},$$

where we remarked that $\psi(y^+) = \psi(y)$, because ψ is right continuous.
   If we let y tend to infinity, we get

$$\gamma = \lim_{y \to \infty} \left(\sum_{n \leqslant y} \frac{1}{n} - \log y\right) = \frac{\psi(1^-)}{1} - \int_{1}^{\infty} \psi(u) \frac{du}{u^2},$$

hence it remains to prove that

$$\int_y^\infty \psi(u)\frac{du}{u^2} = O\left(\frac{1}{y^2}\right).$$

As $\psi(u)$ is an oscillating function, we shall integrate by parts. We set

$$\Psi(u) = \int_0^u \psi(v)dv.$$

As $\Psi(1) = 0$, the function Ψ is periodic with period 1, and continuous hence bounded. Then

$$\int_y^\infty \psi(u)\frac{du}{u^2} = \left[\frac{\Psi(u)}{u^2}\right]_y^\infty + 2\int_y^\infty \Psi(u)\frac{du}{u^3} = O\left(\frac{1}{y^2}\right).$$

Proposition 8.20 *For every real number $x \geqslant 1$,*

$$\Delta(x) = -2\sum_{n\leqslant\sqrt{x}} \psi\left(\frac{x}{n}\right) + O(1).$$

Proof We use the hyperbola method. We write

$$\sum_{\ell\leqslant x} \tau(\ell) = \sum_{mn\leqslant x} 1$$

$$= \sum_{\substack{mn\leqslant x \\ m\leqslant\sqrt{x}}} 1 + \sum_{\substack{mn\leqslant x \\ n\leqslant\sqrt{x}}} 1 - \sum_{\substack{m\leqslant\sqrt{x} \\ n\leqslant\sqrt{x}}} 1.$$

The first two sums are equal, namely to

$$\sum_{n\leqslant\sqrt{x}} \left\lfloor\frac{x}{n}\right\rfloor = \sum_{n\leqslant\sqrt{x}} \left(\frac{x}{n} - \psi\left(\frac{x}{n}\right) - \frac{1}{2}\right)$$

$$= x\left(\log\sqrt{x} + \gamma - \frac{\psi(\sqrt{x})}{\sqrt{x}} + O\left(\frac{1}{x}\right)\right) - \sum_{n\leqslant\sqrt{x}} \psi\left(\frac{x}{n}\right) - \frac{1}{2}\lfloor\sqrt{x}\rfloor$$

$$= \tfrac{1}{2}x\log x + \gamma x - \sqrt{x}\,\psi(\sqrt{x}) - \sum_{n\leqslant\sqrt{x}} \psi\left(\frac{x}{n}\right) - \tfrac{1}{2}\sqrt{x} + O(1).$$

We have

$$\sum_{\substack{m \leqslant \sqrt{x} \\ n \leqslant \sqrt{x}}} 1 = \lfloor \sqrt{x} \rfloor^2 = \left(\sqrt{x} - \psi(\sqrt{x}) - \tfrac{1}{2} \right)^2 = x - 2\sqrt{x}\, \psi(\sqrt{x}) - \sqrt{x} + O(1).$$

Thus

$$\sum_{\ell \leqslant x} \tau(\ell) = 2 \sum_{n \leqslant \sqrt{x}} \left\lfloor \frac{x}{n} \right\rfloor - \lfloor \sqrt{x} \rfloor^2 = x \log x + (2\gamma - 1)x - 2 \sum_{n \leqslant \sqrt{x}} \psi\left(\frac{x}{n}\right) + O(1).$$

Remark 8.21 Dirichlet's proof consists simply in replacing $\psi(x/n)$ with $O(1)$ in what precedes, and using the simplest asymptotic expansion

$$\sum_{n \leqslant y} \frac{1}{n} = \log y + \gamma + O\left(\frac{1}{y}\right),$$

and we get immediately

$$\Delta(x) = O(\sqrt{x}).$$

Lemma 8.22 *Let f be a twice continuously differentiable function on an interval I containing $N > 0$ integers. We suppose there exist $\lambda_1 > 0$, $\lambda_2 > 0$, $\alpha \geqslant 1$, such that $\lambda_1 \asymp N\lambda_2$ and*

$$\lambda_1 \leqslant |f'(x)| \leqslant \alpha\lambda_1, \qquad \lambda_2 \leqslant |f''(x)| \leqslant \alpha\lambda_2 \qquad (x \in I).$$

Then we have

$$\sum_{n \in I} \psi(f(n)) = O\left(\alpha N \lambda_2^{1/3} + \lambda_1^{-1}\right)$$

Proof We can suppose $\lambda_2 \leqslant 1$ as otherwise the result is trivial. Applying Vaaler's lemma, we get

$$\left| \sum_{n \in I} \psi(f(n)) \right| \leqslant \frac{N}{2H + 2} + \sum_{1 \leqslant |h| \leqslant H} \left(\frac{1}{2\pi |h|} + \frac{1}{2H + 2} \right) \left| \sum_{n \in I} e(hf(n)) \right|$$

$$\leqslant \frac{N}{2H} + \sum_{1 \leqslant h \leqslant H} \frac{2}{h} \left| \sum_{n \in I} e(hf(n)) \right|.$$

The exponential sum on the right-hand side is bounded through the van der Corput lemma (improved version in the remark). We get

$$\sum_{1 \leqslant h \leqslant H} \frac{1}{h} \left| \sum_{n \in I} e(hf(n)) \right| = \sum_{1 \leqslant h \leqslant H} \frac{1}{h} O\left(\alpha_2 N \sqrt{\lambda_2 h} + \frac{1}{\lambda_1 h} \right)$$

$$= O\left(\alpha_2 N \sqrt{\lambda_2 H} + \frac{1}{\lambda_1} \right),$$

so that

$$\left| \sum_{n \in I} \psi(f(n)) \right| = O\left(\frac{N}{H} + \alpha_2 N \sqrt{\lambda_2 H} + \frac{1}{\lambda_1} \right).$$

Taking $H = \left\lfloor \lambda_2^{-1/3} \right\rfloor$, we get the expected result.

We are now able to complete the proof of the Voronoi theorem. We write (with Vinogradov's notation)

$$\Delta(x) \ll \left| \sum_{n \leqslant \sqrt{x}} \psi\left(\frac{x}{n} \right) \right| \ll \sum_{r=1}^{R} \left| \sum_{n \in I_r} \psi\left(\frac{x}{n} \right) \right|$$

with, for $r = 1, \ldots, R$,

$$I_r =]2^{-r} \sqrt{x}, 2^{-r+1} \sqrt{x}], \quad 2^{R-1} \leqslant x^{1/2} < 2^R.$$

Hence we apply the previous lemma with

$$\lambda_1 \asymp \frac{x}{(2^{-r} \sqrt{x})^2} \asymp 2^{2r}, \qquad \lambda_2 \asymp \frac{x}{(2^{-r} \sqrt{x})^3} \asymp \frac{2^{3r}}{x^{1/2}},$$

hence

$$\Delta(x) \ll \sum_{r=1}^{R} \left(2^{-r} \sqrt{x} \frac{2^r}{x^{1/6}} + 2^{-2r} \right) \ll x^{1/3} \log x.$$

Remark 8.23 The van der Corput theorem above is useful to bound an exponential sum through the second derivative, particularly when we have $|f'| > 1$ and $|f''| < 1$ on I. When $|f''| > 1$, we use the following lemma to lower the order of magnitude of the derivatives.

Lemma 8.24 *Let N and Q be integers $\geqslant 1$, and z_n complex numbers. Then we have:*

$$\left| \sum_{1 \leqslant n \leqslant N} z_n \right|^2 \leqslant \left(1 + \frac{N-1}{Q} \right) \sum_{|q| < Q} \left(1 - \frac{|q|}{Q} \right) \sum_{\substack{1 \leqslant n \leqslant N \\ 1 \leqslant n+q \leqslant N}} z_{n+q} \overline{z_n}.$$

Proof The statement uses only z_n for $1 \leqslant n \leqslant N$. For convenience, we assume $z_n = 0$ for $n \leqslant 0$ and for $n \geqslant N + 1$.

We have the equalities

$$Q \sum_{n \in \mathbb{Z}} z_n = \sum_{1 \leqslant q \leqslant Q} \sum_{n \in \mathbb{Z}} z_{n+q} = \sum_{n \in \mathbb{Z}} \sum_{1 \leqslant q \leqslant Q} z_{n+q}.$$

The integers n which may have a nonzero contribution in the last sum satisfy $1 - Q \leqslant n \leqslant N - 1$. Hence there are at most $N - 1 + Q$ of them.

Applying the Cauchy-Schwarz inequality we get:

$$Q^2 \left| \sum_{n \in \mathbb{Z}} z_n \right|^2 \leqslant (N - 1 + Q) \sum_{n \in \mathbb{Z}} \left| \sum_{1 \leqslant q \leqslant Q} z_{n+q} \right|^2$$

$$\leqslant (N - 1 + Q) \sum_{1 \leqslant q_1 \leqslant Q} \sum_{1 \leqslant q_2 \leqslant Q} \sum_{n \in \mathbb{Z}} z_{n+q_1} \overline{z_{n+q_2}}$$

$$\leqslant (N - 1 + Q) \sum_{1 \leqslant q_1 \leqslant Q} \sum_{1 \leqslant q_2 \leqslant Q} \sum_{m \in \mathbb{Z}} z_{m+q_1-q_2} \overline{z_m}$$

$$\leqslant (N - 1 + Q) \sum_{-Q \leqslant q \leqslant Q} r(q) \sum_{m \in \mathbb{Z}} z_{m+q} \overline{z_m},$$

where $r(q) = \#\{(q_1, q_2), 1 \leqslant q_1 \leqslant Q, 1 \leqslant q_2 \leqslant Q, q_1 - q_2 = q\}$.

It is clear that $r(-q) = r(q)$.

For $0 \leqslant q \leqslant Q$, $r(q)$ is the number of q_1 such that $1 \leqslant q_1 \leqslant Q$ and $1 + q \leqslant q_1 \leqslant Q + q$, i.e. such that $1 + q \leqslant q_1 \leqslant Q$, hence $r(q) = Q - q$.

Consequently $r(q) = Q - |q|$ for $-Q \leqslant q \leqslant Q$. Thus we have

$$Q^2 \left| \sum_{n \in \mathbb{Z}} z_n \right|^2 \leqslant (N - 1 + Q) \sum_{-Q \leqslant q \leqslant Q} (Q - |q|) \sum_{m \in \mathbb{Z}} z_{m+q} \overline{z_m},$$

and we get the expected inequality after dividing by Q^2.

Remark 8.25 If $z_n = e(f(n))$, then $z_{n+q} \overline{z_n} = e(f(n+q) - f(n))$ and under good hypotheses (Q much smaller than N), the derivatives of $g(x) = f(x+q) - f(x)$ are much smaller than those of f.

Theorem 8.26 (van der Corput) *Let R be an integer $\geqslant 2$. We suppose that f has R continuous derivatives on an interval $I \subseteq [N + 1, 2N]$. Furthermore we suppose that there exists a constant F such that*

$$\forall x \in I, \quad FN^{-r} \ll |f^{(r)}(x)| \ll FN^{-r},$$

for $r = 1, \ldots, R$. Then

$$\sum_{n \in I} e(f(n)) \ll (FN^{-R})^{1/(2^R - 2)} N + F^{-1}N.$$

Proof cf. [1, Theorem 2.9].

Complement: Generalized van der Corput Inequality

Theorem 8.27 (Rivat-Sargos) *Let $z_1, \ldots, z_N \in \mathbb{C}$ and $x_1, \ldots, x_N \in \mathbb{R}$. We define*

$$\rho(t) = \left(\frac{\sin \pi t}{\pi t}\right)^2, \quad \rho(0) = 1, \quad \widehat{\rho}(u) = \max(0, 1 - |u|).$$

For all $\delta > 0$, we have

$$\left|\sum_{n=1}^{N} z_n\right|^2 \leqslant \sum_{k \in \mathbb{Z}} \rho(\delta k) \sum_{i=1}^{N} \sum_{j=1}^{N} z_i \overline{z_j} \, e(k(x_i - x_j))$$

$$\leqslant \frac{1}{\delta} \sum_{i=1}^{N} \sum_{j=1}^{N} z_i \overline{z_j} \sum_{k \in \mathbb{Z}} \widehat{\rho}\left(\frac{x_i - x_j + k}{\delta}\right).$$

If furthermore we suppose $\max_{1 \leqslant i,j \leqslant N} |x_i - x_j| \leqslant 1 - \delta$ (hence $0 < \delta \leqslant 1$), then

$$\left|\sum_{n=1}^{N} z_n\right|^2 \leqslant \frac{1}{\delta} \sum_{i=1}^{N} \sum_{j=1}^{N} \widehat{\rho}\left(\frac{x_i - x_j}{\delta}\right) z_i \overline{z_j}.$$

Proof Denote

$$\rho_\delta(t) = \delta \rho(\delta t),$$

$$\widehat{\rho_\delta}(\xi) = \int_{\mathbb{R}} \delta \rho(\delta t) \, e(-t\xi) \, dt = \int_{\mathbb{R}} \rho(u) \, e(-u\xi/\delta) \, du = \widehat{\rho}(\xi/\delta).$$

We have

$$\left| \sum_{n=1}^{N} z_n \right|^2 \leqslant \frac{1}{\delta} \sum_{k \in \mathbb{Z}} \rho_\delta(k) \left| \sum_{n=1}^{N} z_n \, e(kx_n) \right|^2$$

and expanding the square of the right-hand side we get the first inequality. Applying Poisson's summing formula to the right-hand side we get

$$\left| \sum_{n=1}^{N} z_n \right|^2 \leqslant \frac{1}{\delta} \sum_{k \in \mathbb{Z}} \int_{\mathbb{R}} \rho_\delta(t) \left| \sum_{n=1}^{N} z_n \, e(tx_n) \right|^2 e(tk) \, dt.$$

We expand the square and exchange the sums:

$$\frac{1}{\delta} \sum_{i=1}^{N} \sum_{j=1}^{N} z_i \bar{z}_j \sum_{k \in \mathbb{Z}} \int_{\mathbb{R}} \rho_\delta(t) \, e(t(x_i - x_j + k)) \, dt = \frac{1}{\delta} \sum_{i=1}^{N} \sum_{j=1}^{N} z_i \bar{z}_j \sum_{k \in \mathbb{Z}} \widehat{\rho_\delta}(x_i - x_j + k)$$

and get the second inequality because $\widehat{\rho_\delta}(x) = \widehat{\rho}(x/\delta)$.

We have $\widehat{\rho_\delta}(x) = 0$ for $|x| \geqslant \delta$. When $|x_i - x_j| \leqslant 1 - \delta$, we have $|x_i - x_j + k| \geqslant \delta$ for all $k \neq 0$, and thus $\widehat{\rho_\delta}(x_i - x_j + k) = 0$ for all $k \neq 0$, which proves the third inequality.

Remark 8.28 Taking $\delta = \left(1 + \dfrac{N-1}{Q} \right)^{-1} = \dfrac{Q}{N-1+Q}$ and $x_n = \dfrac{n\delta}{Q}$ we have

$$|x_i - x_j| = \frac{|i - j| \, \delta}{Q} \leqslant \frac{(N-1)\delta}{Q} = \frac{N-1}{N-1+Q} = 1 - \delta,$$

and get from applying Theorem 8.27:

$$\left| \sum_{n=1}^{N} z_n \right|^2 \leqslant \left(1 + \frac{N-1}{Q} \right) \sum_{i=1}^{N} \sum_{j=1}^{N} \max \left(0, 1 - \frac{|i-j|}{Q} \right) z_i \bar{z}_j$$

$$= \left(1 + \frac{N-1}{Q} \right) \sum_{|q| < Q} \left(1 - \frac{|q|}{Q} \right) \sum_{\substack{1 \leqslant n \leqslant N \\ 1 \leqslant n+q \leqslant N}} z_{n+q} \bar{z}_n,$$

which is the van der Corput inequality.

Reference

1. S. Graham, G. Kolesnik, *Van der Corput's Method of Exponential Sums*. London Mathematical Society Lecture Note Series, vol. 126 (Cambridge University Press, Cambridge, 1991)

Part II
Interactions Between Arithmetic and Dynamics

Chapter 9
A Brief Guide to Reversing and Extended Symmetries of Dynamical Systems

M. Baake

9.1 Introduction

Symmetries of dynamical systems are important objects to study, as they help in understanding the orbit structure and many other properties. Moreover, the group of symmetries is a topological invariant that can be useful for distinguishing between different dynamical systems. Naturally, this invariant is generally weaker than other invariants (such as those from (co-)homology or homotopy theory), but often easier to access.

For both aspects, studying properties and defining invariants, one is clearly interested in effective generalisations or extensions of the symmetry group. Inspired by the time-reversal symmetry of many fundamental equations in physics, one obvious step in this direction is provided by the *reversing symmetry group* of a dynamical system, which—in the case of reversibility—is an index-2 extension of the symmetry group.

Traditionally, the majority of the studies has concentrated on concrete dynamical systems, where the space is usually simple, but the mapping(s) might be complicated. Even for toral automorphism, the answer is amazingly rich. There is a complementary picture, which arises through the coding of itineraries and leads to the analogous questions in symbolic dynamics [59]. Here, the mapping(s) are simple, but the space (usually a closed shift space) is complicated, and this is particularly so when going to higher-dimensional shifts.

M. Baake (✉)
Faculty of Mathematics, Bielefeld University, Bielefeld, Germany
e-mail: mbaake@math.uni-bielefeld.de

© Springer International Publishing AG, part of Springer Nature 2018
S. Ferenczi et al. (eds.), *Ergodic Theory and Dynamical Systems in their Interactions with Arithmetics and Combinatorics*, Lecture Notes in Mathematics 2213, https://doi.org/10.1007/978-3-319-74908-2_9

In this brief introductory review, we recall the basic definitions and notions, and present some results from the large body of literature that has accumulated. Clearly, the exposition cannot be complete in any way, whence the references will provide further directions.

After some examples from the classic theory of concrete dynamical systems, we shall stroll through some more recent results on the complementary picture from symbolic dynamics.

9.2   General Setting and Notions

A convenient starting point is a topological space $\mathbb{X}$, which is usually (but not always) assumed to be compact, and a mapping $T \in \mathrm{Aut}(\mathbb{X})$, where the automorphism group is understood in the Smale sense, meaning that it is the group of *all* homeomorphisms of $\mathbb{X}$. The pair $(\mathbb{X}, T)$ then defines a (topological) *dynamical system*, and the group $\langle T \rangle \subset \mathrm{Aut}(\mathbb{X})$ is important. Now, we define the *symmetry group* of $(\mathbb{X}, T)$ as

$$S(\mathbb{X}, T) := \{G \in \mathrm{Aut}(\mathbb{X}) : G \circ T = T \circ G\} = \mathrm{cent}_{\mathrm{Aut}(\mathbb{X})}(\langle T \rangle) = \mathrm{Aut}(\mathbb{X}, T). \tag{9.1}$$

This group plays an important role in the analysis of $(\mathbb{X}, T)$, for instance in the context of periodic orbits and dynamical zeta functions. Its is also a useful tool in the classification of dynamical systems, because it is a topological invariant.

Remark 9.1 The group $\mathrm{Aut}(\mathbb{X}, T)$ is often used as a starting point for algebraic considerations, and then simply called the automorphism group of the dynamical system, but this is—as we shall see later—a use of the word that is too restrictive, and effectively excludes many natural mappings from the consideration. We will thus not use this notation, and rather view $S(\mathbb{X}, T)$ as a subgroup of $\mathrm{Aut}(\mathbb{X})$ in the Smale sense.

In some cases, the group $\mathrm{Aut}(\mathbb{X})$ might be too big a 'universe' to consider, and some subgroup of it is a more natural choice, for instance when some additional structure of $\mathbb{X}$ should be preserved. This is particularly so if some general results are available that imply $S(\mathbb{X}, T)$ and $R(\mathbb{X}, T)$ to be subgroups of some group $\mathcal{U} \subset \mathrm{Aut}(\mathbb{X})$. In this case, one can start with $\mathcal{U}$, and simplify the algebraic derivations considerably. The above point simply is that $\mathcal{U}$ should generally *not* be chosen as $\mathrm{Aut}(\mathbb{X}, T)$, as this is too restrictive.

Since we will not consider the case that T is not invertible, a natural extension of $S(\mathbb{X}, T)$ is given by

$$R(\mathbb{X}, T) := \{G \in \mathrm{Aut}(\mathbb{X}) : G \circ T \circ G^{-1} = T^{\pm 1}\}, \tag{9.2}$$

which is motivated by the time-reversal symmetry of many fundamental equations of physics; see [53, 71] and references therein for background. From now on, we write GT instead of $G \circ T$ etc. for ease of notation. The relation between $S(\mathbb{X}, T)$ and $\mathcal{R}(\mathbb{X}, T)$ can be summarised as follows; see [12] and references therein.

Theorem 9.2 *If* $(\mathbb{X}, T)$ *is a topological dynamical system,* $\mathcal{R}(\mathbb{X}, T) \subset \operatorname{Aut}(\mathbb{X})$ *is a group, with* $\langle T \rangle$ *and* $S(\mathbb{X}, T)$ *as normal subgroups. Moreover, one either has* $\mathcal{R}(\mathbb{X}, T) = S(\mathbb{X}, T)$ *or* $[\mathcal{R}(\mathbb{X}, T) : S(\mathbb{X}, T)] = 2$. *In the latter case, the systems is reversible.*

Further, if $T^2 \neq \operatorname{Id}$ *and if there is an involution* H *with* $HTH^{-1} = T^{-1}$, *one has*

$$\mathcal{R}(\mathbb{X}, T) \;=\; S(\mathbb{X}, T) \rtimes \langle H \rangle \;\simeq\; S(\mathbb{X}, T) \rtimes C_2,$$

which is the standard form of reversibility.

An element that conjugates T into its inverse (where we assume $T^2 \neq \operatorname{Id}$) is called a *reversor*. An elementary observation is the fact that a reversor cannot be of odd order, so it is either of even or of infinite order. When the order is finite, hence of the form $2^{\ell}(2m + 1)$ for some $\ell \geqslant 1$, there exists another reversor of order 2^{ℓ}. When T possesses an involutory reversor, R say, one has $T = TR^2 = (TR)R$, where $(TR)^2 = TRTR = TT^{-1} = \operatorname{Id}$, so T is the product of two involutions. This is a frequently used approach in the older literature, before the group-theoretic setting showed [38, 52] that the more general approach is natural and helpful; see [12, 53] and references therein for details.

As is implicit from our formulation so far, reversibility is not an interesting concept when T itself is an involution. More generally, when T has finite order, the structure of $\mathcal{R}(\mathbb{X}, T)$ is a group-theoretic problem, and of independent interest; see [60] for a concise exposition. However, in the context of dynamical systems, one is mainly interest in the case that $\langle T \rangle \simeq \mathbb{Z}$. Then, one can slightly change the point of view by considering T as defining a continuous group action of $\mathbb{Z}$ on $\mathbb{X}$, which is often reflected by the modified notation $(\mathbb{X}, \mathbb{Z})$ for the topological dynamical system. From now on, unless explicitly stated otherwise, we shall adopt this point of view here, too. The following result is elementary.

Fact 9.3 *When* T *is not of finite order, one has* $\mathcal{R}(\mathbb{X}, T) = \operatorname{norm}_{\operatorname{Aut}(\mathbb{X})}(\langle T \rangle)$.

It is thus the interplay between the (topological) centraliser and normaliser that is added in the extension from $S(\mathbb{X}, T)$ to $\mathcal{R}(\mathbb{X}, T)$. One simple (but frequently useful) instance of this is given by the following result, where C_{∞} and $D_{\infty} = C_{\infty} \rtimes C_2$ denote the infinite cyclic and dihedral group, respectively.

Theorem 9.4 ([12, Thm. 1 and Cor. 1]) *Let* $T \in \operatorname{Aut}(\mathbb{X})$ *be of infinite order. If one has* $S(\mathbb{X}, T) \simeq C_{\infty}$ *and if* T *is reversible, one has* $\mathcal{R}(\mathbb{X}, T) = S(\mathbb{X}, T) \rtimes C_2 \simeq D_{\infty}$, *and all reversors of* T *are involutions.*

Conversely, if all reversors of T *are involutions, the symmetry group* $S(\mathbb{X}, T)$ *is Abelian.*

Clearly, in the setting of dynamical systems, one could equally well consider the analogous questions for the measure-theoretic centraliser and normaliser, and this is indeed frequently done in the literature; compare [35, 39, 69] and references therein. Since, in many relevant cases, the measure-theoretic symmetry groups turn out to be topological (see [69] for results in this direction), we concentrate on the latter situation in this overview.

In what follows, we shall meet two rather different general situations, as briefly indicated in the introduction. On the one hand, there are many systems from nonlinear dynamics where the space is simple, but the map is complicated. In this case, we will write $S(T)$ instead of $S(\mathbb{X}, T)$ to emphasise the mapping. Likewise, when we are in the complementary situation (of symbolic dynamics, say) with a simple map acting on a more complicated space, we will use $S(\mathbb{X})$ instead to highlight the difference. This also matches the widely used conventions in these two directions.

9.3 Concrete Systems from Nonlinear Dynamics

In this section, we will describe, in a somewhat informal manner, how symmetries and reversing symmetries arise in three particular families of dynamical systems, namely trace maps, toral automorphisms, and polynomial automorphisms of the plane. Clearly, there are many other relevant examples, some of which can be found in [53, 60, 71] and references therein.

9.3.1 Trace Maps

This class of dynamical system arises in the study of one-dimensional Schrödinger operators with aperiodic potentials of substitutive origin, compare [31] and references therein, and provide a powerful tool for the study of their spectra and transport properties. The paradigmatic *Fibonacci trace map* in 3-space is given by

$$(x, y, z) \longmapsto (y, z, 2yz - x)$$

and is reversible, with involutory reversor $(x, y, z) \mapsto (z, y, x)$; see [65] and references given there. The group-theoretic 'universe' to consider here is given by the group of 3-dimensional invertible polynomial mappings that preserve the Fricke–Vogt invariant

$$I(x, y, z) = x^2 + y^2 + z^2 - 2xyz - 1$$

and fix the point $(1, 1, 1)$; see [9, 14, 65] and references therein for more. This group of mappings is isomorphic with $PGL(2, \mathbb{Z})$, and can thus be analysed by classic methods, including the theory of binary quadratic forms.

In other words, the analysis of (reversing) symmetries of trace maps is equivalent to the determination of $\mathcal{S}(M)$ and $\mathcal{R}(M)$ for matrices $M \in PGL(2, \mathbb{Z})$. Since

$$PGL(2, \mathbb{Z}) \simeq GL(2, \mathbb{Z})/\{\pm 1\},$$

the following result is obvious.

Fact 9.5 *Let $M \in PGL(2, \mathbb{Z})$ and M' be either of the two corresponding matrices in $GL(2, \mathbb{Z})$. Then, the symmetry group $\mathcal{S}(M)$ is given by*

$$\mathcal{S}(M) = \mathrm{cent}_{PGL(2,\mathbb{Z})}(\langle M \rangle) = \mathrm{cent}_{GL(2,\mathbb{Z})}(\langle M' \rangle)/\{\pm 1\}.$$

The symmetry groups can thus be derived from the analysis of general (two-dimensional) toral automorphisms, which we will review in Sect. 9.3.2. For the reversing symmetry group, the role of $\{\pm 1\}$ changes. Let $M \in PGL(2, \mathbb{Z})$ be given, and view it as a $GL(2, \mathbb{Z})$-matrix. Then, we have to find all solutions H to

$$HMH^{-1} = \pm M^{-1},$$

where the calculation modulo ± 1 means that we get *more* cases with reversibility than in $GL(2, \mathbb{Z})$. For instance, $M = \left(\begin{smallmatrix} 1 & 1 \\ 1 & 0 \end{smallmatrix}\right)$ is reversible in $PGL(2, \mathbb{Z})$, with an involutory reversor, but not within $GL(2, \mathbb{Z})$, while M^2 (known as Arnold's cat map [5, Ex. 1.15]) is reversible in both groups. Within $GL(2, \mathbb{Z})$, this phenomenon is called 2-reversibility; see [9] for details.

9.3.2 Toral Automorphisms

These systems, which are also known as 'cat maps', are much studied examples in chaotic dynamics and ergodic theory. Here, in order to preserve the linear structure of $\mathbb{T}^d$, the d-dimensional torus, one usually works within $\mathcal{U} = GL(d, \mathbb{Z}) \subset \mathrm{Aut}(\mathbb{T}^d)$; see [1, 2, 5, 62] for background.

In the planar case ($d = 2$), one thus has to deal with the group $GL(2, \mathbb{Z})$. Here, if M is an element of infinite order, one always finds $\mathcal{S}(M) \simeq C_2 \times C_\infty$, where $C_2 = \{\pm 1\}$. This follows for any parabolic element by a simple calculation, and, for the hyperbolic elements, is a consequence of Dirichlet's unit theorem for real quadratic number fields; see [24] for background.

Remark 9.6 Reversible cases among elements of infinite order are of three possible types: When all reversors are involutions, one has $\mathcal{R}(M) \simeq C_2 \times D_\infty$, again with $D_\infty = C_\infty \rtimes C_2$; when all reversors are of fourth order, one has $\mathcal{R}(M) \simeq C_\infty \rtimes C_4$;

finally, when reversors both of order 2 and 4 exist, one has $\mathcal{R}(M) \simeq (C_2 \times C_\infty) \rtimes C_2$. All three types occur; see [12, Thm. 2 and Ex. 4] and references given there for more.

In this context, it is certainly a valid and interesting question how the concepts can be extended to cover toral endomorphisms, or what happens when one restricts to rational sublattices. This is connected with looking at the related questions over finite fields and residue class rings; see [16, 18] and references therein for some results.

The situation becomes more complex, and also more interesting, in higher dimensions. In a first step, one has to analyse the symmetry group of a toral automorphism, $M \in GL(d, \mathbb{Z})$ say, within this matrix group. In the generic case, where M is simple (meaning that its eigenvalues are distinct) one can employ Dirichlet's unit theorem again. Let us first look at the case that the characteristic polynomial $P(x) = \det(M - x\mathbb{1})$ of M is irreducible over $\mathbb{Z}$, and hence also over $\mathbb{Q}$. Then, if λ is any of the d eigenvalues of M, it is an algebraic integer of degree $d = n_1 + 2n_2$, where n_1 is the number of real algebraic conjugates of λ and n_2 the number of complex conjugate pairs among the algebraic conjugates.

Now, if O is the maximal order in the algebraic number field $\mathbb{Q}(\lambda)$, Dirichlet's unit theorem states that the unit group $O^\times$ is of the form

$$O^\times \simeq T \times \mathbb{Z}^{n_1 + n_2 - 1} \tag{9.3}$$

with $T = O \cap \{\text{roots of unity}\}$ being a finite cyclic group. The latter is known as the *torsion subgroup* of $O^\times$. Due to the isomorphism of $\mathbb{Z}[\lambda]$ with the ring $\mathbb{Z}[M]$ under our irreducibility assumption on P, one then has the following result [10, Prop. 1 and Cor. 1].

Theorem 9.7 *Let $M \in GL(d, \mathbb{Z})$ have an irreducible characteristic polynomial, $P(x)$, of degree $d = n_1 + 2n_2$, with n_1 and n_2 as above. Then, $S(M)$ is isomorphic with a subgroup of $O^\times$ of maximal rank, so*

$$S(M) \simeq T' \times \mathbb{Z}^{n_1 + n_2 - 1},$$

where T' is a subgroup of the torsion group T from Eq. (9.3).

Moreover, whenever $P(x)$ has a real root, which includes all cases with d odd, one simply has $T' = \{\pm 1\} \simeq C_2$.

For our previous example, $M = \left(\begin{smallmatrix} 1 & 1 \\ 1 & 0 \end{smallmatrix}\right)$, one finds $S(M) = \{\pm 1\} \times \langle M \rangle \simeq C_2 \times \mathbb{Z}$. Note that, in general, the generators of the free part of $S(M)$ can correspond to powers of fundamental units, which is related with the question of the existence of matrix roots within $GL(d, \mathbb{Z})$; see [10] for more. One can quite easily extend Theorem 9.7 to the case that M is simple. This is done by factoring P over $\mathbb{Z}$ and treating the factors separately [9, Thm. 1].

Let us look at the reversibility of a matrix $M \in GL(2, \mathbb{Z})$. A necessary condition clearly is that M and M^{-1} have the same spectrum (including multiplicities). In other words, if P is the characteristic polynomial of M with integer coefficients, it must satisfy the self-reciprocity condition

$$P(x) = \frac{(-1)^d x^d}{\det(M)} P(\tfrac{1}{x}). \tag{9.4}$$

Now, if d is odd or if $\det(M) = -1$, this relation implies that 1 or -1 is a root, and P is reducible over $\mathbb{Z}$. In particular, d odd and P irreducible immediately excludes reversibility. This means that, generically, reversible cases can only occur when d is even and $\det(M) = 1$.

Note that, even if Eq. (9.4) is satisfied, the reversibility still depends on the underlying integer matrix M, and the class number of $\mathbb{Z}[\lambda]$ enters. It is then clear that deciding on reversibility is a problem that increases with growing d; we refer to the discussion in [10] for more. However, for any given characteristic polynomial that is self-reciprocal according to the condition of Eq. (9.4), there is at least one reversible class of matrices, and this can be represented by the Frobenius companion matrix [10, Thm. 3].

A natural extension of symmetries can be considered in the setting of matrix rings rather than groups, such as $\mathrm{Mat}(d, K)$ instead of $\mathrm{GL}(d, K)$, where K can itself be a ring (such as $\mathbb{Z}$) or a field (such as $\mathbb{Q}$). Then, one can define

$$S(M) = \{G \in \mathrm{Mat}(d, K) : [M, G] = 0\}.$$

Concretely, if M is an integer matrix with irreducible characteristic polynomial, and λ is any of its roots, one finds $S(M)$ to be isomorphic with an order O in the number field $\mathbb{Q}(\lambda)$ that satisfies $\mathbb{Z}[\lambda] \subseteq O \subseteq O_{\max}$, where $O_{\max}$ denotes the maximal order in $\mathbb{Q}(\lambda)$; see [41, Ch. III] as well as [10, Sec. 3.3] and references given there for more.

9.3.3 Polynomial Automorphisms of the Plane

Let K be a field and consider the group $\mathcal{U}_K = \mathrm{GA}_2(K)$ of polynomial automorphisms of the affine plane over K. Consequently, we have $\mathbb{X} = K^2$ in this case, which need not be compact. $\mathcal{U}_K$ consists of all mappings of the form

$$\begin{pmatrix} x \\ y \end{pmatrix} \longmapsto \begin{pmatrix} P(x, y) \\ Q(x, y) \end{pmatrix}$$

with $P, Q \in K[x, y]$, subject to the condition that the inverse exists and is also polynomial. Note that, over general fields, different polynomials might actually define the same mapping on K^2, but we will distinguish them on the level of the polynomials.

In nonlinear dynamics, where $GA_2(\mathbb{R})$ and $GA_2(\mathbb{C})$ have received considerable attention, a common alternative notation is

$$x' = P(x, y), \quad y' = Q(x, y).$$

Frequently studied examples include the Hénon quadratic map family, defined by $P(x, y) = y$ and $Q(x, y) = -\delta x + y^2 + c$ with constants $c, \delta \in \mathbb{C}$ and $\delta \neq 0$. Quite often, for instance in the context of area-preserving mappings, the starting point is a polynomial automorphism in *generalised standard form*,

$$x' = x + P_1(y), \quad y' = y + P_2(x'),$$

with single-variable polynomials P_1 and P_2; compare [37, 66] and references therein. Here, the inverse is simply given by $y = y' - P_2(x')$ together with $x = x' - P_1(y)$.

In a certain sense, such particular normal forms are important, but do not exhaust the full power of the algebraic setting. Let us explain this a little in the context of combinatorial group theory. We begin by defining three subgroups of $GA_2(K)$ as follows. First,

$$\mathcal{A} := \{(\underline{a}, M) : \underline{a} \in K^2, M \in GL(2, K)\}$$

is the group of *affine* transformations, where $(\underline{a}, M)$ encodes the mapping $\underline{x} \mapsto M\underline{x} + \underline{a}$. We write $\underline{a}$ for a column vector, and tacitly identify the elements of $\mathcal{A}$ with the canonically corresponding elements of $GA_2(K)$. Multiplication is defined by

$$(\underline{a}, A)(\underline{b}, B) = (\underline{a} + A\underline{b}, AB),$$

whence $\mathcal{A}$ is a semi-direct product, namely $\mathcal{A} = K^2 \rtimes GL(2, K)$. The inverse of an element is $(\underline{a}, A)^{-1} = (-A^{-1}\underline{a}, A^{-1})$.

The second group, $\mathcal{E}$, is known as the group of *elementary* transformations. It consists of all mappings of the form

$$\begin{pmatrix} x \\ y \end{pmatrix} \longmapsto \begin{pmatrix} \alpha x + P(y) \\ \beta y + v \end{pmatrix}$$

with P a single-variable polynomial and $\alpha, \beta, v \in K$ subject to the condition $\alpha\beta \neq 0$. It is easy to check that the inverse exists and it of the same form. Transformations of this kind map lines with constant y-coordinate to lines of the same type. It is a well-known fact that the group $GA_2(K)$ is generated by $\mathcal{A}$ and $\mathcal{E}$; see [42, 72] as well as [73, Sec. 1.5].

Finally, our third group, $\mathcal{B}$, is defined as the intersection $\mathcal{B} = \mathcal{A} \cap \mathcal{E}$, with obvious meaning as subgroups of $GA_2(K)$. The elements of $\mathcal{B}$ are called *basic transformations*, and are mappings of the form

$$\begin{pmatrix} x \\ y \end{pmatrix} \longmapsto \begin{pmatrix} \alpha & \gamma \\ 0 & \beta \end{pmatrix} \begin{pmatrix} x \\ y \end{pmatrix} + \begin{pmatrix} u \\ v \end{pmatrix}$$

with $\alpha, \beta, \gamma, u, v \in K$ and $\alpha\beta \neq 0$. Clearly, also $\mathcal{B}$ is a semi-direct product, namely $\mathcal{B} = K^2 \rtimes \mathcal{T}$, where $\mathcal{T}$ denotes the subgroups of $GL(2, K)$ that consists of all invertible upper triangular matrices over K.

Now, the following result [68, 73] is fundamental to the classification of (reversing) symmetries of polynomial automorphisms.

Lemma 9.8 *The group* $GA_2(K)$ *is the free product of the groups* $\mathcal{A}$ *and* $\mathcal{E}$, *amalgamated along their intersection,* $\mathcal{B}$, *which is abbreviated as* $GA_2(K) = \mathcal{A} *_{\mathcal{B}} \mathcal{E}$.

Through this result, the problem has been reset in a purely algebraic way, and one can now explore the subgroup structure [43] of the amalgamated free product. In particular, one can classify the Abelian subgroups of $GA_2(K)$, which has trivial centre. Naturally, $S(T)$ for a given $T \in GA_2(K)$ is more complex, and need no longer be Abelian. When K has characteristic 0, one can derive restrictions on the order of other symmetries, which gives access to the finite subgroups of $S(T)$; for details, the reader is referred to [11].

For an important subclass of transformations known as CR elements, one can say a lot more. In particular, if K is a field of characteristic 0, all reversors must be of finite order. If, in addition, the roots of unity in K are just $\{\pm 1\}$, any reversor is an involution or an element of order 4, which makes their detection feasible. The possible reversing symmetry groups in this case are then the same three types we saw earlier, in Remark 9.6, for 2-dimensional toral automorphisms of infinite order. Since further details in this setting of combinatorial group theory tend to be a bit technical, we refer to [11] and references therein for more.

9.4   Shift Spaces with Faithful $\mathbb{Z}$-action

All examples in the previous section shared the feature that the space $\mathbb{X}$ is simple, but the map T on it is not. This is the standard situation in most dynamical systems that arise from concrete problems, for instance in nonlinear dynamics. However, it has long been known [59] that there is a complementary picture, which arises by coding orbits in such systems by symbolic sequences, for instance via itineraries. The latter keep track of a coarse-grained structure in such a way that the full dynamics can be recovered from them—at least almost surely in some suitable measure-theoretic sense.

This leads to *symbolic dynamics*, where the space $\mathbb{X}$ is 'replaced' by a closed shift space $\mathbb{Y}$ (often over a finite alphabet), and T by the action of the left shift, S. More precisely, one constructs a conjugacy, a semi-conjugacy, or (typically) a measure-theoretic isomorphism that makes the diagram

$$
\begin{array}{ccc}
\mathbb{X} & \xrightarrow{\ T\ } & \mathbb{X} \\
{\scriptstyle\phi}\downarrow & & \downarrow{\scriptstyle\phi} \\
\mathbb{Y} & \xrightarrow{\ S\ } & \mathbb{Y}
\end{array}
$$

commutative and ϕ as 'invertible as possible'. This motivates to also consider symmetries and reversing symmetries of shift spaces, where we shall always assume that the action of $\mathbb{Z}$ on the shift space is *faithful* in order to exclude degenerate situations. We refer to [50, 56] for general background, and to [49, 55] for the study of topological Markov chains in this context.

One immediate problem that arises is the fact that the symmetry group of a shift space (now called $\mathbb{X}$ again) is generally huge, in the sense that it contains a copy of the free group of two generators—and is thus not amenable [56]. This turns a potential classification into a wild problem, and not much has been done in this direction. On the other hand, as has long been known, it is also possible that one simply gets $S(\mathbb{X}) = \langle S \rangle \simeq \mathbb{Z}$, in which case one speaks of a *trivial centraliser*, or of a *minimal symmetry group*. This is a form of *rigidity*, for which different mechanisms are possible. Interestingly, rigidity is not a rare phenomenon [23], but actually generic in some sense [40], which makes it rather relevant also in practice.

To explore the possibilities a little, let us assume that $\mathcal{A}$ is a finite set, called the *alphabet*, and that $\mathbb{X} \subseteq \mathcal{A}^{\mathbb{Z}}$ is a closed and shift-invariant set, which is then automatically compact. Such a space is called a *shift space*, or *subshift* for short.

A special role has the 'canonical' reversor R defined by

$$
(Rx)_n := x_{-n} \tag{9.5}
$$

or any combination of R with a power of the shift S. It is clear that R conjugates S into its inverse on the full shift, $\mathbb{X} = \mathcal{A}^{\mathbb{Z}}$. More generally, one has the following property.

Lemma 9.9 *Let $\mathbb{X}$ be a shift space with faithful shift action. If $\mathbb{X}$ is reflection-invariant, which means $R(\mathbb{X}) = \mathbb{X}$ with the mapping R from Eq. (9.5), the system is reversible, with $\mathcal{R}(\mathbb{X}) = S(\mathbb{X}) \rtimes C_2$, where $C_2 = \langle R \rangle$.*

Let us collect a few examples of reversible subshifts, in an informal manner; see [20] and references therein for precise statements and proofs, and [3, 7, 26, 63, 64] for general background on substitution generated subshifts. Among these examples are

1. the *full* shift [50, 56], $\mathbb{X} = \mathcal{A}^{\mathbb{Z}}$, where $\mathcal{S}(\mathbb{X})$ is huge (and not amenable);
2. any *Sturmian* shift [27], which is always palindromic [33] and hence reversible, with symmetry group $\mathcal{S}(\mathbb{X}) \simeq \mathbb{Z}$;
3. the *period doubling* shift, defined by the primitive substitution rule $a \mapsto ab$, $b \mapsto aa$, again with $\mathcal{S}(\mathbb{X}) \simeq \mathbb{Z}$;
4. the *Thue–Morse* (TM) shift, defined by $a \mapsto ab$, $b \mapsto ba$, here with $\mathcal{S}(\mathbb{X}) \simeq \mathbb{Z} \times C_2$, where the extra symmetry is the letter exchange map defined by $a \leftrightarrow b$;
5. the *square-free* shift, obtained as the orbit closure of the characteristic function of the square-free integers, also with $\mathcal{S}(\mathbb{X}) \simeq \mathbb{Z}$.

In fact, the last example is quite remarkable, as its rigidity mechanism relies on the heredity of the shift, as was recently shown by Mentzen [57]. Note that the square-free shift has positive topological entropy, but nevertheless possesses minimal centraliser. Though this is not surprising in view of known results from Toeplitz sequences [25], it does show that rigidity as a result of low complexity, as studied in [28–30, 32], is only one of *several* mechanisms. We shall see more in Sect. 9.5. The square-free shift is a prominent example from the class of $\mathcal{B}$-free shifts, see [22, 34] and references therein, and also of interest in the context of Sarnak's conjecture on the statistical independence of the Möbius function from deterministic sequences (as discussed at length in other contributions to this volume).

Of course, things are generally more subtle than in these examples. First of all, a subshift can be irreversible, as happens for the one defined by the binary substitution $a \mapsto aba, b \mapsto baa$, where $\mathcal{R}(\mathbb{X}) = \mathcal{S}(\mathbb{X}) \simeq \mathbb{Z}$. Next, consider the subshift $\mathbb{X}_{k,\ell}$ defined by the primitive substitution

$$ a \longmapsto a^k b^\ell, \quad b \longmapsto b^k a^\ell $$

with $k, \ell \in \mathbb{N}$, which is reversible if and only if $k = \ell$. This is an extension of the TM shift (which is the case $k = \ell = 1$), in the spirit of [17, 45]. The symmetry group is $\mathcal{S}(\mathbb{X}_{k,\ell}) \simeq \mathbb{Z} \times C_2$ in all cases, where C_2 is once again the group generated by the letter exchange map.

Going to larger alphabets, $\mathcal{A} = \{a_0, a_1, \ldots, a_{N-1}\}$ say, one can look at a cyclic extension of the TM shift, as defined by the substitution $a_i \mapsto a_i a_{i+1}$ with the index taken modulo N. This shift is reflection invariant only for $N = 2$, but nevertheless reversible for any N, and even with an involutory reversor. The symmetry group is $\mathbb{Z} \times C_N$.

The quaternary Rudin–Shapiro shift shows another phenomenon. Its symmetry group is $\mathbb{Z} \times C_2$, and it is reversible, but no reversor is an involution. Instead, there is a reversor of order 4 (and all reversors have this order), and the reversing symmetry group is $\mathbb{Z} \rtimes C_4$, where the square of the generating element of the cyclic group C_4 is the extra (involutory) symmetry; see [20] for details on this and the previous examples.

9.5 Shift Spaces with Faithful $\mathbb{Z}^d$-action

It is more than natural to also consider higher-dimensional shift actions. Here, given some alphabet $\mathcal{A}$, a subshift is any closed subspace $\mathbb{X} \subseteq \mathcal{A}^{\mathbb{Z}^d}$ that is invariant under the shift in each of the d directions. With $\underline{n} = (n_1, \dots, n_d)^T \in \mathbb{Z}^d$ as well as $x_{\underline{n}} = (x_{n_1}, \dots, x_{n_d})$, one defines the shift in direction i by

$$(S_i x)_{\underline{n}} := x_{\underline{n} + \underline{e}_i},$$

where $\underline{e}_i$ is the standard unit vector in direction i. The individual shifts commute with one another, $S_i S_j = S_j S_i$, for all $1 \leqslant i, j \leqslant d$. Now, we define the *symmetry group* of $\mathbb{X}$ as

$$S(\mathbb{X}) = \mathrm{cent}_{\mathrm{Aut}(\mathbb{X})}(\mathcal{G}),$$

where $\mathcal{G} := \langle S_1, \dots, S_d \rangle$ is a subgroup of $\mathrm{Aut}(\mathbb{X})$.

As before, we are only interested in subshifts with faithful shift action, which means $\mathcal{G} = \langle S_1 \rangle \times \dots \times \langle S_d \rangle \simeq \mathbb{Z}^d$, where the direct product structure is a consequence of the commutativity of the individual shifts. In this case, we define the *group of extended symmetries* as

$$\mathcal{R}(\mathbb{X}) = \mathrm{norm}_{\mathrm{Aut}(\mathbb{X})}(\mathcal{G}),$$

which is the obvious extension of the one-dimensional case. As we shall see shortly, many of the obvious 'symmetries' of $\mathbb{X}$ are only captured by this extension step.

Unlike before, the structure of the normaliser is generally much richer now, which also means that $\mathcal{R}(\mathbb{X})$ is a considerably better topological invariant than $S(\mathbb{X})$. Indeed, the normaliser can even be an *infinite* extension of the centraliser when $d > 1$, as can be seen from the full shift as follows; see [20, Lemma 4].

Fact 9.10 *Let $d \in \mathbb{N}$ and let $\mathbb{X} = \mathcal{A}^{\mathbb{Z}^d}$ be the full d-dimensional shift over the (finite or infinite) alphabet $\mathcal{A}$. Then, the group of extended symmetries is given by $\mathcal{R}(\mathbb{X}) = S(\mathbb{X}) \rtimes \mathrm{GL}(d, \mathbb{Z})$.*

The reasoning behind this observation is simple. Each element of $\mathcal{R}(\mathbb{X})$ must map generators of $\mathcal{G} = \langle S_1, \dots, S_d \rangle \simeq \mathbb{Z}^d$ onto generators of $\mathcal{G}$, and thus induces a mapping into $\mathrm{GL}(d, \mathbb{Z})$, which is the automorphism group of the free Abelian group of rank d. Now, one checks that, for any $M \in \mathrm{GL}(d, \mathbb{Z})$, the mapping h_M defined by

$$(h_M x)_{\underline{n}} = x_{M^{-1} \underline{n}},$$

with $\underline{n}$ considered as a column vector, defines an automorphism of the full shift. This leads to the semi-direct product structure as stated.

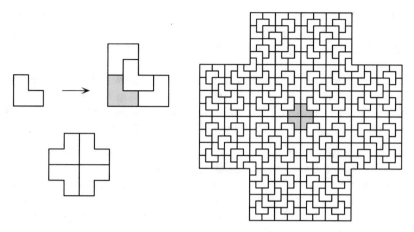

Fig. 9.1 The chair inflation rule (upper left panel; rotated tiles are inflated to rotated patches), a legal patch with full D_4 symmetry (lower left) and a level-3 inflation patch generated from this legal seed (shaded; right panel). Note that this patch still has the full D_4 point symmetry (with respect to its centre), as will the infinite inflation tiling fixed point emerging from it

9.5.1 Tiling Dynamical Systems as Subshifts

Substitution tilings of constant block size are a generalisation of substitutions of constant length, and admit an alternative description as subshifts, for instance via a suitable symbolic coding. Classic examples include the chair and the table tiling [67], but many more are known [8, 36].

Here, we take a look at the chair tiling, which is illustrated in Fig. 9.1; see [7] for more. Its geometric realisation makes it particularly obvious that any reasonable notion of a group of full symmetries must somehow contain the elementary symmetries of the square, simply because the inflation tiling (whose orbit closure under the translation action of $\mathbb{Z}^2$ defines the tiling dynamical system, with compact space $\mathbb{X}$) is invariant under a fourfold rotation and a reflection in the horizontal axis. These two operations generate a group that is isomorphic with the dihedral group D_4, a maximal finite subgroup of $GL(2, \mathbb{Z})$.

Now, none of these orthogonal transformations occur in the centraliser of the shift group, which was shown to be minimal in [61]. This is a rigidity phenomenon of *topological* origin, due to the fibre structure of $\mathbb{X}$ over its maximal equicontinuous factor (MEF). Consequently, this example provides ample evidence that one also needs to consider the normaliser. The general result reads as follows; see [20] for the details.

Theorem 9.11 *Let $\mathbb{X}$ be the hull of the chair tiling, and $(\mathbb{X}, \mathbb{Z}^2)$ the corresponding dynamical system. It is topologically conjugate to a subshift of $\{0, 1, 2, 3\}^{\mathbb{Z}^2}$ with faithful shift action. Moreover, one has $\mathcal{S}(\mathbb{X}) \simeq \mathbb{Z}^2$ and $\mathcal{R}(\mathbb{X}) \simeq \mathbb{Z}^2 \rtimes D_4$, where D_4 is the symmetry group of the square, and a maximal finite subgroup of $GL(2, \mathbb{Z})$.*

Proof (Sketch) It is well known that $\mathbb{X}$ is a.e. one-to-one over its MEF, which is a two-dimensional odometer here. The orbits of non-singleton fibres over the MEF create the topological rigidity that enforce the centraliser to agree with the group generated by the lattice translations.

The extension by D_4 is constructive, via the symmetries of the inflation fixed point. Any further extension would require the inclusion of a $\mathrm{GL}(2, \mathbb{Z})$-element of infinite order (because D_4 is a maximal finite subgroup of $\mathrm{GL}(2, \mathbb{Z})$), which is impossible by the geometric structure (and rigidity) of the prototiles.

Similar results will occur for other tiling dynamical system, also in higher dimensions. For instance, it is clear that the d-dimensional chair (with $d \geqslant 2$; see [7]) will have $\mathcal{S} = \mathbb{Z}^d$ and $\mathcal{R} = \mathbb{Z}^d \rtimes W_d$, where W_d is the symmetry group of the d-dimensional cube, also known as the hyperoctahedral group [6].

Let us note that there is no general reason why the extended symmetry group should be a semi-direct product (though this will be the most frequent case to encounter in the applications). In fact, in (periodic) crystallography, the classification of space groups in dimensions $d \geqslant 2$ contains so-called *non-symmorphic* cases that do not show a semi-direct product structure between translations and linear isometries [70]. It will be an interesting question to identify or construct planar shift spaces that show the planar wallpaper groups as their extended symmetry groups. This and similar results would emphasise once more that and how the extension from $\mathcal{S}(\mathbb{X})$ to $\mathcal{R}(\mathbb{X})$ is relevant to capture the full symmetry of faithful shift actions.

9.5.2   Shifts of Algebraic Origin

There is a particularly interesting and important class of subshifts that has attracted a lot of attention. They are known as subshifts of algebraic origin; see [69] and references therein. The important point here is that such a subshift is also an Abelian group under pointwise addition, and thus carries the corresponding Haar measure as a canonical invariant measure.

Here, we take a look at one of the paradigmatic examples from this class, the Ledrappier shift [54]. This is the subshift $\mathbb{X}_L \subset \{0, 1\}^{\mathbb{Z}^2}$ defined as

$$\mathbb{X}_L = \ker(1 + S_1 + S_2) = \left\{ x \in \{0, 1\}^{\mathbb{Z}^2} : x_{\underline{n}} + x_{\underline{n}+\underline{e}_1} + x_{\underline{n}+\underline{e}_2} = 0 \text{ for all } \underline{n} \in \mathbb{Z}^2 \right\}, \tag{9.6}$$

where the sums are pointwise, and to be taken modulo 2. This definition highlights the special role of elementary lattice triangles, whose vertices are supporting the local variables that need to sum to 0; see Fig. 9.2 for an illustration. The symmetry group is known to be minimal, which can be seen as a rigidity phenomenon of *algebraic* type. More generally, one has the following result.

Theorem 9.12 *The symmetry group of Ledrappier's shift* $\mathbb{X}_L$ *from Eq.* (9.6) *is*

$$S(\mathbb{X}_L) = \langle S_1, S_2 \rangle \simeq \mathbb{Z}^2,$$

while the group of extended symmetries is given by

$$\mathcal{R}(\mathbb{X}) = \langle S_1, S_2 \rangle \rtimes H \simeq \mathbb{Z}^2 \rtimes D_3,$$

where H is the finite group generated by the autormorphisms h_A and h_B, with $A = \begin{pmatrix} -1 & -1 \\ 1 & 0 \end{pmatrix}$ and $B = \begin{pmatrix} 0 & 1 \\ 1 & 0 \end{pmatrix}$. This group is isomorphic with the dihedral group $D_3 \subset \mathrm{GL}(2, \mathbb{Z})$ that is generated by the corresponding matrices, A and B.

Proof (Sketch) The triviality of the centraliser is a consequence of the group structure, which heavily restricts the homeomorphisms between irreducible subshifts that commute with the translations [23, 51, 69].

For the extension to the normaliser, the presence of D_3 is again constructive, and evident from Fig. 9.2. One then excludes any element of order 6 that would complete D_3 to D_6, and finally any element of infinite order that could extend the group D_3. Both types of extensions are impossible because any such additional element would change the defining condition by deforming the elementary triangles.

This example is of interest for a number of reasons. First of all, it shows the phenomenon of rank-1 entropy, which is to say that the number of circular configurations grows exponentially in the *radius* of the patch, but not in the area. While this means that the topological entropy still vanishes, Ledrappier's shift is not an example of low complexity. Second, the spectral structure displays a

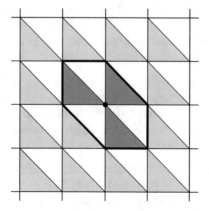

Fig. 9.2 Central configurational patch for Ledrappier's shift condition, indicating the relevance of the triangular lattice. Equation (9.6) implies a condition for the values at the three vertices of all elementary *L*-triangles (shaded). The overall pattern of these triangles is preserved by all (extended) symmetries. The group D_3 from Theorem 9.12 can now be viewed as the colour-preserving symmetry group of the 'distorted' hexagon as indicated around the origin

mixture of trivial point spectrum with further absolutely continuous (Lebesgue) components [13], which highlights the fact that the inverse problem of structure determination, in the presence of mixed spectra, is really a lot more complex than in the case of pure point spectra. Once again, capturing the full extended symmetry group is an important first step in this analysis, as is well-known from classical crystallography [70].

Let us consider the planar point set

$$V := \{(x, y) \in \mathbb{Z}^2 : \gcd(x, y) = 1\} \subset \mathbb{Z}^2,$$

which is known as the set of *visible* (or primitive) *lattice points*; see the cover page of [4] for an illustration. The set V has numerous fascinating properties, both algebraically and geometrically. In particular, it fails to be a Delone set, because it has holes of arbitrary size that even repeat lattice-periodically. Nevertheless, the natural density exists and equals $6/\pi^2 = 1/\zeta(2)$. Moreover, the set V is invariant under the group $\mathrm{GL}(2, \mathbb{Z})$, which acts transitively on V; see [15] and references therein.

The corresponding subshift $\mathbb{X}_V$ is defined as the orbit closure of the characteristic function 1_V under the shift action of $\mathbb{Z}^2$, which turns $(\mathbb{X}_V, \mathbb{Z}^2)$ into a topological dynamical system with faithful shift action and positive topological entropy. This system, like the square-free shift from above, is *hereditary*, which implies rigidity for the symmetry group. On the other hand, due to the way that $\mathrm{GL}(2, \mathbb{Z})$-matrices act on it, the normaliser is the maximal extension of the centraliser in this case [21]. In fact, there is no reason to restrict to the planar case here, as the visible lattice points can be defined for $\mathbb{Z}^d$ with any $d \geqslant 2$ (the case $d = 1$ gives a finite set that is not of interest). Thus, one has the following result.

Theorem 9.13 *Let $\mathbb{X}_V$ be the subshift defined by the visible lattice points of $\mathbb{Z}^d$, where $d \geqslant 2$. Then, $\mathbb{X}_V$ has faithful shift action with minimal symmetry group, $\mathcal{S}(\mathbb{X}_V) = \mathbb{Z}^d$, while the extended symmetry group emerges as the maximal extension of it, $\mathcal{R}(\mathbb{X}_V) = \mathbb{Z}^2 \rtimes \mathrm{GL}(2, \mathbb{Z})$.*

Proof (Sketch) Here, the triviality of the centraliser, as in the earlier example of the square-free shift, is a consequence of the *heredity* of the subshift [21], and really follows from a mild generalisation of Mentzen's approach [57]. The extension to the normaliser, as explained above, is by all of $\mathrm{GL}(d, \mathbb{Z})$, where the semi-direct product structure is the same as for the full shift in Fact 9.10.

More generally, one can study systems of this kind as defined from primitive lattice systems, for instance in the spirit of [19]. This also covers subshifts that are generated from rings of integers in general algebraic number fields subject to certain freeness conditions. This gives a huge class of examples that can be viewed as multi-dimensional generalisations of $\mathcal{B}$-free systems. Interestingly, they are also examples of weak model sets [19], which gives access to a whole new range of tools from the interplay of dynamical systems and algebraic number theory [44, 46–48], in the spirit of the original approach by Meyer [58].

Acknowledgements A substantial part of this exposition is based on a joint work (both past and ongoing) with John Roberts, who introduced me to the concepts over 25 years ago. More recent activities also profited a lot from the interaction and cooperation with Christian Huck, Mariusz Lemańczyk and Reem Yassawi. It is a pleasure to thank the CIRM in Luminy for its support and the stimulating atmosphere during the special program in the framework of the Jean Morlet semester 'Ergodic Theory and Dynamical Systems in their Interactions with Arithmetic and Combinatorics', where part of this work was done.

References

1. R.L. Adler, R. Palais, Homeomorphic conjugacy of automorphisms of the torus. Proc. Am. Math. Soc. **16**, 1222–1225 (1965)
2. R.L. Adler, B. Weiss, *Similarity of Automorphisms of the Torus*. Memoirs AMS, vol. 98 (American Mathematical Society, Providence, RI, 1970)
3. J.-P. Allouche, J. Shallit, *Automatic Sequences* (Cambridge University Press, Cambridge, 2003)
4. T.M. Apostol, *Introduction to Analytic Number Theory* (Springer, New York, 1976)
5. V.I. Arnold, A. Avez, *Ergodic Problems of Classical Mechanics* (Addison-Wesley, Redwood City, CA, 1989)
6. M. Baake, Structure and representations of the hyperoctahedral group. J. Math. Phys. **25**, 3171–3182 (1984)
7. M. Baake, U. Grimm, *Aperiodic Order. Vol. 1: A Mathematical Invitation* (Cambridge University Press, Cambridge, 2013)
8. M. Baake, U. Grimm, Squirals and beyond: substitution tilings with singular continuous spectrum. Ergodic Theory Dyn. Syst. **34**, 1077–1102 (2014). arXiv:1205.1384
9. M. Baake, J.A.G. Roberts, Reversing symmetry group of GL(2, Z) and PGL(2, Z) matrices with connections to cat maps and trace maps. J. Phys. A: Math. Gen. **30**, 1549–1573 (1997)
10. M. Baake, J.A.G. Roberts, Symmetries and reversing symmetries of toral automorphisms. Nonlinearity **14**, R1–R24 (2001). arXiv:math.DS/0006092
11. M. Baake, J.A.G. Roberts, Symmetries and reversing symmetries of polynomial automorphisms of the plane. Nonlinearity **18**, 791–816 (2005). arXiv:math.DS/0501151
12. M. Baake, J.A.G. Roberts, The structure of reversing symmetry groups. Bull. Aust. Math. Soc. **73**, 445–459 (2006). arXiv:math.DS/0605296
13. M. Baake, T. Ward, Planar dynamical systems with pure Lebesgue diffraction spectrum. J. Stat. Phys. **140**, 90–102 (2010). arXiv:1003.1536
14. M. Baake, U. Grimm, D. Joseph, Trace maps, invariants, and some of their applications. Int. J. Mod. Phys. B **7**, 1527–1550 (1993). arXiv:math-ph/9904025
15. M. Baake, R.V. Moody, P.A.B. Pleasants, Diffraction from visible lattice points and k-th power free integers. Discrete Math. **221**, 3–42 (2000). arXiv:math.MG/9906132
16. M. Baake, J.A.G. Roberts, A. Weiss, Periodic orbits of linear endomorphisms on the 2-torus and its lattices. Nonlinearity **21**, 2427–2446 (2008). arXiv:0808.3489
17. M. Baake, F. Gähler, U. Grimm, Spectral and topological properties of a family of generalised Thue–Morse sequences. J. Math. Phys. **53**, 032701:1–24 (2012). arXiv:1201.1423
18. M. Baake, N. Neumärker, J.A.G. Roberts, Orbit structure and (reversing) symmetries of toral endomorphisms on rational lattices. Discrete Contin. Dyn. Syst. A **33**, 527–553 (2013). arXiv:1205.1003
19. M. Baake, C. Huck, N. Strungaru, On weak model sets of extremal density. Ind. Math. **28**, 3–31 (2017). arXiv:1512.07129
20. M. Baake, J.A.G. Roberts, R. Yassawi, Reversing and extended symmetries of shift spaces. Discrete Contin. Dyn. Syst. A **38**, 835–866 (2018). arXiv:1611.05756
21. M. Baake, C. Huck, M. Lemańczyk, Positive entropy shifts with small centraliser and large normaliser (in preparation)

22. A. Bartnicka, Automorphisms of Toeplitz $\mathcal{B}$-free systems. Preprint. arXiv:1705.07021
23. S. Bhattacharya, K. Schmidt, Homoclinic points and isomorphism rigidity of algebraic $\mathbb{Z}^d$-actions on zero-dimensional compact Abelian groups. Isr. J. Math. **137**, 189–209 (2003)
24. Z.I. Borevich, I.R. Shafarevich, *Number Theory* (Academic Press, New York, 1966)
25. W. Bulatek, J. Kwiatkowski, Strictly ergodic Toeplitz flows with positive entropy and trivial centralizers. Stud. Math. **103**, 133–142 (1992)
26. E.M. Coven, Endomorphisms of substitution minimal sets. Z. Wahrscheinlichkeitsth. verw. Geb. **20**, 129–133 (1971/1972)
27. E.M. Coven, G.A. Hedlund, Sequences with minimal block growth. Math. Syst. Theory **7**, 138–153 (1973)
28. E.M. Coven, A. Quas, R. Yassawi, Computing automorphism groups of shifts using atypical equivalence classes. Discrete Anal. **2016**(3) (28pp). arXiv:1505.02482
29. V. Cyr, B. Kra, The automorphism group of a shift of linear growth: beyond transitivity. Forum Math. Sigma **3**, e5 27 (2015). arXiv:1411.0180
30. V. Cyr, B. Kra, The automorphism group of a shift of subquadratic growth. Proc. Am. Math. Soc. **2**, 613–621 (2016). arXiv:1403.0238
31. D. Damanik, A. Gorodetski, W. Yessen, The Fibonacci Hamiltonian. Invent. Math. **206**, 629–692 (2016). arXiv:1403.7823
32. S. Donoso, F. Durand, A. Maass, S. Petite, On automorphism groups of low complexity shifts. Ergodic Theory Dyn. Syst. **36**, 64–95 (2016). arXiv:1501.0051
33. X. Droubay, G. Pirillo, Palindromes and Sturmian words. Theor. Comput. Sci. **223**, 73–85 (1999)
34. E.H. El Abdalaoui, M. Lemańczyk, T. de la Rue, A dynamical point of view on the set of $\mathcal{B}$-free integers. Int. Math. Res. Not. **2015**(16), 7258–7286 (2015). arXiv:1311.3752
35. K. Frączek, J. Kułaga, M. Lemańczyk, Non-reversibility and self-joinings of higher orders for ergodic flows. J. Anal. Math. **122**, 163–227 (2014). arXiv:1206.3053
36. N.P. Frank, Multi-dimensional constant-length substitution sequences. Topology Appl. **152**, 44–69 (2005)
37. A. Gómez, J. Meiss, Reversors and symmetries for polynomial automorphisms of the complex plane. Nonlinearity **17**, 975–1000 (2004). arXiv:nlin.CD/0304035
38. G.R. Goodson, Inverse conjugacies and reversing symmetry groups. Am. Math. Mon. **106**, 19–26 (1999)
39. G. Goodson, A. del Junco, M. Lemańczyk, D. Rudolph, Ergodic transformation conjugate to their inverses by involutions. Ergodic Theory Dyn. Syst. **16**, 97–124 (1996)
40. M. Hochman, Genericity in topological dynamics. Ergodic Theory Dyn. Syst. **28**, 125–165 (2008)
41. N. Jacobson, *Lectures in Abstract Algebra. II. Linear Algebra* (Springer, New York, 1953)
42. H.W.E. Jung, Über ganze irrationale Transformationen der Ebene. J. Reine Angew. Math. (Crelle) **184**, 161–174 (1942)
43. A. Karrass, D. Solitar, The subgroups of a free product of two groups with an amalgamated subgroup. Trans. Am. Math. Soc. **150**, 227–255 (1970)
44. S. Kasjan, G. Keller, M. Lemańczyk, Dynamics of $\mathcal{B}$-free sets: a view through the window. Int. Math. Res. Not. (2017, in press). arXiv:1702.02375
45. M. Keane, Generalized Morse sequences. Z. Wahrscheinlichkeitsth. verw. Geb. **10**, 335–353 (1968)
46. G. Keller, Generalized heredity in $\mathcal{B}$-free systems. Preprint. arXiv:1704.04079
47. G. Keller, C. Richard, Dynamics on the graph of the torus parametrisation. Ergodic Theory Dyn. Syst. (in press). arXiv:1511.06137
48. G. Keller, C. Richard, Periods and factors of weak model sets. Preprint. arXiv:1702.02383
49. Y.-O. Kim, J. Lee, K.K. Park, A zeta function for flip systems. Pac. J. Math. **209**, 289–301 (2003)
50. B.P. Kitchens, *Symbolic Dynamics* (Springer, Berlin, 1998)
51. B. Kitchens, K. Schmidt, Isomorphism rigidity of irreducible algebraic $\mathbb{Z}^d$-actions. Invent. Math. **142**, 559–577 (2000)

52. J.S.W. Lamb, Reversing symmetries in dynamical systems. J. Phys. A: Math. Gen. **25**, 925–937 (1992)
53. J.S.W. Lamb, J.A.G. Roberts, Time-reversal symmetry in dynamical systems: a survey. Physica D **112**, 1–39 (1998)
54. F. Ledrappier, Un champ markovien peut être d'entropie nulle et mélangeant. C. R. Acad. Sci. Paris Sér. A-B **287**, A561–A563 (1978)
55. J. Lee, K.K. Park, S. Shin, Reversible topological Markov shifts. Ergodic Theory Dyn. Syst. **26**, 267–280 (2006)
56. D.A. Lind, B. Marcus, *An Introduction to Symbolic Dynamics and Coding* (Cambridge University Press, Cambridge, 1995)
57. M.K. Mentzen, Automorphisms of shifts defined by $\mathcal{B}$-free sets of integers. Coll. Math. **147**, 87–94 (2017)
58. Y. Meyer, *Algebraic Numbers and Harmonic Analysis* (North Holland, Amsterdam, 1972)
59. M. Morse, G.A. Hedlund, Symbolic dynamics II. Sturmian trajectories. Am. J. Math. **62**, 1–42 (1940)
60. A.G. O'Farrel, I. Short, *Reversibility in Dynamics and Group Theory* (Cambridge University Press, Cambridge, 2015)
61. J. Olli, Endomorphisms of Sturmian systems and the discrete chair substitution tiling system. Discrete Contin. Dyn. Syst. A **33**, 4173–4186 (2013)
62. K. Petersen, *Ergodic Theory* (Cambridge University Press, Cambridge, 1983)
63. N. Pytheas Fogg, *Substitutions in Dynamics, Arithmetics and Combinatorics*. Lecture Notes in Mathematics, vol. 1794 (Springer, Berlin, 2002)
64. M. Queffélec, *Substitution Dynamical Systems – Spectral Analysis*. Lecture Notes in Mathematics, vol. 1294, 2nd edn. (Springer, Berlin, 2010)
65. J.A.G. Roberts, M. Baake, Trace maps as 3D reversible dynamical systems with an invariant. J. Stat. Phys. **74**, 829–888 (1994)
66. J.A.G. Roberts, M. Baake, Symmetries and reversing symmetries of area-preserving polynomial mappings in generalised standard form. Physica A **317**, 95–112 (2002). arXiv:math.DS/0206096
67. E.A. Robinson, On the table and the chair. Indag. Math. **10**, 581–599 (1999)
68. J.-P. Serre, *Trees*, 2nd. corr. printing (Springer, Berlin, 2003)
69. K. Schmidt, *Dynamical Systems of Algebraic Origin* (Birkhäuser, Basel, 1995)
70. R.L.E. Schwarzenberger, *N-Dimensional Crystallography* (Pitman, San Francisco, 1980)
71. M.B. Sevryuk, *Reversible Systems*. Lecture Notes in Mathematics, vol. 1211 (Springer, Berlin, 1986)
72. W. van der Kulk, On polynomial rings in two variables. Nieuw Arch. Wisk. **1**, 33–41 (1953)
73. D. Wright, Abelian subgroups of $\mathrm{Aut}_k(k[X, Y])$ and applications to actions on the affine plane. Ill. J. Math. **23**, 579–634 (1979)

Chapter 10
Kloosterman Sums, Disjointness, and Equidistribution

M. Einsiedler and M. Luethi

10.1 Introduction to Homogeneous Dynamics

Homogeneous dynamics is concerned with the action of a subgroup $A < G$ of a locally compact group G on quotient spaces of the form $X = \Gamma \backslash G$ (or more generally $Y = \Gamma \backslash G / M$), where usually and for this discussion always $\Gamma < G$ is a discrete subgroup (and $M < G$ is a compact subgroup such that A normalizes M). Since Γ is discrete, $X = \Gamma \backslash G$ is locally homeomorphic to G. A similar statement also holds for the dynamics in question, as we will now explain. For the sake of illustration, let $x = \Gamma g \in X$ and for $\varepsilon \in G$ let $\varepsilon \cdot x = \Gamma g \varepsilon^{-1}$. The natural action of an element $a \in A$ maps x and $\varepsilon \cdot x$ respectively to $a \cdot x = \Gamma g a^{-1}$ and to

$$a \cdot (\varepsilon \cdot x) = \Gamma g \varepsilon^{-1} a^{-1} = \Gamma g a^{-1}(a\varepsilon^{-1}a^{-1}) = (a\varepsilon a^{-1}) \cdot (a \cdot x).$$

Assuming that ε is close to the identity in G, it will represent the "minimal displacement" between x and $\varepsilon \cdot x$. At the same time, $a\varepsilon a^{-1}$ will represent the "minimal displacement" between $a \cdot x$ and $a \cdot (\varepsilon \cdot x)$. This shows that the dynamics of a acting on X is closely linked to conjugation on G by a. If G is a Lie group, then the latter is locally conjugated via the exponential map to the adjoint representation Ad_a on the Lie algebra $\mathfrak{g}$ of G. In order to develop some understanding for the number theoretical and geometric meaning and for the dynamical behavior of such systems, we need to discuss a few examples of this setup.

M. Einsiedler · M. Luethi (✉)
Departement Mathematik, ETH Zürich, Zürich, Switzerland
e-mail: manfred.einsiedler@math.ethz.ch; manuel.luethi@math.ethz.ch

© Springer International Publishing AG, part of Springer Nature 2018
S. Ferenczi et al. (eds.), *Ergodic Theory and Dynamical Systems in their
Interactions with Arithmetics and Combinatorics*, Lecture Notes
in Mathematics 2213, https://doi.org/10.1007/978-3-319-74908-2_10

Example 10.1 Probably one of the easiest groups to consider is $G = \mathbb{R}$ and $\Gamma = \{0\}$. In this case the action of a on $\Gamma \backslash G \cong G$ is just translation on $\mathbb{R}$ by a, which does not lend itself to examination using dynamical tools, as this action does not even satisfy Poincaré recurrence.

This example suggests, that we should require Γ to be such that $X = \Gamma \backslash G$ is not too big. One could for example require that X is compact, in which case the discrete subgroup Γ is called a *uniform lattice*. It turns out that this requirement would exclude many natural and important examples which can be treated dynamically. Rather, the correct requirement is that X carries a G-invariant probability measure, in which case Γ is called a *lattice* in G.

Example 10.2 (Circle Rotation) If $G = \mathbb{R}$ and $\Gamma = \mathbb{Z}$, then $X = \mathbb{Z} \backslash \mathbb{R} = \mathbb{T}$ and every real number $a \in \mathbb{R}$ induces a circle rotation $x \bmod 1 \mapsto x + a \bmod 1$.

We can also get a fundamentally different algebraic type of action, namely multiplication by a prime p on $\mathbb{T}$, fit into the framework of homogeneous dynamics, at least if we consider its invertible extension.

Example 10.3 ($\times 2$-Map) In what follows, we will make use of the p-adic numbers $\mathbb{Q}_p$ for $p = 2$, where $\mathbb{Q}_p$ is the completion of $\mathbb{Q}$ with respect to the topology induced by the absolute value $|p^k \frac{r}{s}|_p = p^{-k}$, where $p \nmid rs$ and $k \in \mathbb{Z}$. Vaguely speaking, $\mathbb{Q}_p$ can be considered as the field of Laurent series in "the variable p", and from this perspective the closure $\mathbb{Z}_p$ of $\mathbb{Z}$ in $\mathbb{Q}_p$ corresponds to the subring of power series. The set $\mathbb{Z}_p$ is a compact, open subring of $\mathbb{Q}_p$. For background on the p-adic numbers in general, we refer the reader to [37]. Let $G = (\mathbb{R} \times \mathbb{Q}_2) \rtimes \mathbb{Z}$, where $\mathbb{Z}$ acts by multiplication by 2 on the coordinates of $\mathbb{R} \times \mathbb{Q}_2$. Formally this is realized as

$$G = \left\{ \begin{pmatrix} 2^m & (x_\infty, x_2) \\ 0 & 1 \end{pmatrix} \;\middle|\; m \in \mathbb{Z}, x_\infty \in \mathbb{R}, x_2 \in \mathbb{Q}_2 \right\} \subseteq \mathrm{Mat}_2(\mathbb{R} \times \mathbb{Q}_2),$$

where we view the top left coordinate (in $\mathbb{Z}[\frac{1}{2}]$) as an element in $\mathbb{R} \times \mathbb{Q}_2$ by embedding it diagonally, and the group operation is matrix multiplication

$$\begin{pmatrix} 2^m & (x_\infty, x_2) \\ 0 & 1 \end{pmatrix} \begin{pmatrix} 2^n & (y_\infty, y_2) \\ 0 & 1 \end{pmatrix} = \begin{pmatrix} 2^{n+m} & (x_\infty + 2^m y_\infty, x_2 + 2^m y_2) \\ 0 & 1 \end{pmatrix}$$

for $m, n \in \mathbb{Z}$, $x_\infty, y_\infty \in \mathbb{R}$ and $x_2, y_2 \in \mathbb{Q}_2$. Consider the subgroups $\Gamma = \mathbb{Z}[\frac{1}{2}] \rtimes \mathbb{Z}$ and $H = \mathbb{R} \times \mathbb{Q}_2$ of G, where again $\mathbb{Z}[\frac{1}{2}]$ is diagonally embedded in H. Since the generator of $\mathbb{Z}$ in the product $G = H \rtimes \mathbb{Z}$ belongs to Γ, we have $\Gamma \backslash G \cong H/\Gamma \cap H$. In fact, we claim that

$$X := \Gamma \backslash G \cong H / \Gamma \cap H = \mathbb{R} \times \mathbb{Q}_2 / \mathbb{Z}[\frac{1}{2}] \cong \mathbb{R} \times \mathbb{Z}_2 / \mathbb{Z}$$

where $\mathbb{Z}$ is diagonally embedded in $\mathbb{R} \times \mathbb{Z}_2$. Let us assume for the moment that we are given the isomorphism $\mathbb{R} \times \mathbb{Q}_2 / \mathbb{Z}[\frac{1}{2}] \cong \mathbb{R} \times \mathbb{Z}_2 / \mathbb{Z}$. We will give the construction of this isomorphism below. Once we know that $\mathbb{Z}$ is a lattice in $\mathbb{R} \times \mathbb{Z}_2$, one deduces that Γ is a cocompact lattice in G. Discreteness of $\mathbb{Z} \subseteq \mathbb{R} \times \mathbb{Z}_2$ is obtained as follows: The set $(-\varepsilon, \varepsilon) \times \mathbb{Z}_2$ is an open neighbourhood of 0 inside $\mathbb{R} \times \mathbb{Z}_2$, and if $\varepsilon < 1$, then any x contained in the intersection of the diagonally embedded copy of $\mathbb{Z}$ and this open neighbourhood is in fact 0. The cocompactness follows from compactness of $\mathbb{Z}_2$ together with the fact that the set $[0, 1] \times \mathbb{Z}_2$ is surjective for the quotient map $\mathbb{R} \times \mathbb{Z}_2 \to \mathbb{R} \times \mathbb{Z}_2/\mathbb{Z}$. Using the isomorphism indicated above, we obtain a canonical projection

$$X = {}_{\Gamma}\backslash^{G} \cong \mathbb{R} \times \mathbb{Z}_2/\mathbb{Z} \to \mathbb{T} = \mathbb{R}/\mathbb{Z},$$

where we simply forget the $\mathbb{Z}_2$-component of the representative.

We will now construct the isomorphism $\mathbb{R} \times \mathbb{Q}_2/\mathbb{Z}[\frac{1}{2}] \cong \mathbb{R} \times \mathbb{Z}_2/\mathbb{Z}$. Using the interpretation of $\mathbb{Q}_2$ as Laurent series in the variable 2 with coefficients equal to 0 or 1, every $x_2 \in \mathbb{Q}_2$ is of the form $x_2 = [x_2] + \{x_2\}$ with $[x_2] \in \mathbb{Z}_2$ and $\{x_2\} \in \mathbb{Z}[\frac{1}{2}]$ (say with $\{x_2\} \in [0, 1)$), so that for $x_\infty \in \mathbb{R}$ and $x_2 \in \mathbb{Q}_2$ we obtain the following equality of cosets

$$(x_\infty, x_2) + \mathbb{Z}[\tfrac{1}{2}] = (x_\infty - \{x_2\}, [x_2]) + \mathbb{Z}[\tfrac{1}{2}].$$

This shows that $(x_\infty, x_2) + \mathbb{Z}[\frac{1}{2}] \in \mathbb{R} \times \mathbb{Z}_2 + \mathbb{Z}[\frac{1}{2}]$, i.e. the orbit of the identity coset $\mathbb{Z}[\frac{1}{2}]$ under $\mathbb{R} \times \mathbb{Z}_2$ is the full space. As $\mathbb{Z}_2 \cap \mathbb{Z}[\frac{1}{2}] = \mathbb{Z}$, the stabilizer of the identity coset in $\mathbb{R} \times \mathbb{Z}_2$ is the diagonally embedded $\mathbb{Z}$. From this it follows that, up to choice of the basepoint, a canonical isomorphism is given by the map

$$x + \mathbb{Z} \in \mathbb{R} \times \mathbb{Z}_2/\mathbb{Z} \mapsto x + \mathbb{Z}[\tfrac{1}{2}] \in \mathbb{R} \times \mathbb{Q}_2/\mathbb{Z}[\tfrac{1}{2}].$$

Note that this isomorphism is evidently $\times 2$-equivariant.

We now consider the element $a = \begin{pmatrix} 2 & 0 \\ 0 & 1 \end{pmatrix} \in \Gamma < G$. Given $x_\infty \in \mathbb{R}$, $x_2 \in \mathbb{Q}_2$ and an element $h = \begin{pmatrix} 1 & (x_\infty, x_2) \\ 0 & 1 \end{pmatrix} \in G$, then

$$a \cdot (h\Gamma) = (aha^{-1}) \cdot \Gamma = \begin{pmatrix} 1 & (2x_\infty, 2x_2) \\ 0 & 1 \end{pmatrix} \cdot \Gamma$$

and using the isomorphism described above and the stated canonical projection, we see that the action of a projects to $T_2 : \mathbb{T} \to \mathbb{T}$, the multiplication by 2 on $\mathbb{T}$. In fact, X is the invertible extension of $T_2 : \mathbb{T} \to \mathbb{T}$ and $\mathbb{Q}_2$ can be interpreted as the field that is needed to make the invertible extension of T_2 a diagonalizable map.

For the reader who is not accustomed with the p-adic numbers, we want to point out that $\mathbb{Z}_2$ is the set underlying the *simple dyadic odometer*, whose dynamics then is realized as addition of the multiplicative unit, and hence is a rotation on the compact group $\mathbb{Z}_2$. This connection to odometers is discussed in greater detail for example in [8].

Moreover, it might be helpful to compare the previous construction to the following realization of another common dynamical system as a homogeneous dynamical system. Consider the toral automorphism defined by the matrix $A = \left(\begin{smallmatrix} 0 & 1 \\ 1 & 1 \end{smallmatrix}\right)$ and the associated (invertible) dynamical system. We can realize this system as a homogeneous dynamical system, by considering the group $G = \mathbb{R}^2 \rtimes \mathbb{Z}$ embedded in $\mathrm{Mat}_3(\mathbb{R})$ via $(v, m) \mapsto \left(\begin{smallmatrix} A^m & v \\ 0 & 1 \end{smallmatrix}\right)$. As before, we consider the subgroups $\Gamma = \mathbb{Z}^2 \rtimes \mathbb{Z}$ and $H = \mathbb{R}^2$. Then the homogeneous space $X = \Gamma \backslash G$ is $X \cong H/_{H \cap \Gamma} \cong \mathbb{T}^2$ and under the isomorphism the action of the element $a = \left(\begin{smallmatrix} A & 0 \\ 0 & 1 \end{smallmatrix}\right)$ on X maps to the action of A on $\mathbb{T}^2$.

We recall that dynamically the two maps arising in the above examples could not be more different from each other.

- For the circle rotation, there are two cases to consider. If $a \in \mathbb{Q}$, then clearly every orbit is periodic. If $a \notin \mathbb{Q}$, then every orbit is dense and equidistributed, as in that case the dynamical system is uniquely ergodic (cf. [15, p. 107]).
- For the $\times 2$ map $T_2 : \mathbb{T} \to \mathbb{T}$, there are infinitely many finite orbits, as every rational point has a finite orbit under T_2. This follows immediately from the fact that the rational numbers are exactly the numbers whose base 2 expansion is eventually periodic and the property that T_2 acts by left-shift on the base 2 expansion. Alternatively we note that in case $2 \nmid q$, the denominator of $T_2(\frac{p}{q})$ is the same as the denominator of $\frac{p}{q}$. As there are only finitely many rational numbers with denominator q in $[0, 1)$, it follows that the orbit of $\frac{p}{q}$ is finite. Note that the image of any rational number will be of this form after finitely may applications of T_2.

  Ergodicity of T_2 implies that the orbit of almost every point is equidistributed, which is an application of Birkhoff's pointwise ergodic theorem. However, contrary to the first example, there are plenty of points with closed fractal Cantor-set-like orbit closures. In fact, T_2 is isomorphic to the one-sided shift on $\{0, 1\}^{\mathbb{N}}$, and it is easy to define closed subshifts by e.g. disallowing sequences containing three 1's in a row.

We introduce two more homogeneous spaces. In terms of generality, the first one is at the opposite relative to the ones discussed above, because it will play the role of a mother-space for almost all homogeneous spaces appearing in number theory. The second will be in between and will reveal the strong interaction between dynamics on homogeneous spaces and Riemannian geometry.

10.1.1 The Space of Lattices

We set $G = \mathrm{SL}_d(\mathbb{R})$ and $\Gamma = \mathrm{SL}_d(\mathbb{Z})$ for some $d \geq 2$ and we consider

$$X_d = \Gamma \backslash G = \mathrm{SL}_d(\mathbb{Z}) \backslash \mathrm{SL}_d(\mathbb{R}).$$

As was already applied during the construction of the invertible extension of the ×2-map, it is an elementary fact from algebra, that given some space Ω equipped with a G-action and some $\omega_0 \in \Omega$ such that $H = \mathrm{Stab}_G(\omega_0)$, the quotient $_H\backslash^G$ identifies with the orbit of ω_0 under G and for continuous or smooth actions, the identification will be of the corresponding category. For the space X_d this is easily achieved by letting

$$\Omega = \{\Lambda \mid \Lambda \text{ is a closed subgroup of } \mathbb{R}^d\}$$

and $\omega_0 = \mathbb{Z}^d$, which leads to

$$X_d =_{\mathrm{SL}_d(\mathbb{Z})}\backslash^{\mathrm{SL}_d(\mathbb{R})} \cong \{\mathbb{Z}^d g \mid g \in \mathrm{SL}_d(\mathbb{R})\}$$
$$= \{\Lambda \mid \Lambda < \mathbb{R}^d \text{ is discrete, cocompact and has covolume 1}\}.$$

Here, the covolume $\mathrm{covol}(\Lambda)$ of the lattice $\Lambda < \mathbb{R}^d$ is the Lebesgue measure $m(F)$ of a fundamental domain F for translation by Λ. Note that $[0, 1)^d$ is a fundamental domain for $\mathbb{Z}^d$ and that for every $g \in \mathrm{SL}_d(\mathbb{Z})$ the set $[0, 1)^d g$ is a fundamental domain for $\Lambda = \mathbb{Z}^d g$, so that $m(F) = m([0, 1)^d g) = \det(g) = 1$ (where the Lebesgue measure considered is the one normalized so that $m([0, 1)^d) = 1$).

Using this identification, it becomes clearer that the space X_d is not compact, as for the sequence of matrices[1]

$$g_n = \begin{pmatrix} \frac{1}{n} & & \\ & n & \\ & & I_{d-2} \end{pmatrix} \in \mathrm{SL}_d(\mathbb{R})$$

the sequence $\mathbb{Z}^d g_n$ in the space X_d does not have any converging subsequence with limit within X_d. Indeed, it converges to the subgroup $\mathbb{R} \times \{0\} \times \mathbb{Z}^{d-2} \in \Omega \setminus X_d$ for the right topology on Ω, which would be the Chabauty topology (see [4]).

Pursuing this line of thought and combining it with Minkowski's geometry of numbers leads to

Theorem 10.1 (Mahler's Compactness Criterion [2, p. 16]) *A subset $K \subset X_d$ has compact closure if and only if the lattices in K are uniformly discrete, i.e. there exists some $\delta > 0$ such that for all $\Lambda \in K$ we have $B_\delta(0) \cap \Lambda = \{0\}$, where $B_\delta(0)$ is the ball of radius δ around 0 in $\mathbb{R}^d$.*

Taking the closure of X_d in Ω with respect to the Chabauty topology, one obtains the lattice compactification of X_d (cf. [3, §III.19]), which reveals some of the complexity of X_d for larger values of d.

---

[1] Here and in what follows matrix entries which are 0 are omitted.

Even though X_d is not compact, it still has finite volume in a natural way, i.e. it carries a natural finite $SL_d(\mathbb{R})$-invariant measure, which we simply call the *Haar measure* on X_d. The fact that this Haar measure is finite, i.e. that the volume of X_d is finite, is essentially a calculation carried out in [2, p. 17f.] for example. This calculation is aided by the Iwasawa decomposition of $SL_d(\mathbb{R})$, which is obtained from the familiar Gram-Schmidt orthonormalization procedure, and, once more, Minkowski's geometry of numbers.

On X_d one can study the action of many different subgroups of $SL_d(\mathbb{R})$, which we categorize as follows:

- If the acting group (in this case denoted U) is unipotent, i.e. all eigenvalues of all elements of the acting group equal 1, then many dynamical questions are completely answered by the celebrated theorems of Ratner concerning unipotent dynamics (cf. [26–28]) and the related theorems in [6] and [25]. This situation can be viewed as a vast and rich generalization of the circle rotation from before. For instance, Ratner proved that for any $x \in X_d$ the orbit closure $\overline{Ux}$ equals the orbit Lx for some intermediate closed subgroup $U < L < SL_d(\mathbb{R})$.

- If the acting group (in this case denoted A) is diagonalizable, there are two subtypes to consider:

  - If $\dim A = 1$, then many different types of fractal orbits appear and it is pretty much impossible to classify its invariant probability measures. The situation is very similar to the situation for the $\times 2$-map considered above.

  - If $\dim A > 1$, then several conjectures of Furstenberg, Margulis, and Katok-Spatzier apply. These conjectures would classify invariant probability measures for some actions as described above. A number of these conjectures have been confirmed (see e.g. [10, 12, 13]), at least under not too strong additional assumptions (as for example positive entropy for the invariant measure). The situation is frequently very similar to the situation in Furstenberg's $\times 2$, $\times 3$-problem on $\mathbb{T}$, which asks what the jointly T_2- and T_3-invariant ergodic probability measures on $\mathbb{T}$ are, where the $\times 3$ map $T_3 : \mathbb{T} \to \mathbb{T}$ is defined in complete analogy by $T_3(x \bmod 1) = 3x \bmod 1$.

- If the acting group H is semisimple without compact factors, then H is generated by unipotents and thus Ratner's theorems apply again. However, this case is from many perspectives easier (see [9]) and in some cases the corresponding results can be made effective (see [16]), which in greater generality would be a very difficult objective when using ergodic theoretic arguments (see also [23]).

We close our discussion of X_d by indicating, why we called it the mother space. Since $SL_d(\mathbb{R})$ has many closed subgroups (unipotent, diagonalizable, and semisimple), all of these could have different types of interesting orbits. In fact, for every connected, semisimple Lie group H with finite center there is some $d \geqslant 2$ such that H is isogeneous to a subgroup of $SL_d(\mathbb{R})$ and has closed orbits in X_d (which have finite volume). Hence we can obtain finite volume quotients of H by considering orbits $Hx \cong H/{\text{Stab}_H(x)}$ in X_d.

10.1.2   The Modular Surface

The special case of $X_2 = {}_{SL_2(\mathbb{Z})}\backslash^{SL_2(\mathbb{R})}$ deserves special attention due to its connection to hyperbolic geometry, which allows us to visualize the global structure of X_2 (which for the 8-dimensional manifold X_3 or even for the 5-dimensional double quotient $_{SL_3(\mathbb{Z})}\backslash^{SL_3(\mathbb{R})}/_{SO_3(\mathbb{R})}$ many people find harder).

In fact, $SL_2(\mathbb{R})$ acts by isometries on the hyperbolic plane $\mathbb{H} = \{z \in \mathbb{C} \mid \Im z > 0\}$, which is equipped with the hyperbolic Riemannian metric $\frac{dx^2+dy^2}{y^2}$. The action of an element $g = \left(\begin{smallmatrix} a & b \\ c & d \end{smallmatrix}\right)$ on $z \in \mathbb{C}$ is defined by the Möbius transformation

$$g \cdot z = \frac{az + b}{cz + d}.$$

It is a straight-forward calculation to show that this indeed defines an action of $SL_2(\mathbb{R})$ on $\mathbb{H}$ and that $z \in \mathbb{H} \mapsto g \cdot z$ is differentiable and preserves the length of tangent vectors with respect to the hyperbolic metric. Moreover, this action is transitive and $\mathrm{Stab}_{SL_2(\mathbb{R})}(i) = SO_2(\mathbb{R})$, so that $_{SL_2(\mathbb{R})}/_{SO_2(\mathbb{R})} \cong \mathbb{H}$.

If we let $SL_2(\mathbb{Z})$ act on $\mathbb{H}$, then it is possible to completely describe a fundamental domain for the action (see Fig. 10.1). The hyperbolic triangle F should be understood as a hyperbolic analog of the fundamental domain $[0, 1)^2 \subseteq \mathbb{R}^2$ for the Euclidean isometries of $\mathbb{R}^2$ corresponding to translation by an element in $\mathbb{Z}^2$. In our case the map defined by $\left(\begin{smallmatrix} 1 & 1 \\ & 1 \end{smallmatrix}\right) \in SL_2(\mathbb{Z})$ acts by $z \in \mathbb{H} \mapsto z + 1$ and glues the left and the right sides of F together. Moreover the map $\left(\begin{smallmatrix} & -1 \\ 1 & \end{smallmatrix}\right)$ acts by $z \mapsto -\frac{1}{z}$, thus mapping the unit circle to itself and glueing the lower boundary of F together. The resulting surface $_{SL_2(\mathbb{Z})}\backslash^{\mathbb{H}}$ has one so-called cusp at ∞, and two special points obtained from i and $\frac{1}{2} + \frac{\sqrt{3}}{2}i$ (which are conical singularities). This surface is called the modular surface. Filling in the cusp point, i.e. taking the one-point compactification, $_{SL_2(\mathbb{Z})}\backslash^{\mathbb{H}} \cup \{\infty\}$ topologically becomes a sphere.

Fig. 10.1 A fundamental domain for the action of $SL_2(\mathbb{Z})$ on $\mathbb{H}$

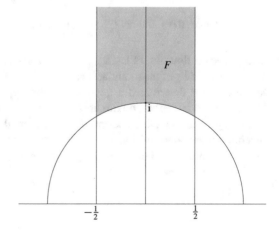

We now wish to return from the study of $\mathbb{H} \cong SL_2(\mathbb{R})/SO_2(\mathbb{R})$ to the study of $SL_2(\mathbb{R})$ or rather the study of $PSL_2(\mathbb{R}) = SL_2(\mathbb{R})/\{\pm I\}$. If we look at the unit tangent bundle $T^1\mathbb{H} \cong \mathbb{H} \times S^1$ of $\mathbb{H}$ consisting of all tangent vectors of unit length, then the induced action of $SL_2(\mathbb{R})$ on this space is transitive but $-I$ acts trivially. One checks, that I and $-I$ are the only elements stabilizing the reference vector (i, i), so that any choice of a reference vector $(z, v) \in T^1\mathbb{H}$ yields an isomorphism $T^1\mathbb{H} \cong PSL_2(\mathbb{R})$ by $g \mapsto g \cdot (z, v)$. It follows that the quotient

$$X_2 \cong {}_{SL_2(\mathbb{Z})}\backslash {}^{SL_2(\mathbb{R})} \cong {}_{PSL_2(\mathbb{Z})}\backslash {}^{PSL_2(\mathbb{R})} \cong {}_{PSL_2(\mathbb{Z})}\backslash {}^{T^1\mathbb{H}}$$

should be thought of as the unit tangent bundle of the surface ${}_{SL_2(\mathbb{Z})}\backslash {}^{\mathbb{H}}$. In what follows, we fix the identification corresponding to the reference vector that points north at i, i.e. the tangent vector $(i, i) \in T^1\mathbb{H}$. In particular, under the resulting isomorphism the identity $I \in PSL_2(\mathbb{R})$ corresponds to $(i, i) \in T^1\mathbb{H}$.

If we multiply $I \in PSL_2(\mathbb{R})$ on the right by a diagonal element $\begin{pmatrix} y \\ & 1/y \end{pmatrix}$ in $SL_2(\mathbb{R})$, then the natural isomorphism $PSL_2(\mathbb{R}) \cong T^1\mathbb{H}$ maps the element $I\begin{pmatrix} y \\ & 1/y \end{pmatrix}$ to the point $\begin{pmatrix} y \\ & 1/y \end{pmatrix} \cdot (i, i) = (y^2 i, y^2 i)$. This shows that the group

$$A = \left\{ \begin{pmatrix} e^{t/2} \\ & e^{-t/2} \end{pmatrix} \,\middle|\, t \in \mathbb{R} \right\} < SL_2(\mathbb{R})$$

moves (i, i) up and down along the imaginary axis. However, the imaginary axis is precisely a geodesic in the hyperbolic plane. Using that the natural left- and right-action of $SL_2(\mathbb{R})$ on $PSL_2(\mathbb{R})$ commute, and that on the left $SL_2(\mathbb{R})$ acts by isometries for some left-invariant Riemannian metric on $PSL_2(\mathbb{R})$, it follows that the right action of A on $PSL_2(\mathbb{R})$ corresponds to following the geodesic determined by the given vector in $T^1\mathbb{H}$ (cf. [15, Chapter 9.2]). This flow on $PSL_2(\mathbb{R})$ is called the geodesic flow.

The Riemannian manifold $X_2 = {}_\Gamma\backslash {}^{SL_2(\mathbb{R})}$ carries the topology induced by the metric

$$d_{X_2}(\Gamma g_1, \Gamma g_2) = \inf_{\gamma \in \Gamma} d_{PSL_2(\mathbb{R})}(\gamma g_1, g_2) \quad (g_1, g_2 \in PSL_2(\mathbb{R})).$$

The geodesic flow on $PSL_2(\mathbb{R})$ then descends to the geodesic flow on the unit tangent bundle X_2 of the modular surface, which is illustrated in Fig. 10.2. The initial tangent vector determines a geodesic in the hyperbolic plane, which (typically) leaves the fundamental domain at some point. At this point we apply an isometry from $SL_2(\mathbb{Z})$ to move it back into the fundamental domain. (The translation action of a one-dimensional subspace $V \subset \mathbb{R}^2$ on $\mathbb{T}^2 = \mathbb{R}^2/\mathbb{Z}^2$ is described in the same way using the fundamental domain $[0, 1)^2$ and the isometries $\mathbb{Z}^2 \subset \mathbb{R}^2$.) The geodesic

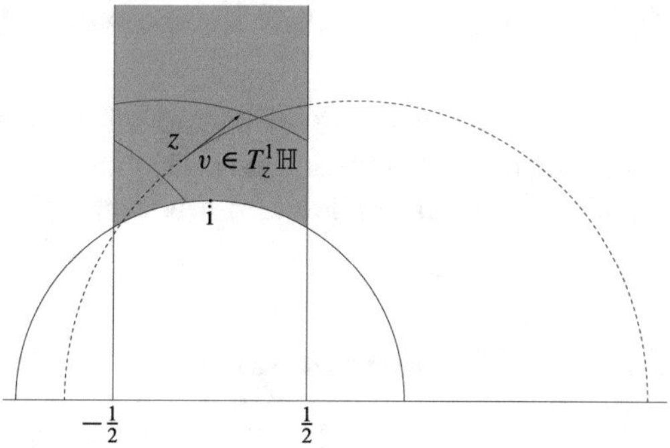

Fig. 10.2 An illustration of
the geodesic flow
on $\mathrm{SL}_2(\mathbb{Z})\backslash\mathbb{H}$ along the
direction determined by the
tangent vector $(z, v) \in T^1\mathbb{H}$

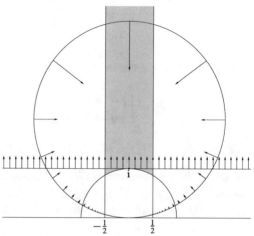

Fig. 10.3 The U-orbit
of (i, i) is a horizontal line
with vectors pointing north.
The image of $U \cdot (i, i)$ under
the Möbius transform
corresponding to $\left(\begin{smallmatrix} 1 & -2 \\ 1/2 & \end{smallmatrix}\right)$ is a
circle tangent to the real axis

flow on the modular surface (or on any other hyperbolic surface) has many invariant subsets and invariant measures. This can for instance be seen via a symbolic coding of the geodesic flow, which in the case of the modular surface can be linked to the continued fraction expansion of real numbers, see [1, 32] and [15, Ch. 9].

The second type of dynamical systems on X_2 that we wish to consider is the horocycle flow. We define $U = \left\{ \left(\begin{smallmatrix} 1 & s \\ & 1 \end{smallmatrix}\right) \mid s \in \mathbb{R} \right\}$ and as with the geodesic flow we first apply U to our reference vector to obtain a horizontal line with tangent vectors pointing north. The Möbius transformations map this orbit to other so-called horocycle orbits. Two such horocycles are illustrated in Fig. 10.3. As in the case of the geodesic flow, the horocycle flow on X_2 can be understood within the fundamental domain by following the horocycle determined by the initial vector until a boundary is reached and then moving the continuation back into the fundamental domain using an isometry from $\mathrm{SL}_2(\mathbb{Z})$.

As was discussed before, since U is unipotent, the dynamics of U on X_2 is better behaved and can be completely understood. This is due to the work of Hedlund [20], Furstenberg [19], Dani [5], and Dani-Smillie [7] (and is today also a very special case of Ratner's theorems).

10.2   Disjointness and Kloosterman Sums

We wish to discuss here an interaction between the classical Kloosterman sums appearing in number theory and a disjointness result for (T_2, T_3) and related systems (which is a partial result towards Furstenberg's conjecture mentioned above).

10.2.1   Kloosterman Sums

Given an integer $q > 1$, we define the modulo q hyperbola

$$\mathcal{H}_q = \left\{ (a, b) \in \left(\mathbb{Z}/q\mathbb{Z} \right)^2 \,\middle|\, ab = 1 \right\}$$

and its image (after division by q) in the 2-torus

$$\mathcal{H}_q^{\mathbb{T}} = \left\{ \left(\frac{a}{q}, \frac{b}{q} \right) \in \mathbb{T}^2 \,\middle|\, a, b \in \mathbb{Z} \text{ and } ab \equiv 1 \bmod q \right\}$$

Note that $|\mathcal{H}_q| = |\mathcal{H}_q^{\mathbb{T}}| = \varphi(q)$ is the Euler φ-function.

Theorem 10.2 (Kloosterman/Weil's Bound) *As $q \to \infty$, the set $\mathcal{H}_q^{\mathbb{T}}$ equidistributes in $\mathbb{T}^2$ with respect to the Lebesgue measure. Moreover, for any $(m, n) \in \mathbb{Z}^2$ satisfying $\gcd(m, n, q) = 1$ we have the estimate*

$$\left| \frac{1}{\varphi(q)} \sum_{(a,b) \in \mathcal{H}_q} e\left(\frac{ma + nb}{q} \right) \right| l \ll q^{-\kappa},$$

where $\kappa = \frac{1}{4}$ (Kloosterman's bound) or $\kappa = \frac{1}{2}$ (Weil's bound).

Here $e(\cdot)$ stands for the function $x \mapsto e^{2\pi i x}$, so that the expression on the left-hand side equals the mean value of the product $\phi(a)\psi(a^{-1})$ for units $a \bmod q$, where ϕ, ψ are characters of the additive group $\mathbb{Z}/q\mathbb{Z}$.

We are not concerned with optimality of the exponents, so we are quite content with using the original Kloosterman bound $\kappa = \frac{1}{4}$, which can be obtained via elementary means, whereas Weil's bound originally was much harder to achieve, even though there is by now a relatively elementary proof thereof, based on a method by Stepanov (see [33]). For a complete discussion, we refer the reader to [21]. In what follows, we denote by

$$K_{m,n}^q = \sum_{(a,b) \in \mathcal{H}_q} e\left(\frac{ma + nb}{q} \right)$$

the so-called Kloosterman sums appearing in the above statement. Using the bound and a simple Fourier series argument, one can obtain the equidistribution statement in the theorem. In fact, the bound even implies an effective equidistribution result, i.e.

$$\forall f \in C^\infty(\mathbb{T}^2) : \left| \frac{1}{\varphi(q)} \sum_{(a,b)\in\mathcal{H}_q^{\mathbb{T}}} f(a,b) - \int_{\mathbb{T}^2} f\,dm \right| \ll q^{-\kappa} S(f)$$

for some Sobolev norm S on $C^\infty(\mathbb{T}^2)$ and some $\kappa > 0$.

10.2.2 A Disjointness Result

Let H be a group that acts measure preservingly on two probability spaces (X, μ) and (Y, ν), hence yielding two separate dynamical systems. A joining between these two systems is defined to be a probability measure ρ on $X \times Y$ that is invariant under the induced diagonal action of H defined by $h \cdot (x, y) = (h \cdot x, h \cdot y)$ for $h \in H$ and $(x, y) \in X \times Y$, and that projects to the original measures on X and Y, i.e.

$$(\pi_X)_* \rho = \mu \text{ and } (\pi_Y)_* \rho = \nu.$$

For example the product measure is a joining and is called the trivial joining (as it always exists). In what follows, we assume that the probability spaces (X, μ) and (Y, ν) are nontrivial, i.e. there exist measurable subsets which neither have measure zero nor full measure. The existence of nontrivial joinings indicates some nontrivial relationship between the two systems. More precisely, disjointness— i.e. the absence of nontrivial joinings—proves (see [15, Ch. 6]) that they can not have a common factor (otherwise one could construct the relatively independent joining over the common factor, which will be trivial if and only if the common factor is trivial) and in particular that the two systems can not be isomorphic. The latter statement is proven very easily, as otherwise the graph would immediately give rise to a joining.

Theorem 10.3 (Disjointness of $\times 2$, $\times 3$ and Its Inverse) *Let $H = \mathbb{Z}^2$ and*

$$X = \mathbb{R} \times \mathbb{Q}_2 \times \mathbb{Q}_3 \big/ \mathbb{Z}[\frac{1}{6}] \cong \mathbb{R} \times \mathbb{Z}_2 \times \mathbb{Z}_3 \big/ \mathbb{Z}$$

be the common invertible extension for T_2 and T_3 acting on $\mathbb{T}$ via $T_p(x) = px \bmod 1$. Consider the action

$$(m, n) : x \in X \mapsto 2^m 3^n x \in X$$

and its inverse

$$(m, n) : x \in X \mapsto 2^{-m}3^{-n}x \in X$$

of H on X and equip X with the Haar measure. The two resulting dynamical systems are disjoint.

This follows from [11] but can also be proven by the TNS-method in [22], and the proof is similar to the slightly easier zero-dimensional analog considered in [14].

10.2.3 The Connection

We will now show that Theorem 10.3 implies the ineffective equidistribution statement in Theorem 10.2. For this we assume first that $2 \nmid q$ and $3 \nmid q$. With this, the projection of $\mathcal{H}_q^{\mathbb{T}}$ to one factor, i.e. the points

$$P_q^{\mathbb{T}} = \left\{ \frac{a}{q} \in \mathbb{T} \,\middle|\, a \in \mathbb{Z}, \gcd(a, q) = 1 \right\},$$

consists of $\varphi(q)$ points periodic both for T_2 and T_3. Moreover, this set can be lifted to the collection

$$P_q^X = \left\{ \left(\frac{a}{q}, \frac{a}{q}, \frac{a}{q} \right) \in X \,\middle|\, a \in \mathbb{Z}, \gcd(a, q) = 1 \right\}$$

of points in $X = \mathbb{R} \times \mathbb{Z}_2 \times \mathbb{Z}_3/_{\mathbb{Z}}$, which consists of $\varphi(q)$ points which are periodic under the lifted realizations of T_2 and T_3.

Lemma 10.4 *As $q \to \infty$, the finite sets P_q^X become equidistributed in X w.r.t. the Haar measure on X.*

Proof As P_q^X is invariant under T_2 and T_3, it suffices to show the similar claim for $P_q^{\mathbb{T}}$. To this end we denote by ν_q the counting measure on P_q^X and by ν any weak* limit for $q \to \infty$ along a sequence of numbers coprime to 2 and 3. The counting measure μ_q on $P_q^{\mathbb{T}}$ is the push-forward of the counting measure on P_q^X under the canonical projection $\pi : X \to \mathbb{T}$, that is $\mu_q = \pi_* \nu_q$. If μ_q converges to the Lebesgue measure $m_{\mathbb{T}}$ on $\mathbb{T}$, then also $\pi_* \nu = m_{\mathbb{T}}$. As the sets P_q^X are invariant under T_2 and T_3, so is the measure ν. Hence $(\mathbb{T}, m_{\mathbb{T}})$ is a factor of (X, ν) for T_2 and T_3 with factor map being the canonical projection. However, the measure on the invertible extension, for which the original system is a factor, is unique, and as X equipped with the Haar measure is an invertible extension also with factor map being the canonical projection, it follows that ν and m_X agree.

It remains to show that the sets $P_q^{\mathbb{T}}$ equidistribute as $q \to \infty$ along a suitable sequence. Using Weyl's criterion, we need to show that for any $m \in \mathbb{Z} \setminus \{0\}$ we have

$$\frac{1}{\varphi(q)} \sideset{}{'}\sum_{a \in \mathbb{Z}/q\mathbb{Z}} e\left(m \frac{a}{q}\right) \xrightarrow{q \to \infty} 0,$$

where $\sum'$ denotes the sum restricted to the congruence classes mod q which are coprime to q. If $\gcd(m, q) > 1$, we can cancel them partially and consider the cancelled expressions. As m is fixed while $q \to \infty$, we can without loss of generality assume that $\gcd(m, q) = 1$. If $q = p^k$ is a power of a prime p, then

$$\left\{ \frac{a}{q} \,\middle|\, \gcd(a, q) = 1 \right\} = \bigsqcup_{j=1}^{p-1} \left\{ \frac{j + pl}{p^k} \,\middle|\, l = 0, \ldots, p^{k-1} - 1 \right\}$$

and

$$\frac{1}{\varphi(q)} \sideset{}{'}\sum_{a \in \mathbb{Z}/q\mathbb{Z}} e\left(m \frac{a}{q}\right) = \frac{1}{\varphi(q)} \sum_{j=1}^{p-1} \sum_{l=0}^{p^{k-1}-1} e\left(m \frac{j + pl}{q}\right)$$

In particular, we see that

$$\frac{1}{\varphi(q)} \sideset{}{'}\sum_{a \in \mathbb{Z}/q\mathbb{Z}} e\left(m \frac{a}{q}\right) = \begin{cases} 0 & \text{if } k > 1 \\ \frac{-1}{p-1} & \text{else} \end{cases}$$

which yields the equidistribution statement along powers of p.

If $q = q_1 \cdots q_n$ is a product of powers $q_1, \ldots, q_n$ of distinct primes, then the Chinese remainder theorem implies multiplicativity of the sum in the sense that it can be written as a product of character sums as above, each factor corresponding to the correct power of a prime divisor of q. If q contains a square, i.e. there is a prime p and $k > 1$ such that $p^k \mid q$, then the above discussion together with the multiplicativity implies that the factor corresponding to p^k vanishes and thus

$$\left| \frac{1}{\varphi(q)} \sideset{}{'}\sum_{a \in \mathbb{Z}/q\mathbb{Z}} e\left(m \frac{a}{q}\right) \right| = 0 < \varepsilon.$$

Otherwise, $q = q_1 \cdots q_n$ for distinct primes $q_1, \ldots, q_n$. Fix some arbitrary $\varepsilon > 0$ and let $p_1 = 2, p_2 = 3, p_3 = 5, \ldots$ be the sequence of prime numbers. Then there is some $N \in \mathbb{N}$ such that $(p_i - 1)^{-1} < \varepsilon$ for all $i \geq N$. Now assume that $q \geq p_1 \cdots p_N$, then we can find some $1 \leqslant i \leqslant m$ so that $q_i \geqslant p_N$, and thus

$$
\left| \frac{1}{\varphi(q)} \sum_{a \in \mathbb{Z}/q\mathbb{Z}}' e\left(m\frac{a}{q}\right) \right| = \prod_{i=1}^{n} \frac{1}{q_i - 1} \leqslant \frac{1}{p_N - 1} < \varepsilon.
$$

Lemma 10.5 *Consider the family of normalized counting measures on*

$$
\mathcal{H}_{q,d}^{X} = \left\{ \left(\left(\frac{a}{q}, \frac{a}{q}, \frac{a}{q}\right), \left(\frac{b}{q}, \frac{b}{q}, \frac{b}{q}\right) \right) \,\middle|\, a, b \in \mathbb{Z}/q\mathbb{Z}, ab = d \bmod q \right\} \subset X^2
$$

for $q \in \mathbb{N}$ with $2 \nmid q$ and $3 \nmid q$ and $\gcd(q, d) = 1$. Then any accumulation point is a joining for the system generated by T_2 and T_3 in the first and their inverses in the second factor.

Proof We note that the projection of the set $\mathcal{H}_{q,d}^{X}$ to the second component is P_q^X, because d is a unit $\bmod\, q$. Recall from the introduction that the maps T_2 and T_3 on $\mathbb{R} \times \mathbb{Q}_2 \times \mathbb{Q}_3/\mathbb{Z}[\frac{1}{6}]$ are realized on $\mathbb{R} \times \mathbb{Z}_2 \times \mathbb{Z}_3/\mathbb{Z}$ also by multiplication by 2 and 3 respectively. Assume that $b \in \mathbb{Z}$. Let $k \in \mathbb{Z}$ such that $2k \equiv 1 \bmod q$, then

$$
\left((2k)\frac{b}{q}, (2k)\frac{b}{q}, (2k)\frac{b}{q} \right) \equiv \left(\frac{b}{q}, \frac{b}{q}, \frac{b}{q} \right) \quad \bmod \mathbb{Z},
$$

so that on points in $P_q^X \subset \mathbb{R} \times \mathbb{Z}_2 \times \mathbb{Z}_3/\mathbb{Z}$ the inverse of $\times 2$ is given by multiplication with $2^{-1} \in \mathbb{Z}/q\mathbb{Z}$. The same argument works for $\times 3$. In particular it follows that $\mathcal{H}_{q,d}^{X}$ is invariant under (T_2, T_2^{-1}) and (T_3, T_3^{-1}).

Let ρ be any weak* limit of the normalized counting measure on $\mathcal{H}_{q,d}^{X}$ where d is allowed to vary with q, then ρ is invariant under (T_2, T_2^{-1}) and under (T_3, T_3^{-1}). Equidistribution of P_q^X implies that ρ projects onto the Haar measure on X in both components. Hence ρ is a joining.

Proof (of the Ineffective Equidistribution) We start with the more general statement that for any sequence $q \to \infty$ satisfying $2 \nmid q$ and $3 \nmid q$ and any sequence d_q so that $\gcd(q, d_q) = 1$, if $\gcd(m, n, q) = 1$, then

$$
K_{m,d_q n}^{q} = \frac{1}{\varphi(q)} \sum_{ab \equiv d_q \bmod q} e\left(\frac{ma + nb}{q} \right) \overset{q \to \infty}{\longrightarrow} 0.
$$

By the preceding lemma, any weak* limit ρ of the normalized counting measures on $\mathcal{H}_{q,d_q}^{X}$ is a joining between the system generated by T_2 and T_3 and the system generated by their inverses. By the disjointness theorem this joining is trivial, i.e. $\rho = m_X \otimes m_X$. In particular the limit is independent of the choice of the subsequence and the whole sequence converges to the same limit $\rho = m_X \otimes m_X$. It follows that the corresponding counting measures on the torus $\mathbb{T}^2$ converge to the uniform measure $m_{\mathbb{T}^2}$. The integral of $\chi_{m,n} : \mathbb{T}^2 \to \mathbb{C}$, $(s,t) \mapsto e(ms + nt)$ with respect to the Haar measure vanishes whenever m, n are not simultaneously 0, so the ineffective statement follows.

If on the other hand $q = 2^j 3^k$ for some $(j, k) \to \infty$, we apply the same argument replacing 2 and 3 by 5 and 7.

The general case can be reduced to these cases in a fashion similar to the proof of Lemma 10.4. To this end assume that q_1, q_2 satisfy $\gcd(q_1, q_2) = 1$ and assume $q = q_1 q_2$. Using the Chinese remainder theorem, we can find $a, b \in \mathbb{Z}$ so that $aq_1 \equiv 1 \mod q_2$ and $bq_2 \equiv 1 \mod q_1$. One calculates

$$K^{q_1}_{bm,bn}K^{q_2}_{am,an} = \sum_{x\in(\mathbb{Z}/q_1\mathbb{Z})^\times} \sum_{y\in(\mathbb{Z}/q_2\mathbb{Z})^\times} e\left(\frac{bmx + bnx^{-1}}{q_1} + \frac{amy + any^{-1}}{q_2}\right)$$

$$= \sum_{x\in(\mathbb{Z}/q_1\mathbb{Z})^\times} \sum_{y\in(\mathbb{Z}/q_2\mathbb{Z})^\times} e\left(\frac{m(aq_1 y + bq_2 x) + n(aq_1 y^{-1} + bq_2 x^{-1})}{q}\right)$$

$$= K^q_{m,n},$$

where the map $\mathbb{Z}/q_1\mathbb{Z} \oplus \mathbb{Z}/q_2\mathbb{Z} \to \mathbb{Z}/q\mathbb{Z}$ given by $(x, y) \mapsto bq_2 x + aq_1 y$ is exactly the isomorphism appearing in the proof of the Chinese remainder theorem and the same theorem implies $(\mathbb{Z}/q\mathbb{Z})^\times \cong (\mathbb{Z}/q_1\mathbb{Z})^\times \times (\mathbb{Z}/q_2\mathbb{Z})^\times$. This yields

$$K^q_{m,n} = K^{q_1}_{bm,bn}K^{q_2}_{am,an} = K^{q_1}_{m,b^2 n}K^{q_2}_{m,a^2 n} \overset{q\to\infty}{\longrightarrow} 0$$

after splitting q according to the cases discussed above.

The ineffective equidistribution statement in Theorem 10.2 now follows from density of the characters inside $C(\mathbb{T}^2)$.

10.3   Kloosterman Sums and Spectral Gap

We return to the study of $X_2 = \mathrm{SL}_2(\mathbb{Z})\backslash\mathrm{SL}_2(\mathbb{R})$. This section, when explicated, strongly connects dynamical properties of the action of subgroups of $\mathrm{SL}_2(\mathbb{R})$ on X_2 and Kloosterman sums, as the effective claims made here could be proven using bounds on Kloosterman sums. For what follows, we recall that the action of $\mathrm{SL}_2(\mathbb{R})$ on X_2 is defined by

$$(g, x) \in \mathrm{SL}_2(\mathbb{R}) \times X_2 \mapsto g \cdot x = xg^{-1} \in X_2.$$

10.3.1 Ineffective Equidistribution

As $SL_2(\mathbb{R})$ acts transitively on X_2, it also acts ergodically. By a special feature of $SL_2(\mathbb{R})$ (or rather of simple Lie groups), any ergodic action of $SL_2(\mathbb{R})$ is actually strongly mixing. In particular,

$$\forall f_1, f_2 \in L^2(X_2) : \langle \pi_g(f_1), f_2 \rangle \overset{g \to \infty}{\longrightarrow} \int f_1 dm_{X_2} \int \overline{f_2} dm_{X_2},$$

where the unitary representation π on $L^2(X_2)$ of $SL_2(\mathbb{R})$ is defined by

$$\pi_g(f)(x) := f(xg) \quad m_{X_2}\text{-a.e.},$$

for all $f \in L^2(X_2)$ and for all $g \in SL_2(\mathbb{R})$. This is the "vanishing of matrix coefficients at ∞" or the Howe-Moore theorem, for which we refer to [15, Ch. 11].

We wish to explain how this proves the following ineffective version of a theorem by Sarnak [29].

Theorem 10.6 *Let $P_1 = U \cdot SL_2(\mathbb{Z}) \subset X_2$ be the periodic horocycle orbit at the identity coset, where $U = \left\{ u_s := \left(\begin{smallmatrix} 1 & s \\ & 1 \end{smallmatrix} \right) \mid s \in \mathbb{R} \right\}$ is the horocycle subgroup and P_1 is equipped with the normalized Lebesgue measure m_{P_1} (note that $P_1 \cong \mathbb{T}$). Then the push-forward $(a_T)_* m_{P_1}$ of m_{P_1} under the diagonal element $a_T = \left(\begin{smallmatrix} T^{-1} & \\ & T \end{smallmatrix} \right)$ equidistributes w.r.t. m_{X_2} as $T \to 0$.*

Geometrically this theorem corresponds to the equidistribution of long periodic horocycle orbits in X_2, see Fig. 10.4. We note that the U-orbit $a_T U \cdot \Gamma$ has volume T^{-2}, which follows from the calculation $a_T u_s a_T^{-1} = u_{T^{-2}s}$.

Proof (Sketch) The idea is to thicken the very thin object P_1 and its one-dimensional Lebesgue measure in the two other directions (with the resulting "banana" having positive Haar measure) and to apply the strong mixing property of $SL_2(\mathbb{R}) \curvearrowright X_2$.

Fig. 10.4 The dashed line is a periodic horocycle orbit, which, if drawn within the fundamental domain, would look very messy, as guaranteed by Sarnak's theorem

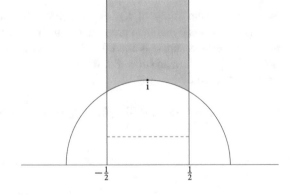

This method commonly goes under the names mixing-trick, Margulis' trick, or banana-trick and has proven useful in many applications (e.g. [18]).

More precisely, fix $f \in C_c(X_2)$ and let $\varepsilon > 0$ arbitrary. By uniform continuity of f, there exists some $\delta > 0$ such that

$$d_{X_2}(x_1, x_2) < \delta \Rightarrow |f(x_1) - f(x_2)| < \varepsilon,$$

where d_{X_2} is the metric on X_2 induced by the left-invariant metric on $\mathrm{SL}_2(\mathbb{R})$. Using basic properties of Lie groups, one can show that the map sending $r, t, s \in \mathbb{R}$ to the product of matrices

$$u_s b(t, r) = \begin{pmatrix} 1 & s \\ & 1 \end{pmatrix} \begin{pmatrix} e^t & \\ r & e^{-t} \end{pmatrix}$$

yields a local coordinate system around the identity. We use this coordinate system to thicken the orbit P_1 to a three-dimensional tube of positive measure. Using the abundant continuity properties of the objects involved, we can find $\eta > 0$ such that $|t| < \eta$ and $|r| < \eta$ implies

$$d_{\mathrm{SL}_2(\mathbb{R})}\left(\begin{pmatrix} e^t & \\ r & e^{-t} \end{pmatrix}, I\right) < \delta$$

If now $|t| < \eta$, $|r| < \eta$, and $s \in [0, 1]$, then (for $T \leqslant 1$)

$$d_{X_2}\left(\Gamma \begin{pmatrix} 1 & s \\ & 1 \end{pmatrix} \begin{pmatrix} e^t & \\ r & e^{-t} \end{pmatrix} a_T^{-1}, \Gamma \begin{pmatrix} 1 & s \\ & 1 \end{pmatrix} a_T^{-1}\right)$$

$$= d_{X_2}\left(\Gamma \begin{pmatrix} 1 & s \\ & 1 \end{pmatrix} a_T^{-1} a_T \begin{pmatrix} e^t & \\ r & e^{-t} \end{pmatrix} a_T^{-1}, \Gamma \begin{pmatrix} 1 & s \\ & 1 \end{pmatrix} a_T^{-1}\right)$$

$$\leqslant d_{\mathrm{SL}_2(\mathbb{R})}\left(a_T \begin{pmatrix} e^t & \\ r & e^{-t} \end{pmatrix} a_T^{-1}, I\right) = d_{\mathrm{SL}_2(\mathbb{R})}\left(\begin{pmatrix} e^t & \\ T^2 r & e^{-t} \end{pmatrix}, I\right) < \delta,$$

which in particular implies that

$$\left| f\left(\Gamma \begin{pmatrix} 1 & s \\ & 1 \end{pmatrix} \begin{pmatrix} e^t & \\ r & e^{-t} \end{pmatrix} a_T^{-1}\right) - f\left(\Gamma \begin{pmatrix} 1 & s \\ & 1 \end{pmatrix} a_T^{-1}\right) \right| < \epsilon.$$

As discussed in [15, Lemma 11.31], we can locally express the Haar measure in this coordinate system. Integration of the above expression with respect to $r, t \in (-\eta, \eta)$ and $s \in [0, 1]$ implies that

$$\left| \frac{1}{m_{X_2}(B)} \int f(a_T \cdot x) \chi_B(x) dm_{X_2}(x) - \int f(a_T \cdot x) dm_{P_1}(x) \right| < \varepsilon$$

where χ_B is the indicator function of some neighbourhood B of $I \in SL_2(\mathbb{R})$. However, for small enough T (or rather large enough T^{-1}) the first expression is ε-close to

$$\int f dm_{X_2} \int \frac{1}{m_{X_2}(B)} \chi_B dm_{X_2} = \int f dm_{X_2}$$

due to strong mixing.

10.3.2   Effective Decay of Matrix Coefficients

The mixing property mentioned above has the following significant strengthening, which follows from the work of Selberg [31].

Theorem 10.7 (Effective Mixing on X_2) *For $f_1, f_2 \in C_c^\infty(X_2)$ and for all $g \in G$ holds*

$$\left| \langle \pi_g(f_1), f_2 \rangle - \int f_1 dm_{X_2} \int \overline{f_2} dm_{X_2} \right| \ll \|g\|^{-\kappa} S(f_1) S(f_2)$$

for some absolute implicit constant and some absolute $\kappa > 0$, where $S(f)$ denotes some fixed Sobolev norm for $f \in C_c^\infty(X_2)$.

Using this theorem, one can prove an effective version of the equidistribution of (pieces of) long periodic horocycles from Theorem 10.6. We refer to [36] for similar arguments. In the case of the whole periodic horocycle orbit, much stronger results appear already in [29].

Corollary 10.8 *For $f \in C_c^\infty(X_2)$ we have*

$$\left| \int f d(a_T)_* m_{P_1} - \int f dm_{X_2} \right| \ll T^\kappa S(f)$$

for some absolute implicit constant, some fixed $\kappa > 0$, and some fixed Sobolev norm.

10.4   Sparse Equidistribution of Primitive Points

In what follows, we want to consider subsets of the orbits $a_T \cdot P_1$, and examine equidistribution properties of these subsets. The question here is twofold: First we are in some sense interested in how much of the U-orbit we actually need in order to equidistribute in X_2, and second we ask for the equidistribution of arithmetically defined sets, i.e. whether the arithmetic nature of a subset forces the possibly small collection to still equidistribute effectively.

10.4.1   The Discrete Periodic Horocycle Orbit: Ineffective Discussion

Recall from the preceding section that the long periodic orbits

$$\Gamma U \begin{pmatrix} y^{1/2} & \\ & y^{-1/2} \end{pmatrix} = \Gamma \begin{pmatrix} y^{1/2} & \\ & y^{-1/2} \end{pmatrix} U$$

equidistribute in X_2 as their volume y^{-1} goes to infinity. We also recall, that we denote $a_{y-1/2} = \begin{pmatrix} \sqrt{y} & \\ & 1/\sqrt{y} \end{pmatrix}$. Assume for the moment, that $y = n^{-1}$ for some $n \in \mathbb{N}$. In this case we consider the discrete orbit $\Gamma a_{y-1/2} U(\mathbb{Z})$ consisting of n points, where

$$U(\mathbb{Z}) = \left\{ \begin{pmatrix} 1 & k \\ & 1 \end{pmatrix} \,\middle|\, k \in \mathbb{Z} \right\}.$$

We can use equidistribution of long periodic horocycles to deduce the following discrete version.

Corollary 10.9 *The discrete periodic orbit*

$$\left\{ \frac{k}{n} + \frac{i}{n} \,\middle|\, k = 0, \ldots, n - 1 \right\}$$

equidistributes inside $\Gamma \backslash \mathbb{H}$ as $n \to \infty$ and so does the discrete periodic orbit

$$\Gamma \begin{pmatrix} n^{-\frac{1}{2}} & \\ & n^{\frac{1}{2}} \end{pmatrix} U(\mathbb{Z})$$

in the tangent bundle $\Gamma \backslash \mathrm{SL}_2(\mathbb{R}) \cong \Gamma \backslash T^1(\mathbb{H})$ of the modular surface.

The above corollary and in particular Corollary 10.10 are in spirit related to a conjecture by Nimish Shah, a much harder problem. The question is, whether certain sequences along horocycles in compact quotients equidistribute in the full space. Note that in the context of compact quotients, horocycle orbits are equidistributed. We refer the reader to the work of Venkatesh, Tanis and Tanis-Vishe as well as Sarnak-Ubis (cf. [30, 34–36]).

Proof (of Corollary 10.9) Let $\delta \in (0, 1)$ and thicken the discrete periodic orbit to obtain

$$P_{n,\delta} = \Gamma a_{n^{1/2}} U(\mathbb{Z}) \left\{ \begin{pmatrix} 1 & s \\ & 1 \end{pmatrix} \,\middle|\, s \in [0, \delta] \right\}.$$

Consider now the normalized Lebesgue measure $m_{n,\delta}$ on this set and take the weak* limit along some subsequence to obtain a limit point μ_δ. Recall from above that the normalized probability measure on the full periodic orbit converges to m_{X_2} as $n \to \infty$. Hence the normalized measure on the complement of $P_{n,\delta}$ inside $a_{n-1/2} \cdot P_1$ converges as well, say to the probability measure μ'_δ. We note that by construction μ_δ and $\mu_{\delta'}$ are both invariant probability measures for the action of $\left(\begin{smallmatrix} 1 & 1 \\ & 1 \end{smallmatrix}\right)$. It follows that

$$\delta\mu_\delta + (1 - \delta)\mu'_\delta = m_{X_2}.$$

However, one of the characterizations of ergodicity is extremality of the probability measure in question—in our case the measure m_{X_2}—within the convex set of invariant probability measures. Since $\left(\begin{smallmatrix} 1 & 1 \\ & 1 \end{smallmatrix}\right)$ acts ergodically, we obtain $\mu_\delta = m_{X_2}$.

If now $f \in C_c(X_2)$ and $\varepsilon > 0$, then we can choose $\delta > 0$ satisfying the uniform continuity estimate for f and this given ε. This implies

$$\left| \int f \, dm_{n,\delta} - \frac{1}{n} \sum_{k=0}^{n-1} f\left(\Gamma \begin{pmatrix} n^{-\frac{1}{2}} & \\ & n^{\frac{1}{2}} \end{pmatrix} \begin{pmatrix} 1 & k \\ & 1 \end{pmatrix} \right) \right| < \varepsilon,$$

so that the corollary follows by first letting $n \to \infty$, then $\varepsilon \to 0$ (and with it $\delta \to 0$).

10.4.2   The Discrete Periodic Horocycle: Effective Discussion

Like in the discussion of the full periodic orbit, we wish to upgrade the equidistribution claim in the above corollary and give an error rate, which will be needed in the following discussions. Indeed, using the effective version of equidistribution of long periodic horocycles (10.8) instead of just equidistribution and the mean value theorem instead of uniform continuity, one can effectivize the preceding proof in order to obtain an error rate in terms of the function f and the covolume n.

Corollary 10.10 *The discrete periodic orbit $\Gamma a_{n^{1/2}} U(\mathbb{Z})$ equidistributes effectively: There exists some $\kappa > 0$ such that for any $f \in C_c^\infty(X_2)$ we have*

$$\left| \frac{1}{n} \sum_{k=0}^{n-1} f\left(\Gamma \begin{pmatrix} n^{-\frac{1}{2}} & \\ & n^{\frac{1}{2}} \end{pmatrix} \begin{pmatrix} 1 & k \\ & 1 \end{pmatrix} \right) - \int f \, dm_{X_2} \right| \ll n^{-\kappa} S(f),$$

for some L^2-Sobolev norm S on $C_c^\infty(X_2)$.

10.4.3 Selecting the Primitive Rational Points and Retaining Effective Equidistribution

We want to choose an even sparser set from the discrete periodic orbit and want to discuss its dynamics. To select this sparse set, we start with a short calculation. Let $n > 1$ be any positive number, for the moment not necessarily an integer. We will show below, that n in fact has to be an integer for the question we are interested in. In addition to the periodic orbit (the so-called stable horocycle orbit) discussed above, there is also the periodic orbit P_1' of Γ under the subgroup $V = \left\{ v_r = \left(\begin{smallmatrix} 1 & \\ r & 1 \end{smallmatrix} \right) \mid r \in \mathbb{R} \right\}$, called the unstable horocycle. In complete analogy to the above discussion, one can show that the sequence of measures $(a_T)_* m_{P_1'}$ supported on one-dimensional periodic orbits of volume T^2 equidistributes as T goes to ∞. It is an interesting question to find out, what the intersection points, if any, of stretched one-dimensional V and U orbits of Γ are. To answer this question, it is sufficient, to consider the intersections of a stretched U orbit with the orbit $V \cdot \Gamma$ of volume 1. Explicitly, we wish to describe the set

$$\left\{ \Gamma \begin{pmatrix} 1 & s \\ & 1 \end{pmatrix} \,\middle|\, \exists s \in [0,1) \exists r \in [0,1) : \Gamma \begin{pmatrix} 1 & s \\ & 1 \end{pmatrix} \begin{pmatrix} n^{-1} & \\ & n \end{pmatrix} = \Gamma \begin{pmatrix} 1 & \\ r & 1 \end{pmatrix} \right\}$$

and understand the relationship between $s \in [0,1)$ and $r \in [0,1)$ appearing in this definition. So suppose that $\gamma = \left(\begin{smallmatrix} a & -b \\ c & d \end{smallmatrix} \right) \in \Gamma$ and $s, r \in [0,1)$ satisfy

$$\begin{pmatrix} a & -b \\ c & d \end{pmatrix} \begin{pmatrix} 1 & s \\ & 1 \end{pmatrix} \begin{pmatrix} n^{-1} & \\ & n \end{pmatrix} = \begin{pmatrix} 1 & \\ r & 1 \end{pmatrix}$$

or equivalently

$$\begin{pmatrix} a & as - b \\ c & cs + d \end{pmatrix} = \begin{pmatrix} n & \\ rn & n^{-1} \end{pmatrix}.$$

As $\gamma \in \mathrm{SL}_2(\mathbb{Z})$, it follows that $a = n \in \mathbb{N}$, $s = \frac{b}{n}$, $r = \frac{c}{n}$, where in particular we have $bc = 1 - ad \equiv 1 \bmod (n)$. In other words, the set of pairs (r, s) obtained from the geometric intersection above, is precisely the modulo-n hyperbola $\mathcal{H}_n^{\mathbb{T}}$ that we considered in Sect. 10.2. If we now consider the collection of points

$$\Gamma \begin{pmatrix} 1 & \frac{b}{n} \\ & 1 \end{pmatrix} \begin{pmatrix} t & \\ & t^{-1} \end{pmatrix}$$

for $b \in \mathbb{Z}$ coprime to n and $t > 0$, we observe the following behaviour:

- For $t = 1$ the points equidistribute on the periodic orbit $\Gamma U \cong \mathbb{T}$ as $n \to \infty$.
- If $t > 1$, the points move closer together, sitting on the periodic orbit of length t^{-2} which moves to the cusp, as $t \to \infty$. For fixed t, the points equidistribute inside the periodic orbit $\Gamma \left({}^{t}_{\ t^{-1}} \right) U \cong \mathbb{T}$ as $n \to \infty$.
- For $t < 1$, the points (originally at distance about $\frac{1}{n}$ from each other) move apart (along the U-orbit) as we shrink t. There are two special cases which help us understand the dynamics as $t \to 0$:

  - When $t = \frac{1}{\sqrt{n}}$, the points, lying on the orbit of volume n, are at least at distance 1 apart from the perspective of $U \cong \mathbb{R}$ and form a potentially thin subset of the set of points considered in Sect. 10.4.1.
  - When $t = \frac{1}{n}$, the points have moved apart at least to distance n, when viewed along the U-orbit. However, by the above geometric-algebraic miracle, we know that the points get to lie on the periodic orbit ΓV of volume 1, so in fact they move closer together again (but in a different order).
  - As $t < \frac{1}{n}$ goes to 0, the points remain on the periodic orbit $\Gamma \left({}^{\tau}_{\ \tau^{-1}} \right) V$ of volume τ^2 with $\tau = nt$, which uniformly diverges into the cusp as $\tau \to 0$.

This above description of the rational points on ΓU serves as a motivation for understanding the intermediate case $t = \frac{1}{\sqrt{n}}$ better.

Theorem 10.11 *The normalized counting measure on*

$$\left\{ \Gamma \begin{pmatrix} 1 & \frac{b}{n} \\ & 1 \end{pmatrix} \begin{pmatrix} \frac{1}{\sqrt{n}} & \\ & \sqrt{n} \end{pmatrix} \,\middle|\, \gcd(b, n) = 1 \right\}$$

equidistributes to m_{X_2} as $n \to \infty$.

The proof (of an effective version) of this theorem is the content of [17]. Here, let us provide a sketch of the proof of the theorem in increasingly harder cases:

- If $n = p$ is a prime, then the theorem is an immediate consequence of Corollary 10.9, since the two sets appearing in Theorem 10.11 and Corollary 10.9 differ only by one point (as $\varphi(p) = p - 1$).
- If n is a product of many prime powers but $\frac{\varphi(n)}{n} \geq \frac{1}{100}$ say, then by the multiplicativity of the totient function and using $\lim \inf_{n \to \infty} \frac{\varphi(n)}{n} = 0$, there is one prime p from a list of finitely many primes (independent of n) such that n is coprime to p. This p is a unit mod n, so that the set $\left(\mathbb{Z}/n\mathbb{Z} \right)^{\times} = \left\{ b \in \mathbb{Z}/n\mathbb{Z} \mid \gcd(b, n) = 1 \right\}$ of units mod n is invariant under multiplication by p. This multiplication can also be constructed in a homogeneous space X_p (as indicated in the introduction), so that under this multiplication $a_p : X_p \to X_p$ our realization of the above points is invariant and a_p acts ergodically with respect to the Haar measure on X_p. However, the ergodicity of a_p together with equidistribution of the discrete periodic orbit in Corollary 10.9 extended to X_p then gives Theorem 10.11. The

correct space to use in this context is $X_p = {}_{\mathrm{SL}_2(\mathbb{Z}[1/p])}\backslash^{\mathrm{SL}_2(\mathbb{R} \times \mathbb{Q}_p)}$, and the map $a_p : X_p \to X_p$ is given by

$$a_p : x = \Gamma_p(g_\infty, g_p) \mapsto x \left(\begin{pmatrix} p^{-1} & \\ & p \end{pmatrix}, \begin{pmatrix} p^{-1} & \\ & p \end{pmatrix} \right),$$

which commutes with the diagonal maps in $\mathrm{SL}_2(\mathbb{R})$ considered above and normalizes the upper unipotent group in $\mathrm{SL}_2(\mathbb{R} \times \mathbb{Q}_p)$. In particular it fixes the periodic orbit

$$\Gamma_p \left\{ \left(\begin{pmatrix} 1 & s_\infty \\ & 1 \end{pmatrix}, \begin{pmatrix} 1 & s_p \\ & 1 \end{pmatrix} \right) \;\middle|\; s_\infty \in \mathbb{R}, \, s_p \in \mathbb{Q}_p \right\} \cong {}_{\mathbb{Z}[1/p]}\backslash^{\mathbb{R} \times \mathbb{Q}_p}.$$

As a_p can be realized as multiplication by a group element which is not contained in a compact subgroup of $\mathrm{SL}_2(\mathbb{R} \times \mathbb{Q}_p)$, we can use the Howe-Moore theorem to deduce ergodicity of a_p. We now can argue as in the proof of Corollary 10.9 using the assumption $\frac{\varphi(n)}{n} \geqslant \frac{1}{100}$ to obtain the desired equidistribution.

- In the general case, we have

$$\frac{\varphi(n)}{n} \gg \frac{1}{\log \log n},$$

so the expression can go to zero along a subsequence, but only very slowly. In this case one uses the smallest prime p coprime to n and effectivizes the argument outlined above. Here an effective argument is required, because we can not use a fixed dynamical system (p depends on n) and also because $\frac{\varphi(n)}{n}$ may go to zero. Spectral gap in the form of effective mixing (Theorem 10.7) is again the key to the proof, together with a discrepancy argument.

We point out that the proof of the corresponding statement for Hilbert modular surfaces is essentially the same but easier, i.e. consider the group $\mathbb{G} = \mathrm{SL}_2$ defined over k, where k is a totally real number field, and formulate the corresponding statement on $X = {}_{\mathrm{Res}_{k/\mathbb{Q}}\mathbb{G}(\mathbb{Z})}\backslash^{\mathrm{Res}_{k/\mathbb{Q}}\mathbb{G}(\mathbb{R})}$. Then this can be proven without having to consider a p-adic extension but instead making use of the existence of a similar non-compact element $a \in \mathrm{Res}_{k/\mathbb{Q}}\mathbb{G}(\mathbb{Z})$ which normalizes the horospherical subgroups. This case is examined in [24].

10.4.4   Another Application of the Disjointness Result

We wish to mention one more application of a disjointness result, which yields a relatively soft proof of equidistribution of sparse, arithmetically defined subsets. Instead of asking for equidistribution of the primitive points $\{\frac{i}{n} + \frac{j}{n} \mid (j, n) = 1\}$, we can ask the same question for a polynomial version thereof. Consider the set

$$\left\{ \left(\tfrac{i}{n} + \tfrac{p_1(j)}{n}, \ldots, \tfrac{i}{n} + \tfrac{p_k(j)}{n} \right) \,\middle|\, (j, n) = 1 \right\} \subset \left(\Gamma \backslash \mathbb{H} \right)^k,$$

where $p_1, \ldots, p_k \in \mathbb{Z}[X]$ are polynomials. The disjointness result in [13] allows a proof of the equidistribution of these sets as $n \to \infty$ along any sequence for which $p, q \nmid n$, for the special case $p_l = X^l$, where $1 \leqslant l \leqslant k$. The method of proof is similar to our discussion in Sect. 10.2.3, using the invariance of the sets lifted to the p, q-adic cover of the unit tangent bundle under suitable diagonalizable elements. To our knowledge, there is no proof of this using more classical number theoretic methods.

Acknowledgements We thank Shahar Mozes, Cagri Sert, Uri Shapira, and Andreas Wieser for helpful comments on an earlier draft of this article. M.E. has learned a lot from Akshay Venkatesh and wants to thank him for many discussions related to these topics.

References

1. E. Artin, Ein mechanisches system mit quasiergodischen bahnen. Abh. Math. Semin. Univ. Hambg. **3**(1), 170–175 (1924)
2. A. Borel, *Introduction aux groupes arithmétiques*, Publications de l'Institut de Mathématique de l'Université de Strasbourg, XV. Actualités Scientifiques et Industrielles, vol. 1341 (Hermann, Paris, 1969)
3. A. Borel, L. Ji, *Compactifications of Symmetric and Locally Symmetric Spaces*. Mathematics: Theory and Applications (Birkhäuser, Boston, MA, 2006)
4. C. Chabauty, Limite d'ensembles et géométrie des nombres. Bull. Soc. Math. Fr. **78**, 143–151 (1950)
5. S.G. Dani, Invariant measures and minimal sets of horospherical flows. Invent. Math. **64**(2), 357–385 (1981)
6. S.G. Dani, G.A. Margulis, Limit distributions of orbits of unipotent flows and values of quadratic forms, in *I. M. Gel'fand Seminar* (1993), pp. 91–137
7. S.G. Dani, J. Smillie, Uniform distribution of horocycle orbits for Fuchsian groups. Duke Math. J. **51**(1), 185–194 (1984)
8. T. Downarowicz, Survey of odometers and Toeplitz flows, in *Algebraic and Topological Dynamics – Proceedings of the Conference on Algebraic and Topological Dynamics*, 1 May–31 July, 2004 (2005), pp. 7–38
9. M. Einsiedler, Ratner's theorem on SL(2,ℝ)-invariant measures. Jahresber. Dtsch. Math.-Ver. **108**(3), 143–164 (2006)
10. M. Einsiedler, A. Katok, E. Lindenstrauss, Invariant measures and the set of exceptions to Littlewood's conjecture. Ann. Math. (2) **164**(2), 513–560 (2006)
11. M. Einsiedler, E. Lindenstrauss, Rigidity properties of $\mathbb{Z}^d$-actions on tori and solenoids. Electron. Res. Announc. Am. Math. Soc. **9**, 99–110 (2003)
12. M. Einsiedler, E. Lindenstrauss, Joinings of higher-rank diagonalizable actions on locally homogeneous spaces. Duke Math. J. **138**(2), 203–232 (2007)
13. M. Einsiedler, E. Lindenstrauss, Joinings of higher rank torus actions on homogeneous spaces (2017, preprint), https://arxiv.org/pdf/1502.05133.pdf
14. M. Einsiedler, T. Ward, Entropy geometry and disjointness for zero-dimensional algebraic actions. J. Reine Angew. Math. **584**, 195–214 (2005)

15. M. Einsiedler, T. Ward, *Ergodic Theory with a View towards Number Theory* (Springer, New York, 2010)
16. M. Einsiedler, G. Margulis, A. Venkatesh, Effective equidistribution for closed orbits of semisimple groups on homogeneous spaces. Invent. Math. **177**(1), 137–212 (2009)
17. M. Einsiedler, M. Luethi, N. Shah, Effective equidistribution of primitive rational points in the modular surface (2017, preprint)
18. A. Eskin, C. McMullen, Mixing, counting, and equidistribution in Lie groups. Duke Math. J. **71**(1), 181–209 (1993)
19. H. Furstenberg, The unique ergodicity of the horocycle flow, in *Recent Advances in Topological Dynamics: Proceedings of the Conference on Topological Dynamics*, held at Yale University, 19–23 June 1972, in honor of Professor Gustav Arnold Hedlund on the occasion of his retirement (Springer, Berlin, 1973), pp. 95–115
20. G.A. Hedlund, Fuchsian groups and transitive horocycles. Duke Math. J. **2**(3), 530–542 (1936)
21. H. Iwaniec, E. Kowalski, *Analytic Number Theory*. Colloquium Publications, vol. 53 (American Mathematical Society, Providence, RI, 2004)
22. B. Kalinin, A. Katok, Measurable rigidity and disjointness for $\mathbb{Z}^k$ actions by toral automorphisms. Ergodic Theory Dyn. Syst. **22**(2), 507–523 (2002)
23. E. Lindenstrauss, G. Margulis, Effective estimates on indefinite ternary forms. Isr. J. Math. **203**(1), 445–499 (2014)
24. M. Luethi, Effective equidistribution of primitive rational points in the Hilbert modular surface (2017, preprint)
25. S. Mozes, N. Shah, On the space of ergodic invariant measures of unipotent flows. Ergodic Theory Dyn. Syst. **15**, 149–159 (1995)
26. M. Ratner, On measure rigidity of unipotent subgroups of semisimple groups. Acta Math. **165**, 229–309 (1990)
27. M. Ratner, Strict measure rigidity for unipotent subgroups of solvable groups. Invent. Math. **101**, 449–482 (1990)
28. M. Ratner, On Raghunathan's measure conjecture. Ann. Math. **134**(3), 545–607 (1991)
29. P. Sarnak, Asymptotic behavior of periodic orbits of the horocycle flow and Eisenstein series. Commun. Pure Appl. Math. **34**(6), 719–739 (1981)
30. P. Sarnak, A. Ubis, The horocycle flow at prime times. J. Math. Pures Appl. **103**(2), 575–618 (2015)
31. A. Selberg, On the estimation of Fourier coefficients of modular forms, in *Proceedings of Symposia in Pure Mathematics*, vol. VIII (1965), pp. 1–15
32. C. Series, The modular surface and continued fractions. J. Lond. Math. Soc. **s2-31**(1), 69–80 (1985)
33. S.A. Stepanov, On the number of points of a hyperelliptic curve over a finite prime field. Izv. Akad. Nauk SSSR Ser. Mat. **3**(5), 1103 (1969)
34. J. Tanis, The cohomological equation and invariant distributions for horocycle maps. Ergodic Theory Dyn. Syst. **34**(1), 299–340 (2014)
35. J. Tanis, P. Vishe, Uniform bounds for period integrals and sparse equidistribution. Int. Math. Res. Not. **2015**(24), 13728–13756 (2015)
36. A. Venkatesh, Sparse equidistribution problems, period bounds and subconvexity. Ann. Math. **172**(2), 989–1094 (2010)
37. A. Weil, *Basic Number Theory*. Classics in Mathematics (Springer, Berlin, 1995)

Chapter 11
Sarnak's Conjecture: What's New

Sébastien Ferenczi, Joanna Kułaga-Przymus, and Mariusz Lemańczyk

11.1 Introduction

11.1.1 Möbius Disjointness

Assume that T is a continuous map[1] of a compact metric space X. Following Peter Sarnak [148, 150], we will say that T, or, more precisely, the topological dynamical system (X, T) *is Möbius disjoint* (or *Möbius orthogonal*)[2] if:

$$\lim_{N \to \infty} \frac{1}{N} \sum_{n \leqslant N} f(T^n x) \boldsymbol{\mu}(n) = 0 \text{ for each } f \in C(X) \text{ and } x \in X. \tag{11.1}$$

---

[1] Most often, however not always, T will be a homeomorphism.

[2] $\boldsymbol{\mu}$ stands for the arithmetic Möbius function, see next sections for explanations of notions that appear in Introduction.

S. Ferenczi
Aix-Marseille Université, CNRS, Centrale Marseille, Institut de Mathématiques de Marseille, I2M – UMR 7373, Marseille, France

J. Kułaga-Przymus
Aix-Marseille Université, CNRS, Centrale Marseille, Institut de Mathématiques de Marseille, I2M – UMR 7373, Marseille, France

Faculty of Mathematics and Computer Science, Nicolaus Copernicus University, Toruń, Poland

M. Lemańczyk (✉)
Faculty of Mathematics and Computer Science, Nicolaus Copernicus University, Toruń, Poland
e-mail: mlem@mat.umk.pl

© Springer International Publishing AG, part of Springer Nature 2018
S. Ferenczi et al. (eds.), *Ergodic Theory and Dynamical Systems in their Interactions with Arithmetics and Combinatorics*, Lecture Notes in Mathematics 2213, https://doi.org/10.1007/978-3-319-74908-2_11

In 2010, Sarnak [148, 150] formulated the following conjecture[3]:

> Each zero entropy continuous map T of a compact metric space X is
> Möbius disjoint.   (11.2)

Note that if f is constant then convergence (11.1) takes place in an arbitrary topological system (X, T); indeed, $\frac{1}{N} \sum_{n \leqslant N} \mu(n) \to 0$ is equivalent to the Prime Number Theorem (PNT), e.g. [89, 160]. We can also interpret this statement as the equivalence of the PNT and the Möbius disjointness of the one-point dynamical system. The Prime Number Theorem in arithmetic progressions (Dirichlet's theorem) can also be viewed similarly: it is equivalent to the Möbius disjointness of the system (X, T), where $Tx = x + 1$ on $X = \mathbb{Z}/k\mathbb{Z}$ for each $k \geqslant 1$. Note also that the classical Davenport's [33] estimate: for each $A > 0$, we have

$$\max_{t \in \mathbb{T}} \left| \sum_{n \leqslant N} e^{2\pi i n t} \mu(n) \right| \leqslant C_A \frac{N}{\log^A N} \text{ for some } C_A > 0 \text{ and all } N \geqslant 2, \quad (11.3)$$

yields the Möbius disjointness of irrational rotations.[4]

The present article is concentrated on an overview of research done during the last 7 years[5] on Sarnak's conjecture (11.2) from the ergodic theory point of view. It is also rather aimed at the readers with a good orientation in dynamics, especially in ergodic theory. It means that we assume that the reader is familiar with at least basics of ergodic theory, but often more than that is required, monographs [28, 50, 76, 78, 165] are among best sources to be consulted. In contrast to that, we included in the article a selection of some basics of analytic number theory. Those which appear here, in principle, are not contained in [146] and, as we hope, allow one for a better understanding of dynamical aspects of some number-theoretic results. We should however warn the reader that some number-theoretic results will be presented in their simplified (typically, non-quantitative) forms, sufficient for some ergodic interpretations but not putting across the whole complexity and depth of the results. In particular, this remark applies to recent break-through results of Matomäki and

---

[3] To be compared with Möbius Randomness Law by Iwaniec and Kowalski [99], p. 338, that any "reasonable" sequence of complex numbers is orthogonal to μ.

[4] In order to establish Möbius disjointness, we need to show convergence (11.1) (for all $x \in X$) only for a set of functions linearly dense in $C(X)$, so, for the rotations on the (additive) circle $\mathbb{T} = [0, 1)$, we only need to consider characters. Note also that if the topological system (X, T) is uniquely ergodic then we need to check (11.1) (for all $x \in X$) only for a subset of $C(X)$ which is linearly dense in L^1.

In what follows, for inequalities (as (11.3)), we will also use notation $\ll$ or $O(\cdot)$, or $\ll_A$ or $O_A(\cdot)$ if we need to emphasize a role of $A > 0$.

[5] For a presentation of a part of it, see [37].

Radziwiłł [126] and some related concerning a behavior of multiplicative functions on short intervals.[6]

11.1.2 Ergodic Theory Viewpoint on Sarnak's Conjecture

Sarnak's conjecture (11.2) is formulated as a problem in topological dynamics. However, for each topological system (X, T) the set $M(X, T)$ of (Borel, probability) T-invariant measures is non-empty and we can study dynamical properties of (X, T) by looking at all measure-theoretic dynamical systems $(X, \mathcal{B}, \mu, T)$ for $\mu \in M(X, T)$. Via the Variational Principle, Sarnak's conjecture can be now formulated as Möbius disjointness of the topological systems (X, T) whose measure-theoretic systems $(X, \mathcal{B}, \mu, T)$ for all $\mu \in M(X, T)$ have zero Kolmogorov-Sinai entropy. But one of main motivations for (11.2) in [150] was that this condition is weaker than a certain (open since 1965) pure number-theoretic result, known as the Chowla conjecture (see Sect. 11.4.1). Since the Chowla conjecture has its pure ergodic theory interpretation (Sect. 11.4.1), the approach through invariant measures allows one to see the implication[7]

Chowla conjecture $\Rightarrow$ Sarnak's conjecture

as a consequence of some disjointness (in the sense of Furstenberg) results in ergodic theory. While the Chowla conjecture remains open, some recent break-through results in number theory find their natural interpretation as particular instances of the validity of Sarnak's conjecture. Samples of such results are (see Sects. 11.4.4.1 and 11.4.5):

1. The result of Matomäki, Radziwiłł and Tao [127]:

$$\sum_{h \leqslant H} \left| \sum_{m \leqslant M} \mu(m)\mu(m+h) \right| = \mathrm{o}(HM)$$

(when $H, M \to \infty$, $H \ll M$) implies that each system (X, T) for which all invariant measures yield measure-theoretic systems with discrete spectrum is Möbius disjoint.[8]

---

[6]For a detailed account of these results, we refer the reader to [153].

[7]As proved by Tao [156], the logarithmic averages version of the Chowla conjecture is equivalent to the logarithmic version of Sarnak's conjecture. We will see later in Sect. 11.4 that once the logarithmic Chowla conjecture holds for the Liouville function λ, we have that all configurations of ± 1s appear in λ (infinitely often).

[8]The same argument applied to the Liouville function λ implies that the subshift X_λ generated by λ is uncountable, see Sect. 11.4.

2. The result of Tao [155]:

$$\sum_{n \leqslant N} \frac{\mu(n)\mu(n+h)}{n} = o(\log N)$$

(when $N \to \infty$) for each $h \neq 0$ implies that each system (X, T) for which all invariant measures yield measure-theoretic systems with singular spectrum are logarithmically Möbius disjoint.

This is done by:

- interpreting the number theoretic results as ergodic properties of the dynamical systems given by the invariant measures of the subshift X_μ for which μ is quasi-generic,
- using classical disjointness results in ergodic theory.

It is surprising and important that the ergodic theoretical methods of the last decades that led to new non-conventional ergodic theorems and showed a particular role of nil-systems, also appear in the context of Sarnak's conjecture, and again the role of nil-systems seems to be decisive. Together with some new disjointness results in ergodic theory, it pushes forward significantly our understanding of Möbius disjointness, at least on the level of logarithmic version of Sarnak's conjecture. The most spectacular achievement here is the recent result of Frantzikinakis and Host [71] (see Sect. 11.4.5) who proved that each zero entropy topological system (X, T) with only countably many ergodic measures is logarithmically Möbius disjoint.

The proofs reflect the "local" nature of all the aforementioned results. In other words, regardless the total entropy of the system, to obtain (11.1) for a FIXED $x \in X$ (and all $f \in C(X)$), we only need to look at ergodic properties of the dynamical systems given by measures "produced" by x itself (the limit points of the empiric measures given by x). So, if all such measures yield zero entropy systems, the Chowla conjecture implies (11.1) (for the fixed x and all $f \in C(X)$). When all such measures yield systems with discrete spectrum/singular spectrum/countably many ergodic components then the relevant Möbius disjointness holds (at x). Points with one of the listed properties may appear in (X, T) having positive entropy. In fact, a positive entropy system can be Möbius disjoint [45]. To distinguish between zero and positive entropy systems it is natural to expect that in the zero entropy case the behavior of sums in (11.1) is homogenous in x (for a fixed $f \in C(X)$). Indeed, the uniform convergence (in $x \in X$, under the Chowla conjecture) of sums (11.1) has been proved in [58] (see Sect. 11.5); in fact (11.2) is equivalent to Sarnak's conjecture in its uniform form and also in a uniform short interval form. Moreover (still under the Chowla conjecture), for the Liouville function, no positive entropy system satisfies (11.1) in its uniform short interval form.

The problem of uniform convergence turns out to be closely related to the general question whether Möbius disjointness is stable under our ergodic theory approach. More precisely, suppose that the topological dynamical systems (X, T) and (X', T') are such that the dynamical systems obtained from invariant measures are the same for each of them (up to measure-theoretic isomorphism). Does the Möbius disjointness of (X, T) imply the Möbius disjointness of (X', T')? Although the answer in general seems unknown, in case of uniquely ergodic models of the same measure-theoretic system a satisfactory (positive) answer can be given [58].

11.1.3 Content of the Article

We include the following topics:

- Sarnak's conjecture a.e., Sarnak's conjecture versus Prime Number Theorem in dynamics—see Introduction and Sect. 11.2.
- Brief introduction to multiplicative functions, Prime Number Theorem, Kátai-Bourgain-Sarnak-Ziegler criterion—see Sect. 11.3.
- Results of Matomäki, Radziwiłł and Matomäki, Radziwiłł, Tao on multiplicative functions and some of their ergodic interpretations—see Sect. 11.4.
- Chowla conjecture, logarithmic Chowla and logarithmic Sarnak conjectures (Tao's results and Frantzikinakis and Host's results)—see Sect. 11.4.
- Frantzikinakis' theorem on some consequences of ergodicity of measures for which μ is quasi-generic—see Sect. 11.4.
- Ergodic criterion for Sarnak's conjecture—the AOP and MOMO properties (uniform convergence in (11.1)), Sarnak's conjecture in topological models—see Sect. 11.5.
- Glimpses of results on Sarnak's conjecture: systems of algebraic origin (horo-cycle flows, nilflows); systems of measure-theoretic origin (finite rank systems, distal systems), interval exchange transformations, systems of number-theoretic origin (automatic sequences and related)—see Sect. 11.6.
- Related research: $\mathcal{B}$-free systems, applications to ergodic Ramsey theory—see Sect. 11.7.

11.1.4 Sarnak's Conjecture a.e.

Before we really get into the subject of Sarnak's conjecture, let us emphasize that this is the requirement "for each $f \in C(X)$ and $x \in X$" in (11.1) that makes Sarnak's conjecture deep and difficult to establish. As it has been already noticed in [150], the a.e. version of (11.2) is always true regardless of the entropy assumption:

Proposition 11.1 ([150]) *Let T be an automorphism of a standard Borel probability space $(X, \mathcal{B}, \mu)$ and let $f \in L^1(X, \mathcal{B}, \mu)$. Then, for a.e. $x \in X$, we have*

$$\frac{1}{N} \sum_{n \leqslant N} f(T^n x) \mu(n) \xrightarrow[N \to \infty]{} 0.$$

For a complete proof, see [57]. The main ingredient is the Spectral Theorem which replaces $\left\| \frac{1}{N} \sum_{n \leqslant N} f(T^n x) \mu(n) \right\|_2$ by $\left\| \frac{1}{N} \sum_{n \leqslant N} z^n \mu(n) \right\|_{L^2(\sigma_f)}$,[9] together with Davenport's estimate (11.3) (for $A = 2$) which yields

$$\left\| \frac{1}{N} \sum_{n \leqslant N} f(T^n x) \mu(n) \right\|_2 \ll \frac{1}{\log^2 N}, \quad N \geqslant 2.$$

The latter shows that, for $\rho > 1$, the function $\sum_{k \geqslant 1} \left| \frac{1}{\rho^k} \sum_{n \leqslant \rho^k} f(T^n \cdot) \mu(n) \right|$ is in $L^2(X, \mu)$, which, letting $\rho \to 1$ allows one to conclude for $f \in L^\infty(X, \mu)$. The general case $f \in L^1(X, \mu)$ follows from the pointwise ergodic theorem.

As shown in [51], a use of Davenport's type estimate proved in [83] for the nil-case, yields a polynomial version of Proposition 11.1. See also [29] for the pointwise ergodic theorem for other arithmetic weights.

11.2   From a PNT in Dynamics to Sarnak's Conjecture

The content of this section can be viewed as a kind of motivation for Sarnak's conjecture (and is written on the base of Tao's post [157] and Sarnak's lecture given at CIRM [149]).

We denote by $\mathbb{N} := \{1, 2, \ldots\}$ the set of positive integers. Given $N \in \mathbb{N}$, we let $\pi(N) := \{p \leqslant N : p \in \mathbb{P}\}$. The classical Prime Number Theorem states that

$$\lim_{N \to \infty} \frac{\pi(N)}{N / \log N} = 1. \tag{11.4}$$

We will always refer to this theorem as the (classical) PNT.

Assume that T is a continuous map of a compact metric space X. Assume moreover that (X, T) is uniquely ergodic, that is, the set $M(X, T)$ of T-invariant probability Borel measures is reduced to one measure, say μ. By unique ergodicity, the ergodic averages go to zero (even uniformly) for zero mean continuous functions:

$$\frac{1}{N} \sum_{n \leqslant N} f(T^n x) \xrightarrow[N \to \infty]{} 0$$

for each $f \in C(X)$, $\int_X f \, d\mu = 0$, and $x \in X$. Hence, the statement that *a PNT holds in (X, T)* "should" mean

$$\lim_{N \to \infty} \frac{1}{\pi(N)} \sum_{\mathbb{P} \ni p \leqslant N} f(T^p x) = 0 \tag{11.5}$$

---

[9]σ_f stands for the spectral measure of f.

for all zero mean $f \in C(X)$ and $x \in X$ (in what follows, instead of $\sum_{\mathbb{P} \ni p}$, we write simply $\sum_p$ if no confusion arises).[10] Let us see how to arrive at (11.5) differently.

Recall that the von Mangoldt function Λ is defined by $\Lambda(n) = \log p$ if $n = p^k$ for a prime number p (and $k \geqslant 1$) and $\Lambda(n) = 0$ otherwise. Contrary to most of arithmetic functions considered in this article, Λ is not multiplicative. It is not bounded either and its support has zero density. The (classical) PNT is equivalent to

$$\frac{1}{N} \sum_{n \leqslant N} \Lambda(n) \xrightarrow[N \to \infty]{} 1.$$

A given sequence $(a_n) \subset \mathbb{C}$ can be said *to satisfy a PNT* whenever we can give an asymptotic estimate on

$$\sum_{n \leqslant N} a_n \Lambda(n)$$

when N tends to infinity; thus the classical PNT is a PNT for the sequence $a_n = 1$. In particular, a sequence (a_n) also satisfies a PNT if

$$\sum_{n \leqslant N} a_n \Lambda(n) = \sum_{n \leqslant N} a_n + o(N), \tag{11.6}$$

and, if additionally (a_n) has zero mean, i.e. if $\frac{1}{N} \sum_{n \leqslant N} a_n \xrightarrow[N \to \infty]{} 0$, then (a_n) satisfies a PNT if

$$\frac{1}{N} \sum_{n \leqslant N} a_n \Lambda(n) \xrightarrow[N \to \infty]{} 0. \tag{11.7}$$

An interesting special case is $a_n = (-1)^n$, which has zero mean. Here, we do have estimates of the sums of $\Lambda(n)$ over the odd numbers smaller than N, but they are of the order of N, thus (11.7) is not satisfied. Beyond this point, we will not be studying such particular cases and we shall always write that *the sequence (a_n) satisfies a PNT* whenever (11.6) holds.

Zero mean sequences are easily "produced" in uniquely ergodic systems. We will say that a *uniquely ergodic topological dynamical system (X, T) satisfies a PNT* if

$$\frac{1}{N} \sum_{n \leqslant N} f(T^n x) \Lambda(n) \xrightarrow[N \to \infty]{} 0 \tag{11.8}$$

---

[10]We recall that Bourgain in [16–18], proved that for each $\alpha \geqslant (1 + \sqrt{3})/2$, each automorphism T of a probability standard Borel space $(X, \mathcal{B}, \mu)$ and each $f \in L^\alpha(X, \mathcal{B}, \mu)$ the sums in (11.5) converge for a.e. $x \in X$. The result has been extended by Wierdl in [170] for all $\alpha > 1$.

for all zero mean $f \in C(X)$ and $x \in X$. We have

$$\frac{1}{N} \sum_{n \leqslant N} f(T^n x) \Lambda(n) = \frac{1}{N} \sum_{p \leqslant N} f(T^p x) \log p + \frac{1}{N} \sum_{p^k \leqslant N, k \geqslant 2} f(T^{p^k} x) \log p.$$

Now, in the second sum if $p^k \leqslant N$ then $p \in [1, \sqrt{N}]$; the largest value of $\log p$ is bounded by $\frac{1}{2} \log N$, therefore, the second sum is of order $O(\sqrt{N} \cdot \log N / N)$, hence of order $N^{-\frac{1}{2}+\varepsilon}$ for each $\varepsilon > 0$. Thus, a PNT in (X, T) means that

$$\frac{1}{N} \sum_{p \leqslant N} f(T^p x) \log p \xrightarrow[N \to \infty]{} 0 \tag{11.9}$$

for all zero mean $f \in C(X)$ and $x \in X$. Note that by the classical PNT to prove (11.9), we need to show it for a linearly dense set of functions.[11]

Let us now write

$$\frac{1}{N} \sum_{p \leqslant N} f(T^p x) \log p = \frac{1}{N} \sum_{p \leqslant N / \log N} f(T^p x) \log p + \frac{1}{N} \sum_{N / \log N \leqslant p \leqslant N} f(T^p x) \log p.$$

We have $\frac{1}{N} \sum_{p \leqslant N / \log N} f(T^p x) \log p = O(1 / \log N)$ (by $\frac{1}{M} \sum_{p \leqslant M} \log p \to 1$ when $M \to \infty$). Moreover, write $f = f_+ - f_-$ and then we have

$$\frac{\log N - \log \log N}{N} \sum_{N / \log N \leqslant p \leqslant N} f_+(T^p x)$$

$$\leqslant \frac{1}{N} \sum_{N / \log N \leqslant p \leqslant N} f_+(T^p x) \log p \leqslant \frac{\log N}{N} \sum_{N / \log N \leqslant p \leqslant N} f_+(T^p x)$$

as $\log N - \log \log N \leqslant \log p \leqslant \log N$ for the p in the considered interval. Now, $\pi(N) / (N / (\log N - \log \log N)) \xrightarrow[N \to \infty]{} 1$ and $\pi(N) / (N / \log N) \xrightarrow[N \to \infty]{} 1$, whence

$$\left| \frac{1}{N} \sum_{p \leqslant N} f_+(T^p x) \log p - \frac{1}{\pi(N)} \sum_{p \leqslant N} f_+(T^p x) \right| \xrightarrow[N \to \infty]{} 0.$$

---

[11] Indeed, we have

$$\left| \frac{1}{N} \sum_{p \leqslant N} f(T^p x) \log p - \frac{1}{N} \sum_{p \leqslant N} g(T^p x) \log p \right|$$

$$\leqslant \frac{1}{N} \sum_{p \leqslant N} |f(T^p x) - g(T^p x)| \log p \leqslant \|f - g\| \frac{1}{N} \sum_{p \leqslant N} \log p = O(\|f - g\|),$$

as condition $\frac{1}{N} \sum_{n \leqslant N} \Lambda(n) \xrightarrow[N \to \infty]{} 1$ is equivalent to $\frac{1}{N} \sum_{p \leqslant N} \log p \xrightarrow[N \to \infty]{} 1$.

Repeating the same reasoning with f_+ replaced by f_- and by (11.9), we obtain that the statement *a PNT holds in* (X, T) is equivalent to (11.5) for all zero mean $f \in C(X)$ and $x \in X$.

Remark 11.2 By replacing Λ in (11.8) by μ, we come back to Sarnak's conjecture. The identity $\Lambda = \mu * \log$ (see (11.10) below), i.e. $\Lambda(n) = \sum_{d|n} \mu(d) \log(n/d) = -\sum_{d|n} \mu(d) \log d$ suggests some other connections between the simultaneous validity of a PNT and Möbius disjointness in (X, T) but no rigorous theorem toward a formal equivalence of the two conditions has been proved. Actually, such an equivalence taken literally does not hold. Indeed, the fact that the support of Λ is of zero upper Banach density makes a PNT vulnerable under zero density replacements of the observable $(f(T^n x))$. On the other hand, Möbius orthogonality is stable under such replacements. We illustrate this using the following simple example.

Consider the classical case $a_n = 1$ for all $n \in \mathbb{N}$. This is the same as to consider a PNT in a uniquely ergodic model[12] of the one-point system. One can now ask if we have a PNT in all uniquely ergodic models of the one-point system (it is an exercise to prove that all such models are Möbius disjoint). Take any sequence $(c_{p^k})_{p^k} \in \{-1, 1\}^{\mathbb{N}}$ and define b_n as a_n when $n \neq p^k$ and $b_{p^k} = c_{p^k}$. We can see that

$$\frac{1}{N} \sum_{n \leqslant N} b_n \Lambda(n) = \frac{1}{N} \sum_{p^k \leqslant N} c_{p^k} \log p.$$

Now, the subshift $X_b \subset \{-1, 1\}^{\mathbb{N}}$ generated by b (cf. (11.27)) has only one invariant measure $\delta_{11...}$, so it is a uniquely ergodic model of the one-point system and if we take $f(z) = 1 - z(1)$ ($z \in X_b$) as our continuous function, we can see that f has zero mean but neither (11.8) nor (11.5) are satisfied if the sequence c is badly behaving. It follows that we can expect a PNT to hold only in some classes of "natural" dynamical systems, samples of which we will see in Sect. 11.6.

Returning to our discussion on a PNT, in any such situation, given a bounded sequence $(f(n)) \subset \mathbb{C}$, we can write

$$\sum_{n \leqslant N} f(n) \Lambda(n) = -\sum_{n \leqslant N} f(n) \sum_{d|n} \mu(d) \log(d) = -\sum_{d \leqslant N} \mu(d) \log d \sum_{e \leqslant N/d} f(ed).$$

Then a further decomposition of the second sum into a structured part and a remainder leads to two sums and allows one for an application of Möbius Randomness Law to the second sum in order to predict the correct main term value of $\sum_{n \leqslant N} f(n) \Lambda(n)$, see [149].

---

[12]We recall that if $(Z, \mathcal{D}, \kappa, R)$ is a measure-preserving system then by its *uniquely ergodic model* we mean a uniquely ergodic system (X, T) with the unique (Borel) T-invariant measure μ such that $(Z, \mathcal{D}, \kappa, R)$ is measure-theoretically isomorphic to $(X, \mathcal{B}(X), \mu, T)$.

11.3 Multiplicative Functions

11.3.1 Definition and Examples

An arithmetic function $u : \mathbb{N} \to \mathbb{C}$ is called *multiplicative* if $u(1) = 1$ and $u(mn) = u(m)u(n)$ whenever $(m, n) = 1$. If $u(mn) = u(m)u(n)$ without the coprimeness restriction on m, n, then u is called *completely multiplicative*. Clearly, each multiplicative function is entirely determined by its values at p^α, where $p \in \mathbb{P}$ is a prime number and $\alpha \in \mathbb{N}$ (for completely multiplicative functions $\alpha = 1$). A prominent example of a multiplicative function is the Möbius function μ determined by $\mu(p) = -1$ and $\mu(p^\alpha) = 0$ for $\alpha \geq 2$. Note that μ^2 (which is obviously also multiplicative) is the characteristic function of the set of square-free numbers. The Liouville function $\lambda : \mathbb{N} \to \mathbb{C}$ is completely multiplicative and is given by $\lambda(p) = -1$. Clearly, $\mu = \lambda \cdot \mu^2$ and we will see soon some more relations between μ and λ. Many other classical arithmetic functions are multiplicative, for example: the Euler function ϕ; the function $n \mapsto (-1)^{n+1}$ is a periodic multiplicative function which is not completely multiplicative; $d(n) :=$ number of divisors of n, $n \mapsto 2^{\omega(n)}$, where $\omega(n)$ stands for the number of different prime divisors of n; $\sigma(n) = \sum_{d|n} d$. Recall that given $q \geq 1$, a function $\chi : \mathbb{N} \to \mathbb{C}$ is called a *Dirichlet character of modulus q* if:

1. χ is q-periodic and completely multiplicative,
2. $\chi(n) \neq 0$ if and only if $(n, q) = 1$.

It is not hard to see that Dirichlet characters are determined by the ordinary characters of the multiplicative group (of order $\phi(q)$) $(\mathbb{Z}/q\mathbb{Z})^*$ of invertible (under multiplication) elements in $\mathbb{Z}/q\mathbb{Z}$. The Dirichlet character $\chi_1(n) := 1$ iff $(n, q) = 1$ is called the *principal character* of modulus q. Moreover, each periodic, completely multiplicative function is a Dirichlet character (of a certain modulus). Another class of important (completely) multiplicative functions is given by Archimedean characters $n \mapsto n^{it} = e^{it \log n}$ which are indexed by $t \in \mathbb{R}$.

11.3.2 Dirichlet Convolution, Euler's Product

Recall that given two arithmetic functions $u, v : \mathbb{N} \to \mathbb{C}$, by their *Dirichlet convolution* $u * v$ we mean the arithmetic function

$$u * v(n) := \sum_{d|n} u(d)v(n/d), \ n \in \mathbb{N}. \tag{11.10}$$

If by A we denote the set of arithmetic functions then $(A, +, *)$ is a ring which is an integral domain and the unit $e \in A$ is given by $\mathbb{1}_{\{1\}}$.[13] There is a natural ring isomorphism between A and the ring D of (formal)[14] Dirichlet series

$$A \ni \boldsymbol{u} \mapsto U(s) := \sum_{n=1}^{\infty} \frac{\boldsymbol{u}(n)}{n^s} \in D, \ s \in \mathbb{C},$$

under which

$$U(s)V(s) = \sum_{n=1}^{\infty} \frac{\boldsymbol{u} * \boldsymbol{v}(n)}{n^s}.$$

When $\boldsymbol{u} = \mathbb{1}_{\mathbb{N}}$ then the Dirichlet series defines the Riemann ζ function[15]:

$$\zeta(s) = \sum_{n=1}^{\infty} \frac{1}{n^s} \text{ for } \mathrm{Re}\, s > 1.$$

It is classical that if $\boldsymbol{u}$ and $\boldsymbol{v}$ are multiplicative then so is their Dirichlet convolution. The importance of multiplicativity can be seen in the representation of the Dirichlet series of a multiplicative function $\boldsymbol{u}$ as an Euler's product. Indeed, a general term of $\prod_{p \in \mathbb{P}}(1 + \boldsymbol{u}(p)p^{-s} + \boldsymbol{u}(p^2)p^{-2s} + \ldots)$ has the form $\frac{\boldsymbol{u}(p_{i_1}^{\alpha_1}) \cdot \ldots \cdot \boldsymbol{u}(p_{i_r}^{\alpha_r})}{(p_{i_1}^{\alpha_1} \cdot \ldots \cdot p_{i_r}^{\alpha_r})^s} = \frac{\boldsymbol{u}(p_{i_1}^{\alpha_1} \cdot \ldots \cdot p_{i_r}^{\alpha_r})}{(p_{i_1}^{\alpha_1} \cdot \ldots \cdot p_{i_r}^{\alpha_r})^s}$, i.e. equals $\frac{\boldsymbol{u}(n)}{n^s}$ for some n. It easily follows that

$$\sum_{n \geq 1} \frac{\boldsymbol{u}(n)}{n^s} = \prod_{p \in \mathbb{P}}(1 + \boldsymbol{u}(p)p^{-s} + \boldsymbol{u}(p^2)p^{-2s} + \ldots).$$

If additionally $\boldsymbol{u}$ is completely multiplicative (and $|\boldsymbol{u}| \leq 1$), then $\boldsymbol{u}(p^k) = \boldsymbol{u}(p)^k$ and

$$\sum_{n=1}^{\infty} \frac{\boldsymbol{u}(n)}{n^s} = \prod_{p \in \mathbb{P}}(1 - \boldsymbol{u}(p)p^{-s})^{-1}.$$

---

[13] The Möbius Inversion Formula is given by $\mu * \mathbb{1}_{\mathbb{N}} = e$.

[14] We will not discuss here the problem of convergence of Dirichlet series, see [146].

[15] An analytic continuation of ζ yields a meromorphic function on $\mathbb{C}$ (with one pole at $s = 1$) satisfying the functional equation

$$\zeta(s) = 2^s \pi^{s-1} \sin\left(\frac{\pi s}{2}\right) \Gamma(1-s)\zeta(1-s). \tag{11.11}$$

Because of the sine, $\zeta(-2k) = 0$ for all integers $k \geq 1$—these are so called trivial zeros of ζ ($\zeta(2k) \neq 0$ since Γ has simple poles at $0, -1, -2, \ldots$). In $\mathrm{Re}\, s > 1$ there are no zeros of ζ (ζ is represented by a convergent infinite product), so except of $-2k$, $k \geq 1$, there are no zeros for $s \in \mathbb{C}$, $\mathrm{Re}\, s < 0$ (as $\mathrm{Re}(1-s) > 1$). The Riemann Hypothesis asserts that all nontrivial zeros of ζ are on the line $x = \frac{1}{2}$. See [146].

Note that if $u = \mu$, we obtain

$$\sum_{n \geqslant 1} \frac{\mu(n)}{n^s} = \prod_{p \in \mathbb{P}} (1 - p^{-s})$$

since $\mu(p) = -1$ and $\mu(p^r) = 0$ whenever $r \geqslant 2$. Since for the Riemann ζ function, we have $\zeta(s) = \prod_{p \in \mathbb{P}} (1 - p^{-s})^{-1}$ for $\operatorname{Re} s > 1$, we obtain the following.

Corollary 11.3 *We have* $\frac{1}{\zeta(s)} = \sum_{n \geqslant 1} \frac{\mu(n)}{n^s}$ *whenever* $\operatorname{Re} s > 1$.

We could have derived the above assertion in a different way. Indeed, $\mu * \mathbb{1}_{\mathbb{N}} = e$. If $G(s) := \sum_{n=1}^{\infty} \frac{\mu(n)}{n^s}$ stands for the Dirichlet series of the Möbius function, then

$$G(s) \cdot \zeta(s) = \sum_{n=1}^{\infty} \frac{(\mu * \mathbb{1}_{\mathbb{N}})(n)}{n^s} = \sum_{n=1}^{\infty} \frac{e(n)}{n^s} = 1.$$

11.3.3 Distance Between Multiplicative Functions

Denote by

$$\mathcal{M} := \{u \colon \mathbb{N} \to \mathbb{C} : u \text{ is multiplicative and } |u| \leqslant 1\}. \tag{11.12}$$

Let $u, v \in \mathcal{M}$. Define the "distance" function D on $\mathcal{M}$ by setting

$$D(u, v) := \left(\sum_{p \in \mathbb{P}} \frac{1}{p} \left(1 - \operatorname{Re} \left(u(p) \overline{v(p)} \right) \right) \right)^{1/2}. \tag{11.13}$$

For each $u, v, w \in \mathcal{M}$, we have:

- $D(u, u) \geqslant 0$; $D(u, u) = 0$ iff $\sum_{p \in \mathbb{P}} \frac{1}{p}(1 - |u(p)|^2) = 0$ iff $|u(p)| = 1$ for all $p \in \mathbb{P}$, so $D(n^{it}, n^{it}) = 0$ for each $t \in \mathbb{R}$, $D(\lambda, \lambda) = D(\mu, \mu) = 0$. Of course, if $u(p) = 0$ for each $p \in \mathbb{P}$ then $D(u, u) = +\infty$. Moreover, $\phi(n)/n \in \mathcal{M}$ and $D(\phi(n)/n, \phi(n)/n) = \sum_{p \in \mathbb{P}} \frac{1}{p}(1 - \frac{(1-p)^2}{p^2})$ is positive and finite.
- $D(u, v) = D(v, u)$.
- $D(u, v) \leqslant D(u, w) + D(w, v)$, see [81].

When $D(u, v) < +\infty$ then one says that u *pretends to be* v. For example, μ^2 and $\phi(n)/n$ pretend to be $\mathbb{1}$ (as $\sum_{p \in \mathbb{P}} \frac{1}{p}(1 - \frac{p-1}{p}) = \sum_{p \in \mathbb{P}} \frac{1}{p^2} < +\infty$).

Lemma 11.4 ([81]) *For each $u, v, w, w' \in \mathcal{M}$, we have*

(i) $D(uw, vw') \leqslant D(u, v) + D(w, w')$.

Moreover, by (i) and a simple induction,

(ii) $mD(\boldsymbol{u}, \boldsymbol{v}) \geqslant D(\boldsymbol{u}^m, \boldsymbol{v}^m)$ *for all* $m \in \mathbb{N}$.

If we fix $t \neq 0$ and $k \geqslant k_0$ then the number of $p \in \mathbb{P}$ satisfying

$$\exp\left(\frac{2\pi}{t}(k + \frac{1}{3})\right) \leqslant p \leqslant \exp\left(\frac{2\pi}{t}(k + \frac{2}{3})\right)$$

is (by the PNT) at least $C\frac{\exp(2\pi k/t)}{k/t}$ (for a constant $C > 0$), whence

$$\left|\left\{p \in \mathbb{P} : k + \frac{1}{3} \leqslant \frac{t \log p}{2\pi} \leqslant k + \frac{2}{3}\right\}\right| \geqslant C\frac{\exp(2\pi k/t)}{k/t}.$$

It follows that

$$\sum_{\exp(\frac{2\pi}{t}(k+\frac{1}{3}))\leqslant p\leqslant\exp(\frac{2\pi}{t}(k+\frac{2}{3}))} \frac{1}{p}(1 - \cos(t \log p)) \geqslant C'\frac{1}{k} \tag{11.14}$$

for a constant $C' > 0$. Now, using (11.13), (11.14) and summing over k, we obtain the following[16]:

$$D(\mathbb{1}, n^{it}) = \infty \text{ for each } t \neq 0. \tag{11.15}$$

It is not difficult to see that for $t \neq 0$, $D(\chi, n^{it}) = +\infty$ for each Dirichlet character χ, while for $t = 0$, we have $D(\chi, 1) < +\infty$ if and only if χ is principal.

11.3.4 Mean of a Multiplicative Function: The Prime Number Theorem (PNT)

The distance D is useful when we want to compute means of multiplicative functions. Given an arithmetic function $\boldsymbol{u} \colon \mathbb{N} \to \mathbb{C}$, its *mean* $M(\boldsymbol{u})$ is defined as $M(\boldsymbol{u}) := \lim_{N\to\infty} \frac{1}{N} \sum_{n\leqslant N} \boldsymbol{u}(n)$ (if the limit exists).

Theorem 11.5 (Halász; e.g. Thm. 6.3 [60]) *Let* $\boldsymbol{u} \in \mathcal{M}$. *Then* $M(\boldsymbol{u})$ *exists and is non-zero if and only if*

(i) *there is at least one positive integer* k *so that* $\boldsymbol{u}(2^k) \neq -1$, *and*
(ii) *the series* $\sum_{p\in\mathbb{P}} \frac{1}{p}(1 - \boldsymbol{u}(p))$ *converges.*

---

[16]This proof of (11.15) has been shown to us by G. Tenenbaum.

When these conditions are satisfied, we have

$$M(\pmb{u}) = \prod_{p \in \mathbb{P}} \left(1 - \frac{1}{p}\right) \left(1 + \sum_{m=1}^{\infty} p^{-m} \pmb{u}(p^m)\right).$$

The mean value $M(\pmb{u})$ exists and is zero if and only if either

(iii) *there is a real number τ, so that for each positive integer k, $\pmb{u}(2^k) = -2^{ki\tau}$, moreover $D(\pmb{u}, n^{i\tau}) < +\infty$; or*

(iv) *$D(\pmb{u}, n^{it}) = \infty$ for each $t \in \mathbb{R}$.*

Corollary 11.6 (Wirsing's Theorem) *If $\pmb{u} \in \mathcal{M}$ is real-valued then $M(\pmb{u})$ exists.*

Proof Since $\mathrm{Re}(p^{it}) = \mathrm{Re}(p^{-it})$, and $\pmb{u}(p) \in \mathbb{R}$, we have

$$D(\mathbb{1}, n^{2it}) = D(n^{-it}, n^{it}) \leqslant 2D(\pmb{u}, n^{it})$$

by the triangle inequality. By (11.15), it follows that $D(\pmb{u}, n^{it}) = +\infty$ for each $0 \neq t \in \mathbb{R}$. Hence, if $D(\pmb{u}, \mathbb{1}) = +\infty$, then $D(\pmb{u}, n^{it}) = +\infty$ for each $t \in \mathbb{R}$ and then $M(\pmb{u}) = 0$ by Halász's theorem (iv).

If not then $D(\pmb{u}, \mathbb{1}) < +\infty$. Then the series $\sum_{p \in \mathbb{P}} \frac{1}{p}(1 - \pmb{u}(p))$ converges (so (ii) is satisfied) and we check whether or not $\pmb{u}(2^k) = -1$ for all $k \in \mathbb{N}$, that is, either (i) holds or (iii) holds.

Remark 11.7 It follows from (11.15) that in Halász's theorem (iii) and (iv) are two disjoint conditions.

Remark 11.8 Not all functions from $\mathcal{M}$ have mean. Indeed, an Archimedean character n^{it} has mean iff $t = 0$. This can be shown by a direct computation: apply Euler's summation formula to $f(x) = x^{it}$ with $t \neq 0$, to obtain $\frac{1}{N} \sum_{n \leqslant N} n^{it} = \frac{N^{it}}{it+1} + O\left(\frac{\log N}{N}\right)$.

Theorem 11.9 (e.g. [81, 89, 160]) *The PNT is equivalent to $M(\pmb{\mu}) = 0$.*

Remark 11.10 The statement above is an elementary equivalence, see the discussion in Section 4 [42]. For a PNT for a more general f (i.e. not for $f = 1$) the relation between such a disjointness and sums over the primes requires more quantitative estimates than simply o(N).

Remark 11.11 By Halász's theorem, condition $M(\pmb{\mu}) = 0$ is equivalent to $D(\pmb{\mu}, n^{it}) = \infty$ for each $t \in \mathbb{R}$ ($\pmb{\mu}$ does not pretend to be n^{it}), and this can be established similarly to the proof of (11.15).

The PNT tells us about cancelations of $+1$ and -1 for $\pmb{\mu}$. When one requires a behavior similar to random sequences, say "square-root type cancelation", the result is much stronger:

Theorem 11.12 (Littlewood, see [27]) *The Riemann Hypothesis holds if and only if for every $\varepsilon > 0$, we have $\sum_{n \leqslant N} \pmb{\mu}(n) = O_{\varepsilon}(N^{\frac{1}{2}+\varepsilon})$.*

This result is not hard to establish and we show the sufficiency: By Corollary 11.3, we have

$$\frac{1}{\zeta(s)} = \sum_{n=1}^{\infty} \frac{\mu(n)}{n^s} = -\sum_{n=1}^{\infty} \mu(n) \int_{n}^{\infty} dx^{-s} = s \sum_{n=1}^{\infty} \mu(n) \int_{n}^{\infty} \frac{dx}{x^{s+1}}.$$

Setting $\mathbf{M}(x) = \sum_{n \leqslant x} \mu(n)$, we obtain

$$\frac{1}{\zeta(s)} = s \int_{1}^{\infty} \frac{\mathbf{M}(x)}{x^{s+1}} dx, \quad \text{Re } s > 1 \tag{11.16}$$

and, by the assumption on $\mathbf{M}(\cdot)$,

$$\int_{1}^{\infty} \left| \frac{\mathbf{M}(x)}{x^{s+1}} \right| dx = \int_{1}^{\infty} \frac{|\mathbf{M}(x)|}{x^{\text{Re } s+1}} dx \ll \int_{1}^{\infty} x^{\frac{1}{2}+\varepsilon-(\text{Re } s+1)} dx = \int_{1}^{\infty} x^{-\text{Re } s - \frac{1}{2}+\varepsilon} dx.$$

It follows that the integral on the RHS of (11.16) is absolutely convergent for Re $s > \frac{1}{2} + \varepsilon$. Hence, (11.16) yields an analytic extension of $\frac{1}{\zeta(\cdot)}$ to $\{s \in \mathbb{C} : \text{Re } s > \frac{1}{2} + \varepsilon\}$. In this domain there are no zeros of ζ and by the functional equation (see (11.11)) on ζ, we obtain the Riemann Hypothesis.

11.3.5 Aperiodic Multiplicative Functions

Denote by

$$\mathbf{M}_{\text{conv}} := \{\boldsymbol{u} \in \mathbf{M} : \lim_{N \to \infty} \frac{1}{N} \sum_{n \leqslant N} \boldsymbol{u}(an + r) \text{ exists for all } a, r \in \mathbb{N}\}.$$

The following is classical.

Lemma 11.13 *Let* $\boldsymbol{u} \in \mathbf{M}$. *Then* $\boldsymbol{u} \in \mathbf{M}_{\text{conv}}$ *if and only if the mean value* $M(\chi \cdot \boldsymbol{u})$ *exists for each Dirichlet character* χ.

An arithmetic function $\boldsymbol{u} : \mathbb{N} \to \mathbb{C}$ is called *aperiodic* if, for all $a, r \in \mathbb{N}$, we have $\lim_{N \to \infty} \frac{1}{N} \sum_{n \leqslant N} \boldsymbol{u}(an + r) = 0$. Similarly to Lemma 11.13, we obtain that $\boldsymbol{u} \in \mathbf{M}$ is aperiodic if and only if $M(\chi \cdot \boldsymbol{u}) = 0$ for each Dirichlet character χ. Delange theorem (see, e.g., [81]) gives necessary and sufficient conditions for $\boldsymbol{u}$ to be aperiodic. In particular, each $\boldsymbol{u} \in \mathbf{M}$ satisfying $D(\boldsymbol{u}, \chi \cdot n^{it}) = 0$ for all Dirichlet characters χ and all $t \in \mathbb{R}$, is aperiodic. Classical multiplicative functions as μ or λ are aperiodic.

Frantzikinakis and Host in [70] prove a deep structure theorem for multiplicative functions from $\mathbf{M}$. One of the consequences of it is the following characterization of aperiodic functions: $\boldsymbol{u} \in \mathbf{M}$ is aperiodic if and only if it is uniform, that is, all

Gowers uniformity seminorms[17] vanish [70]. In [12] (see Theorem 1.3 therein), this result is extended to show that $u \in M_{conv}$ is either uniform or rational.[18] Also, a variation of this result has been proved in [12] (see Theorem A therein):

for each positive density level set $E = \{n \in \mathbb{N} : u(n) = c\}$ of $u \in M$ there is a (unique if density is smaller than 1) rational (i.e. coming from a rational function from M) level set R of $v \in M$ such that $d(R)\mathbb{1}_E - d(E)\mathbb{1}_R$ is Gowers uniform.   (11.17)

For example, for $E = \{n \in \mathbb{N} : \mu(n) = 1\}$ the unique set R is just the set of square-free numbers.

11.3.6   Davenport Type Estimates on Short Intervals

Given $u \in M$, for our purposes we will need additionally the following[19]: for each $(b_n) \subset \mathbb{N}$ with $b_{n+1} - b_n \to \infty$ and any $c \in \mathbb{C}$, $|c| = 1$, we have

$$\frac{1}{b_{K+1}} \sum_{k \leqslant K} \left| \sum_{b_k \leqslant n < b_{k+1}} c^n u(n) \right| \xrightarrow[K \to \infty]{} 0.   \qquad (11.18)$$

It is not hard to see that if $u \in M$ satisfies (11.18) for each (b_n) and c as above, then it must be aperiodic.

---

[17]For $N \in \mathbb{N}$ we write $[N]$ for the set $\{1, 2, \ldots, N\}$. Given $h, N \in \mathbb{N}$ and $f : \mathbb{N} \to \mathbb{C}$, we let $S_h f(n) = f(n + h)$ and $f_N = \mathbb{1}_{[N]} \cdot f$. For $s \in \mathbb{N}$, the *Gowers uniformity seminorm* [80] $\| . \|_{U_{[N]}^s}$ is defined in the following way:

$$\|f\|_{U_{[N]}^1} := \left| \frac{1}{N} \sum_{n=1}^{N} f_N(n) \right|$$

and for $s \geqslant 1$

$$\|f\|_{U_{[N]}^{s+1}}^{2^{s+1}} := \frac{1}{N} \sum_{h=1}^{N} \|f_N S_h \overline{f_N}\|_{U_{[N]}^s}^{2^s}.$$

A bounded function $f : \mathbb{N} \to \mathbb{C}$ is called *uniform* if $\|f\|_{U_{[N]}^s}$ converges to zero as $N \to \infty$ for each $s \geqslant 1$.

[18]An arithmetic function u is *rational* if for each $\varepsilon > 0$ there is a periodic function v such that $\limsup_{N \to \infty} \frac{1}{N} \sum_{n \leqslant N} |u(n) - v(n)| < \varepsilon$. Note that since μ is aperiodic, whence orthogonal to all periodic sequences, it will also be orthogonal to each rational u [12]. An example of rational sequence is given by μ^2. For more examples, see the sets of $\mathcal{B}$-free numbers in the Erdös case in Sect. 11.7.

[19]To be compared with the estimates (11.3), where we drop the sup requirement.

In fact, it follows from a break-through result in [126] and [127] that the class of $u \in \mathcal{M}$ for which (11.18) holds contains all u for which

$$\inf_{|t| \leqslant M, \chi \bmod q, q \leqslant Q} D(u, n \mapsto \chi(n) n^{it}; M)^2 \to \infty, \tag{11.19}$$

when $10 \leqslant H \leqslant M$, $H \to \infty$ and $Q = \min(\log^{1/125} M, \log^5 H)$; here χ runs over all Dirichlet characters of modulus $q \leqslant Q$ and

$$D(u, v; M) := \left(\sum_{p \leqslant M, p \in \mathbb{P}} \frac{1 - \mathrm{Re}(u(p)\overline{v(p)})}{p} \right)^{1/2}$$

for each $u, v \in \mathcal{M}$. Moreover, classical multiplicative functions like μ and λ satisfy (11.19), see [127].

Finally, note that (11.18) true for all (b_n) as above is equivalent to the following statement:

$$\frac{1}{M} \sum_{M \leqslant m < 2M} \left| \sum_{m \leqslant h < m+H} c^h u(h) \right| \xrightarrow[M, H \to \infty, H = o(M)]{} 0 \tag{11.20}$$

(we can also replace the first sum by $\sum_{1 \leqslant m < M}$), see [59] for details. This statement is much closer to the original formulations of (simplified versions of) theorems from [126, 127].

One more consequence of the main result in [126] is the following:

Theorem 11.14 (Thm. 1.1 in [127] and a Corollary for $k = 2$ Therein) *For $H \to \infty$ arbitrarily slowly with $M \to \infty$ ($H \leqslant M$), we have*

$$\sum_{h \leqslant H} \left| \sum_{m \leqslant M} \mu(m) \mu(m + h) \right| = o(HM).$$

11.3.7   The KBSZ Criterion

Sarnak's conjecture is aimed at showing that deterministic sequences (i.e. those given as observable sequences in the zero entropy systems) are orthogonal to μ. In particular, as μ is a multiplicative function, the result[20] below establishes disjointness with μ.

---

[20]The main ideas for this result appeared in [30] and [134]. It was first established in a slightly different form in [106] and then in [21], see also [88] for a proof. The criterion has its origin in the bilinear method of Vinogradov [164] which is a technique to study sums of a over primes in terms of sums over progressions $\sum_{n \leqslant N} a_{dn}$ and sums $\sum_{n \leqslant N} a_{d_1 n} a_{d_2 n}$. If $a_n = f(T^n x)$ then these sums are Birkhoff sums for powers of T and their joinings.

In what follows we will refer to Theorem 11.15 as to the KBSZ criterion.

Theorem 11.15 ([21, 106]) *Assume that (a_n) is a bounded sequence of complex numbers. Assume that for all prime numbers $p \neq q$*

$$\frac{1}{N} \sum_{n \leqslant N} a_{pn} \bar{a}_{qn} \xrightarrow[N \to \infty]{} 0. \tag{11.21}$$

Then, for each multiplicative function $\boldsymbol{u} \in \boldsymbol{M}$, we have

$$\frac{1}{N} \sum_{n \leqslant N} a_n \boldsymbol{u}(n) \xrightarrow[N \to \infty]{} 0. \tag{11.22}$$

For example, see [106], the criterion applies to the sequences of the form $(e^{iP(n)})$, where $P \in \mathbb{R}[x]$ has at least one irrational coefficient (different from the constant term).

In the context of dynamical systems, we use this criterion for $a_n = f(T^n x), n \geqslant 1$. Clearly, this leads us to study the behavior of different (prime) powers of a fixed map T. We should warn the reader that when applying Theorem 11.15, we do not expect to have (11.21) satisfied for all continuous functions, in fact, even in uniquely ergodic systems, in general, it cannot hold for all zero mean functions[21] but we need a subset of $C(X)$ which is linearly dense, cf. footnote 4.

We will also need the following variation of Theorem 11.15, see [59]:

Proposition 11.16 *Assume that (a_n) is a bounded sequence of complex numbers. Assume, moreover, that*

$$\limsup_{\substack{p,q \to \infty \\ \text{different primes}}} \left(\limsup_{N \to \infty} \left| \frac{1}{N} \sum_{n \leqslant N} a_{pn} \bar{a}_{qn} \right| \right) = 0. \tag{11.23}$$

Then, for each multiplicative function $\boldsymbol{u} : \mathbb{N} \to \mathbb{C}, \boldsymbol{u} \in \boldsymbol{M}$, we have

$$\lim_{N \to \infty} \frac{1}{N} \sum_{n \leqslant N} a_n \cdot \boldsymbol{u}(n) = 0. \tag{11.24}$$

Remark 11.17 In contrast to the KBSZ criterion given by Theorem 11.15, condition (11.23) has its ergodic theoretical counterpart—the property called AOP (see Sect. 11.5) which is a measure-theoretic invariant.

---

[21] We can easily see that when $Tx = x + \alpha$ is an irrational rotation on $\mathbb{T} = [0, 1)$, then, by the Weyl criterion on uniform distribution, (11.21) is satisfied for all characters (for all $x \in \mathbb{T}$), but there are continuous zero mean functions for which (11.21) fails [112].

11.4 Chowla Conjecture

In this section we get into the subject of the Chowla conjecture which is the main motivation for Sarnak's conjecture.

11.4.1 Formulation and Ergodic Interpretation

The Chowla conjecture deals with higher order correlations of the Möbius function,[22] that is, the conjecture asserts that

$$\frac{1}{N} \sum_{n \leqslant N} \mu^{j_0}(n) \mu^{j_1}(n+k_1) \ldots \mu^{j_r}(n+k_r) \xrightarrow[N \to \infty]{} 0 \qquad (11.25)$$

whenever $1 \leqslant k_1 < \ldots < k_r, j_s \in \{1, 2\}$ not all equal to 2, $r \geqslant 0$.[23]

We will now explain an ergodic meaning of the Chowla conjecture. Recall that given a dynamical system (X, T) and $\mu \in M(X, T)$, a point $x \in X$ is called *generic for μ* if

$$\frac{1}{N} \sum_{n \leqslant N} f(T^n x) \xrightarrow[N \to \infty]{} \int_X f \, d\mu$$

for each $f \in C(X)$. Equivalently, $\frac{1}{N} \sum_{n \leqslant N} \delta_{T^n x} \xrightarrow[N \to \infty]{} \mu$ (we recall that $M(X, T)$ is endowed with the weak* topology which makes it a compact metrizable space). By compactness, each point is *quasi-generic* for a certain measure $\nu \in M(X, T)$, i.e.

$$\frac{1}{N_k} \sum_{n \leqslant N_k} \delta_{T^n x} \xrightarrow[k \to \infty]{} \nu$$

---

[22] As a matter of fact, in [27], it is formulated for the Liouville function. We follow [150]. For a discussion on an equivalence of the Chowla conjecture with μ and λ, we invite the reader to [143]. As shown in [127], there are non-pretentious (completely) multiplicative functions for which Chowla conjecture fails. For more information, see the discussion on Elliot's conjecture in [127].

[23] The Chowla conjecture is rather "close" in spirit to the Twin Number Conjecture in the sense that the latter is expressed by (*) $\sum_{n \leqslant x} \Lambda(n) \Lambda(n+2) = (2\Pi_2) \cdot x + o(x)$, where $\Pi_2 = \prod_{p>2}(1 - \frac{1}{(p-1)^2}) = 0,66016\ldots$ which can be compared with $\sum_{n \leqslant x} \mu(n) \mu(n+2) = o(x)$ which is "close" to the Chowla conjecture, see e.g. [158]. A recent development shows that it is realistic to claim that the Chowla conjecture with an error term of the form $o((\log N)^{-A})$ for some A large enough (A depending on the number of shifts of μ that are considered) implies (*). (Of course, everywhere Λ is a good approximation of $\mathbb{1}_{\mathbb{P}}$.)

See also [138] for a (conditional) equivalence of (*) with $\sum_{n \leqslant N} \Lambda(n) \mu(n+2) = o(N)$.

for a certain subsequence $N_k \to \infty$. Let[24]

$$Q\text{-gen}(x) := \{v \in M(X, T) : x \text{ is quasi-generic for } v\}. \tag{11.26}$$

Assume now that we have a finite alphabet A. We consider $(A^{\mathbb{Z}}, S)$, so called *full shift*, or more precisely, *two-sided full shift*, where $A^{\mathbb{Z}}$ is endowed with the product topology and $S((x_n)) = (y_n)$ with $y_n = x_{n+1}$ for each $n \in \mathbb{Z}$. Each $X \subset A^{\mathbb{Z}}$ that is closed and S-invariant yields a *subshift*, i.e. the dynamical system (X, S). One way to obtain a subshift is to choose $x \in A^{\mathbb{Z}}$ and consider the closure X_x of the orbit of x via S. If x is given as a one-sided sequence, $x \in A^{\mathbb{N}}$, we still might consider

$$X_x := \{y \in A^{\mathbb{Z}} : \text{each block appearing in } y \text{ appears in } x\} \tag{11.27}$$

to obtain a two-sided subshift. In case when each block appearing in x reappears infinitely often, $X_x = \overline{\{S^n \bar{x} : n \in \mathbb{Z}\}}$, for some $\bar{x}$ for which $\bar{x}(j) = x(j)$ for each $j \geqslant 1$ but, in general, there is no such a good $\bar{x}$. Moreover, we will let ourselves speak about a one-sided sequence x to be generic or quasi-generic for a measure $v \in M(X_x, S)$.

Now take $A = \{-1, 0, 1\}$. For each subshift $X \subset \{-1, 0, 1\}^{\mathbb{Z}}$ let $\theta \in C(X)$ be defined as

$$\theta(y) = y(0), \quad y \in X. \tag{11.28}$$

Note that directly from the Stone-Weierstrass theorem we obtain the following.

Lemma 11.18 *The linear subspace generated by the constants and the family*

$$\{\theta^{j_0} \circ S^{k_0} \cdot \theta^{j_1} \circ S^{k_1} \cdot \ldots \cdot \theta^{j_r} \circ S^{k_r} : k_i \in \mathbb{Z}, j_i \in \{1, 2\}, i = 0, 1, \ldots, r, r \geqslant 0\}$$

of continuous functions is an algebra of functions separating points, hence it is dense in $C(X)$.

The subshift (X_μ, S) is called the *Möbius system* and $X_{\mu^2} \subset \{0, 1\}^{\mathbb{Z}} \subset \{-1, 0, 1\}^{\mathbb{Z}}$ is the *square-free system*.[25] Note that $s: (z(n)) \mapsto (z(n)^2)$ will settle a factor map between the Möbius system and the square-free system. The point μ^2 is a generic point for so called *Mirsky measure* v_{μ^2} [23, 133] (see Sect. 11.7.2). In other words, there are frequencies of blocks on μ^2: for each block $B \in \{0, 1\}^\ell$, the following limit exists:

$$\lim_{N \to \infty} \frac{1}{N} \left| \{1 \leqslant n \leqslant N - \ell : \mu^2(n, n+\ell-1) = B\} \right| =: v_{\mu^2}(B).$$

---

[24]We recall that either x is generic or Q-gen(x) is a connected uncountable set, see Proposition 3.8 in [40].

[25]The point μ^2 is recurrent, so there is a "completion" of μ^2 to a two-sided sequence generating the same subshift.

We can now consider the relatively independent extension[26] $\widehat{v}_{\mu^2}$ of v_{μ^2} which is the measure on $s^{-1}(X_{\mu^2}) \subset \{-1, 0, 1\}^{\mathbb{Z}}$ given by the following condition: for each block $C \in \{-1, 0, 1\}^{\ell}$, we have

$$\widehat{v}_{\mu^2}(C) := \frac{1}{2^k} v_{\mu^2}(C^2),$$

where C^2 is obtained from B by squaring on each coordinate and k is the number of 1 in C^2. A straightforward computation shows that

$$\int_{\{-1,0,1\}^{\mathbb{Z}}} \theta^{j_0} \circ S^{k_0} \cdot \theta^{j_1} \circ S^{k_1} \cdot \ldots \cdot \theta^{j_r} \circ S^{k_r} \, d\widehat{v}_{\mu^2} = 0 \tag{11.29}$$

whenever $\{j_0, \ldots, j_r\} \neq \{2\}$. On the other hand, in view of Lemma 11.18, the values of integrals

$$\int_{\{-1,0,1\}^{\mathbb{Z}}} \theta^2 \circ S^{k_0} \cdot \theta^2 \circ S^{k_1} \cdot \ldots \cdot \theta^2 \circ S^{k_r} \, d\widehat{v}_{\mu^2}$$

for all $k_i \in \mathbb{Z}$ and $r \geqslant 0$ entirely determine the Mirsky measure v_{μ^2}.

Corollary 11.19 *The Chowla conjecture holds if and only if μ is a generic point for $\widehat{v}_{\mu^2}$.*

Proof We consider any extension of μ to a two-sided sequence (for example, we set $\mu(n) = 0$ for each $n \leqslant 0$). Suppose that

$$\frac{1}{N_k} \sum_{n \leqslant N_k} \delta_{S^n \mu} \xrightarrow[k \to \infty]{} \kappa. \tag{11.30}$$

In order to get $\kappa = \widehat{v}_{\mu^2}$, in view of Lemma 11.18, we need to show that

$$\int_{\{-1,0,1\}^{\mathbb{Z}}} \theta^{j_0} \circ S^{k_0} \cdot \theta^{j_1} \circ S^{k_1} \cdot \ldots \cdot \theta^{j_r} \circ S^{k_r} \, d\kappa = 0$$

for any choice of integers $k_0 < k_1 < \ldots < k_r$, $\{j_0, j_1, \ldots, j_r\} \neq \{2\}$ and $r \geqslant 0$. Since the measure v is S-invariant, it is the same as to show that

$$\int_{\{-1,0,1\}^{\mathbb{Z}}} \theta^{j_0} \cdot \theta^{j_1} \circ S^{k_1-k_0} \cdot \ldots \cdot \theta^{j_r} \circ S^{k_r-k_0} \, d\kappa = 0.$$

---

[26]Consider Bernoulli measure $B(1/2, 1/2)$ on $\{-1, 1\}^{\mathbb{Z}}$ and Mirsky measure v_{μ^2} on $\{0, 1\}^{\mathbb{Z}}$. Measure $\widehat{v}_{\mu^2}$ is the image of the product measure $B(1/2, 1/2) \otimes v_{\mu^2}$ via the map

$$(x, y) \mapsto ((x(n) \cdot y(n)))_{n \in \mathbb{Z}} \in \{-1, 0, 1\}^{\mathbb{Z}}.$$

Now, we have $1 \leqslant k_1 - k_0 < \ldots < k_r - k_0$ and the result follows from (11.25) and (11.30).

The Chowla conjecture for $r = 0$ is just the PNT, however, it remains open even for $r = 1$. As in [150], we could consider a weaker version of the Chowla conjecture. Namely, we say that μ satisfies the *topological Chowla conjecture* if $X_\mu = s^{-1}(X_{\mu^2})$.

Remark 11.20 Note that (11.25) holds if

$$|\{0 \leqslant t \leqslant r : j_t = 1\}| = 1.$$

Indeed, it is not hard to see that if t_0 is the only index for which $j_{t_0} = 1$ then the sequence $a(n) := \prod_{t \neq t_0} \mu^2(n + k_t)$ is rational. Hence, μ is orthogonal to $a(\cdot)$, cf. footnote 18.

11.4.2   The Chowla Conjecture Implies Sarnak's Conjecture

Assume that (X, T) is a topological system. Following [101, 167] a point $x \in X$ is called *completely deterministic* if for each measure $\nu \in Q\text{-gen}(x)$ (see (11.26)), the measure theoretic dynamical system $(X, \mathcal{B}(X), \nu, T)$ has zero Kolmogorov-Sinai entropy: $h_\nu(T) = 0$. Of course, if the topological entropy of T is zero, then by the Variational Principle, each $x \in X$ is completely deterministic. On the other hand, (X_{μ^2}, S) has positive topological entropy [55, 140, 150] and $\mu^2 \in X_{\mu^2}$ is completely deterministic, see [23, 57].

Let $f \in C(X)$ and $x \in X$ be completely deterministic. We have

$$\frac{1}{N} \sum_{n \leqslant N} f(T^n x) \mu(n) = \int_{X \times X_\mu} (f \otimes \theta) d \left(\frac{1}{N} \sum_{n \leqslant N} \delta_{(T \times S)^n(x, \mu)} \right).$$

We can assume that

$$\frac{1}{N_k} \sum_{n \leqslant N_k} \delta_{(T \times S)^n(x, \mu)} \xrightarrow[k \to \infty]{} \rho \text{ in the space } M(X \times X_\mu, T \times S).$$

Under the Chowla conjecture, the projection of ρ on X_μ is equal to $\widehat{\nu}_{\mu^2}$ (since, by Corollary 11.19, μ is a generic point for $\widehat{\nu}_{\mu^2}$), while the projection of ρ on X is some T-invariant measure κ and $h_\kappa(T) = 0$ (since x is completely deterministic).

Note that ρ is a joining[27] of the (measure-theoretic) dynamical systems (X, κ, T) and $(X_\mu, \widehat{v}_{\mu^2}, S)$. Moreover, the latter automorphism has the so called relative Kolmogorov property with respect to the factor $(X_{\mu^2}, v_{\mu^2}, S)$. We then consider the restriction of the joining $\rho|_{X \times X_{\mu^2}}$ and $\rho|_{X_\mu}$ to obtain two systems that have a common factor (namely X_{μ^2}) relatively to which the first one has zero entropy and the second being relatively Kolmogorov. Since the function θ is orthogonal to $L^2(X_{\mu^2}, v_{\mu^2})$, the relative disjointness theorem on zero entropy and Kolmogorov property yields the following (see also Remark 11.23):

Theorem 11.21 ([57]) *The Chowla conjecture implies*

$$\frac{1}{N} \sum_{n \leqslant N} f(T^n x) \mu(n) \to 0$$

for each dynamical system $(X, T), f \in C(X)$ and $x \in X$ completely deterministic. In particular, the Chowla conjecture implies Sarnak's conjecture.[28]

Remark 11.22 It is also proved in [57] that this seemingly stronger statement of the validity of Sarnak's conjecture at completely deterministic points is in fact equivalent to the Möbius disjointness of all zero entropy systems.

Remark 11.23 A word for word repetition of the above proof[29] yields the same result when we replace μ by another generic point of $\widehat{v}_{\mu^2}$ in which we control the relative Kolmogorov property over the maximal factor with zero entropy, so called Pinsker factor. In particular, we can replace μ by λ (for which the Pinsker factor will be just the one-point dynamical system).

As a matter of fact, it is expected that each aperiodic real-valued multiplicative function satisfies the Chowla type result (and hence satisfies the Sarnak type result), see the conjectures by Frantzikinakis and Host formulated after Theorem 11.47.

Remark 11.24 The original proof of Sarnak of the implication "Chowla conjecture $\Rightarrow$ Sarnak's conjecture" used some combinatorial arguments and probabilistic methods, see [158].

---

[27]Recall that if R_i is an automorphism of a probability standard Borel space $(Z_i, \mathcal{D}_i, v_i)$, $i = 1, 2$, then each $R_1 \times R_2$-invariant measure λ on $(Z_1 \times Z_2, \mathcal{D}_1 \otimes \mathcal{D}_2)$ having the projections v_1 and v_2, respectively is called a *joining* of R_1 and R_2: we write $\lambda \in J(R_1, R_2)$. If R_1, R_2 are ergodic then the set $J^e(R_1, R_2)$ of ergodic joinings between R_1 and R_2 is non-empty. A fundamental notion here is the *disjointness* (in sense of Furstenberg) [73]: R_1 and R_2 are disjoint if $J(R_1, R_2) = \{v_1 \otimes v_2\}$: we write $R_1 \perp R_2$. For example, zero entropy automorphisms are disjoint with automorphisms having completely positive entropy (Kolmogorov automorphisms) and also a relativized version of this assertion holds.

[28]We will see later that some special cases of validity of convergence in (11.25) also have their ergodic interpretations and they imply Möbius disjointness for restricted classes of dynamical systems of zero entropy; in particular, see Corollaries 11.37 and 11.42.

[29]The above proof was already suggested by Sarnak in [150].

Sarnak's conjecture (11.2) is formulated for the Möbius function. But of course one can consider other multiplicative functions.[30] Below, we show that if we use the Liouville function then nothing changes.

Corollary 11.25 *Sarnak's conjecture with respect to* μ *is equivalent to Sarnak's conjecture with respect to* λ.

Proof Let us recall the basic relation between these two functions: $\lambda(n) = \sum_{d^2|n} \mu(n/d^2)$.

Assume that (X, T) is a dynamical system with $h(T) = 0$. As the zero entropy class is closed under taking powers, we assume Möbius disjointness for all powers of T. Then

$$\frac{1}{N} \sum_{n \leqslant N} f(T^n x) \lambda(n) = \frac{1}{N} \sum_{n \leqslant N} f(T^n x) \left(\sum_{d^2|n} \mu(n/d^2) \right)$$

$$= \frac{1}{N} \sum_{n \leqslant N} \sum_{d^2|n} \mu(n/d^2) f((T^{d^2})^{n/d^2} x)$$

$$= \sum_{d \leqslant \sqrt{N}} \frac{1}{d^2} \cdot \frac{1}{N/d^2} \sum_{n \leqslant N/d^2} \mu(n) f((T^{d^2})^n x).$$

Take $\varepsilon > 0$ and select $M \geqslant 1$ so that $\sum_{d \geqslant M} \frac{1}{d^2} < \varepsilon$. Consider $T, T^2, T^3, \ldots, T^M$. We have

$$\left| \frac{1}{N} \sum_{n \leqslant N} f(T^{kn} x) \mu(n) \right| < \varepsilon$$

for all $k = 1, \ldots, M$ whenever $N \geqslant N_0$. It follows that

$$\left| \frac{1}{N/d^2} \sum_{n \leqslant N/d^2} \mu(n) f(T^{d^2 n} x) \right| < \varepsilon$$

for all $d = 1, \ldots, M$ if $N > MN_0$. Otherwise we estimate such a sum by $\|f\|_\infty$.

To obtain the other direction, we first recall that μ^2 is a completely deterministic point. Then use Theorem 11.21 for λ (see Remark 11.23), write $\lambda(n)\mu^2(n) = \mu(n)$ for each $n \geqslant 1$ and we obtain

$$\frac{1}{N} \sum_{n \leqslant N} f(T^n x) \mu^2(n) \lambda(n) = \frac{1}{N} \sum_{n \leqslant N} (f \otimes \theta)((T \times S)^n (x, \mu^2)) \lambda(n) \to 0$$

as the point (x, μ^2) is completely deterministic.

---

[30]If Möbius disjointness in a dynamical system is shown through the KBSZ criterion then we obtain orthogonality with respect to all multiplicative functions.

11.4.3 The Logarithmic Versions of Chowla and Sarnak's Conjectures

An intriguing problem arises whether the Chowla and Sarnak's conjecture are equivalent. An intuition from ergodic theory would say that this is rather not the case as the class of systems that are disjoint (in the Furstenberg sense) from all zero entropy measure-theoretic systems is the class of Kolmogorov automorphisms and not only Bernoulli automorphisms (and a relative version of this result persists).[31]

From that point of view a recent remarkable result of Terence Tao [156] about the equivalence of logarithmic versions of the Chowla and Sarnak's conjectures is quite surprising. We will formulate some versions[32] of three (out of five) conjectures from [156].

Conjecture A We have

$$\frac{1}{\log N} \sum_{n \leqslant N} \frac{\mu^{j_0}(n) \mu^{j_1}(n+k_1) \ldots \mu^{j_r}(n+k_r)}{n} \xrightarrow[N\to\infty]{} 0$$

whenever $1 \leqslant k_1 < \ldots < k_r, j_s \in \{1, 2\}$ not all equal to 2, $r \geqslant 0$.

Remark 11.26 It should be noted that passing to such logarithmic averages moves one away from questions about primes, twin primes and subtleties such as the parity problem. For example, the statement $\sum_{n \leqslant N} \frac{\mu(n)}{n} = o(\log N)$ is easy to establish,[33] while the PNT is equivalent to much stronger statement $\sum_{n=1}^{\infty} \frac{\mu(n)}{n} = 0$ (as conditionally convergent series).

On the other hand, the logarithmically averaged Chowla conjecture implies that all "admissible" configurations do appear on μ, see Corollary 11.30 below (the topological Chowla conjecture for λ implies that all blocks of ± 1 appear in λ).

Conjecture B We have

$$\frac{1}{\log N} \sum_{n \leqslant N} \frac{f(T^n x) \mu(n)}{n} \xrightarrow[N\to\infty]{} 0$$

---

[31]If we consider general sequences $z \in \{-1, 0, 1\}^{\mathbb{N}}$ then we can speak about the Sarnak and Chowla properties on a more abstract level: for example the Chowla property of z means (11.25) with μ replaced by z. See Example 5.1 and Remark 5.3 in [57] for sequences orthogonal to all deterministic sequences but not satisfying the Chowla property. However, arithmetic functions in these examples are not multiplicative.

However, an analogy between disjointness results in ergodic theory and disjointness of sequences is sometimes accurate. For example, a measure-theoretic dynamical system has zero entropy if and only if it is disjoint with all Bernoulli automorphisms. As pointed out in [57] (Prop. 5.21), a sequence $t \in \{-1, 1\}^{\mathbb{N}}$ is completely deterministic if and only if it is disjoint with any sequence $z \in \{-1, 0, 1\}^{\mathbb{N}}$ satisfying the Chowla property.

[32]See Remark 1.9. Also, in [156] the Liouville function λ is considered, see page 2 in [156] how to replace λ by μ.

[33]In fact, $|\sum_{n \leqslant N} \mu(n)/n| \leqslant 1$.

whenever (X, T) is a topological system of zero topological entropy, $f \in C(X)$ and $x \in X$.

To formulate the third conjecture, we need to recall the definition of a nilrotation. Let G be a connected, simply connected Lie group and $\Gamma \subset G$ a lattice (a discrete, cocompact subgroup). For any $g_0 \in G$ we define $T_{g_0}(gH) := g_0gH$. Then the topological system $(G/\Gamma, T_{g_0})$ is called a *nilrotation*.

Conjecture C Let $f \in C(G/\Gamma)$ be Lipschitz continuous and $x_0 \in G$. Then (for $H \leqslant N$)

$$\sum_{n \leqslant N} \frac{\sup_{g \in G} \left| \sum_{h \leqslant H} f(T_g^{h+n}(x_0 \Gamma)) \boldsymbol{\mu}(n+h) \right|}{n} = o(H \log N).$$

Theorem 11.27 ([156]) *Conjectures A, B and C are equivalent.*

Remark 11.28 Tao also shows that if instead of logarithmic averages we come back to Cesàro averages, then

$$\text{Conjecture A} \;\Rightarrow\; \text{Conjecture B} \;\Rightarrow\; \text{Conjecture C}$$

and it is the implication Conjecture C $\Rightarrow$ Conjecture A that requires logarithmic averages.

Remark 11.29 Let us consider the Cesàro version of Conjecture C with $H = o(N)$ and we drop the assumption on the sup (which is inside), i.e.: for each $g \in G$, we have

$$\frac{1}{N} \sum_{n \leqslant N} \left| \sum_{h \leqslant H} f(T_g^{h+n}(x_0 \Gamma)) \boldsymbol{\mu}(n+h) \right| \xrightarrow[H,N \to \infty, H=o(N)]{} 0.$$

This is a particular case of what we will see in Sect. 11.5, where we introduce the strong MOMO notion (hence, the validity of Sarnak's conjecture on (typical) short interval).

Corollary 11.30 (A Letter of W. Veech in June 2016) *Sarnak's conjecture implies topological Chowla conjecture. Equivalently, Sarnak's conjecture implies that each block $B \in \{-1, 0, 1\}^\ell$ for which B^2 appears in $\boldsymbol{\mu}^2$ appears in $\boldsymbol{\mu}$ (and the entropy of $(X_{\boldsymbol{\mu}}, S)$ equals $\frac{6}{\pi^2} \log 3$).*

Proof Indeed, Sarnak's conjecture implies its logarithmic version which, by Theorem 11.27, implies logarithmic Chowla conjecture, that is, $\frac{1}{\log N} \sum_{n \leqslant N} \frac{\delta_{S^n \boldsymbol{\mu}}}{n} \to \widehat{\nu}_{\boldsymbol{\mu}^2}$. However, the logarithmic averages of the Dirac measures are convex combinations of the consecutive Cesàro averages[34] $\frac{1}{n}\sum_{j\leqslant n}\delta_{S^j\mu}$, so if we take a block $B \in s^{-1}(X_{\mu^2})$, we have $\widehat{\nu}_{\mu^2}(B) > 0$ and therefore there exists n such that $\frac{1}{n}\sum_{j\leqslant n}\delta_{S^j\mu}(B) > 0$, which means that B appears in μ.

Remark 11.31 (Added in October 2017) As a matter of fact, as shown in [79], Sarnak's conjecture implies the existence of a subsequence (N_k) along which we have $\frac{1}{N_k}\sum_{n\leqslant N_k}\delta_{S^n\mu} \to \widehat{\nu}_{\mu^2}$. This follows from a general observation that, given a topological system (X, T), whenever an ergodic measure ν is a limit of a subsequence (M_k) of logarithmic averages of Dirac measures: $\nu = \lim_{k\to\infty} \frac{1}{\log M_k}\sum_{m\leqslant M_k}\frac{\delta_{T^m x}}{m}$, then there exists a subsequence (N_k) for which $\nu = \lim_{k\to\infty}\frac{1}{N_k}\sum_{n\leqslant N_k}\delta_{T^n x}$. We apply this to the measure $\widehat{\nu}_{\mu^2}$ which is ergodic.

In [155], Tao proves the logarithmic version of Chowla conjecture for the correlations of order 2 (which we formulate for the Liouville function):

Theorem 11.32 ([155]) *For each $0 \neq h \in \mathbb{Z}$, we have*

$$\frac{1}{\log N}\sum_{n\leqslant N}\frac{\lambda(n)\lambda(n+h)}{n} \xrightarrow[N\to\infty]{} 0.$$

See also [126], where it is proved that for each integer $h \geqslant 1$ there exists $\delta(h) > 0$ such that $\limsup_{N\to\infty}\frac{1}{N}\left|\sum_{n\leqslant N}\lambda(n)\lambda(n+h)\right| \leqslant 1 - \delta(h)$ and [128], where it is proved that for the Liouville function the eight patterns of length 3 of signs occur with positive lower density, and the density result with lower density replaced by upper density persists for $k + 5$ patterns (out of total 2^k) for each $k \in \mathbb{N}$.

For a proof of a function field Chowla's conjecture, see [22].

Remark 11.33 See also [159], where, given $k_0, \ldots k_\ell \in \mathbb{Z}$ and $u_0, \ldots, u_\ell \in \mathcal{M}$, one studies sequences of the form

$$n \mapsto u_0(n + ak_0) \cdot \ldots \cdot u_\ell(n + ak_\ell), \ a \in \mathbb{Z}.$$

---

[34] Assume that (a_n) is a bounded sequence and set $A_n = a_1 + \ldots + a_n$. Then, we have by summation by parts

$$\frac{1}{\log N}\sum_{n\leqslant N}\frac{a_n}{n} = \frac{1}{\log N}\sum_{n\leqslant N}(A_{n+1} - A_n)\frac{1}{n}$$

$$= \frac{1}{\log N}\sum_{n\leqslant N}A_n\left(\frac{1}{n} - \frac{1}{n+1}\right) + o(1) = \frac{1}{\log N}\sum_{n\leqslant N}\frac{A_n}{n}\frac{1}{n+1} + o(1). \quad (11.31)$$

It follows that:

- If the Cesàro averages of (a_n) converge, so do the logarithmic averages of (a_n).
- The converse does not hold (see e.g. [14] in $\mathcal{B}$-free case, Sect. 11.7.1).
- If the Cesàro averages converge along a subsequence (N_k) then not necessarily the logarithmic averages do the same. Indeed, by (11.31), $\frac{1}{\log N_k}\sum_{n\leqslant N_k}\frac{a_n}{n}$ is (up to a small error) a convex combination of the Cesàro averages for all $n \leqslant N_k$.

By considering their logarithmic averages, one obtains a sequence $(f(a))$. The main result of [159] is a structure theorem (depending on whether or not the product $u_0 \cdot \ldots u_\ell$ weakly pretends to be a Dirichlet character) for the sequences $(f(a))$. As a corollary, the logarithmically averaged Chowla conjecture is proved for any odd number of shifts.

11.4.4 Frantzikinakis' Theorem

Tao's approach from [156] is continued in [69]. Before we formulate Frantzikinakis' results, let us interpret some arithmetic properties, especially the role of a "good behavior" on (typical) short interval of a multiplicative function in the ergodic theory language.

11.4.4.1 Ergodicity of Measures for Which μ Is Quasi-Generic

In this subsection we summarize ergodic consequences of some recent, previously mentioned number-theoretic results, cf. [68]. By that we mean that we consider all measures $\kappa \in Q\text{-gen}(\mu)$ and we study ergodic properties of the dynamical systems (X_μ, κ, S).

Let $\kappa \in Q\text{-gen}(\mu)$, i.e. $\frac{1}{M_k} \sum_{m \leqslant M_k} \delta_{S^m \mu} \xrightarrow[k \to \infty]{} \kappa \in M(X_\mu, S)$ for some increasing sequence (M_k). As usual, $\theta(x) = x(0)$ $(\theta \in C(X_\mu))$. We have

$$\int_{X_\mu} \theta \, d\kappa = 0, \tag{11.32}$$

as the integral equals $\lim_{k \to \infty} \frac{1}{M_k} \sum_{n \leqslant M_k} \theta(S^n \mu) = 0$ (by the PNT). Denoting by Inv the σ-algebra of S-invariant (modulo the measure κ) subsets of X_μ, we recall that

$$\frac{1}{H} \sum_{h \leqslant H} \theta \circ S^h \xrightarrow[H \to \infty]{} \mathbb{E}(\theta | Inv) \text{ in } L^2(X_\mu, \kappa)$$

(by the von Neumann ergodic theorem). We want to show that

$$\theta \perp L^2(X_\mu, Inv, \kappa)$$

(i.e. κ must be "slightly" ergodic). In other words, we want to show that

$$\int_{X_\mu} \left| \frac{1}{H} \sum_{h \leqslant H} \theta \circ S^h \right|^2 d\kappa \xrightarrow[H \to \infty]{} 0.$$

But such integrals can be computed:

$$\frac{1}{M_k} \sum_{m \leqslant M_k} \left| \frac{1}{H} \sum_{h \leqslant H} \theta \circ S^h (S^m \mu) \right|^2 \xrightarrow[k \to \infty]{} \int_{X_\mu} \left| \frac{1}{H} \sum_{h \leqslant H} \theta \circ S^h \right|^2 d\kappa.$$

Putting things together, given $\varepsilon > 0$, for $H \geqslant 1$ large enough, we want to see

$$\limsup_{k \to \infty} \frac{1}{M_k} \sum_{m \leqslant M_k} \left| \frac{1}{H} \sum_{h \leqslant H} \mu(m+h) \right|^2 \leqslant \varepsilon.$$

The latter is true because of [126]: for a "typical" m the sum $\left| \frac{1}{H} \sum_{m \leqslant h < m+H} \mu(h) \right|$ is small.

Remark 11.34 As the calculation above shows, the fact that

$$\frac{1}{M} \sum_{m \leqslant M} \left| \frac{1}{H} \sum_{h \leqslant H} \mu(m+h) \right|^2 \to 0$$

when $H \to \infty$ and $H = o(M)$ is equivalent to $\theta \perp L^2(X_\mu, Inv, \kappa)$ for each $\kappa \in$ Q-gen(μ). In particular, the Chowla conjecture implies the above short interval behavior.

However, remembering that $\kappa|_{X_{\mu^2}} = \nu_{\mu^2}$, one can ask now whether θ is measurable with respect to the factor given by the Mirsky measure. As this factor has rational discrete spectrum [23], to show that this is not the case, we need to prove that $\theta \perp L^2(\Sigma_{rat})$, where Σ_{rat} stands for the factor given by the whole rational spectrum of (X_μ, κ, S). To do it, we need to show that for each $r \geqslant 1$, we have

$$\frac{1}{N} \sum_{n \leqslant N} \theta \circ S^{rn} \xrightarrow[N \to \infty]{} 0 \text{ in } L^2(X_\mu, \kappa).$$

This convergence can be shown by using the strong MOMO property (which we will consider in Sect. 11.5) for the rotation $j \mapsto j+1$ on $\mathbb{Z}/r\mathbb{Z}$. We skip this argument here and show still a stronger consequence.

Assume that $\kappa \in$ Q-gen(μ) and that we want to show that the spectral measure of $\theta \in L^2(X_\mu, \kappa)$ is continuous. Hence, we need to show that

$$\frac{1}{H} \sum_{h \leqslant H} |\widehat{\sigma}_\theta(h)| \xrightarrow[H \to \infty]{} 0$$

when $H \to \infty$. Equivalently, we need to show that

$$\frac{1}{H} \sum_{h \leqslant H} \left| \int_{X_\mu} \theta \circ S^h \cdot \theta \, d\kappa \right| \xrightarrow[H \to \infty]{} 0.$$

If we fix $H \geqslant 1$ then

$$\int_{X_\mu} \theta \circ S^h \cdot \theta \, d\kappa = \lim_{k \to \infty} \frac{1}{M_k} \sum_{m \leqslant M_k} \theta \circ S^h(S^m \boldsymbol{\mu}) \cdot \theta(S^m \boldsymbol{\mu})$$

$$= \frac{1}{M_k} \sum_{m \leqslant M_k} \boldsymbol{\mu}(m + h) \boldsymbol{\mu}(m).$$

It follows that we need to show that

$$\frac{1}{H} \sum_{h \leqslant H} \left| \frac{1}{M_k} \sum_{m \leqslant M_k} \boldsymbol{\mu}(m + h) \boldsymbol{\mu}(m) \right| \to 0$$

when $H, M_k \to \infty$; to be precise, given $\varepsilon > 0$ we want to show that for $H > H_\varepsilon$, we have $\limsup_{k \to \infty} \frac{1}{H} \sum_{h \leqslant H} \left| \frac{1}{M_k} \sum_{m \leqslant M_k} \boldsymbol{\mu}(m + h) \boldsymbol{\mu}(m) \right| < \varepsilon$. Hence, directly from Theorem 11.14, we obtain the following.

Corollary 11.35 *The spectral measure of θ is continuous for each $\kappa \in Q\text{-gen}(\boldsymbol{\mu})$.*

While it is obvious that the subshift X_μ is uncountable (indeed, it is the subshift X_{μ^2} which is already uncountable, see Sect. 11.7), it is not clear whether X_λ is uncountable. However, if a subshift (Y, S) is countable, all its ergodic measures are given by periodic orbits, hence there are only countably many of them and it easily follows that each $\kappa \in M(Y, S)$ yield a system with discrete spectrum. Hence, immediately from Corollary 11.35, we obtain that:

Corollary 11.36 *The subshift X_λ is uncountable.*[35]

From Corollary 11.35 we derive immediately the Möbius disjointness of all dynamical systems with "trivial" invariant measures (see also [93]). This kind of problems will be the main subject of our discussion in Sect. 11.5.

Corollary 11.37 *Let (X, T) be any topological dynamical system such that, for each measure $v \in M(X, T)$, (X, v, T) has discrete spectrum (not necessarily ergodic, of course). Then (X, T) is Möbius disjoint. In particular, the result holds if $M^e(X, T)$ is countable with each member of $M^e(X, T)$ yielding a discrete spectrum dynamical system.*

---

[35]The result has been observed in [71], cf. also [95].

Proof Fix $x \in X$ and consider

$$\frac{1}{M_k} \sum_{m \leqslant M_k} \delta_{(T^m x, S^m \mu)} \xrightarrow[k \to \infty]{} \rho.$$

We have $\rho|_{X_\mu} =: \kappa \in \text{Q-gen}(\mu)$ and $\rho|_X =: \nu$. Now, we fix $f \in C(X)$ and we need to show that $\int f \otimes \theta \, d\rho = 0$. But

$$\int_{X \times X_\mu} f \otimes \theta \, d\rho = \int_{X \times X_\mu} (f \otimes 1) \cdot (1 \otimes \theta) \, d\rho = 0. \tag{11.33}$$

Indeed, the spectral measure of $f \otimes 1$ with respect to ρ is the same as the spectral measure of f with respect to ν and the spectral measure of $1 \otimes \theta$ with respect to ρ is the same as the spectral measure of θ with respect to κ. Therefore, these spectral measures are mutually singular by assumption and Corollary 11.35. Hence, the functions $f \otimes 1$ and $1 \otimes \theta$ are orthogonal, i.e. (11.33) holds.[36]

If we have all ergodic measures giving discrete spectrum but we have too many ergodic measures then the argument above does not go through. Consider[37]

$$(x, y) \mapsto (x, x + y) \text{ on } \mathbb{T}^2. \tag{$*$}$$

Question 1 (Frantzikinakis (2016)) Can we obtain $\kappa \in \text{Q-gen}(\lambda)$, so that (X_λ, κ, S) is isomorphic to $(*)$?

Of course, the answer to Question 1 is expected to be negative.

11.4.4.2 Frantzikinakis' Results

We now follow [69] and formulate results for the Liouville function, although, up to some obvious modifications, they also hold for μ.

Theorem 11.38 ([69]) *Assume that $N_k \to \infty$ and let $\frac{1}{\log N_k} \sum_{n \leqslant N_k} \frac{\delta_{S^n \lambda}}{n} \xrightarrow[k \to \infty]{} \kappa$. If κ is ergodic then the Chowla conjecture (and Sarnak's conjecture) holds along (N_k) for the logarithmic averages.*

Taking into account footnote 34, we cannot deduce a similar statement for ordinary averages along (N_k) but in view of [79], see Remark 11.31, the Chowla

---

[36] We use here the standard result in the theory of unitary operators that mutual singularity of spectral measures implies orthogonality. Recall also the classical result in ergodic theory that spectral disjointness implies disjointness.

[37] Consider $X_1 = X_2 = \mathbb{T}^2$ with $\mu_1 = \mu_2 = Leb_{\mathbb{T}^2}$, the diagonal joining Δ on $X_1 \times X_2$ and $f(x, y) = \overline{\theta(x, y)}$ with $\theta(x, y) = e^{2\pi i y}$. The spectral measure of θ is Lebesgue, and all ergodic components of the measure μ_1 have discrete spectra.

conjecture holds along another subsequence. The situation becomes clear when (N_k) is the sequence of all natural numbers and we assume genericity.

Corollary 11.39 ([69]) *If λ is generic for an ergodic measure then the Chowla conjecture holds.*

Let us say a few words on the proof. Recall that given a bounded sequence $(a(n)) \subset \mathbb{C}$ admitting correlations,[38] one defines its local uniformity seminorms (see Host and Kra [91]) in the following manner:

$$\|a\|_{U^1(\mathbb{N})}^2 = \mathbb{E}_{h \in \mathbb{N}} \mathbb{E}_{n \in \mathbb{N}} a(n+h)\overline{a(n)}, \tag{11.34}$$

$$\|a\|_{U^{s+1}(\mathbb{N})}^{2^{s+1}} = \mathbb{E}_{h \in \mathbb{N}} \|S_h\, a \cdot \overline{a}\|_{U^s(\mathbb{N})}^{2^s}, \quad s \geqslant 2, \tag{11.35}$$

where, for each bounded sequence $(b(n))$, $(S_h\, b)(n) := b(h+n)$ and $\mathbb{E}_{n \in \mathbb{N}} b(n) = \lim_{N \to \infty} \frac{1}{N} \sum_{n \leqslant N} b(n)$. (Similar definitions are considered along a subsequence (N_k).)

The following result has been proved by Tao:

Theorem 11.40 ([156]) *Assume that λ is generic. The Chowla conjecture holds if and only if $\|\lambda\|_{U^s(\mathbb{N})} = 0$ for each $s \geqslant 1$.*[39]

Remark 11.41 We have assumed in the statement of Theorem 11.40 that λ is generic but we would like also to note that, without this latter (strong) assumption, Tao obtained the equivalence in Theorem 11.40 for the logarithmic averages, see Conjecture 1.6 and Theorem 1.9 in [156] (however, one has to modify the definition of seminorms [156]).

Hence, under the assumption of Corollary 11.39, we need to prove that all local uniform seminorms of λ vanish. The inverse theorem for seminorms reduces this problem to the statement: for every basic nilsequence $(a(n))$[40] on an $s-1$-step nilmanifold G/Γ and every $s-2$-step manifold H/Λ, we have

$$\lim_{N \to \infty} \mathbb{E}_{m \in \mathbb{N}} \sup_{b \in \Psi_{H/\Lambda}} \left| \mathbb{E}_{n \in [m,m+N]} \lambda(n)a(n)b(n) \right| = 0,$$

where $\Psi_{H/\Lambda}$ is a special class of basic nil-sequences (coming from Lipschitz functions). The latter is then proved using a deep induction argument.

---

[38]I.e., we assume the existence of the limits of sequences $\left(\frac{1}{N} \sum_{n \leqslant N} a'(n)a'(n+k_1)\ldots a'(n+k_r) \right)_{N \geqslant 1}$ for every $r \in \mathbb{N}$ and $k_1, \ldots, k_r \in \mathbb{N}$ (not necessarily distinct) with $a' = a$ or $\overline{a}$. It is not hard to see that a admits correlations if and only if it is generic, cf. Sect. 11.4.1.

[39]We have $\|\lambda\|_{U^1(\mathbb{N})} = 0$ by Matomäki and Radziwiłł [126], moreover $\|\lambda\|_{U^2(\mathbb{N})} = 0$ is equivalent to $\lim_{N \to \infty} \mathbb{E}_{m \in \mathbb{N}} \sup_{\alpha \in [0,1)} \left| \mathbb{E}_{n \in [m,m+N]} \lambda(n)e^{2\pi i n\alpha} \right| = 0$ (cf. Conjecture C) and remains open. For a subsequence version of Theorem 11.40 for logarithmic averages, see [156].

[40]By that we mean $a(n) = f(g^n \Gamma)$ for some continuous $f \in C(G/\Gamma)$ and $g \in G$.

11.4.5   Dynamical Properties of Furstenberg Systems Associated to the Liouville and Möbius Functions

We now continue considerations about logarithmic version of Sarnak's conjecture, cf. Conjecture B, Theorem 11.38. Consider all measures κ for which λ is logarithmically quasi-generic, i.e. $\frac{1}{\log N_k} \sum_{n \leqslant N_k} \frac{\delta_{S^n \lambda}}{n} \to \kappa$ for some $N_k \to \infty$. We denote the set of all such measures by $\log -Q\text{-gen}(\lambda)$. Following [71], for each $\kappa \in \log -Q\text{-gen}(\lambda)$ the corresponding measure-theoretic dynamical system (X_λ, κ, S) will be called a *Furstenberg system of* λ. Before we get closer to the results of [71], let us see first some consequence of Theorem 11.32 for the logarithmic Sarnak's conjecture:

> For each Furstenberg system (X_λ, κ, S), the spectral measure σ_θ of
> θ is Lebesgue.
> (11.36)

Indeed, assuming $\frac{1}{\log N_k} \sum_{n \leqslant N_k} \frac{\delta_{S^n \lambda}}{n} \xrightarrow[k \to \infty]{} \kappa$, Theorem 11.32 tells us that for each $h \in \mathbb{Z} \setminus \{0\}$, we have

$$\widehat{\sigma}_\theta(h) = \int_{X_\lambda} \theta \circ S^h \cdot \theta \, d\kappa = \lim_{k \to \infty} \frac{1}{\log N_k} \sum_{n \leqslant N_k} \frac{\lambda(n+h)\lambda(n)}{n} = 0.$$

Using (11.36) and repeating the proof of Corollary 11.37, we obtain the following.

Corollary 11.42 *Let (X, T) be a topological system such that each of its Furstenberg's systems has singular spectrum. Then (X, T) is logarithmically Liouville disjoint.*

The starting point of the paper [71] is a surprising Tao's identity (implicit in [155]) for general sequences which in its ergodic theory language (cf. Sect. 11.4.4.1) takes the following form:

Theorem 11.43 (Tao's Identity, [71]) *Let $\kappa \in \log -Q\text{-gen}(\lambda)$. Then*

$$\int_{X_\lambda} \left(\prod_{j=1}^{\ell} \theta \circ S^{k_j} \right) d\kappa = (-1)^{\ell} \lim_{N \to \infty} \frac{\log N}{N} \sum_{\mathbb{P} \ni p \leqslant N} \int_{X_\lambda} \left(\prod_{j=1}^{\ell} \theta \circ S^{pk_j} \right) d\kappa$$

for all $\ell \in \mathbb{N}$ and $k_1, \ldots, k_\ell \in \mathbb{Z}$.

Now, the condition in Theorem 11.43 is purely abstract (indeed, the function θ generates the Borel σ-algebra), and the strategy to cope with logarithmic Sarnak's conjecture is to describe the class of measure-theoretic dynamical systems satisfying the assertion of Theorem 11.43 and then to obtain Liouville disjointness for all systems which are disjoint (in the Furstenberg sense) from all members of the class.

In fact, Frantzikinakis and Host deal with extensions of Furstenberg systems of λ, so called *systems of arithmetic progressions with prime steps*.[41] They prove the following result.

Theorem 11.44 ([71]) *For each system of arithmetic progressions with prime steps, its "typical" ergodic component is isomorphic to the direct product of an infinite-step nilsystem and a Bernoulli automorphism.*[42] *In particular, each Furstenberg system (X_λ, κ, S) of λ is a factor of a system which:*

(i) has no irrational spectrum and
(ii) has ergodic components isomorphic to the direct product of an infinite-step nilsystem and a Bernoulli automorphism.

Remark 11.45 All the above results are also true when we replace λ by μ.

Then, some new disjointness results in ergodic theory are proved (for example, all totally ergodic automorphisms are disjoint from an automorphism satisfying (i) and (ii) in Theorem 11.44) and the following remarkable result is obtained:

Theorem 11.46 ([71]) *Let (X, T) be a topological dynamical system of zero entropy with countably many ergodic invariant measures. Then Conjecture B holds for (X, T).*

In particular, logarithmic Sarnak's conjecture holds for all zero entropy uniquely ergodic systems. As a matter of fact, some new[43] consequences are derived:

Theorem 11.47 ([71]) *Let (X, T) be a topological dynamical system with zero entropy. Assume that $x \in X$ is generic for a measure v with only countably many ergodic components all of which yield totally ergodic systems. Then, for every $f \in C(X), \int_X f \, dv = 0$, we have*

$$\lim_{N \to \infty} \frac{1}{\log N} \sum_{n \leqslant N} \frac{f(T^n x) \prod_{j=1}^{\ell} \mu(n + k_j)}{n} = 0$$

for all $\ell \in \mathbb{N}$ and $k_1, \ldots, k_\ell \in \mathbb{Z}$.

---

[41] Given a measure-theoretic dynamical system $(Z, \mathcal{D}, \rho, R)$, its system of arithmetic progressions with prime steps is of the form $(Z^{\mathbb{Z}}, \mathcal{B}(Z^{\mathbb{Z}}), \widetilde{\rho}, S)$, where S is the shift and the (shift invariant) measure $\widetilde{\rho}$ is determined by

$$\int_{Z^{\mathbb{Z}}} \prod_{j=-m}^{m} f_j(z_j) \, d\widetilde{\rho}(z) = \lim_{N \to \infty} \frac{\log N}{N} \sum_{p \leqslant N} \int_Z \prod_{j=-m}^{m} f_j \circ R^{pj} \, d\rho$$

for all $m \geqslant 0, f_{-m}, \ldots, f_m \in L^\infty(Z, \rho)$ (here $z = (z_j)$). It is proved that such shift systems have no irrational spectrum. One of key observations is that each Furstenberg system of the Liouville function is a factor of the associated system of arithmetic progressions with prime steps.

[42] The product decomposition depends on the component.

[43] They are new even for irrational rotations. Cf. the notions of (S)-strong and (S$_0$)-strong and their equivalence to the Chowla type condition in [57].

New conjectures are proposed in [71]:

1. Every real-valued $u \in \mathcal{M}$ has a unique Furstenberg system (i.e. u is generic) which is ergodic and isomorphic to the direct product of a Bernoulli automorphism and an odometer.
2. If, additionally, $u \in \mathcal{M}$ takes values ± 1 then its Furstenberg system is either Bernoulli or it is an odometer.

Finally, it is noticed in [71] that the complexity of the Liouville function has to be superlinear, that is

$$\lim_{N \to \infty} \frac{1}{N} \left| \{ B \in \{-1, 1\}^N : B \text{ appears in } \lambda \} \right| = \infty. \qquad (11.37)$$

The reason is that, as shown in [71], for transitive systems having linear block growth we have only finitely many ergodic measures (and clearly systems with linear block growth have zero topological entropy). Hence, by Theorem 11.46, such systems are Liouville disjoint. As X_λ is not Liouville disjoint, λ cannot have linear block growth, i.e. (11.37) holds.

11.5 The MOMO and AOP Properties

11.5.1 The MOMO Property and Its Consequences

We will now consider Sarnak's conjecture from the ergodic theory point of view. We ask whether (already) measure-theoretic properties of a measurable system $(Z, \mathcal{D}, \kappa, R)$ imply the validity of (11.1) for any (X, T), $f \in C(X)$ provided that $x \in X$ is a generic point for a measure μ such that the measure-theoretic system $(X, \mathcal{B}(X), \mu, T)$ is measure-theoretically isomorphic to $(Z, \mathcal{D}, \kappa, R)$. More specifically, we can ask whether some measure-theoretic properties of $(Z, \mathcal{D}, \kappa, R)$ can imply Möbius disjointness of all its uniquely ergodic models.[44] We recall that the Jewett-Krieger theorem implies the existence of a uniquely ergodic model of each ergodic system.[45] As a matter of fact, there are plenty of such models and they

---

[44]Note that the answer is positive in all uniquely ergodic models of the one-point system: each such a model has a unique fixed point that attracts each orbit on a subset of density 1, cf. the map $e^{2\pi ix} \mapsto e^{2\pi ix^2}$, $x \in [0, 1)$. This argument is however insufficient already for uniquely ergodic models of the exchange of two points: in this case we have a density 1 attracting 2-periodic orbit $\{a, b\}$, but we do not control to which point a or b the orbit returns first. Quite surprisingly, it seems that already in this case we need [126] to obtain Möbius disjointness of all uniquely ergodic models.

[45]If all uniquely ergodic systems were Möbius disjoint, then as noticed by T. Downarowicz, we would get that the Chowla conjecture fails in view of the result of B. Weiss [169] Thm. 4.4' on approximation of generic points of ergodic measures by uniquely ergodic sequences.

can have various additional topological properties including topological mixing[46]
[117]. Here is another variation of the approach to view Möbius disjointness as a
measure-theoretic invariant:

Question 2 Does the Möbius disjointness in a certain uniquely ergodic model of an
ergodic system yield the Möbius disjointness in all its uniquely ergodic models?

To cope with these questions we need a definition. Let $u \colon \mathbb{N} \to \mathbb{C}$ be an arithmetic
function.[47]

Definition 1 (Strong MOMO[48] Property [58]) We say that (X, T) satisfies the
strong MOMO property (relatively to u) if, for any increasing sequence of integers
$0 = b_0 < b_1 < b_2 < \cdots$ with $b_{k+1} - b_k \to \infty$, for any sequence (x_k) of points in
X, and any $f \in C(X)$, we have

$$\frac{1}{b_K} \sum_{k<K} \left| \sum_{b_k \leqslant n < b_{k+1}} f(T^{n-b_k} x_k) u(n) \right| \xrightarrow[K \to \infty]{} 0. \qquad (11.38)$$

Remark 11.48 The property (11.38) looks stronger than the condition on Möbius
disjointness. The idea behind it is to look at the pieces of orbits (of different points)
in one system as a single orbit of a point in a different, larger but "controllable"
(from measure-theoretic point of view) system.

Remark 11.49 One can easily show (as in Sect. 11.4.4.1) that the strong MOMO
property (relative to μ) implies $f \otimes \theta \perp L^2(Inv, \rho)$ for each $\rho \in$ Q-gen$((x, \mu)$,
$T \times S)$.[49]

By taking $f = 1$ in Definition 1, we obtain that whenever strong MOMO holds,
u has to satisfy:

$$\frac{1}{b_K} \sum_{k<K} \left| \sum_{b_k \leqslant n < b_{k+1}} u(n) \right| \xrightarrow[K \to \infty]{} 0 \qquad (11.39)$$

for every sequence $0 = b_0 < b_1 < b_2 < \cdots$ with $b_{k+1} - b_k \to \infty$. In particular,
$\frac{1}{N} \sum_{n<N} u(n) \xrightarrow[N \to \infty]{} 0$. This is to be compared with (11.18), (11.20) and (11.19) to
realize that we require a special behavior of u on a typical short interval.

---

[46]Topological mixing for example excludes the possibility of having eigenfunctions continuous.

[47]Our objective is of course the Möbius function μ, however the whole approach can be developed
for an arithmetic function satisfying some additional properties.

[48]The acronym comes from Möbius Orthogonality of Moving Orbits.

[49]Inv stands here for the σ-algebra of $T \times S$-invariant sets modulo ρ.

Theorem 11.50 ([58]) *Let $(Z, \mathcal{D}, \kappa, R)$ be an ergodic dynamical system. Let $u: \mathbb{N} \to \mathbb{C}$ be an arithmetic function. The following conditions are equivalent:*

(a) *There exist a topological system (Y, S) enjoying the strong MOMO property (relative to u) and $\nu \in M^e(Y, S)$ such that the measurable systems $(Y, \mathcal{B}(Y), \nu, S)$ and $(Z, \mathcal{D}, \kappa, R)$ are isomorphic.*

(b) *For any topological dynamical system (X, T) and any $x \in X$, if there exists a finite number of T-invariant measures μ_j, $1 \leqslant j \leqslant t$, such that*

- *$(X, \mathcal{B}(X), \mu_j, T)$ is measure-theoretically isomorphic to $(Z, \mathcal{D}, \kappa, R)$ for each j,*
- *any measure for which x is quasi-generic is a convex combination of the measures μ_j, i.e. Q-gen$(x) \subset \text{conv}(\mu_1, \dots, \mu_t)$,*

then $\frac{1}{N} \sum_{n \leqslant N} f(T^n x) u(n) \xrightarrow[N \to \infty]{} 0$ for each $f \in C(X)$.

(c) *All uniquely ergodic models of $(Z, \mathcal{D}, \kappa, R)$ enjoy the strong MOMO property (relative to u).*

The proof of implication $(a) \Rightarrow (b)$ borrows some ideas from [93] and the proof of implication $(b) \Rightarrow (c)$ uses some ideas from [59].

Remark 11.51 It can be easily shown that any minimal (hence uniquely ergodic) rotation on a compact Abelian group satisfies the strong MOMO property (say, relatively to μ). It follows from Theorem 11.50 (and the Halmos-von Neumann theorem) that in each uniquely ergodic model of an ergodic automorphism with discrete spectrum, we also have the strong MOMO property (in particular, the Möbius disjointness).

We now list three consequences of Theorem 11.50:

Corollary 11.52 ([58])

(a) *If Sarnak's conjecture holds then the strong MOMO property (relative to μ) holds for every zero entropy dynamical system.[50]*

(b) *If Sarnak's conjecture holds then it holds uniformly, that is, the convergence in (11.1) is uniform in x.[51]*

(c) *Fix $\delta_{(\dots 0.00\dots)} \neq \kappa \in M^e((\mathbb{D}_L)^{\mathbb{Z}}, S)$, where $\mathbb{D}_L = \{z \in \mathbb{C} : |z| \leqslant L\}$. Let (X, T) be any uniquely ergodic model of $((\mathbb{D}_L)^{\mathbb{Z}}, \kappa, S)$. Then for any $u \in (\mathbb{D}_L)^{\mathbb{Z}}$ for which Q-gen$(u) \subset \text{conv}(\kappa_1, \dots, \kappa_m)$, where $((\mathbb{D}_L)^{\mathbb{Z}}, \kappa_j, S)$ for $j = 1, \dots, m$ is*

---

[50]That is, Sarnak's conjecture and the strong MOMO property (relatively to μ) for all deterministic systems are equivalent statements.

[51]It is not hard to see that the MOMO property implies the relevant uniform convergence. As a matter of fact, the strong MOMO property is equivalent to the uniform convergence (in x, for a fixed $f \in C(X)$) on short intervals: $\frac{1}{M} \sum_{1 \leqslant m < M} \left| \frac{1}{H} \sum_{m \leqslant h < m+H} f(T^h x) \mu(n) \right| \to 0$ (when $H, M \to \infty$ and $H = o(M)$). It follows that we have equivalence of: Sarnak's conjecture (11.2), Sarnak's conjecture in its uniform form, Sarnak's conjecture in its short interval uniform form and the strong MOMO property. Moreover, each of these conditions is implied by the Chowla conjecture.

measure-theoretically isomorphic to $((\mathbb{D}_L)^{\mathbb{Z}}, \kappa, S)$, *the system* (X, T) *does not satisfy the strong MOMO property (relative to* $\boldsymbol{u}$*).*[52]

Remark 11.53 Let us come back to Theorem 11.21 and Remark 11.23, i.e. to the reformulation of Sarnak's conjecture using completely deterministic sequences. We intend to show that a natural generalization of Corollary 11.52 (b) to the completely deterministic case fails. Indeed, consider the square-free system (X_{μ^2}, S). In Remark 11.20, we have already noticed that whenever $k_j, j = 1, \ldots, r$ are different non-negative integers, then

$$\sum_{n \leqslant N} \mu^2(n + k_1) \ldots \mu^2(n + k_{r-1}) \mu(n + k_r) = \mathrm{o}(N). \tag{$*$}$$

It follows that for each $f \in C(X_{\mu^2})$, for each $k \in \mathbb{Z}$, we have

$$\frac{1}{N} \sum_{n \leqslant N} f(S^{n+k} \mu^2) \mu(n) \to 0. \tag{$**$}$$

On the other hand, the convergence in $(**)$ cannot be uniform in $k \in \mathbb{Z}$. Indeed, if it were then the whole square-free system would be Möbius disjoint. This is however impossible since (X_{μ^2}, S) is hereditary, see Remark 11.77. Indeed, we can find $y \in X_{\mu^2}$ such $y(n) = 1$ if and only if $\mu(n) = 1$ and $y(n) = 0$ otherwise (then $y \leqslant \mu^2$) and if we set $\theta(z) := z(0)$ then $\lim_{N \to \infty} \frac{1}{N} \sum_{n \leqslant N} \theta(S^n y) \mu(n) = \frac{3}{\pi^2}$.
    See also [137], where a quantitative version of $(*)$ has been proved.

Note that Theorem 11.50 does not fully answer Question 2. In certain situations the following general (lifting) lemma of Downarowicz and Lemańczyk can be helpful:

Lemma 11.54 ([44, 56]) *Assume that an ergodic automorphism R is coalescent.*[53] *Let* $(\widetilde{X}, \widetilde{T})$ *and* (X, T) *be uniquely ergodic models of R. Assume that T is a topological factor of $\widetilde{T}$, i.e. there exists* $\pi : \widetilde{X} \to X$ *which is continuous and onto and which satisfies* $\pi \circ \widetilde{T} = T \circ \pi$. *If T is Möbius disjoint then also $\widetilde{T}$ is Möbius disjoint.*

11.5.2   Möbius Disjointness and Entropy

Sarnak's conjecture deals with deterministic systems but Möbius disjointness, a priori, does not exclude the possibility of positive (topological) entropy systems

---

[52]This result means that there must be an observable sequence in (X, T) which significantly correlates with $\boldsymbol{u}$.

[53]This means that each measure-preserving transformation commuting with R must be invertible. Finite multiplicity of the Koopman operator associated to R guarantees coalescence. In particular, all ergodic rotations are coalescent.

which are Möbius disjoint.[54] The first "natural" trial would be to take the square-free system (X_{μ^2}, S) which has positive entropy (see Sect. 11.7.2) and clearly μ^2 is orthogonal to μ. However, in spite of the orthogonality of the two sequences, as we have noticed in Remark 11.53, the square-free system is not Möbius disjoint.

Recently, Downarowicz and Serafin [45] constructed Möbius disjoint positive entropy homeomorphisms of arbitrarily large entropy. On the other hand, see [104], in the subshift of finite type case we do not have Möbius disjointness. Using Katok's horseshoe theorem, it follows that $C^{1+\delta}$-diffeomorphisms of surfaces are not Möbius disjoint but the following question seems to be open:

Question 3 Is there a positive entropy diffeomorphism of a compact manifold which is Möbius disjoint?

Viewed all this above, another natural question arises:

Question 4 Does there exist an ergodic positive entropy measure-theoretic system all uniquely ergodic models of which are Möbius disjoint?

Using Theorem 11.50, Sinai's theorem on Bernoulli factors (see e.g. [78]) and B. Weiss' theorem [168] on strictly ergodic models of some diagrams a partial answer to Question 4 is given by the following result:

Corollary 11.55 ([58]) *Assume that $u \in (\mathbb{D}_L)^{\mathbb{Z}}$ is generic for a Bernoulli measure κ. Let $v \in (\mathbb{D}_L)^{\mathbb{Z}}$, u and v correlate. Then for each dynamical system (X, T) with $h(X, T) > h((\mathbb{D}_L)^{\mathbb{Z}}, \kappa, S)$, we do not have the strong MOMO property relatively to v.*

By substituting $u = \lambda$, $v = \mu$ and assuming the Chowla conjecture for λ, we obtain that no system (X, T) with entropy $> \log 2$ satisfies the strong MOMO relatively to μ. When μ is replaced by λ, we still have a stronger result.

Proposition 11.56 ([58]) *Assume that the Chowla conjecture holds for λ. Then no topological system (X, T) with positive entropy satisfies the strong MOMO property relatively to λ.*

Remark 11.57 The proof of Theorem 11.50 tells us that when $(Z, \mathcal{D}, \kappa, R)$ is ergodic and has positive entropy then there exists a system (X, T), which is not Liouville disjoint, with at most three ergodic measures and all of these measures yield a measurable system isomorphic to R. Therefore, it seems reasonable to conjecture that the answer to Question 4 is negative.

We now have a completely clear picture for the Liouville function: it follows from Theorem 11.21 (for λ) and Proposition 11.56 that if the Chowla conjecture holds for λ then the strong MOMO property (relatively to λ) holds for (X, T) if and

---

[54]Sarnak in [150] mentions that Bourgain has constructed a positive entropy system which is Möbius disjoint but this construction has never been published.

only if $h(X, T) = 0$. Using footnote 51, we immediately obtain Proposition 11.56 in its equivalent form:

Corollary 11.58 *Assume that the Chowla conjecture holds for* λ. *Then, the short interval uniform convergence in* (11.1) *(with* μ *replaced by* λ*) takes place if and only if* $h(X, T) = 0$.

11.5.3 The AOP Property and Its Consequences

We need an ergodic criterion to establish the strong MOMO property in models of an automorphism. This turns out to be a natural ergodic counterpart of the KBSZ criterion (Theorem 11.15). Following [59] an ergodic automorphism R is said to have *asymptotically orthogonal powers* (AOP) if for each $f, g \in L_0^2(Z, \mathcal{D}, \kappa)$, we have

$$\lim_{\mathbb{P} \ni p, q \to \infty, p \neq q} \sup_{\kappa \in J^e(R^p, R^q)} \left| \int_{X \times X} f \otimes g \, d\kappa \right| = 0. \tag{11.40}$$

Rotation $Rx = x + 1$ acting on $\mathbb{Z}/k\mathbb{Z}$ with $k \geqslant 2$ has no AOP property because of Dirichlet's theorem on primes in arithmetic progressions. Hence, AOP implies total ergodicity (clearly, AOP is closed under taking factors). The AOP property implies zero entropy [59].

Clearly, if the powers of R are pairwise disjoint[55] then R enjoys the AOP property. In order to see a less trivial example of an AOP automorphism, consider any totally ergodic discrete spectrum automorphism R on $(Z, \mathcal{D}, \kappa)$. For f, g take eigenfunctions corresponding to eigenvalues c, d, respectively. Now, take $\rho \in J^e(R^p, R^q)$ and consider

$$\int_{Z \times Z} f \otimes g \, d\rho = \int_{Z \times Z} (f \otimes \mathbb{1}_Z) \cdot (\mathbb{1}_Z \otimes g) \, d\rho.$$

Notice that $f \otimes \mathbb{1}_Z$ and $\mathbb{1}_Z \otimes g$ are eigenfunctions of $(Z \times Z, \rho, R^p \times R^q)$ corresponding to c^p and d^q, respectively. If $c^p \neq d^q$ (and this is the case for all but one pair (p, q) because of total ergodicity) then these eigenfunctions are orthogonal and we are done. We will see more examples in Sect. 11.6.

Remark 11.59 For an AOP automorphism the powers need not be disjoint. As a matter of fact, we can have an AOP automorphism with all of its non-zero powers isomorphic.[56]

---

[55]This is a "typical" property of an automorphism of a probability standard Borel space [38]. Möbius disjointness for uniquely ergodic models for this case is already noticed in [21].

[56]Take an ergodic rotation with the group of eigenvalues $\{e^{2\pi i \alpha m/n} : m, n \in \mathbb{Z}, n \neq 0, \alpha \notin \mathbb{Q}\}$.

Theorem 11.60 ([58, 59]) *Let $u \in \mathcal{M}$. Suppose that $(Z, \mathcal{D}, \kappa, R)$ satisfies AOP. Then the following are equivalent:*

- *u satisfies (11.39);*
- *The strong MOMO property relatively to u is satisfied in each uniquely ergodic model (X, T) of R.*

In particular, if the above holds, for each $f \in C(X)$, we have

$$\frac{1}{N} \sum_{n \leqslant N} f(T^n x) u(n) \xrightarrow[N \to \infty]{} 0 \text{ uniformly in } X.$$

Corollary 11.61 *Assume that $(Z, \mathcal{D}, \kappa, R)$ enjoys the AOP property. Then, in each uniquely ergodic model (X, T) of R, we have*

$$\frac{1}{M} \sum_{M \leqslant m < 2M} \left| \frac{1}{H} \sum_{m \leqslant h < m+H} f(T^n x) \mu(n) \right| \xrightarrow[H, M \to \infty, H = o(M)]{} 0 \qquad (11.41)$$

for all $f \in C(X)$, $x \in X$.

The AOP property can be defined for actions of locally compact (second countable) groups. Then, for induced actions this property lifts [67], and in particular (by taking the induced $\mathbb{R}$-action), if we have an automorphism then the corresponding suspension flow[57] has this lifted property. In particular, using induced $\mathbb{Z}$-actions (for $a\mathbb{Z} \subset \mathbb{Z}$), one can derive easily that for uniquely ergodic systems (X, T) with the measure-theoretic AOP property we not only have Möbius disjointness but also

$$\frac{1}{N} \sum_{n \leqslant N} f(T^n x) \mu(an + b) \xrightarrow[N \to \infty]{} 0 \qquad (11.42)$$

for each $a, b \in \mathbb{N}, f \in C(X)$ and the convergence is uniform in x [67].[58]

11.6 Glimpses of Results on Sarnak's Conjecture

The cases for which the Möbius disjointness has been proved, depend on the complexity of the deterministic system. They fit into two basic types. The first comes with sufficiently quantitative estimates for the disjointness sums which

---

[57]By the *suspension flow* of R we mean the special flow over R under the constant function (equal to 1).

[58]The same argument shows that if Sarnak's conjecture holds then (11.42) holds for each zero entropy (X, T), $a, b \in \mathbb{N}, f \in C(X)$ uniformly in $x \in X$.

makes possible an analysis of the sums on primes yielding a PNT. This group includes Kronecker systems (Vinogradov [163]), nilsystems (Green and Tao [83]) and, perhaps the most striking, the Thue-Morse system (Mauduit and Rivat [129]) which resolved a conjecture of Gelfond [77]. When the systems are more complex, such as horocycles flows,[59] then at least to date they do not come with a PNT,[60] and for them the KBSZ criterion is used, in other words, the disjointness (perhaps in its weaker form, see Sect. 11.5) is achieved.

We now review most of important cases in which Möbius disjointness has been proved.

11.6.1 Systems of Algebraic Origin

11.6.1.1 Horocycle Flows

Let $\Gamma \subset PSL_2(\mathbb{R})$ be a discrete subgroup with finite covolume.[61] Then the homogeneous space $X = \Gamma \backslash PSL_2(\mathbb{R})$ is the unit tangent bundle of a surface M of constant negative curvature. Let us consider the corresponding *horocycle flow*[62] $(h_t)_{t \in \mathbb{R}}$ and the *geodesic flow* $(g_s)_{s \in \mathbb{R}}$ on X. Since

$$g_s h_t g_s^{-1} = h_{e^{-2s}t} \text{ for all } s, t \in \mathbb{R}, \qquad (11.43)$$

the flows $(h_t)_{t \in \mathbb{R}}$ and $(h_{e^{-2s}t})_{t \in \mathbb{R}}$ are measure-theoretically isomorphic for each $s \in \mathbb{R}$. In order to show that $T := h_1$ is Möbius disjoint, the KBSZ criterion is used, and, given $x \in PSL_2(\mathbb{R})$, one studies limit points of $\frac{1}{N} \sum_{n \leqslant N} \delta_{(T^{pn}\Gamma x, T^{qn}\Gamma x)}$, $N \geqslant 1$. Now, the celebrated Ratner's rigidity theorem [145] tells us two important things: the point $(\Gamma x, \Gamma x)$ is generic for a measure ρ (which must be a joining by unique ergodicity: $\rho \in J(T^p, T^q)$) and moreover this joining is ergodic.[63] Again using Ratner's theory (cf. [144]) such joinings are determined by the commensurator $Com(\Gamma)$ of the lattice Γ:

$$Com(\Gamma) := \{z \in PSL_2(\mathbb{R}) : z^{-1}\Gamma z \cap \Gamma \text{ has finite index in both } \Gamma \text{ and } z^{-1}\Gamma z\}.$$

---

[59] Horocycle flows are mixing of all orders, see [123].

[60] In case of horocycle flows (Bourgain, Sarnak and Ziegler [21]) Ratner's theorems on joinings are used and these provide no rate.

[61] We will tacitly assume that Γ is cocompact, so that the homogenous space $\Gamma \backslash PSL_2(\mathbb{R})$ is compact and the system is uniquely ergodic by Furstenberg in [74]; otherwise, as in the modular case when $\Gamma = PSL_2(\mathbb{Z})$ we need to compactify our space. The proof of Theorem 11.62 in the modular case is slightly different than what we describe below.

[62] We have $h_t(\Gamma x) = \Gamma \cdot \left(x \cdot \begin{bmatrix} 1 & t \\ 0 & 1 \end{bmatrix} \right)$ and $g_s(\Gamma x) = \Gamma \cdot \left(x \cdot \begin{bmatrix} e^{-s} & 0 \\ 0 & e^s \end{bmatrix} \right)$; we identify g_s and h_t with the relevant matrices.

[63] The measure ρ depends on p, q and x and it is so called algebraic measure, i.e. a Haar measure.

Set $x_{p,q} := xg_{\frac{1}{2}\log(\frac{p}{q})}x^{-1}(\infty)$. The intersection of the stabilizer of $x_{p,q}$ with $Com(\Gamma)$ yields the correlator of $x_{p,q}$: it is a subgroup $C(\Gamma, x_{p,q}) \subset \mathbb{R}_+^*$ and if ρ is not the product measure then $\frac{p}{q} \in C(\Gamma, x_{p,q})$. The careful analysis of the arithmetic and non-arithmetic cases done in [21] shows that given $x \in PSL_2(\mathbb{R})$, $\frac{p}{q} \in C(\Gamma, x_{p,q})$ only for finitely many different primes p, q. Hence, the joining ρ has to be product measure for all but finitely many pairs $(p, q) \in \mathbb{P}^2$ with $p \neq q$ which, by Theorem 11.15, yields the following:

Theorem 11.62 ([21]) *All time-automorphisms of horocycle flows are Möbius disjoint.*

Remark 11.63 As noticed in [59], this is (11.43) which yields the absence[64] of AOP and makes the following questions of interest.

Question 5 Do we have the MOMO property for horocycle flows? Are all uniquely ergodic models of horocycle flows Möbius disjoint? Do we have uniform convergence in (11.1)?

Since the method to prove Möbius disjointness is through the KBSZ criterion (hence offers no rate of convergence), the following question is still open:

Question 6 (Sarnak) Do we have a PNT for horocycle flows?

For a partial answer, see [151], where it is proved that if Γx is a generic point for Haar measure μ_X of X then any limit point of $\left(\frac{1}{\pi(N)} \sum_{p \leqslant N} \delta_{T^p \Gamma x}\right)$ is a measure which is absolutely continuous with respect to μ_X.

Question 7 (Ratner) Are smooth time changes for horocycle flows Möbius disjoint?

As smooth time changes of horocycle flows enjoy so called Ratner's property, the above question can be asked in the larger context of flows possessing Ratner's property.

Added in September 2017

In the recent paper [102], a new criterion (of Ratner's type) for disjointness of different time-automorphisms of flows has been proved. The criterion applies for some classes of flows with Ratner's property, namely, in case of so called Arnold flows and for non-trivial smooth time changes of horocycle flows (in particular, the answer to Question 7 is positive).

---

[64]To be compared with Remark 11.59; the difference however is that when the ratio of p and q is close to 1, we can choose graph joinings in a compact set.

11.6.1.2 Nilrotations, Affine Automorphisms

Green and Tao in [83] proved Möbius disjointnes in the following strong form:

Theorem 11.64 ([83]) *Let G be a simply-connected nilpotent Lie group with a discrete and cocompact subgroup Γ. Let $p: \mathbb{Z} \to G$ be any is polynomial sequence[65] and $f: G/\Gamma \to \mathbb{R}$ a Lipschitz function. Then*

$$\left| \frac{1}{N} \sum_{n \leqslant N} f(p(n)\Gamma)\mu(n) \right| = O_{f,G,\Gamma,A}\left(\frac{N}{\log^A N} \right)$$

for all $A > 0$.

In particular, by considering $T_g(x\Gamma) = gx\Gamma$, we see that all nilrotations are Möbius disjoint with uniform Davenport's estimate (11.3).

Also, a PNT holds for nilrotations: Let $2 = p_1 < p_2 < \ldots$ denote the sequence of primes.

Theorem 11.65 ([83], Theorem 7.1) *Assume that a nil-rotation T_g is ergodic.[66] Then, for every $x \in G$, we have*

$$\lim_{N \to \infty} \frac{1}{N} \sum_{n \leqslant N} f(T_g^{p_n} x\Gamma) = \int_{G/\Gamma} f \, d\lambda_{G/\Gamma}$$

for all continuous functions $f: G/\Gamma \to [-1, 1]$.

In [67], it is proved that all nil-rotations enjoy the AOP property (hence all uniquely ergodic models of nil-rotations are Möbius disjoint). In fact, the result is proved for all nil-affine automorphisms whose Möbius disjointness has been established earlier in [122]. Earlier, AOP has been proved for all quasi-discrete spectrum automorphism in [59], that is (following [85]) for all unipotent affine automorphisms $Tx = Ax + b$ of compact Abelian groups (A is a continuous group automorphism and b is an element of the group). The Möbius disjointness of the latter automorphisms has been established still earlier in [122].

The proof of the following corollary in [59] shows that Furstenberg's proof [72] (see e.g. [50]) of Weyl's uniform distribution theorem can be adapted to the short interval version.

---

[65]I.e. $p(n) = d_1^{p_1(n)} \ldots d_k^{p_k(n)}$, where $p_j: \mathbb{N} \to \mathbb{N}$ is a polynomial, $j = 1, \ldots, k$. See, Section 6 in [84] for the equivalence with the classical definition of polynomials sequences in nilpotent Lie groups.

[66]We assume that G is connected.

Corollary 11.66 ([59]) *Assume that* $u \colon \mathbb{N} \to \mathbb{C}$, $u \in \mathcal{M}$. *Then, for each non constant polynomial* $P \in \mathbb{R}[x]$ *with irrational leading coefficient, we have*[67]

$$\frac{1}{M} \sum_{M \leqslant m < 2M} \frac{1}{H} \left| \sum_{m \leqslant n < m+H} e^{2\pi i P(n)} u(n) \right| \xrightarrow[H,M \to \infty, H=o(M)]{} 0.$$

Recall that a sequence $(a_n) \subset \mathbb{C}$ is called a *nilsequence* if it is a uniform limit of basic nilsequences, i.e. of sequences of the form $(f(T_g^n x \Gamma))$, where $f \in C(G/\Gamma)$ (here, we do not assume that G/Γ is connected, neither that T_g is ergodic).

Corollary 11.67 ([67]) *We have*

$$\frac{1}{M} \sum_{M \leqslant m < 2M} \frac{1}{H} \left| \sum_{m \leqslant n < m+H} a_n u(n) \right| \xrightarrow[H,M \to \infty, H=o(M)]{} 0.$$

It has been proved by Leibman [119] that all polynomial multicorrelation sequences[68] are limits in the Weyl pseudo-metric of nil-sequences, all such polynomial sequences are orthogonal to μ on typical short interval, cf. Sect. 11.7.

The main problem connected with nilsequences is to prove the uniform version of convergence on short intervals as it is made precise in Conjecture C of Tao (see Sect. 11.4.3 and also Frantzikinakis' proofs [69]).

11.6.1.3 Other Algebraic Systems

For a more general zero entropy algebraic systems and their Möbius disjointness we refer the reader to [139], where in particular the Ad-unipotent translation case is treated.

---

[67] For degree 1 polynomials, the result is already in [127].

[68] More precisely, given an automorphism T of a probability standard Borel space $(X, \mathcal{B}, \mu)$, we consider

$$a_n = \int_X g_1 \circ T^{p_1(n)} \cdot \ldots \cdot g_k \circ T^{p_k(n)} \, d\mu,$$

where $g_i \in L^\infty(X, \mu)$, $p_i \in \mathbb{Z}[x]$, $i = 1, \ldots, k$ $(k \geqslant 1)$.

11.6.2 Systems of Measure-Theoretic Origin: Substitutions and Interval Exchange Transformations

11.6.2.1 Systems Whose Powers Are Disjoint

We are interested in ergodic automorphisms $(Z, \mathcal{D}, \kappa, R)$ for which (sufficiently large) prime powers R^p are pairwise disjoint. Clearly, such automorphisms enjoy the AOP property. A typical automorphism has this property [38] but there are also large classes of rank one (we detail on this class below) automorphisms with this property [20, 54, 147]. Also minimal self-joining automorphisms [39] enjoy this property. Chaika and Eskin in [25] show that for a.e. 3-interval exchange transformation (we detail on interval exchange transformations below) there are sufficiently many prime powers that are disjoint. It follows that all uniquely ergodic models of these automorphisms are Möbius disjoint.

11.6.2.2 Adic Systems and Bourgain's Criterion

Let $(Z, \mathcal{D}, \kappa, R)$ be a measure-theoretic system.

Definition 2 In $(Z, \mathcal{D}, \kappa, R)$, a Rokhlin *tower* is a collection of disjoint measurable sets called *levels* $F, RF, \ldots, R^{h-1}F$. If Z is equipped with a partition P such that each level $R^r F$ is contained in one atom $P_{w(r)}$, the *name* of the tower is the word $w(0) \ldots w(h-1)$.

Definition 3 A system $(Z, \mathcal{D}, \kappa, R)$ is of *rank one* if there exists a sequence of Rokhlin towers $(F_n, \ldots, R^{h_n-1}F_n)$, $n \geqslant 1$, such that the whole σ-algebra is generated by the partitions $\{F_n, RF_n, \ldots, R^{h_n-1}F_n, X \setminus \bigcup_{j=0}^{h_n-1} R^j F_n\}$.

For topological systems, there is no canonical notion of rank, but the useful notion is that of adic presentation [162], which we translate here from the original vocabulary into the one of Rokhlin towers.

Definition 4 An *adic presentation* of a topological system (X, T) is given, for each $n \geqslant 0$, by a finite collection $\mathcal{Z}_n$ of Rokhlin towers such that:

- the levels of the towers in $\mathcal{Z}_n$ partition X,
- each level of a tower in $\mathcal{Z}_n$ is a union of levels of towers in $\mathcal{Z}_{n+1}$,
- the levels of the towers in $\bigcup_{n \geqslant 0} \mathcal{Z}_n$ form a basis of the topology of X.

In that case, the towers of $\mathcal{Z}_{n+1}$ are built from the towers of $\mathcal{Z}_n$ by cutting and stacking, following recursion rules: a given tower in $\mathcal{Z}_{n+1}$ can be built by taking columns of successive towers in $\mathcal{Z}_n$ and stacking them successively one above another. These rules are best seen by looking at the partition P into levels of the towers in $\mathcal{Z}_0$; possibly replacing $\mathcal{Z}_0$ by some $\mathcal{Z}_k$, we can always assume P has at least two atoms. The names of the towers in $\mathcal{Z}_n$ form sets of words $\mathcal{W}_n$, and the cutting and stacking of towers gives a canonical decomposition of every $W \in \mathcal{W}_n$:

$$W = W_1^{k_1} \cdots W_r^{k_r}$$

for r words $W_i \in \mathcal{W}_{n-1}$, $1 \leqslant i \leqslant r$, integers $k_1, \ldots, k_r$; all these parameters depend on the word W. These decompositions are called the *rules of cutting and stacking* of the system.

The following result is an improvement on Theorem 3.1 of [64], which itself can be found in [20], though it is not completely explicit in that paper (it is stated in full only in a particular case, as Theorem 3, and its proof is understated). The following effective bound stems from a closer reading of [20]:

Theorem 11.68 *Let (X, T) be a topological dynamical system admitting an adic presentation, as in Definition 4 and the comment just after.*
Suppose that for any n and W in $\mathcal{W}_n$, we have:

- *in the rules of cutting and stacking $r \leqslant C$, with $C \geqslant 2$,*
- *if we decompose W into words $W_\ell \in \mathcal{W}_{n-s}$ by iteration of the rules of cutting and stacking then for all ℓ and s large enough, we have*

$$|W| > C^{200s}|W_\ell|.$$

Then (X, T) is Möbius disjoint.
If such a system is uniquely ergodic and weakly mixing for its invariant probability, it satisfies also the following PNT: for any word $W = w_1 \ldots w_N$ which is a factor of a word in any $\mathcal{W}_n$, we have

$$\sum_{i=1}^{N} \Lambda(i)w_i = \sum_{i=1}^{N} w_i + o(N).$$

Proof We look at Theorem 2 of [20]. It requires a stronger assumption, denoted by relations (2.2) and (2.3) in p. 119 of [20], which is indeed the assumption of the present theorem with the estimate C^{200s} replaced by $\beta(s)$ for some function satisfying $\frac{\log \beta(s)}{s} \to \infty$ when $s \to \infty$ (note that the assumption in [20] that the words W_n are on the alphabet $\{0, 1\}$ is not used in the proof, which works for any finite alphabet). Then this theorem gives, for any word $w_1 \cdots w_N$ in some $\mathcal{W}_m$ and N large enough, an estimate for

$$\int \left| \sum_{n \leqslant N} w_n e^{2\pi i n\theta} \right| \left| \sum_{n \leqslant N} \mu(n) e^{2\pi i n\theta} \right| d\theta,$$

and this, through the relation (1.62) on p. 118, implies that $\sum_{n \leqslant N} w_n \mu(n) = o(N)$.

Lacking space to rewrite the extensive computations in [20], we explain how to weaken the hypothesis. First, as suggested in the remark at the beginning of Section 2, p. 119, of that paper, we replace $\beta(s)$ by C_0^s for some constant C_0, as yet unknown (the C_0s written in the same p. 119 is a misprint). The relations (2.2) and (2.3) are used twice in the course of the proof: first, to get the relation (2.15), namely

$$\left(C \frac{\log |W|}{n} \right)^s < |W|^\epsilon$$

for a word W in $\mathcal{W}_n$, and then to get the estimate (2.42), which states that

$$\left(C \frac{\log K}{s} \right)^s < K^\epsilon,$$

where s is the number of stages such that a word of length N in $\mathcal{W}_n$ is divided into words of $\mathcal{W}_{n-s}$, of lengths in the order of $\frac{N}{K}$. Under our hypothesis, in the first case, $|W|$ is in C_0^n, and in the second case K is in C_0^s. Thus both (2.15) and (2.42) are implied by the relation

$$\frac{\log \log C_0 + \log C}{\log C_0} < \epsilon.$$

The value of ϵ is dictated by relation (2.49), which requires $Q^\epsilon K^\epsilon (Q + K)^{-\frac{1}{4}} \leqslant (Q + K)^{-\frac{1}{5}}$ for some large numbers Q and K, thus we can take $\epsilon = \frac{1}{20}$. Then $\frac{\log \log C_0}{\log C_0}$ will be bounded if C_0 is large enough independently of C, while to bound $\frac{\log C}{\log C_0}$ we need to take $C_0 = C^a$; as $C \geqslant 2$, we see that $a = 200$ is convenient for the sum of the two terms.

Now, if we replace w_n by $u(n) = f(T^n(x_0))$, because of Definition 4 above, we can first assume that f is constant on all levels of the towers of some stage m, and then conclude by approximation. Such an f is also constant on all levels of all towers at stages $q > m$; fixing x_0 and N, except for some initial values $u(1)$ to $u(N_0)$ where N_0 is much smaller than N, we can replace $u(n)$ by w'_n, where w'_n is the value of f on the n-th level of some tower with name W in some $\mathcal{W}_q$ for $q \geqslant m$. Then the $w'_1 \cdots w'_N$ are built by the same induction rules as the $w_1 \cdots w_N$, and the estimates using the w'_n are computed as those using the w_n in the proof of Theorem 2 of [20], thus we get the same result.

The PNT is in (3.4), (3.7), (3.14) of [20] ((3.14) is proved for the particular case of 3-interval exchanges but holds in the same way for the more general case).

Of course, the value of C_0 could be improved, but we need it to be at least some power of C.

11.6.2.3 Substitutions

We start with some basic notions.

Definition 5 A *substitution* σ is an application from an alphabet $\mathcal{A}$ into the set $\mathcal{A}^\star$ of finite words on $\mathcal{A}$; it extends to a morphism of $\mathcal{A}^\star$ for the concatenation. A *fixed point* of σ is an infinite sequence u with $\sigma u = u$. The associated symbolic dynamical system (X_σ, S) is (X_u, S) for a fixed point u.

Substitution σ has *constant length* q if $|\sigma a| = q$ for all a in $\mathcal{A}$.

The *Perron-Frobenius eigenvalue* is the largest eigenvalue of the matrix giving the number of occurrences of j in σa. A substitution σ is *primitive* if a power of this matrix has strictly positive entries.

For the class of constant length substitutions, there have been a lot of partial results on Möbius orthogonality:

- First for the most famous example, the Thue-Morse substitution $0 \to 01, 1 \to 10$, with Indlekofer and Kátai [98], Dartyge and Tenenbaum [31], Mauduit and Rivat [129], El Abdalaoui, Kasjan and Lemańczyk [56].[69]
- The case of the Rudin-Shapiro substitution $0 \to 01, 1 \to 02, 2 \to 31, 3 \to 32$ was solved by Mauduit and Rivat [130], Deshouillers, Drmota and Müllner [41].
- Then Drmota [46], Ferenczi, Kułaga-Przymus, Lemańczyk and Mauduit [66] proved Möbius disjointness for the dynamical systems given by bijective substitutions, while [41] proved it for the opposite case, the so-called synchronized substitutions.[70]

See also [124, 125] for a PNT for some digital functions.

But all this was superseded by the general result of Müllner [136], whose proof uses the arithmetic techniques of [130] together with a new structure theorem on the underlying *automata*:

Theorem 11.69 ([136]) *For any substitution of constant length, the associated symbolic system is Möbius disjoint. Moreover, a PNT holds if the substitution is primitive.*

The substitutions which are not of constant length are much less known:

- The most famous example is the Fibonacci substitution, $0 \to 01, 1 \to 0$: in that case, the associated symbolic system is a coding of an irrational rotation, hence it is Möbius disjoint as a uniquely ergodic model of a discrete spectrum automorphism, see Sect. 11.6.3.1.

---

[69]In [31, 98, 129] it is proved that the sequence $(-1)^{u(n)}$, $n \geqslant 1$ is orthogonal to μ.

[70]As noted in [13], this leads to dynamical systems given by rational sequences and such systems are Möbius disjoint. Note also that for the synchronized case, once the system is uniquely ergodic, it is automatically a uniquely ergodic model of an automorphism with discrete spectrum, cf. Corollary 11.37 and Remark 11.51.

- Drmota, Müllner and Spiegelhofer [47] have just shown Möbius disjointness for a new example, a substitution which generates $(-1)^{s_\phi(n)}$, where $s_{\phi(n)}$ is the Zeckendorf sum-of-digits function.[71]
- Also, we can exhibit a small subclass of examples which are Möbius disjoint, by a straightforward translation of Bourgain's criterion above:

Theorem 11.70 *Suppose that σ is a primitive substitution satisfying*

- *for all $i \in \mathcal{A}$, $\sigma i = (j_1(i))^{a_1(i)} \ldots (j_{q_i}(i))^{a_{q_i}(i)}, , a_1(i) \in \mathcal{A}, \ldots, a_{q_i}(i) \in \mathcal{A}$, $q_i \leqslant C$ (this can be expressed as: the multiplicative length of σ is smaller than C),*
- *the Perron-Frobenius eigenvalue of σ is larger than C^{200};*

then the associated symbolic dynamical system is Möbius disjoint. If (X_σ, S) is weakly mixing, the fixed points satisfy a PNT.

Proof If all fixed points are periodic, the result is trivial. If σ has a non-periodic fixed point, it is well known (and proved by the methods of [142] together with the recognizability result of [135]) that the system has an adic presentation, where the names of the towers in Z_n are the words $\sigma^n a$, $a \in \mathcal{A}$. Thus the results come from Theorem 11.68 above and the properties of the matrix of σ.

Example 1 Here are some substitutions for which the above theorem applies, with a PNT: $0 \to 0^{k+1}12$, $1 \to 12$, $2 \to 0^k 12$, $k + 2 > 3^{200}$.

Question 8 Are dynamical systems associated to substitutions Möbius disjoint?[72]

11.6.2.4 Interval Exchanges

Definition 6 A *k-interval exchange* with probability vector $(\alpha_1, \alpha_2, \ldots, \alpha_k)$, and permutation π is defined by

$$Tx = x + \sum_{\pi^{-1}(j) < \pi^{-1}(i)} \alpha_j - \sum_{j<i} \alpha_j.$$

when $x \in \Delta_i = \left[\sum_{j<i} \alpha_j, \sum_{j \leqslant i} \alpha_j \right)$.

Exchanges of 2 intervals are just rotations, thus Möbius disjointness holds for them by the Prime Number Theorem (on arithmetic progressions when the rotation is rational and from a result of Davenport [33]—using a result of Vinogradov [164]—when the rotation is irrational, cf. (11.3) in Introduction).

Then [20] exhibits exchanges of 3 intervals which are Möbius disjoint, with a PNT if weak mixing holds: these use the criterion developed in Theorem 11.68 above, together with the adic presentation built in [65]. Generalizing these methods,

---

[71]This example has partly continuous spectrum.

[72]One can also ask about Möbius disjointness of related systems as tiling systems.

it is shown in [64] that Möbius disjointness holds for examples of exchanges of k intervals for every $k \geqslant 2$ and every *Rauzy class*, with a PNT in the weak mixing case. A breakthrough came with [25], for a large subclass of exchange of 3 intervals:

Theorem 11.71 ([25]) *For (Lebesgue)-almost all* (α_1, α_2), *Sarnak's conjecture holds for exchanges of* 3 *intervals with permutation* $\pi i = 3 - i$ *and probability vector* $(\alpha_1, \alpha_2, 1 - \alpha_1 - \alpha_2)$.

To prove this, Chaika and Eskin use first the well-known fact that such an exchange of 3 intervals, denoted by T, is the induced map of the rotation of angle $\alpha = \frac{1-\alpha_1}{1+\alpha_2}$ on the interval $[0, x)$ where $x = \frac{1}{1+\alpha_2}$. This approach, of course, does not generalize to 4 intervals or more.

In fact, in [25] two different results are proved. In the easier one, they deduce Möbius disjointness from the disjointness of powers of T; they give a sufficient condition for T^m to be disjoint from T^n for all $m \neq n$, which is satisfied by almost all these T. Namely, if we take $(a_1, \ldots)$ to be the continued fraction of α and $(b_1, \ldots)$ the α-*Ostrowski* expansion of x, then it is enough that, for any ordered k-tuple of pairs $((c_1, d_1), \ldots (c_k, d_k))$ of natural numbers such that $d_i \leqslant c_i - 1$, there are infinitely many i with $a_i = c_1, \ldots, a_{i+k-1} = c_k, b_i = d_1, \ldots, b_{i+k-1} = d_k$.

Then most of the paper is used to give an explicit Diophantine condition on α and x, which implies a slightly weaker property than the disjointness of powers. Under that condition, there exists a constant C such that for all n, and $0 \leqslant m \leqslant n$, T^m is disjoint from T^n except maybe when m belongs to a sequence $m_i(n)$ in which any two consecutive terms satisfy $m_{i+1}(n) > Cm_i(n)$, and this is proved to imply Möbius disjointness. The Diophantine condition holds for almost all T, and, as it is long, we refer the reader to Theorem 1.4 of [25]; it expresses the fact that the geodesic ray from a certain flat torus with two marked points, defined naturally from T and its inducing rotation, spends significant time in compact subsets of the space of such tori.

11.6.2.5 Systems of Rank One

These systems form a measure-theoretic class defined in Definition 3 above. It is well known, but has been shown explicitly for all cases only in the recent [1], that each system of rank-one is measure-theoretically isomorphic to one of the topological systems we define now.

Definition 7 A *standard model of rank one* is the shift on the orbit closure of the sequence u which, for each $n \geqslant 0$, begins with the word B_n defined recursively by concatenation as follows. We take sequences of positive integers $q_n, n \geqslant 0$, with $q_n > 1$ for infinitely many n, and $a_{n,i}, n \geqslant 0, 0 \leqslant i \leqslant q_n - 1$, such that, if h_n are defined by $h_0 = 1, h_{n+1} = q_n h_n + \sum_{j=0}^{q_n-1} a_{n,i}$, then

$$\sum_{n=0}^{\infty} \frac{h_{n+1} - q_n h_n}{h_{n+1}} < \infty.$$

We define $B_0 = 0$,

$$B_{n+1} = B_n 1^{a_{n,0}} B_n \ldots B_n 1^{a_{n,q_n-1}}$$

for $n \geqslant 0$.

In [20], Bourgain proved Möbius disjointness for a standard model of rank one if both the $q_n, n \in \mathbb{N} \cup \{0\}$ and $a_{n,i}, n \in \mathbb{N} \cup \{0\}$, are bounded by some constant C (we will refer to this as to a *bounded rank one construction*).

Note however that, in the same paper, the half-hidden criterion deduced from Theorem 2 or 3, see Theorem 11.68 above, is much more than an auxiliary to prove the supposedly main Theorem 1 of [20]; it applies to a much wider class of systems, and even for some famous rank one systems this criterion works while Theorem 1 does not apply.

Bourgain's result was improved in [54], where so called *recurrent rank one constructions* are considered with a stabilizing bounded subsequence of spacers (that is, of a subsequence of $(a_{n,i})$).[73] One of main tools in [54] is a representation of each rank one transformation as an integral automorphism over an odometer with so called Morse-type roof function which goes back to [92]. See also [147] for a simpler proof of a generalization of Bourgain's result to a class of partially bounded rank one constructions.

Spectral Approach

In order to prove Möbius disjointness for standard models of rank one transformations, both papers [20] and [54] use a spectral approach. In [54], unitary operators U (of separable Hilbert spaces) are considered and weak limits of powers $(U^{p m_k})$ (for different primes p) are studied. Once such limits yield sufficiently different (for different p) analytic functions (of U), the powers U^p and U^q are spectrally disjoint.[74] If for a positive real number a we set $s_a(x) = ax \mod 1$ on the additive circle $\mathbb{T} = [0, 1)$, then the above spectral disjointness means that

$$\sigma^{(p)} := (s_p)_*(\sigma) \text{ are mutually singular for different } p \in \mathbb{P}, \qquad (11.44)$$

where $\sigma = \sigma_U$ stands for the maximal spectral type of U.

In [20], a different spectral approach (sufficient for a use of the KBSZ criterion, hence, sufficient for Möbius disjointness) is used. Namely, if $r \geqslant 1$ is an integer, then by σ_r, we will denote the measure which is obtained first by taking the image

---

[73]Moreover, Möbius disjointness is established for some other famous classes of rank one transformations such as: Katok's α-weak mixing class (these are a special case of three interval exchange maps) or rigid generalized Chacon's maps.

[74]Hence, T^p and T^q are disjoint in Furstenberg's sense, and, in fact, we even have AOP.

of σ under the map $x \mapsto \frac{1}{r}x$, i.e. the measure $\sigma^{(1/r)}$, and then repeating this new measure periodically in intervals $[\frac{j}{r}, \frac{j+1}{r})$, that is:

$$\sigma_r := \frac{1}{r} \sum_{j=0}^{r-1} \sigma^{1/r} * \delta_{j/r}.$$

Bourgain [20] uses a representation of the maximal spectral type of a rank one transformation as a Riesz product and then shows the mutual disjointness of measures σ_p and σ_q for different $p, q \in \mathbb{P}$ (for more information about the measures σ_r, see e.g. [142], p. 196). Although, there seems not to be too much relation between the measures $\sigma^{(r)}$ and σ_r, the following observation[75] explains some equivalence of these both spectral approaches:

Lemma 11.72 *Assume that σ and η are two probability measures on the circle. Then:*

(a) if $\sigma^{(r)} \perp \eta^{(s)}$ then $\sigma_s \perp \eta_r$;
(b) if $(r, s) = 1$ then $\sigma^{(r)} \perp \eta^{(s)}$ if and only if $\sigma_s \perp \eta_r$.

11.6.2.6 Rokhlin Extensions

Let T be a uniquely ergodic homeomorphism of a compact metric space X and let $f : X \to \mathbb{R}$ be continuous. Set $T_f(x, t) := (Tx, f(x) + t)$ to obtain a skew product homeomorphism on $X \times \mathbb{R}$. Note that the latter space is not compact. But, if we take any continuous flow $\mathcal{S} = (S_t)_{t \in \mathbb{R}}$ acting on a compact metric space Y then the skew product $T_{f,\mathcal{S}}$ acting on $X \times Y$ by the formula:

$$T_{f,\mathcal{S}}(x, y) = (Tx, S_{f(x)}(y)), \quad (x, y) \in X \times Y$$

is a homeomorphism of the compact space $X \times Y$ and it is called a *Rokhlin extension* of T. To get a good theory, usually one has to put some further assumptions on f (considered as a cocycle taking values in a locally compact but not compact group, see e.g. [120, 152]). It is proved in [113] that there are irrational rotations $Tx = x + \alpha$ and continuous $f : \mathbb{T} \to \mathbb{R}$ (even smooth) such that $T_{f,\mathcal{S}}$ has the AOP property for each uniquely ergodic $\mathcal{S}$.[76]

We would like to emphasize that the Rokhlin skew product construction are usually relatively weakly mixing [120], so the class we consider here is drastically different from the distal class which is our next object to give account.

---

[75]This has been proved, e.g. in an unpublished preprint of El Abadalaoui, Kułaga-Przymus, Lemańczyk and de la Rue.

[76]If $\mathcal{S}$ preserves a measure ν then $T_{f,\mathcal{S}}$ preserves measure $\mu \otimes \nu$, the AOP property is considered with respect to this measure.

This approach leads in [113] to so called random sequences[77] $(a_n) \subset \mathbb{R}$ such that

$$\frac{1}{N} \sum_{n \leqslant N} g(S_{a_n} y) \mu(n) \to 0$$

for each uniquely ergodic flow S acting on a compact metric space Y, each $g \in C(Y)$ and (due to [58]) uniformly in $y \in Y$.

11.6.3 Distal Systems

Assume that R is an ergodic automorphism of a probability standard Borel space $(Z, \mathcal{D}, \kappa)$. R is called (measurably) *distal* if it can be represented as transfinite sequence of consecutive isometric extensions, where in case of a limit ordinal, we take the corresponding inverse limit (i.e. we start with the one-point dynamical system, the first isometric extension is a rotation and then we take a further isometric extension of it etc.). Recall that by a *separating sieve* we mean a sequence

$$Z \supset A_1 \supset A_2 \supset \ldots \supset A_n \supset \ldots$$

of sets of positive measure such that $\mu(A_n) \to 0$ and there exists $Z_0 \subset Z, \mu(Z_0) = 1$, such that for each $z, z' \in Z_0$ if for each $n \geqslant 1$ there is $k_n \in \mathbb{Z}$ such that $R^{k_n} z, R^{k_n} z' \in A_n$, then $z = z'$. A theorem by Zimmer [172] says that T is distal if and only if it has a separating sieve.

Distal automorphisms play a special role in ergodic theory: each automorphism has a maximal distal factor and is relatively weakly mixing over it [76, 171, 172]. Hence, many problems in ergodic theory can be reduced to study the two opposite cases: the distal and the weak mixing one.[78] Recall that distal automorphisms have entropy zero.

There is also a notion of distality in topological dynamics. A homeomorphism T of a compact metric space X is called *distal* if the orbit $(T^n x, T^n x')$, $n \in \mathbb{Z}$, is bounded away from the diagonal in $X \times X$ for each $x \neq x'$. Some of topologically distal classes already appeared in previous sections. Indeed, zero entropy (minimal) affine transformations are examples of distal homeomorphisms. Another natural class of distal (uniquely ergodic) homeomorphisms is given by nil-translations and, more generally, affine unipotent diffeomorphisms of nilmanifolds. A theorem by

---

[77] Such a sequence (a_n) is of the form $(\varphi^{(n)}(x))$ with $\varphi^{(n)}(x) = \varphi(x) + \varphi(Tx) + \ldots + \varphi(T^{n-1} x)$, $n \geqslant 0$.

[78] See the most prominent example of such a reduction, namely, Furstenberg's ergodic proof of Szemerédi theorem on the existence of arbitrarily long arithmetic progressions in subsets of integers of positive upper Banach density [76].

Lindenstrauus [121] says that a measurably distal automorphism R has a minimal[79] model (X, T) together with $\mu \in M^e(X, T)$ of full support (and (X, μ, T) is isomorphic to (Z, κ, R)) in which T is topologically distal.

The following (still open) question seems to be a natural and important step in proving Sarnak's conjecture:

Question 9 (Liu and Sarnak [122]) Are all topologically distal systems Möbius disjoint?

As transformations with discrete spectrum are measurably distal and Theorem 11.73 holds, we can of course ask whether given a measurably distal automorphism, all of its uniquely ergodic models are Möbius disjoint.[80]

We now focus on the famous class of Anzai skew products. This is the class of transformations defined on $\mathbb{T}^2$ by the formula:

$$T_\varphi : \mathbb{T}^2 \to \mathbb{T}^2, \ T_\varphi(x, y) = (x + \alpha, \varphi(x) + y).$$

In other words, Anzai skew products are given by $Tx = x + \alpha$ an irrational rotation on the (additive) circle, and a measurable $\varphi : \mathbb{T} \to \mathbb{T}$; the skew product T_φ preserves the Lebesgue measure. If φ is continuous, T_φ is a homeomorphism of $\mathbb{T}^2$. If we cannot solve the functional equations

$$k\varphi(x) = \xi(x) - \xi(Tx) \tag{11.45}$$

($k \in \mathbb{N}$) in continuous functions $\xi : \mathbb{T} \to \mathbb{T}$, then T_φ is minimal, but if for one $k \in \mathbb{N}$ we have a measurable solution then T_φ is not uniquely ergodic. In [122], we find examples of Anzai skew products which are minimal not uniquely ergodic but are Möbius disjoint,[81] moreover it is proved that if φ is analytic with an additional condition on the decay (from below) of Fourier coefficients then T_φ is Möbius disjoint for each irrational α. In [112], it is proved that if φ is of class $C^{1+\delta}$ then for a typical (in topological sense) α, we have Möbius disjointness of T_φ.[82] A remarkable result is proved by Wang [166]: all analytic Anzai skew products are Möbius disjoint. The proofs in all these papers are using Fourier analysis techniques but in [166], it is also a short interval argument from [127] used in one crucial case.

---

[79]In general, there is no uniquely ergodic model (X, T) of R with T topologically distal.

[80]As a matter of fact, such a question remains open even for 2-point extensions of irrational rotations.

[81]As a matter of fact, in [58] it is proved that if a uniquely ergodic homeomorphism T satisfies the strong MOMO property (see Definition 1 on page 198) and (continuous) $\varphi : X \to G$ (G is a compact Abelian group) satisfies $\varphi := \xi - \xi \circ T$ has a measurable solution $\xi : X \to G$, then the homeomorphism T_φ of $X \times G$ is Möbius disjoint. This applies if (11.45) has a measurable solution for $k = 1$. It is however an open question whether we have Möbius disjointness when there is no measurable solution for $k = 1$ but there is such a solution for some $k \geqslant 2$.

[82]It follows from a subsequent paper [113] that the Anzai skew products considered in [112] enjoy the AOP property.

Nothing seems to be proved about a PNT in the class of distal systems (except for rotations).

11.6.3.1 Discrete Spectrum Automorphisms

The simplest examples of (measurably) distal automorphisms are those with discrete spectrum. Recall that a measure-theoretic system $(Z, \mathcal{D}, \kappa, R)$ is said *to have discrete spectrum* if the L^2-space is generated by the eigenfunctions of the Koopman operator $Tf := f \circ T$. The classical Halmos-von Neumann theorem tells us that each ergodic automorphism with discrete spectrum has a uniquely ergodic model being a rotation on a compact Abelian (monothetic) group.

Theorem 11.73 *All uniquely ergodic models of automorphisms with discrete spectrum are Möbius disjoint.*

This result was first proved in [59] for totally ergodic discrete spectrum automorphisms (as they have the AOP property) and in full generality by Huang, Wang and Zhang in [93]. In fact, the latter result is stronger:

Theorem 11.74 ([93]) *Let (X, T) be a dynamical system, $x \in X$ and $N_i \to \infty$. Assume that $\frac{1}{N_i} \sum_{n \leqslant N_i} \delta_{T^n x} \xrightarrow[i \to \infty]{} \mu$. Assume that μ is a convex combination of countably many ergodic measures, each of which yields a system with discrete spectrum. Then $\lim_{i \to \infty} \frac{1}{N_i} \sum_{n \leqslant N_i} f(T^n x) \boldsymbol{\mu}(n) = 0$ for each $f \in C(X)$.*

Note that Theorem 11.73 also follows from Theorem 11.50 because ergodic rotations enjoy the strong MOMO property [58] (see Remark 11.51). As a matter of fact, as we have already noticed in Corollary 11.37, Theorem 11.73 follows from [127].

11.6.4 Sub-polynomial Complexity

Let T be a homeomorphism of a compact metric space (X, d) and let $\mu \in M(X, T)$. Assume also that $a \colon \mathbb{N} \to \mathbb{R}$ is increasing with $\lim_{n \to \infty} a(n) = \infty$. In the spirit of [63], we say that the measure complexity of μ is weaker than a if

$$\liminf_{n \to \infty} \frac{\min\{m \geqslant 1 : \mu(\bigcup_{j=1}^{m} B_{d_n}(x_j, \varepsilon)) > 1 - \varepsilon \text{ for some } x_1, \ldots, x_m \in X\}}{a(n)} = 0$$

for each $\varepsilon > 0$ (here $d_n(y, z) = \frac{1}{n} \sum_{j=1}^{n} d(T^j y, T^j z)$).
The main result of the recent article [95] states the following:

Theorem 11.75 ([95]) *If (X, T) is a topological system for which all its invariant measures have sub-polynomial complexity, i.e. their complexity is weaker than n^δ for each $\delta > 0$, then (X, T) is Möbius disjoint.*

As shown in [95], Theorem 11.75 applies to: topological systems whose all invariant measures yield systems with discrete spectrum (cf. Corollary 11.37), Anzai skew products of C^∞-class (over each irrational rotation), $K(\mathbb{Z})$-sequences introduced by Veech [161] and tame systems.[83]

11.6.5 Systems of Number-Theoretic Origin

Recall that a sequence $x \in \{0, 1\}^{\mathbb{N}}$ is called a *generalized Morse sequence* [107] if

$$x = b^0 \times b^1 \times \dots \tag{11.46}$$

with $b^i \in \{0, 1\}^{\ell_i}$, $\ell_i \geq 2$, $b^i(0) = 0$ for each $i \geq 0$.[84] The following question still remains open.

Question 10 (Mauduit (2014)) Are dynamical systems arising from generalized Morse sequences Möbius disjoint?

Consider the simplest subclass of the class of generalized Morse sequences, for which in (11.46) we have $|b^i| = 2$ for all $i \geq 0$ (in other words, either $b^i = 01$ or $b^i = 00$). Such sequences are called *Kakutani sequences* [116]. A particular case of Sarnak's conjecture, namely:

$$\frac{1}{N} \sum_{n=1}^{N} (-1)^{x(n)} \mu(n) \to 0, \tag{11.47}$$

for the classical Thue-Morse sequence $x = 01 \times 01 \times \dots$ follows from [98, 106] (see also [31] where, additionally, the speed of convergence to zero is given and [129], where, additionally, a PNT has been proved). Then (11.47) has been proved for some subclass of Kakutani sequences in [82]. As a matter of fact, in [82], the problem whether $\frac{1}{N} \sum_{n=1}^{N} (-1)^{s_E(n)} \mu(n) \to 0$ is considered. Here $E \subset \mathbb{N}$ is fixed and $s_E(n) := \sum_{i \in E} n_i$, where $n = \sum_{i=0}^{\infty} n_i 2^i$ ($n_i \in \{0, 1\}$). To see a relationship with Kakutani sequences define a Kakutani sequence $x = b^0 \times b^1 \times \dots$ with $b^n = 01$ iff $n + 1 \in E$; it is now not hard to see that $s_E(n) = x(n) \bmod 2$. Finally, using some methods from [129], Bourgain [19] completed the result from [82] so that (11.47) holds in the whole class of Kakutani sequences (moreover, in [19, 82] a relevant PNT has been proved). One can show that the methods used in the aforementioned papers allow us to have (11.47) with x replaced by every $y \in \overline{O(x)}$ (as shown in [66] in Lemma 6.5 therein, this can be sufficient to show Möbius disjointness for

---

[83]For the latter two classes all invariant measures yield discrete spectrum.

[84]If $B \in \{0, 1\}^k$ and $C = C(0)C(1)\dots C(\ell-1) \in \{0, 1\}^{\ell}$ then we define $B \times C := (B+C(0))(B+C(1))\dots(B+C(\ell-1))$.

the simple spectrum case; for example, this approach works for the Thue-Morse system).

The problem of Möbius disjointness is also studied (and solved) in the class of (generalized) Kakutani sequences taking values in compact (even non-Abelian) groups, see [161].

11.6.6   Other Research Around Sarnak's Conjecture

As all periodic observable sequences are orthogonal to μ, one could think that a limit of periodic constructions of type of Toeplitz sequences[85] also yields systems that are Möbius disjoint.[86] However, in [56] (and then [44]) there are examples of Toeplitz systems which are not Möbius orthogonal. These examples have positive entropy [44, 57]. Karagulyan in [103] shows Möbius disjointness of zero entropy continuous maps of the interval and (orientation preserving) homeomorphisms of the circle. In [51], Eisner proposes to study a polynomial version of Sarnak's conjecture (in the minimal case). See also [36, 43, 52, 53, 62, 94].

11.7   Related Research: $\mathcal{B}$-free Numbers

11.7.1   Introduction

11.7.1.1   Sets of Multiples

We have already seen that some properties of the Möbius function μ can be investigated by looking at its square μ^2, i.e. the characteristic function of the set of square-free numbers $Q := \{n \in \mathbb{Z} : p^2 \nmid n \text{ for all primes } p\}$. A natural generalization comes when we study sets of integers that are not divisible by elements of a given set. Let $\mathcal{B} \subset \mathbb{N}$ and let $\mathcal{M}_{\mathcal{B}}$ be the corresponding set of multiples, i.e. $\mathcal{M}_{\mathcal{B}} = \bigcup_{b \in \mathcal{B}} b\mathbb{Z}$ and the associated set of $\mathcal{B}$-free numbers $\mathcal{F}_{\mathcal{B}} := \mathbb{Z} \setminus \mathcal{M}_{\mathcal{B}}$ (for convenience, we will deal now with subsets of $\mathbb{Z}$ instead of subsets of $\mathbb{N}$— the Möbius function μ is not defined for negative arguments, but its square has a natural extension to negative integers). By $\eta = \eta_{\mathcal{B}}$ we will denote the characteristic function of $\mathcal{F}_{\mathcal{B}}$. It is not hard to see that a subset $F \subset \mathbb{Z}$ is a $\mathcal{B}$-free set (for some $\mathcal{B}$) if F is closed under taking divisors.

---

[85]A sequence $x \in A^{\mathbb{N}}$ is called Toeplitz if for each $n \in \mathbb{N}$ there is $q_n \in \mathbb{N}$ such that $x(n+jq_n) = x(n)$ for each $j = 0, 1, \ldots$

[86]So called regular Toeplitz sequences are treated in [56] and [44], these are however uniquely ergodic models of odometers.

11.7.1.2 Historical Remarks

Sets of multiples were an object of intensive studies already in the 1930s [15, 26, 32, 61]. The basic motivating example there was the set of *abundant numbers* ($n \in \mathbb{Z}$ is *abundant* if $|n|$ is smaller than the sum of its (positive) proper divisors, i.e. $|n| < \sigma(|n|)$), see also more recent [97, 100, 110] on that subject. Also many natural questions on general $\mathcal{B}$-free sets emerged. Besicovitch [14] showed that the asymptotic density of $\mathcal{M}_{\mathcal{B}}$ may fail to exist. It turned out that it was more natural to use the notion of logarithmic density (denoted by δ) which always exists in this case and equals the lower density. More precisely, we have the following result of Davenport and Erdös:

Theorem 11.76 ([34, 35]) *For any* $\mathcal{B}$, *the logarithmic density* $\delta(\mathcal{M}_{\mathcal{B}})$ *of* $\mathcal{M}_{\mathcal{B}}$ *exists. Moreover,* $\delta(\mathcal{M}_{\mathcal{B}}) = \underline{d}(\mathcal{M}_{\mathcal{B}}) = \lim_{n\to\infty} d(\mathcal{M}_{\{b\in\mathcal{B}:b\leqslant n\}})$.

In the so-called *Erdös case* when $\mathcal{B}$ consists of pairwise coprime elements whose sum of reciprocals converges, the density does exist, cf. [86] (in particular, $\mathbb{1}_{\mathcal{F}_{\mathcal{B}}}$ is rational). We refer the reader to [86, 87] for a coherent, self-contained introduction to the theory of sets of multiples from the analytic and probabilistic number theory viewpoint.

11.7.1.3 Dynamics Comes Into Play

Sarnak in [150], suggested to study μ^2 from the dynamical viewpoint and he announced the following results:

(i) μ^2 is *generic* for an ergodic S-invariant measure ν_{μ^2} on $\{0, 1\}^{\mathbb{Z}}$ such that the measure-theoretical dynamical system $(X_{\mu^2}, \nu_{\mu^2}, S)$ has zero measure-theoretic entropy[87];
(ii) the topological entropy of (X_{μ^2}, S) is equal to $\frac{6}{\pi^2}$;
(iii) $X_{\mu^2} = X_{\{p^2 : p \in \mathbb{P}\}}$ (see the definition of admissibility below);
(iv) (X_{μ^2}, S) is *proximal*.

This triggered intensive research in analogous direction for dynamical systems given by other $\mathcal{B}$-free sets. In [55], El Abdalaoui, Lemańczyk and de la Rue developed the necessary tools in the Erdös case and covered (i)–(iii) from the above list. Given $\mathcal{B} = \{b_k : k \geqslant 1\}$, In particular, they defined a function $\varphi : G = \prod_{k\geqslant 1} \mathbb{Z}/b_k\mathbb{Z} \to \{0, 1\}^{\mathbb{Z}}$ given by

$$\varphi(g)(n) = 1 \iff g_k + n \not\equiv 0 \bmod b_k \text{ for all } k \geqslant 1.$$

---

[87]This is clearly a refinement of the fact that the asymptotic density of square-free integers exists (it is given by $6/\pi^2 = 1/\zeta(2)$). It follows that μ^2 is a completely deterministic point.

Note that $\eta_{\mathcal{B}} = \varphi(0)$ and φ is the coding of points under the translation by $(1, 1, \dots)$ on G with respect to a two-set partition $\{W, W^c\}$, where

$$W = \{h \in G : h_b \neq 0 \text{ for all } b \in \mathcal{B}\}. \tag{11.48}$$

This study was continued in a general setting in [48] and the first obstacle was that it was no longer clear which subshift to study—it turned out that the most important role is played by the following three subshifts, which coincide in the Erdös case (for the square-free, case see [140] by Peckner and for the Erdös case, see [55]):

- X_η is the closure of the orbit of $\eta_{\mathcal{B}}$ under S ($\mathcal{B}$-free subshift),
- $\widetilde{X}_\eta$ is the smallest hereditary subshift containing X_η (a subshift (X, S) is *hereditary*, whenever $x \in X$ and $y \leqslant x$ coordinatewise, then $y \in X$),
- $X_{\mathcal{B}}$ is the set of $\mathcal{B}$-admissible sequences, i.e. of $x \in \{0, 1\}^{\mathbb{Z}}$ such that, for each $b \in \mathcal{B}$, the support $\text{supp}\, x := \{n \in \mathbb{Z} : x(n) = 1\}$ of x taken modulo b is a proper subset of $\mathbb{Z}/b\mathbb{Z}$ ($\mathcal{B}$-admissible subshift).

Remark 11.77 As $X_{\mathcal{B}}$ is hereditary, we have $X_\eta \subset \widetilde{X}_\eta \subset X_{\mathcal{B}}$. In the Erdös case, we have $X_\eta = X_{\mathcal{B}}$ [55] (for the square-free system [150]).

Also the group G turned out to be too large for the studies—it is natural to consider its closed subgroup

$$H := \overline{\{(n, n, \dots) \in G : n \in \mathbb{Z}\}}. \tag{11.49}$$

In the Erdös case we have $H = G$. Certain special cases more general than the Erdös one were considered in [48]:

- we say that $\mathcal{B}$ is *taut* whenever $\delta(\mathcal{F}_{\mathcal{B}}) < \delta(\mathcal{F}_{\mathcal{B}\setminus\{b\}})$ for each $b \in \mathcal{B}$;
- we say that $\mathcal{B}$ has *light tails*, i.e. $\overline{d}(\sum_{b>K} b\mathbb{Z}) \to 0$ as $K \to \infty$.

Following [87], we also say that $\mathcal{B}$ is *Besicovitch* if $d(\mathcal{M}_{\mathcal{B}})$ exists (equivalently, $d(\mathcal{F}_{\mathcal{B}})$ exists). A set $\mathcal{B} \subset \mathbb{N} \setminus \{1\}$ is called *Behrend* if $\delta(\mathcal{M}_{\mathcal{B}}) = 1$. Throughout, we will tacitly assume that $\mathcal{B}$ is *primitive*, i.e. does not contain $b \neq b'$ with $b|b'$. Recall that $\mathcal{B}$ is taut if and only if $\mathcal{B}$ does not contain $d\mathcal{A}$, where $\mathcal{A} \subset \mathbb{N} \setminus \{1\}$ is Behrend and $d \in \mathbb{N}$.

11.7.1.4 Further Generalizations

Several further generalizations of $\mathcal{B}$-free integers were discussed in the literature from the dynamical viewpoint. Let us briefly recall them here:

- Pleasants and Huck [141] considered *k-free lattice points* $\mathcal{F}_k = \mathcal{F}_k(\Lambda) := \Lambda \setminus \bigcup_{p \in \mathbb{P}} p^k \Lambda$, where Λ is a lattice in $\mathbb{R}^d$ (the corresponding dynamical system given by the orbit closure of $\mathbb{1}_{\mathcal{F}_k} \in \{0, 1\}^{\Lambda}$ under the multidimensional shift).

- Cellarosi and Vinogradov [24] considered *k-free integers in number fields* $\mathscr{F}_k =$ $\mathscr{F}_k(O_K) := O_K \setminus \bigcup_{\mathfrak{p} \in \mathfrak{P}} \mathfrak{p}^k$. Here K is a finite extension of $\mathbb{Q}$, $O_K \subset K$ is the ring of integers, $\mathfrak{P}$ stands for the family of all prime ideals in O_K and $\mathfrak{p}^k$ stands for $\mathfrak{p} \ldots \mathfrak{p}$ ($\mathfrak{p}$ is taken k times).
- Baake and Huck in their survey [5] considered $\mathscr{B}$-*free lattice points* $\mathscr{F}_\mathscr{B} =$ $\mathscr{F}_\mathscr{B}(\Lambda) := \Lambda \setminus \bigcup_{b \in \mathscr{B}} b\Lambda$. Here Λ is a lattice in $\mathbb{R}^d$ and $\mathscr{B} \subseteq \mathbb{N} \setminus \{1\}$ is an infinite pairwise coprime set with $\sum_{b \in \mathscr{B}} 1/b^d < \infty$.
- Finally, one can consider $\mathscr{B}$-free integers $\mathscr{F}_\mathscr{B}$ in number fields as suggested in [5]. Here K is a finite extension of $\mathbb{Q}$, $O_K \subset K$ is the ring of integers and $\mathscr{B}$ is a family of pairwise coprime ideals in O_K such that the sum of reciprocals of their norms converges.

We will recall some of the main results from the above papers in the relevant sections below.

11.7.2 Invariant Measures and Entropy

11.7.2.1 Mirsky Measure

Cellarosi and Sinai proved (i) in [23]: they showed that v_{μ^2} is generic for a shift-invariant measure v_{μ^2} on $\{0, 1\}^{\mathbb{Z}}$, and that $(X_{\mu^2}, v_{\mu^2}, S)$ is isomorphic to a rotation on the compact Abelian group $\prod_{p \in \mathbb{P}} \mathbb{Z}/p^2\mathbb{Z}$. In particular, $(X_{\mu^2}, v_{\mu^2}, S)$ is of zero Kolmogorov entropy.[88] In case of k-free lattice points and k-free integers in number fields an analogous result can be found in [141] and [24], respectively and for $\mathscr{B}$-free lattice points it was announced in [5]. Recently, Huck [96] showed that in case of $\mathscr{B}$-free integers in number fields, the logarithmic density of $\mathscr{F}_\mathscr{B}$ always exists and equals the lower density, thus extending Theorem 11.76 in the (1-dimensional) Erdös case.

Since $\mathscr{F}_\mathscr{B}$ may fail to have asymptotic density, the more η may fail to be a generic point. However (Proposition E in [48]), for any $\mathscr{B} \subset \mathbb{N}$, η is always a quasi-generic point for a natural ergodic S-invariant measure v_η on $\{0, 1\}^{\mathbb{Z}}$ (the relevant Mirsky measure). Moreover, $\mathscr{B}$ is Besicovitch if and only if η is generic for v_η. Now, if we additionally assume that $\mathscr{B}$ is taut, then (X_η, v_η, S) is isomorphic to an ergodic rotation on a compact metric group (Theorem F in [48]).[89] In particular, (X_η, v_η, S) has zero entropy.

Finally, for a generalization to so-called weak model sets, see [6], and for some results related to the distribution of $\mathscr{B}$-free integers, see [2, 3].

---

[88]The frequencies of blocks on μ^2 were first studied by Mirsky [132, 133] and that is why we refer to v_{μ^2} (and the analogous measure in case of general $\mathscr{B}$-free systems) as the *Mirsky measure*.

[89]More precisely, it is isomorphic to $(H, \mathbb{P}, T)$, where H is the closure of $\{(n \bmod b_k)_{k \geqslant 1} : n \in \mathbb{Z}\}$ in $\prod_{k \geqslant 1} \mathbb{Z}/b_k\mathbb{Z}$ and $Tg = g + (1, 1, \ldots)$, cf. (11.49).

11.7.2.2 Entropy

The topological entropy of X_{μ^2} is positive and equals $6/\pi^2 = \prod_{p \in \mathbb{P}}(1 - 1/p^2) = d(\mathcal{F}_\mathcal{B})$ for $\mathcal{B} = \{p^2 : p \in \mathbb{P}\}$, see [140]. This extends to the Erdös case, where the topological entropy of $X_\eta = \widetilde{X}_\eta = X_\mathcal{B}$ equals $\prod_{b \in \mathcal{B}}(1 - 1/b) = d(\mathcal{F}_\mathcal{B})$, see [55]. In the general case of $\mathcal{B}$-free systems, we have $h_{top}(\widetilde{X}_\eta, S) = h_{top}(X_\mathcal{B}, S) = \delta(\mathcal{F}_\mathcal{B})$ (see Proposition K in [48]). The formula for the topological entropy of k-free lattice points is provided in [141].

In view of the variational principle, the positivity of the topological entropy evokes two problems: whether the system under consideration is intrinsically ergodic (i.e. whether there is a unique measure of maximal entropy) and to describe the set of all invariant measures. We address them next.

11.7.2.3 Maximal Entropy Measure

In the square-free case, the intrinsic ergodicity is proved by Peckner in [140]. This extends to the Erdös case, see [114] by Kułaga-Przymus, Lemańczyk and Weiss. Finally, for any $\mathcal{B} \subset \mathbb{N}$, the subshift $(\widetilde{X}_\eta, S)$ is intrinsically ergodic, see Theorem J in [48]. In particular, if $\mathcal{B}$ has light tails and contains an infinite pairwise coprime subset then $(X_\mathcal{B}, S)$ is intrinsically ergodic.

11.7.2.4 All Invariant Measures

Notice that for each $\mathcal{B}$, the map $M: X_\eta \times \{0, 1\}^\mathbb{Z} \to \widetilde{X}_\eta$ given by the coordinatewise multiplication of sequences is well-defined and each $S \times S$-invariant measure ρ on $X_\eta \times \{0, 1\}^\mathbb{Z}$ yields an S-invariant measure on $\widetilde{X}_\eta$. In particular, this applies to those ρ whose projection on the first coordinate is ν_η. It turns out that the converse is also true: for any S-invariant measure ν on $\widetilde{X}_\eta$ there exists an $S \times S$-invariant measure ρ on $X_\eta \times \{0, 1\}^\mathbb{Z}$ whose projection on the first coordinate is ν_η and such that $M_*(\rho) = \nu$. For the Erdös case see [114] and for general $\mathcal{B}$-free systems, see Theorem I in [48] (for further generalizations of $\mathcal{B}$-free systems listed before (see page 222) no analogous description of the set of all invariant measures is known).

It turns out that a special role is played by $\mathcal{B}$ that are taut. We have the following: for any $\mathcal{B}$, there exists a unique taut set $\mathcal{B}' \subset \mathbb{N}$ such that $\mathcal{F}_{\mathcal{B}'} \subset \mathcal{F}_\mathcal{B}$, $\widetilde{X}_{\eta'} \subset \widetilde{X}_\eta$ and all S-invariant measures on $\widetilde{X}_\eta$ are in fact supported on $\widetilde{X}_{\eta'}$ (Theorem C in [48]).

More subtle properties of the simplex of invariant measures of the $\mathcal{B}$-shift have been studied in [115] by Kułaga-Przymus, Lemańczyk and Weiss—it was shown that in the positive entropy case the simplex of S-invariant measures on $\widetilde{X}_\eta$ is *Poulsen*, i.e. the ergodic measures are dense. In particular, if we additionally know that X_η is hereditary (and has positive entropy), then its simplex of invariant measures is Poulsen. However, this is no longer true for a general (not necessarily $\mathcal{B}$-free) hereditary system. On the other hand, Konieczny, Kupsa and Kwietniak [111] showed that the set of ergodic invariant measures of a hereditary shift is always arcwise connected (when endowed with the d-bar metric).

11.7.3 Topological Results

A lot can be said about the topological properties of (X_η, S). E.g. for any $\mathcal{B} \subset \mathbb{N}$ the subshift X_η has a unique minimal subset that is the orbit closure of a Toeplitz system (Theorem A in [48]). In particular, X_η is minimal if and only if X_η is a Toeplitz system.[90] In fact, η itself can be a Toeplitz sequence (see Example 3.1 in [48]) and it was shown in [105] that η is a Toeplitz sequence different from $\ldots 0.00 \ldots$ if and only if $\mathcal{B}$ does not contain a subset of the form $d\mathcal{A}$, where $d \in \mathbb{N}$ and $\mathcal{A} \subset \mathbb{N} \setminus \{1\}$ is infinite and pairwise coprime. Moreover, if η is Toeplitz then $\mathcal{B}$ is necessarily taut [105].

On the other hand, the proximality of X_η is equivalent to $\{\ldots 0.00 \ldots\}$ being the unique minimal subset of X_η. Moreover, X_η is proximal if and only if $\mathcal{B}$ contains an infinite pairwise coprime subset (Theorem B in [48]).

Some of the properties of the $\mathcal{B}$-free subshift X_η can be characterized via properties of a set W called the *window*: $W = \{h \in H : h_b \neq 0 \text{ for all } b \in \mathcal{B}\}$, cf. (11.48). This name has its origins in the theory of weak model sets (for more details see [4]); $\mathcal{F}_\mathcal{B}$ is an example of such a set. Again a special role is played by sets $\mathcal{B}$ that are taut. In [105], Kasjan, Keller and Lemańczyk show the following:

- $\mathcal{B}$ is taut if and only if W is Haar regular, i.e. the topological support of Haar measure restricted to W is the whole W;
- if $\mathcal{B}$ is primitive then X_η is a Toeplitz system if and only if W is topologically regular;
- X_η is proximal if and only if W has empty interior.

In [105] there is also a detailed description of the maximal equicontinuous factor of X_η (with no extra assumptions on $\mathcal{B}$). See also [109].

Clearly, if X_η is hereditary, i.e. $X_\eta = \widetilde{X}_\eta$ then $(\ldots 0.00 \ldots) \in X_\eta$ and hence X_η is proximal. If we assume that $\mathcal{B}$ has light tails and contains an infinite pairwise coprime subset then the converse is true: proximality yields heredity (Theorem D in [48]). However, $\widetilde{X}_\eta = X_\mathcal{B}$ may fail to hold, even under quite strong assumptions on $\mathcal{B}$. Indeed, the set of abundant numbers $\mathbb{A}$ is the corresponding set of multiples $\mathcal{M}_\mathcal{B}$ for a certain set $\mathcal{B}$ with the property that $\sum_{b \in \mathcal{B}} 1/b < \infty$. Here, $\widetilde{X}_\eta \neq X_\mathcal{B}$, see Section 11 in [48].

More subtle results on heredity were recently obtained by Keller in [108]. He shows that whenever X_η is proximal then it is contained in a slightly larger subshift that is hereditary (there is no need to make extra assumptions on $\mathcal{B}$). He also generalizes the concept of heredity to the non-proximal case.

---

[90]This has been recently improved in [105] and by A. Bartnicka: X_η is minimal if and only if η is Toeplitz.

It is also interesting to ask about the (invertible) centralizer of (S, X_η). In the Erdös case it was proved by Mentzen[91] in [131] that the group of homeomorphisms commuting with the shift (S, X_η) consists only of the powers of S. In case of some Toeplitz $\mathcal{B}$-free systems an analogous result was proved by Bartnicka in [49].

Question 11 Is the invertible centralizer trivial for each $\mathcal{B}$-free subshift?

11.7.4 Ergodic Ramsey Theory

We will now see some connections of the theory of $\mathcal{B}$-free sets with the theory uniform distribution and ergodic Ramsey theory.

11.7.4.1 Polynomial Recurrence and Divisibility

Recall that Szemerédi showed [154] that any set $S \subset \mathbb{N}$ with positive upper density contains arbitrarily long arithmetic progressions and Furstenberg [75, 76] introduced an ergodic approach to this result that proved very fruitful from the point of view of various generalizations. In particular, it allowed one to prove the following: for any probability space $(X, \mathcal{B}, \mu)$, invertible measure preserving transformation $T: X \to X$, $A \in \mathcal{B}$ with $\mu(A) > 0$ and any polynomials $p_i \in \mathbb{Q}[t]$ satisfying $p_i(\mathbb{Z}) \subset \mathbb{Z}$ and $p_i(0) = 0$, $1 \leqslant i \leqslant \ell$, there exists arbitrarily large $n \in \mathbb{N}$ such that

$$\mu\left(A \cap T^{-p_1(n)}A \cap \ldots \cap T^{-p_\ell(n)}A\right) > 0. \tag{11.50}$$

In fact, we have

$$\lim_{N \to \infty} \frac{1}{N} \sum_{n=1}^{N} \mu\left(A \cap T^{-p_1(n)}A \cap \ldots \cap T^{-p_\ell(n)}A\right) > 0$$

[9, 90, 118]. One can now restrict attention to a specific subset R of $n \in \mathbb{N}$ for which we ask whether (11.50) holds or even demand

$$\lim_{N \to \infty} \frac{1}{|R \cap [1, N]|} \sum_{n=1}^{N} \mathbb{1}_R(n)\mu\left(A \cap T^{-p_1(n)}A \cap \ldots \cap T^{-p_\ell(n)}A\right) > 0. \tag{11.51}$$

If (11.51) holds for any invertible measure preserving system $(X, \mathcal{B}, \mu, T)$, $A \in \mathcal{B}$ with $\mu(A) > 0$, $\ell \in \mathbb{N}$ and any polynomials $p_i \in \mathbb{Q}[t]$, $i = 1, \ldots, \ell$, with $p_i(\mathbb{Z}) \subset \mathbb{Z}$ and $p_i(0) = 0$ for all $i \in \{1, \ldots, \ell\}$, we say (cf. [8, Definition 1.5]) that $R \subset \mathbb{N}$ is

---

[91]Mentzen's result is extended in [7] to every hereditary $\mathcal{B}$-free subshift.

averaging set of polynomial multiple recurrence. If $\ell = 1$, we speak of an *averaging set of polynomial single recurrence*.

We will be interested in polynomial recurrence for $\mathcal{B}$-free sets. Before we get there, let us direct our attention to so-called rational sets. Recall that $R \subset \mathbb{N}$ is rational if it can be approximated in density by finite unions of arithmetic progressions, cf. footnote 18. Note that the rationality of $\mathcal{F}_{\mathcal{B}}$ is equivalent to $\mathcal{B}$ being Besicovitch. An easy necessary condition for $R \subset \mathbb{N}$ to be an averaging set of polynomial recurrence is that the density of $R \cap u\mathbb{N}$ exists and is positive for any $u \in \mathbb{N}$ (indeed, otherwise consider the cyclic rotation on $\mathbb{Z}/u\mathbb{Z}$ to see that even usual recurrence fails). If the latter holds, we will say that R is *divisible*. It turns out that in case of rational sets, divisibility is not only necessary but also sufficient for polynomial recurrence. More precisely, we have the following:

Theorem 11.78 ([13]) *Let $R \subset \mathbb{N}$ be rational and of positive density. The following conditions are equivalent:*

(a) R is divisible.
(b) R is an averaging set of polynomial single recurrence.
(c) R is an averaging set of polynomial multiple recurrence.

Recall that it was proved in [10] that every self-shift $Q - r$, $r \in Q$, of the set of square-free numbers Q is divisible and these are the only divisible shifts of Q. For general $\mathcal{B}$-free sets the situation is more complicated and we have the following result:

Theorem 11.79 ([13]) *Given $\mathcal{B} \subset \mathbb{N}$ that is Besicovitch, there exists a set $D \subset \mathcal{F}_{\mathcal{B}}$ with $d(\mathcal{F}_{\mathcal{B}} \setminus D) = 0$ such that the set $\mathcal{F}_{\mathcal{B}} - r$ is an averaging set of polynomial multiple recurrence if and only if $r \in D$. Moreover, $D = \mathcal{F}_{\mathcal{B}}$ if and only if the set $\mathcal{B}$ is taut.*

This can be generalized to $\mathcal{B}$ that are not Besicovitch by considering divisibility and recurrence along a certain subsequence $(N_k)_{k \geqslant 1}$. As a combinatorial application, one obtains in [13] the following result: Suppose that $(N_k)_{k \geqslant 1}$ is such that the density of $\mathcal{F}_{\mathcal{B}}$ along $(N_k)_{k \geqslant 1}$ exists and is positive. Then there exists $D \subset \mathcal{F}_{\mathcal{B}}$ which equals $\mathcal{F}_{\mathcal{B}}$ up to a set of zero density along $(N_k)_{k \geqslant 1}$ such that for all $r \in D$ and for all $E \subset \mathbb{N}$ with positive upper density, for any polynomials $p_i \in \mathbb{Q}[t]$, $i = 1, \ldots, \ell$, which satisfy $p_i(\mathbb{Z}) \subset \mathbb{Z}$ and $p_i(0) = 0$, for all $1 \leqslant i \leqslant \ell$, there exists $\beta > 0$ such that the set

$$\left\{ n \in \mathcal{F}_{\mathcal{B}} - r : \overline{d}\Big(E \cap (E - p_1(n)) \cap \ldots \cap (E - p_\ell(n))\Big) > \beta \right\}$$

has positive lower density along $(N_k)_{k \geqslant 1}$. If, additionally, $\mathcal{B}$ is taut then one can take $D = \mathcal{F}_{\mathcal{B}}$.

Results of similar flavor as above have been also obtained in [12] in the context of level sets of multiplicative functions. In particular, if E is a level set of a multiplicative function and has positive density then every self-shift of E is an averaging set of polynomial multiple recurrence (Corollary C in [12]). The key tool here is (11.17) that provides an important link between level sets of multiplicative functions and rational sets. See also [11].

Acknowledgements The research resulting in this survey was carried out during the Research in Pairs Program of CIRM, Luminy, France, 15-19.05.2017. J. Kułaga-Przymus and M. Lemańczyk also acknowledge the support of Narodowe Centrum Nauki grant UMO-2014/15/B/ST1/03736. J. Kułaga-Przymus was also supported by the European Research Council (ERC) under the European Union's Horizon 2020 research and innovation programme (grant agreement No 647133 (ICHAOS)) and by the Foundation for Polish Science (FNP).

The authors special thanks go to N. Frantzikinakis and P. Sarnak for a careful reading of the manuscript, numerous remarks and suggestions to improve presentation. We also thank M. Baake, V. Bergelson, B. Green, D. Kwietniak, C. Mauduit and M. Radziwiłł for some useful comments on the subject.

References

1. T. Adams, S. Ferenczi, K. Petersen, Constructive symbolic presentations of rank one measure-preserving systems. Coll. Math. **150**(2), 243–255 (2017)
2. M. Avdeeva, Variance of $\mathcal{B}$-free integers in short intervals. Preprint (2015). https://arxiv.org/abs/1512.00149
3. M. Avdeeva, F. Cellarosi, Y.G. Sinai, Ergodic and statistical properties of $\mathcal{B}$-free numbers. Teor. Veroyatnost. i Primenen. **61**, 805–829 (2016)
4. M. Baake, U. Grimm, *Aperiodic Order. Vol 1. A Mathematical Invitation, Encyclopedia of Mathematics and Its Applications*, vol.149 (Cambridge University Press, Cambridge, 2013)
5. M. Baake, C. Huck, Ergodic properties of visible lattice points. Proc. Steklov Inst. Math. **288**(1), 165–188 (2015) (English)
6. M. Baake, C. Huck, N. Strungaru, On weak model sets of extremal density. Indag. Math. **28**(1), 3–31 (2017). Special Issue on Automatic Sequences, Number Theory, and Aperiodic Order
7. M. Baake, C. Huck, M. Lemańczyk, Positive entropy shifts with small centraliser and large normaliser (in preparation)
8. V. Bergelson, I.J. Håland, Sets of recurrence and generalized polynomials, in *Convergence in Ergodic Theory and Probability (Columbus, OH, 1993)*. Ohio State University Mathematical Research Institute Publications, vol. 5 (de Gruyter, Berlin, 1996), pp. 91–110
9. V. Bergelson, A. Leibman, Polynomial extensions of van der Waerden's and Szemerédi's theorems. J. Am. Math. Soc. **9**(3), 725–753 (1996)
10. V. Bergelson, I. Ruzsa, Squarefree numbers, IP sets and ergodic theory, in *Paul Erdős and His Mathematics, I (Budapest, 1999)*. Bolyai Society Mathematical Studies, vol. 11 (János Bolyai Mathematical Society, Budapest, 2002), pp. 147–160
11. V. Bergelson, J. Kułaga-Przymus, M. Lemańczyk, F.K. Richter, A generalization of Kátai's orthogonality criterion with applications. Preprint (2017). https://arxiv.org/abs/1705.07322
12. V. Bergelson, J. Kułaga-Przymus, M. Lemańczyk, F.K. Richter, A structure theorem for level sets of multiplicative functions and applications. Int. Math. Res. Not. (2017, to appear). https://arxiv.org/abs/1708.02613
13. V. Bergelson, J. Kułaga-Przymus, M. Lemańczyk, F.K. Richter, Rationally almost periodic sequences, polynomial multiple recurrence and symbolic dynamics. Erg. Theory Dyn. Syst. (to appear). https://arxiv.org/abs/1611.08392
14. A.S. Besicovitch, On the density of certain sequences of integers. Math. Ann. **110**(1), 336–341 (1935)
15. E. Bessel-Hagen, *Zahlentheorie* (Teubner, Leipzig, 1929)

16. J. Bourgain, An approach to pointwise ergodic theorems, in *Geometric Aspects of Functional Analysis (1986/87)*. Lecture Notes in Mathematics, vol. 1317 (Springer, Berlin, 1988), pp. 204–223
17. J. Bourgain, On the maximal ergodic theorem for certain subsets of the integers. Isr. J. Math. **61**(1), 39–72 (1988)
18. J. Bourgain, On the pointwise ergodic theorem on L^p for arithmetic sets. Isr. J. Math. **61**(1), 73–84 (1988)
19. J. Bourgain, Möbius-Walsh correlation bounds and an estimate of Mauduit and Rivat. J. Anal. Math. **119**, 147–163 (2013)
20. J. Bourgain, On the correlation of the Moebius function with rank-one systems. J. Anal. Math. **120**, 105–130 (2013)
21. J. Bourgain, P. Sarnak, T. Ziegler, Disjointness of Möbius from horocycle flows, in *From Fourier Analysis and Number Theory to Radon Transforms and Geometry*. Developments in Mathematics, vol. 28 (Springer, New York, 2013), pp. 67–83
22. D. Carmon, Z. Rudnick, The autocorrelation of the Möbius function and Chowla's conjecture for the rational function field. Q. J. Math. **65**(1), 53–61 (2014)
23. F. Cellarosi, Y.G. Sinai, Ergodic properties of square-free numbers. J. Eur. Math. Soc. **15**(4), 1343–1374 (2013)
24. F. Cellarosi, I. Vinogradov, Ergodic properties of k-free integers in number fields. J. Mod. Dyn. **7**(3), 461–488 (2013)
25. J. Chaika, A. Eskin, Moebius disjointness for interval exchange transformations on three intervals. Preprint (2016). https://arxiv.org/abs/1606.02357
26. S. Chowla, On abundant numbers. J. Indian Math. Soc. New Ser. **1**, 41–44 (1934) (English)
27. S. Chowla, *The Riemann Hypothesis and Hilbert's Tenth Problem*. Mathematics and Its Applications, vol. 4 (Gordon and Breach Science Publishers, New York, 1965)
28. I.P. Cornfeld, S.V. Fomin, Y.G. Sinaĭ, *Ergodic Theory*. Grundlehren der Mathematischen Wissenschaften [Fundamental Principles of Mathematical Sciences], vol. 245 (Springer, New York, 1982)
29. C. Cuny, M. Weber, Ergodic theorems with arithmetical weights. Isr. J. Math. **217**(1), 139–180 (2017)
30. H. Daboussi, H. Delange, On multiplicative arithmetical functions whose modulus does not exceed one. J. Lond. Math. Soc. (2) **26**(2), 245–264 (1982)
31. C. Dartyge, G. Tenenbaum, Sommes des chiffres de multiples d'entiers. Ann. Inst. Fourier (Grenoble) **55**(7), 2423–2474 (2005)
32. H. Davenport, Über numeri abundantes. Sitzungsber. Preuss. Akad. Wiss. **26/29**, 830–837 (1933)
33. H. Davenport, On some infinite series involving arithmetical functions. II. Q. J. Math. **8**, 313–320 (1937)
34. H. Davenport, P. Erdös, On sequences of positive integers. Acta Arith. **2**(1), 147–151 (1936)
35. H. Davenport, P. Erdös, On sequences of positive integers. J. Indian Math. Soc. (N.S.) **15**, 19–24 (1951)
36. R. de la Breteche, G. Tenenbaum, A remark on Sarnak's conjecture. Preprint (2017). https://arxiv.org/abs/1709.01194
37. T. de la Rue, La fonction de Möbius à la rencontre des systèmes dynamiques. Gaz. Math. (150), 31–40 (2016)
38. A. del Junco, Disjointness of measure-preserving transformations, minimal self-joinings and category, in *Ergodic Theory and Dynamical Systems, I (College Park, MD, 1979–80)*. Progress in Mathematics, vol. 10 (Birkhäuser, Boston, 1981), pp. 81–89
39. A. del Junco, D. Rudolph, On ergodic actions whose self-joinings are graphs. Ergod. Theory Dyn. Syst. **7**(4), 531–557 (1987)
40. M. Denker, C. Grillenberger, K. Sigmund, *Ergodic Theory on Compact Spaces*. Lecture Notes in Mathematics, vol. 527 (Springer, Berlin, 1976)
41. J.-M. Deshouillers, M. Drmota, C. Müllner, Automatic sequences generated by synchronizing automata fulfill the Sarnak conjecture. Stud. Math. **231**(1), 83–95 (2015)

42. H.G. Diamond, Elementary methods in the study of the distribution of prime numbers. Bull. Am. Math. Soc. (N.S.) **7**(3), 553–589 (1982)
43. T. Downarowicz, E. Glasner, Isomorphic extensions and applications. Topol. Methods Nonlinear Anal. **48**(1), 321–338 (2016)
44. T. Downarowicz, S. Kasjan, Odometers and Toeplitz systems revisited in the context of Sarnak's conjecture. Stud. Math. **229**(1), 45–72 (2015)
45. T. Downarowicz, J. Serafin, Almost full entropy subshifts uncorrelated to the Möbius function. Int. Math. Res. Not. (to appear). https://arxiv.org/abs/1611.02084
46. M. Drmota, Subsequences of automatic sequences and uniform distribution, in *Uniform Distribution and Quasi-Monte Carlo Methods*. Radon Series on Computational and Applied Mathematics, vol. 15 (De Gruyter, Berlin, 2014), pp. 87–104
47. M. Drmota, C. Müllner, L. Spiegelhofer, Möbius orthogonality for the Zeckendorf sum-of-digits function. Proc. AMS (to appear). http://arxiv.org/abs/1706.09680
48. A. Dymek, S. Kasjan, J. Kułaga-Przymus, M. Lemańczyk, $\mathscr{B}$-free sets and dynamics. Trans. Am. Math. Soc. (2015, to appear). http://arxiv.org/abs/1509.08010
49. A. Dymek, Automorphisms of Toeplitz $\mathscr{B}$-free systems. Bull. Pol. Acad. Sci. Math. **65**(2), 139–152 (2017)
50. M. Einsiedler, T. Ward, *Ergodic Theory with a View Towards Number Theory*. Graduate Texts in Mathematics, vol. 259 (Springer, London, 2011)
51. T. Eisner, A polynomial version of Sarnak's conjecture. C. R. Math. Acad. Sci. Paris **353**(7), 569–572 (2015)
52. E.H. El Abdalaoui, M. Disertori, Spectral properties of the Möbius function and a random Möbius model. Stoch. Dyn. **16**(1), 1650005, 25 (2016)
53. E.H. El Abdalaoui, X. Ye, A cubic non-conventional ergodic average with multiplicative or von Mangoldt weights. Preprint (2016). https://arxiv.org/abs/1606.05630
54. E.H. El Abdalaoui, M. Lemańczyk, T. de la Rue, On spectral disjointness of powers for rank-one transformations and Möbius orthogonality. J. Funct. Anal. **266**(1), 284–317 (2014)
55. E.H. El Abdalaoui, M. Lemańczyk, T. de la Rue, A dynamical point of view on the set of $\mathscr{B}$-free integers. Int. Math. Res. Not. **2015**(16), 7258–7286 (2015)
56. E.H. El Abdalaoui, S. Kasjan, M. Lemańczyk, 0-1 sequences of the Thue-Morse type and Sarnak's conjecture. Proc. Am. Math. Soc. **144**(1), 161–176 (2016)
57. E.H. El Abdalaoui, J. Kułaga-Przymus, M. Lemańczyk, T. de la Rue, The Chowla and the Sarnak conjectures from ergodic theory point of view. Discrete Contin. Dyn. Syst. **37**(6), 2899–2944 (2017)
58. E.H. El Abdalaoui, J. Kułaga-Przymus, M. Lemańczyk, T. de la Rue, Möbius disjointness for models of an ergodic system and beyond. Isr. J. Math. (2017, to appear). https://arxiv.org/abs/1704.03506
59. E.H. El Abdalaoui, M. Lemańczyk, T. de la Rue, Automorphisms with quasi-discrete spectrum, multiplicative functions and average orthogonality along short intervals. Int. Math. Res. Not. (14), 4350–4368 (2017)
60. P.D.T.A. Elliott, *Probabilistic Number Theory. I Mean-Value Theorems*. Grundlehren der Mathematischen Wissenschaften [Fundamental Principles of Mathematical Science], vol. 239 (Springer, New York, 1979)
61. P. Erdös, On the density of the abundant numbers. J. Lond. Math. Soc. **9**(4), 278–282 (1934)
62. A. Fan, Weighted Birkhoff ergodic theorem with oscillating weights. Ergod. Theory Dyn. Syst. (2017). https://arxiv.org/abs/1705.02501
63. S. Ferenczi, Measure-theoretic complexity of ergodic systems. Isr. J. Math. **100**, 189–207 (1997)
64. S. Ferenczi, C. Mauduit, On Sarnak's conjecture and Veech's question for interval exchanges. J. d'Analyse Math. **134**, 545–573 (2018)
65. S. Ferenczi, C. Holton, L.Q. Zamboni, Structure of three-interval exchange transformations. II. A combinatorial description of the trajectories. J. Anal. Math. **89**, 239–276 (2003)

66. S. Ferenczi, J. Kułaga-Przymus, M. Lemańczyk, C. Mauduit, Substitutions and Möbius disjointness, in *Ergodic Theory, Dynamical Systems, and the Continuing Influence of John C. Oxtoby*. Contemporary Mathematics, vol. 678 (American Mathematical Society, Providence, 2016), pp. 151–173

67. L. Flaminio, K. Frączek, J. Kułaga-Przymus, M. Lemańczyk, Approximate orthogonality of powers for ergodic affine unipotent diffeomorphisms on nilmanifolds. Stud. Math. (to appear). https://arxiv.org/abs/1609.00699

68. N. Frantzikinakis, An averaged Chowla and Elliott conjecture along independent polynomials (2016). https://arxiv.org/abs/1606.08420

69. N. Frantzikinakis, Ergodicity of the Liouville system implies the Chowla conjecture. Discrete Anal. (2016, to appear). https://arxiv.org/abs/1611.09338.

70. N. Frantzikinakis, B. Host, Higher order Fourier analysis of multiplicative functions and applications. J. Am. Math. Soc. **30**(1), 67–157 (2017)

71. N. Frantzikinakis, B. Host, The logarithmic Sarnak conjecture for ergodic weights. Ann. Math. (2017, to appear). https://arxiv.org/abs/1708.00677

72. H. Furstenberg, Strict ergodicity and transformation of the torus. Am. J. Math. **83**, 573–601 (1961)

73. H. Furstenberg, Disjointness in ergodic theory, minimal sets, and a problem in Diophantine approximation. Math. Syst. Theory **1**, 1–49 (1967)

74. H. Furstenberg, *The Unique Ergodicity of the Horocycle Flow*. Lecture Notes in Mathematics, vol. 318 (Springer, Berlin, 1973), pp. 95–115

75. H. Furstenberg, Ergodic behavior of diagonal measures and a theorem of Szemerédi on arithmetic progressions. J. Anal. Math. **31**, 204–256 (1977)

76. H. Furstenberg, *Recurrence in Ergodic Theory and Combinatorial Number Theory*. M. B. Porter Lectures (Princeton University Press, Princeton, 1981)

77. A.O. Gel'fond, Sur les nombres qui ont des propriétés additives et multiplicatives données. Acta Arith. **13**, 259–265 (1967/1968)

78. E. Glasner, *Ergodic Theory via Joinings*. Mathematical Surveys and Monographs, vol. 101 (American Mathematical Society, Providence, 2003)

79. A. Gomilko, D. Kwietniak, M. Lemańczyk, Sarnak's conjecture implies the Chowla conjecture along a subsequence, in *Ergodic Theory and Dynamical Systems in Their Interactions with Arithmetic and Combinatorics*, ed. By S. Ferenczi et al. Lecture Notes in Mathematics, vol. 2213 (Springer, New York, 2018)

80. W.T. Gowers, A new proof of Szemerédi's theorem. Geom. Funct. Anal. **11**(3), 465–588 (2001)

81. K. Granville, A. Soundararajan, *Multiplicative Number Theory: The Pretentious Approach* (book manuscript in preparation)

82. B. Green, On (not) computing the Möbius function using bounded depth circuits. Combin. Probab. Comput. **21**(6), 942–951 (2012)

83. B. Green, T. Tao, The Möbius function is strongly orthogonal to nilsequences. Ann. Math. (2) **175**(2), 541–566 (2012)

84. B. Green, T. Tao, The quantitative behaviour of polynomial orbits on nilmanifolds. Ann. Math. (2) **175**(2), 465–540 (2012)

85. F. Hahn, W. Parry, Some characteristic properties of dynamical systems with quasi-discrete spectra. Math. Syst. Theory **2**, 179–190 (1968)

86. H. Halberstam, K.F. Roth, *Sequences*, 2nd edn. (Springer, New York, 1983)

87. R.R. Hall, *Sets of Multiples*. Cambridge Tracts in Mathematics, vol. 118 (Cambridge University Press, Cambridge, 1996)

88. A.J. Harper, A different proof of a finite version of Vinogradov's ilinear sum inequality (NOTES) (2011). https://www.dpmms.cam.ac.uk/~ajh228/FiniteBilinearNotes.pdf

89. A. Hildebrand, Introduction to analytic number theory. http://www.math.uiuc.edu/~ajh/531.fall05/

90. B. Host, B. Kra, Convergence of polynomial ergodic averages. Isr. J. Math. **149**, 1–19 (2005). Probability in mathematics

91. B. Host, B. Kra, Nonconventional ergodic averages and nilmanifolds. Ann. Math. (2) **161**(1), 397–488 (2005)

92. B. Host, J.-F. Méla, F. Parreau, Nonsingular transformations and spectral analysis of measures. Bull. Soc. Math. France **119**(1), 33–90 (1991)

93. W. Huang, Z. Wang, G. Zhang, Möbius disjointness for topological models of ergodic systems with discrete spectrum. Preprint (2016). https://arxiv.org/abs/1608.08289

94. W. Huang, Z. Lian, S. Shao, X. Ye, Sequences from zero entropy noncommutative toral automorphisms and Sarnak Conjecture. J. Differ. Equ. **263**(1), 779–810 (2017)

95. W. Huang, Z. Wang, X. Ye, Measure complexity and Möbius disjointness (2017). https://arxiv.org/abs/1707.06345

96. C. Huck, On the logarithmic probability that a random integral ideal is relatively $\mathcal{A}$-free, in *Ergodic Theory and Dynamical Systems in Their Interactions with Arithmetic and Combinatorics*, ed. By S. Ferenczi et al. Lecture Notes in Mathematics, vol. 2213 (Springer, New York, 2018)

97. D.E. Iannucci, On the smallest abundant number not divisible by the first k primes. Bull. Belg. Math. Soc. Simon Stevin **12**(1), 39–44 (2005)

98. K.-H. Indlekofer, I. Kátai, Investigations in the theory of q-additive and q-multiplicative functions. I. Acta Math. Hungar. **91**(1–2), 53–78 (2001)

99. H. Iwaniec, E. Kowalski, *Analytic Number Theory*. American Mathematical Society Colloquium Publications, vol. 53 (American Mathematical Society, Providence, 2004)

100. E. Jennings, P. Pollack, L. Thompson, Variations on a theorem of Davenport concerning abundant numbers. Bull. Aust. Math. Soc. **89**(3), 437–450 (2014)

101. T. Kamae, Subsequences of normal sequences. Isr. J. Math. **16**, 121–149 (1973)

102. A. Kanigowski, M. Lemańczyk, C. Ulcigrai, On disjointness properties of some parabolic flows (in preparation)

103. D. Karagulyan, On Möbius orthogonality for interval maps of zero entropy and orientation-preserving circle homeomorphisms. Ark. Mat. **53**(2), 317–327 (2015)

104. D. Karagulyan, On Möbius orthogonality for subshifts of finite type with positive topological entropy. Stud. Math. **237**(3), 277–282 (2017)

105. S. Kasjan, G. Keller, M. Lemańczyk, Dynamics of $\mathcal{B}$-free sets: a view through the window. Int. Math. Res. Not. (to appear). https://arxiv.org/abs/1702.02375

106. I. Kátai, A remark on a theorem of H. Daboussi. Acta Math. Hungar. **47**(1–2), 223–225 (1986)

107. M. Keane, Generalized Morse sequences. Z. Wahrscheinlichkeitstheorie und Verw. Gebiete **10**, 335–353 (1968)

108. G. Keller, Generalized heredity in $\mathcal{B}$-free systems. Preprint (2017). https://arxiv.org/abs/1704.04079

109. G. Keller, C. Richard, Periods and factors of weak model sets. Preprint (2017). https://arxiv.org/abs/1702.02383

110. M. Kobayashi, A new series for the density of abundant numbers. Int. J. Number Theory **10**(1), 73–84 (2014)

111. J. Konieczny, M. Kupsa, D. Kwietniak, Arcwise connectedness of the set of ergodic measures of hereditary shifts. Preprint (2016). https://arxiv.org/abs/1610.00672

112. J. Kułaga-Przymus, M. Lemańczyk, The Möbius function and continuous extensions of rotations. Monatsh. Math. **178**(4), 553–582 (2015)

113. J. Kułaga-Przymus, M. Lemańczyk, Möbius disjointness along ergodic sequences for uniquely ergodic actions. Ergodic Theory Dyn. Syst. (2017, to appear). https://arxiv.org/abs/1703.02347

114. J. Kułaga-Przymus, M. Lemańczyk, B. Weiss, On invariant measures for $\mathcal{B}$-free systems. Proc. Lond. Math. Soc. (3) **110**(6), 1435–1474 (2015)

115. J. Kułaga-Przymus, M. Lemańczyk, B. Weiss, Hereditary subshifts whose simplex of invariant measures is Poulsen, in *Ergodic Theory, Dynamical Systems, and the Continuing Influence of John C. Oxtoby*. Contemporary Mathematics, vol. 678 (American Mathematical Society, Providence, 2016), pp. 245–253

116. J. Kwiatkowski, Spectral isomorphism of Morse dynamical systems. Bull. Acad. Polon. Sci. Sér. Sci. Math. **29**(3–4), 105–114 (1981)
117. E. Lehrer, Topological mixing and uniquely ergodic systems. Isr. J. Math. **57**(2), 239–255 (1987)
118. A. Leibman, Convergence of multiple ergodic averages along polynomials of several variables. Isr. J. Math. **146**, 303–315 (2005)
119. A. Leibman, Multiple polynomial correlation sequences and nilsequences. Ergod. Theory Dyn. Syst. **30**(3), 841–854 (2010)
120. M. Lemańczyk, F. Parreau, Lifting mixing properties by Rokhlin cocycles. Ergod. Theory Dyn. Syst. **32**(2), 763–784 (2012)
121. E. Lindenstrauss, Measurable distal and topological distal systems. Ergod. Theory Dyn. Syst. **19**(4), 1063–1076 (1999)
122. J. Liu, P. Sarnak, The Möbius function and distal flows. Duke Math. J. **164**(7), 1353–1399 (2015)
123. B. Marcus, The horocycle flow is mixing of all degrees. Invent. Math. **46**(3), 201–209 (1978)
124. B. Martin, C. Mauduit, J. Rivat, Théorème des nombres premiers pour les fonctions digitales. Acta Arith. **165**(1), 11–45 (2014)
125. B. Martin, C. Mauduit, J. Rivat, Fonctions digitales le long des nombres premiers. Acta Arith. **170**(2), 175–197 (2015)
126. K. Matomäki, M. Radziwiłł, Multiplicative functions in short intervals. Ann. Math. (2) **183**(3), 1015–1056 (2016)
127. K. Matomäki, M. Radziwiłł, T. Tao, An averaged form of Chowla's conjecture. Algebra Number Theory **9**, 2167–2196 (2015)
128. K. Matomäki, M. Radziwiłł, T. Tao, Sign patterns of the Liouville and Möbius functions. Forum Math. Sigma **4**(e14), 44 (2016)
129. C. Mauduit, J. Rivat, Sur un problème de Gelfond: la somme des chiffres des nombres premiers. Ann. Math. (2) **171**(3), 1591–1646 (2010)
130. C. Mauduit, J. Rivat, Prime numbers along Rudin–Shapiro sequences. J. Eur. Math. Soc. **17**(10), 2595–2642 (2015)
131. M.K. Mentzen, Automorphisms of subshifts defined by $\mathcal{B}$-free sets of integers. Colloq. Math. **147**(1), 87–94 (2017)
132. L. Mirsky, Note on an asymptotic formula connected with r-free integers. Q. J. Math. Oxford Ser. **18**, 178–182 (1947)
133. L. Mirsky, Arithmetical pattern problems relating to divisibility by r th powers. Proc. Lond. Math. Soc. (2) **50**, 497–508 (1949)
134. H.L. Montgomery, R.C. Vaughan, Exponential sums with multiplicative coefficients. Invent. Math. **43**(1), 69–82 (1977)
135. B. Mossé, Puissances de mots et reconnaissabilité des points fixes d'une substitution. Theor. Comput. Sci. **99**(2), 327–334 (1992)
136. C. Müllner, Automatic sequences fulfill the Sarnak conjecture. Duke Math. (to appear). https://arxiv.org/abs/1602.03042.
137. M.R. Murty, A. Vatwani, A remark on a conjecture of Chowla. J. Ramanujan Math. Soc. (2017, to appear)
138. M.R. Murty, A. Vatwani, Twin primes and the parity problem. J. Number Theory **180**, 643–659 (2017)
139. R. Peckner, Two dynamical perspectives on the randomness of the Mobius function. Thesis (Ph.D.), Princeton University. ProQuest LLC, Ann Arbor, 2015
140. R. Peckner, Uniqueness of the measure of maximal entropy for the squarefree flow. Isr. J. Math. **210**(1), 335–357 (2015)
141. P.A.B. Pleasants, C. Huck, Entropy and diffraction of the k-free points in n-dimensional lattices. Discrete Comput. Geom. **50**(1), 39–68 (2013)
142. M. Queffélec, *Substitution Dynamical Systems—Spectral Analysis*, 2nd edn. Lecture Notes in Mathematics, vol. 1294 (Springer, Berlin, 2010)

143. O. Ramaré, From Chowla's conjecture: from the Liouville function to the Moebius function, in *Ergodic Theory and Dynamical Systems in Their Interactions with Arithmetic and Combinatorics*, ed. By S. Ferenczi et al. Lecture Notes in Mathematics, vol. 2213 (Springer, New York, 2018)

144. M. Ratner, Horocycle flows, joinings and rigidity of products. Ann. Math. (2) **118**(2), 277–313 (1983)

145. M. Ratner, On Raghunathan's measure conjecture. Ann. Math. (2) **134**(3), 545–607 (1991)

146. J. Rivat, Analytic number theory, *in Ergodic Theory and Dynamical Systems in Their Interactions with Arithmetic and Combinatorics*, ed. By S. Ferenczi et al. Lecture Notes in Mathematics, vol. 2213 (Springer, New York, 2018)

147. V.V. Ryzhikov, Bounded ergodic constructions, disjointness, and weak limits of powers. Trans. Mosc. Math. Soc. **74**, 165–171 (2013)

148. P. Sarnak, Mobius randomness and dynamics. Not. S. Afr. Math. Soc. **43**(2), 89–97 (2012)

149. P. Sarnak, Möbius randomness and dynamics six years later. http://www.youtube.com/watch?v=LXX0ntxrkb0

150. P. Sarnak, Three lectures on the Möbius function, randomness and dynamics. http://publications.ias.edu/sarnak/

151. P. Sarnak, A. Ubis, The horocycle flow at prime times. J. Math. Pures Appl. (9) **103**(2), 575–618 (2015)

152. K. Schmidt, *Cocycles on Ergodic Transformation Groups*. Macmillan Lectures in Mathematics, vol. 1 (Macmillan Company of India, Ltd., Delhi, 1977)

153. K. Soundararajan, The Liouville function in short intervals [after Matomäki and Radziwiłł]. Séminaire Boubaki 68ème année (2015–2016), no. 1119

154. E. Szemerédi, On sets of integers containing no *k* elements in arithmetic progression. Acta Arith. **27**, 199–245 (1975). Collection of articles in memory of Juriĭ Vladimirovič Linnik

155. T. Tao, The logarithmically averaged Chowla and Elliot conjectures for two-point correlations. Forum Math. Pi **4**, e8 (2016)

156. T. Tao, Equivalence of the logarithmically averaged Chowla and Sarnak conjectures, in *Number Theory – Diophantine Problems, Uniform Distribution and Applications: Festschrift in Honour of Robert F. Tichy's 60th Birthday*, ed. by C. Elsholtz, P. Grabner (Springer International Publishing, Cham, 2017), pp. 391–421

157. T. Tao, The Bourgain-Sarnak-Ziegler orthogonality criterion. What's new. http://terrytao.wordpress.com/2011/11/21/the-bourgain-sarnak-ziegler-orthogonality-criterion/

158. T. Tao, The Chowla and the Sarnak conjecture. What's new. http://terrytao.wordpress.com/2012/10/14/the-chowla-conjecture-and-the-sarnak-conjecture/

159. T. Tao, J. Teräväinen, The structure of logarithmically averaged correlations of multiplicative functions, with applications to the Chowla and Elliott conjectures. Preprint (2017). https://arxiv.org/abs/1708.02610

160. G. Tenenbaum, *Introduction to Analytic and Probabilistic Number Theory*, 3rd edn. Translated from the 2008 French edition by Patrick D. F. Ion. Graduate Studies in Mathematics, vol. 163 (American Mathematical Society, Providence, 2015), xxiv+629 pp.

161. W.A. Veech, Möbius orthogonality for generalized Morse-Kakutani flows. Am. J. Math. **139**(5), 1157–1203 (2017)

162. A.M. Vershik, A.N. Livshits, Adic models of ergodic transformations, spectral theory, substitutions, and related topics, in *Representation Theory and Dynamical Systems*. Advances in Soviet Mathematics, vol. 9 (American Mathematical Society, Providence, 1992), pp. 185–204

163. I.M. Vinogradov, Some theorems concerning the theory of primes. Rec. Math. Moscou **2**(44), 179–195 (1937)

164. I.M. Vinogradov, *The Method of Trigonometrical Sums in the Theory of Numbers* (Dover Publications, Inc., Mineola, 2004). Translated from the Russian, revised and annotated by K.F. Roth and Anne Davenport, Reprint of the 1954 translation

165. P. Walters, *An Introduction to Ergodic Theory*. Graduate Texts in Mathematics, vol. 79 (Springer, New York, 1982)

166. Z. Wang, Möbius disjointness for analytic skew products. Inventiones Math. **209**(1), 175–196 (2017). https://arxiv.org/abs/1509.03183.
167. B. Weiss, Normal sequences as collectives, in *Proceedings of Symposium on Topological Dynamics and Ergodic Theory*, University of Kentucky, 1971
168. B. Weiss, Strictly ergodic models for dynamical systems. Bull. Am. Math. Soc. (N.S.) **13**(2), 143–146 (1985)
169. B. Weiss, *Single Orbit Dynamics*. CBMS Regional Conference Series in Mathematics, vol. 95 (American Mathematical Society, Providence, 2000)
170. M. Wierdl, Pointwise ergodic theorem along the prime numbers. Isr. J. Math. **64**(3), 315–336 (1989)
171. R.J. Zimmer, Ergodic actions with generalized discrete spectrum. Ill. J. Math. **20**(4), 555–588 (1976)
172. R.J. Zimmer, Extensions of ergodic group actions. Ill. J. Math. **20**(3), 373–409 (1976)

Chapter 12
Sarnak's Conjecture Implies the Chowla Conjecture Along a Subsequence

Alexander Gomilko, Dominik Kwietniak, and Mariusz Lemańczyk

12.1 Introduction

For the definitions, basic notation and results concerning the conjecture of Sarnak and the one of Chowla we refer the reader to the survey [5]. We only recall here that the Chowla conjecture implies Sarnak's conjecture [4, 11]. Then the intriguing question arises whether the reverse implication is true. Motivated by some recent results concerning logarithmic autocorrelations of the classical Möbius function $\mu \colon \mathbb{N} \to \{-1, 0, 1\}$ (and the Liouville function $\lambda \colon \mathbb{N} \to \{-1, 1\}$), see [6, 7, 12, 13] and using Tao's result [13] on the equivalence of logarithmic versions of Sarnak's and Chowla conjectures, we give a partial answer to the aforementioned question by showing that:

Theorem 12.1 *If Sarnak's conjecture is satisfied then there exists an increasing sequence (N_k) of positive integers along which the Chowla conjecture holds.*

As a matter of fact, we show that the assertion of Theorem 12.1 follows from the logarithmic version of the Chowla conjecture.

A. Gomilko · M. Lemańczyk (✉)
Faculty of Mathematics and Computer Science, Nicolaus Copernicus University, Toruń, Poland
e-mail: gomilko@mat.umk.pl; mlem@mat.umk.pl

D. Kwietniak
Faculty of Mathematics and Computer Science, Jagiellonian University in Kraków,
Kraków, Poland

Institute of Mathematics, Federal University of Rio de Janeiro, Rio de Janeiro, Brazil
e-mail: dominik.kwietniak@uj.edu.pl

© Springer International Publishing AG, part of Springer Nature 2018
S. Ferenczi et al. (eds.), *Ergodic Theory and Dynamical Systems in their
Interactions with Arithmetics and Combinatorics*, Lecture Notes
in Mathematics 2213, https://doi.org/10.1007/978-3-319-74908-2_12

12.2 Cesàro and Harmonic Limits of Empirical Measures

12.2.1 Ergodic Components of Members of $V(x)$ and $V^{log}(x)$

Let X be a compact metric space. By $M(X)$ we denote the space of probability Borel measures on X, in fact, we will also consider the space of $\widetilde{M}(X)$ of Borel measures μ on X for which $\mu(X) \leqslant 2$. With the weak-∗-topology, $M(X)$ is a compact metrizable space and the metric we will consider is given by

$$d(\mu, \nu) = \sum_{j \geqslant 1} \frac{1}{2^j} \left| \int f_j \, d\mu - \int f_j \, d\nu \right|, \tag{12.1}$$

where $\{f_j : j \geqslant 1\}$ is a linearly dense set of continuous function whose sup norm is $\leqslant 1$. Note that, by (12.1), if $0 \leqslant \alpha \leqslant 1$ then $d(\alpha\mu, \alpha\nu) = \alpha d(\mu, \nu)$. In particular, d is convex:

$$d\left(\int \mu_\gamma \, dP(\gamma), \int \nu_\gamma \, dP(\gamma)\right) \leqslant \int d(\mu_\gamma, \nu_\gamma) \, dP(\gamma), \tag{12.2}$$

where P is a Borel probability measure on the set of indices γ in some Polish metric space, $\gamma \mapsto \mu_\gamma$ is an $M(X)$-valued Borel function, and $\int \mu_\gamma \, dP(\gamma)$ denotes the Pettis integral.

Let (X, T) be a dynamical system given by a continuous map $T : X \to X$ and $M(X, T)$ (respectively, $M^e(X, T)$) stands for the set of T-invariant (respectively, ergodic) measures. Given $x \in X$ and $n \in \mathbb{N}$, we write $\delta_{T^n(x)}$ for the Dirac measure concentrated at the point $T^n(x)$. Let $\Delta(x, N)$ denote the counting measure concentrated on $\{x, T(x), \ldots, T^{N-1}(x)\}$, where $N \geqslant 1$ and let the *empirical measure* $\mathcal{E}(x, N)$ be the normalized counting measure, that is,

$$\Delta(x, N) = \sum_{n=1}^{N} \delta_{T^{n-1}(x)}, \quad \text{and} \quad \mathcal{E}(x, N) = \frac{1}{N}\Delta(x, N).$$

We say that $x \in X$ is quasi-generic for a measure $\nu \in M(X)$ if for some subsequence (N_k) we have $\mathcal{E}(x, N_k) \to \nu$. The Cesàro limit set of x is

$$V(x) := \{\nu \in M(X) : x \text{ is quasi-generic for } \nu\}.$$

The set was studied by many authors and it is well-known (and easy to see) that we always have

$$V(x) \subset M(X, T) \tag{12.3}$$

and $V(x)$ is a nonempty, closed and connected set, see [1], hence

$$\text{either } |V(x)| = 1 \text{ or } V(x) \text{ is uncountable.} \tag{12.4}$$

Choosing different normalization method for the counting measures $\Delta(x, N)$, we arrive at the notion of harmonic limit set of a point. Let

$$\mathcal{E}^{\log}(x, N) = \frac{1}{\log N} \sum_{n=1}^{N} \frac{1}{n} \delta_{T^{n-1}(x)}, \qquad \text{for } N \geqslant 2.$$

Note that $\mathcal{E}^{\log}(x, N)$ is not a probability measure. In order to stay in $M(X)$, we should consider

$$\mathcal{E}_{\text{nrm}}^{\log}(x, N) = \frac{1}{H_N} \sum_{n=1}^{N} \frac{1}{n} \delta_{T^{n-1}(x)}, \text{ where } H_N = \sum_{n=1}^{N} \frac{1}{n}.$$

Note that the limit sets of sequences $(\mathcal{E}^{\log}(x, N))$ and $(\mathcal{E}_{\text{nrm}}^{\log}(x, N))$ as N goes to ∞ coincide, so by abuse of notation we will not distinguish between $M(X)$ and $\widetilde{M}(X)$ and we will often deal with sequences of linear combinations of measures which are not exactly convex (affine), but are closer and closer to be so when we pass with the index N to infinity. We say that $x \in X$ is logarithmically quasi-generic for a measure ν if for some subsequence (N_k) we have $\mathcal{E}^{\log}(x, N_k) \to \nu$ as $k \to \infty$. The harmonic limit set of x is defined as

$$V^{\log}(x) := \{\nu \in M(X) : x \text{ is logarithmically quasi-generic for } \nu\}.$$

It is easy to see that the proofs of (12.3)–(12.4) presented for $V(x)$ in [1] can be easily adapted to harmonic averages, so we have that $V^{\log}(x)$ is nonempty, closed, connected, and consists of T-invariant measures. In particular, (12.4) also holds for $V^{\log}(x)$. If $V(x) = \{\nu\}$, then $V^{\log}(x) = \{\nu\}$, but the converse need not be true. Nevertheless, the measures in the harmonic limit set of x are always members of the closed convex hull of $V(x)$ (note that the latter set need not be convex as in a dynamical system with the specification property every nonempty compact and connected set $V \subset M(X, T)$ is the Cesàro limit set of some point $x \in X$, cf. [1, Proposition 21.14]).

Proposition 12.2 We have $V^{\log}(x) \subset \overline{conv}(V(x))$.

Proof Fix $x \in X$ and let

$$A := \{\mathcal{E}(x, N) : N \geqslant 1\}, \quad B := \{\mathcal{E}^{\log}(x, N) : N \geqslant 2\}.$$

If x is eventually periodic, that is, $T^k(x) = T^\ell(x)$ for some $0 \leqslant k < \ell$, then it is easy to see that $V(x) = V^{\log}(x) = \{\mathcal{E}(T^k(x), \ell - k)\}$. We will assume that x is not an eventually periodic point. It follows that $A \cap M(X, T) = B \cap M(X, T) = \emptyset$. Furthermore, $\overline{A} = V(x) \cup A$ and $\overline{B} = V^{\log}(x) \cup B$, where the summands are disjoint in both cases. Note that

$$\Delta(x, n) - \Delta(x, n - 1) = \delta_{T^{n-1}(x)} \qquad \text{for } n = 2, 3, \ldots.$$

Using the summation by parts trick, we obtain

$$\sum_{n=1}^{N} \frac{1}{n} \delta_{T^{n-1}(x)} = \Delta(x, 1) + \sum_{n=2}^{N} \frac{1}{n}(\Delta(x, n) - \Delta(x, n - 1))$$

$$= \frac{1}{N} \Delta(x, N) + \sum_{n=1}^{N-1} \left(\frac{1}{(n+1)n} \right) \Delta(x, n)$$

$$= \mathcal{E}(x, N) + \sum_{n=1}^{N-1} \frac{\mathcal{E}(x, n)}{n + 1}. \tag{12.5}$$

Fix $\varepsilon > 0$. Then, there exists $K = K_\varepsilon \geqslant 1$ such that

$$d(\mathcal{E}(x, n), V(x)) < \varepsilon \quad \text{for } n \geqslant K_\varepsilon. \tag{12.6}$$

Using (12.5), we get

$$\mathcal{E}_{\text{nrm}}^{\log}(x, N) = \frac{1}{H_N} \sum_{n=1}^{K} \frac{\mathcal{E}(x, n)}{n + 1} + \frac{\mathcal{E}(x, N)}{H_N} + \frac{1}{H_N} \sum_{n=K+1}^{N-1} \frac{\mathcal{E}(x, n)}{n + 1}. \tag{12.7}$$

Now, keeping K fixed, we can assure that the total mass of the first two summands on the RHS of (12.7) is as close to 0 as we want provided that N is large enough. Therefore, for every N large enough, the measure $\mathcal{E}_{\text{nrm}}^{\log}(x, N)$ is ε-close to

$$\xi_N = \frac{1}{H_N - H_K} \sum_{n=K+1}^{N-1} \frac{\mathcal{E}(x, n)}{n + 1}.$$

The latter measure is an affine combination of $\mathcal{E}(x, n)$ for $n > K_\varepsilon$ which are all ε-close to $V(x)$. Using (12.2), we get that ξ_N is ε-close to $\overline{\text{conv}}(V(x))$, thus $d(\xi_N, \mathcal{E}_{\text{nrm}}^{\log}(x, N)) < 2\varepsilon$ for all N large enough. Putting all this together, we obtain that for each $\rho \in V^{\log}(x)$, we have

$$d(\rho, \overline{\text{conv}}(V(x))) = 0$$

which completes the proof of Proposition 12.2.

12.2.2 Ergodic Measures in $V^{log}(x)$

It turns out that if $x \in X$ is logarithmically quasi-generic for an ergodic measure then x is also quasi-generic in the classical (Cesàro) sense.

Corollary 12.3 *If an ergodic measure* $v \in V^{log}(x)$, *then* $v \in V(x)$.

Proof We first recall Milman's theorem ([10], Chapter 1.3, Theorem 3.25): *If K is a compact set in a locally convex space and* $\overline{conv}(K)$ *is compact then* $ex(\overline{conv}(K)) \subset K$. Here, and elsewhere by $ex(L)$ we denote the set of extreme points of a convex set L. We apply Milman's result to $K = V(x)$ and $ex(M(X, T)) = M^e(X, T)$. By Proposition 12.2, we have

$$V^{log}(x) \cap M^e(X, T) \subset \overline{conv}(V(x)) \cap M^e(X, T) \subset ex(\overline{conv}(V(x))) \subset V(x).$$

12.2.3 Ergodic Components of Measures in $V^{log}(x)$

Let us recall that in our setting $M(X, T)$ is a metrizable, compact and convex subset of a locally convex space. It follows that we can use Choquet's representation theorem (see [9], Chapters 3 and 10) to conclude that if L is a closed convex subset of $M(X, T)$ and $\kappa \in L$ then there exists a Borel probability measure P_κ on $M(X, T)$ supported by $ex(L)$ such that

$$\kappa = \int_{ex(L)} \rho \, dP_\kappa(\rho). \tag{12.8}$$

Furthermore, if $L = M(X, T)$, then P_κ satisfying (12.8) is unique and the map $M(X, T) \ni \kappa \mapsto P_\kappa \in M(M^e(X, T))$ is Borel measurable [2] Fact A.2.12 and [9] Proposition 11.1.

Proposition 12.4 *The set of ergodic components of measures in* $V^{log}(x)$ *is contained in the set of ergodic components of measures in* $V(x)$. *More precisely: if a Borel set* $\mathcal{D} \subset M^e(X, T)$ *satisfies*

$$P_\mu(\mathcal{D}) = 1 \text{ for each } \mu \in V(x)$$

then

$$P_\kappa(\mathcal{D}) = 1 \text{ for each } \kappa \in V^{log}(x).$$

In particular, if the set of ergodic components of measures in $V(x)$ *is countable, so is the set of ergodic components of* $V^{log}(x)$.

Proof Fix $\kappa \in V^{\log}(x)$. It follows from Proposition 12.2 that $\kappa \in \overline{\text{conv}}(V(x))$. By Choquet's theorem, there exists $Q \in M(M(X, T))$ with $Q(\text{ex}(\overline{\text{conv}}(V(x)))) = 1$ such that

$$\kappa = \int_{\text{ex}(\overline{\text{conv}}(V(x)))} \mu \, dQ(\mu).$$

Using Milman's theorem, we obtain that $Q(V(x)) = 1$, so that

$$\kappa = \int_{V(x)} \mu \, dQ(\mu).$$

Hence

$$\kappa = \int_{V(x)} \left(\int_{M^e(X,T)} \rho \, dP_\mu(\rho) \right) dQ(\mu) = \int_{M^e(X,T)} \rho \, dR(\rho),$$

where $R \in M(M(X, T))$ is defined by

$$R(C) := \int_{V(x)} P_\mu(C) \, dQ(\mu) \text{ for a Borel subset } C \subset M(X, T).$$

Note that the definition of R is correct as $V(x)$ is Borel and $\mu \mapsto P_\mu$ is Borel measurable. Now, since $M(X, T)$ is a simplex, there is only one measure on $M^e(X, T)$ satisfying (12.8) and we obtain that $R = P_\kappa$. Since $P_\mu(\mathcal{D}) = 1$ for each $\mu \in V(x)$, we have $P_\kappa(\mathcal{D}) = 1$.

12.3   Relations Between Sarnak's Conjecture and the Chowla Conjecture

Thinking of the Möbius function as of a point $\boldsymbol{\mu}$ in the sequence space $\{-1, 0, 1\}^{\mathbb{N}}$, we can consider the Möbius dynamical system $(X_{\boldsymbol{\mu}}, S)$, where $X_{\boldsymbol{\mu}}$ stands for the orbit closure of $\boldsymbol{\mu}$ and S denotes the left shift. We now apply results from the previous sections to the Möbius system and sets $V(\boldsymbol{\mu})$ and $V^{\log}(\boldsymbol{\mu})$.

Proof (Proof of Theorem 12.1) It is obvious that Sarnak's conjecture implies logarithmic Sarnak's conjecture which, by a result of Tao [13], implies the logarithmic version of Chowla conjecture. The logarithmic version of Chowla conjecture for $\boldsymbol{\mu}$ phrased in the language of ergodic theory means that

$$\frac{1}{\log N} \sum_{n=1}^{N} \frac{1}{n} \delta_{S^{n-1}(\boldsymbol{\mu})} \to \widehat{\nu}_{\boldsymbol{\mu}^2}, \quad \text{as } N \to \infty,$$

where $\widehat{v}_{\mu^2}$ is the relatively independent extension of the Mirsky measure v_{μ^2} of the square-free system (X_{μ^2}, S) [4]. In particular, $\widehat{v}_{\mu^2}$ is ergodic. Equivalently, the conjecture says that $\widehat{v}_{\mu^2} \in V^{log}(\mu)$. It follows from Corollary 12.3 that μ is a quasi-generic point (in the sense of Cesàro) for $\widehat{v}_{\mu^2}$, that is, there is a sequence (N_i) such that for each $1 \leqslant a_1 < \ldots < a_k$ and each choice of $j_0, j_1, \ldots, j_k \in \{1, 2\}$ not all equal to 2, we have

$$\lim_{i \to \infty} \frac{1}{N_i} \sum_{n \leqslant N_i} \mu^{j_0}(n) \mu^{j_1}(n + a_1) \cdot \ldots \cdot \mu^{j_k}(n + a_k) = 0,$$

i.e., we obtain the Chowla conjecture along the subsequence (N_i).

Remark 12.5 If instead of the Möbius function μ we consider the Liouville function λ then the Chowla conjecture claims that the limit is the Bernoulli measure $B(1/2, 1/2)$ on $\{-1, 1\}^{\mathbb{N}}$.

In fact, using [6], we have the following:

Corollary 12.6 *Assume that there exists an ergodic measure $\kappa \in V^{log}(\mu)$. Then there exists an increasing sequence (N_i) such that the Chowla conjecture holds along (N_i).*

Proof If there exists $\kappa \in V^{log}(\mu) \cap M^e(X_\mu, S)$, then, reasoning as above, we see that for a subsequence (N_i), we have

$$\frac{1}{N_i} \sum_{1 \leqslant n \leqslant N_i} \delta_{S^{n-1}(\mu)} \to \kappa \qquad \text{as } i \to \infty.$$

Now, by Frantzikinakis [6], we get $\kappa = \widehat{v}_{\mu^2}$. The result follows.

Remark 12.7 Assume that (X, T) is a dynamical system and $x \in X$ is completely deterministic (i.e. each member $\kappa \in V(x)$ yields zero entropy measurable system (X, κ, T)) such that the ergodic components of all measures from $V(x)$ give a countable set. It follows from Proposition 12.4 and [7] (see Remarks after Theorem 1.3 in [7]) that, at x, we obtain Möbius disjointness in the logarithmic sense.

12.4 Examples

We collect here a couple of examples demonstrating that some of our results are optimal and cannot be improved. Our examples are points in the full shift $\{0, 1\}^{\mathbb{N}}$ or $\{0, 1, 2\}^{\mathbb{N}}$ constructed so that the Cesàro and harmonic limit sets are easy to identify. We will routinely omit some easy computations used in our proofs.

Let $\underline{d}(J)$, $\overline{d}(J)$, $\underline{\delta}(J)$, and $\overline{\delta}(J)$ denote, respectively the lower/upper asymptotic and lower/upper logarithmic density of a set $J \subset \mathbb{Z}$. It is well known that we always have $\underline{d}(J) \leqslant \underline{\delta}(J) \leqslant \overline{\delta}(J) \leqslant \overline{d}(J)$. We write $\delta(J)$ for the common value of $\underline{\delta}(J)$ and $\overline{\delta}(J)$ (if such an equality holds).

Our approach is based on the following criterion for logarithmic genericity. It can be proved the same way as for Cesàro averages the only difference is that logarithmic (harmonic) averages replace asymptotic averages.

Proposition 12.8 *A point $x = (x_n) \in \{0, 1\}^{\mathbb{N}}$ is logarithmically generic for a measure μ if and only if for every $k \geqslant 1$ and for every finite block $w \in \{0, 1\}^k$, the set of positions j such that w appears at the position j in w, that is the set $J_w = \{j \in \mathbb{N} : x_j = w_1, \ldots, x_{j+k-1} = w_k\}$ satisfies $\underline{\delta}(J_w) = \overline{\delta}(J_w) = \mu(\{y : y_{[0,k)} = w\})$.*

We will also apply the following observation.

Proposition 12.9 *For every point $x = (x_n) \in \{0, 1\}^{\mathbb{N}}$, $k \geqslant 1$ and finite block $w \in \{0, 1\}^k$, we have*

$$\underline{d}(J_w) = \min\{\nu(\{y : y_{[0,k)} = w\}) : \nu \in V(x)\}$$

and

$$\overline{d}(J_w) = \max\{\nu(\{y : y_{[0,k)} = w\}) : \nu \in V(x)\},$$

where $J_w = \{j \in \mathbb{N} : x_j = w_1, \ldots, x_{j+k-1} = w_k\}$.

Both results are well-known. We now present our examples.

Proposition 12.10 *It can happen that $V^{log}(x) \subsetneq V(x)$.*

Proof Let

$$x = 011000011111111110000000000000000\ldots,$$

that is,

$$x_n = \begin{cases} 0, & \text{if } 2^{2k} \leqslant n+1 < 2^{2k+1} \text{ for some } k \geqslant 0, \\ 1, & \text{if } 2^{2k-1} \leqslant n+1 < 2^{2k} \text{ for some } k \geqslant 1. \end{cases}$$

It is easy to see that $V(x) = \{\alpha\delta_{\overline{0}} + (1 - \alpha)\delta_{\overline{1}} : 1/3 \leqslant \alpha \leqslant 2/3\}$, where $\delta_{\overline{p}}$ denotes the Dirac measure concentrated at the fixed point $ppp\ldots \in \{0, 1\}^{\mathbb{N}}$. We claim that $V^{log}(x) = \{1/2\delta_{\overline{0}} + 1/2\delta_{\overline{1}}\}$. Indeed, it is easy to see that:

1. $\underline{\delta}(J_w) = \overline{\delta}(J_w) = 1/2$ for $w = 0$ and $w = 1$,
2. $\underline{d}(J_w) = \overline{d}(J_w) = 0$ for any block w containing 01 or 10 as a subblock,
3. $\underline{\delta}(J_w) = \overline{\delta}(J_w) = 1/2$ for any $k \geqslant 2$ and $w = 0^k$ or $w = 1^k$.

Our claim is an immediate consequence of these three observations. The proofs of the first two are based on easy computations. To see the third one, fix $k \geqslant 2$ and consider $w = 1^k$ (the case $w = 0^k$ is proved in the same way). Note that $J_1 \setminus J_w$ can be equivalently described as the set of positions j such that the block $x_j x_{j+1} \ldots x_{j+k-1}$ starts with 1 and contains 10 as a subblock. By 2. this set satisfies $d(J_1 \setminus J_w) = 0$, thus $\underline{\delta}(J_w) = \underline{\delta}(J_1)$ and $\overline{\delta}(J_w) = \overline{\delta}(J_1)$. It follows from 1. that $\delta(J_w) = 1/2$.

We are grateful to J. Kułaga-Przymus for the following remark. It has also inspired our next proposition presenting a simpler example of the same phenomenon.

Remark 12.11 (J. Kułaga-Przymus) It is implicit in [3] that for each $\mathscr{B} \subset \mathbb{N} \setminus \{1\}$ which is not Besicovitch, for the subshift (X_η, S), where $\eta := \mathbb{1}_{\mathcal{F}_{\mathscr{B}}}$ ($\mathcal{F}_{\mathscr{B}}$ stands for the set of $\mathscr{B}$-free numbers), we have $V^{\log}(\eta) = \{\nu_\eta\}$ and ν_η (so called Mirsky measure of (X_η, S)) is ergodic, while $V(\eta)$ is uncountable, whence the set of ergodic components of members in $V(\eta)$ is strictly larger than the analogous set for $V^{\log}(\eta)$.

Proposition 12.12 *The set of ergodic measures appearing in the ergodic decompositions of members of $V(x)$ can be strictly bigger than the set of ergodic components of analogous set for $V^{\log}(x)$.*

Proof Let $x = (x_n) \in \{0, 1\}^{\mathbb{N}}$, where

$$x_n = \begin{cases} 1, & \text{if } 2^{k^2-1} \leqslant n + 1 < 2^{k^2}, \text{ for some } k \geqslant 1, \\ 0, & \text{otherwise.} \end{cases}$$

Then it follows either from [8, Lemma 2] or from direct computations that $\overline{\delta}(J_1) = \overline{\delta}(\{0 \leqslant j < 2^{k^2} : x_j = 1\}) = 0$, which implies that $V^{\log}(x) = \{\delta_{\overline{0}}\}$. On the other hand for every $k \geqslant 1$ we have

$$|\{0 \leqslant j < 2^{k^2} : x_j = 1\}| = \sum_{j=1}^{k} 2^{j^2-1}.$$

Therefore $d(\{0 \leqslant j < 2^{k^2} : x_j = 1\}) = 1/2$ and we conclude that $V(x) \neq \{\delta_{\overline{0}}\}$.

Proposition 12.13 *The sets $V(x)$ and $V^{\log}(x)$ can be disjoint.*

Proof Let

$$x = 0011111112222222222222222220000000000000 \ldots,$$

that is,

$$x_n = \begin{cases} 0, & \text{if } 3^{3k} \leqslant n + 1 < 3^{3k+1} \text{ for some } k \geqslant 0, \\ 1, & \text{if } 3^{3k+1} \leqslant n + 1 < 3^{3k+2} \text{ for some } k \geqslant 0, \\ 2, & \text{if } 3^{3k+2} \leqslant n + 1 < 3^{3k+3} \text{ for some } k \geqslant 0. \end{cases}$$

We claim that $V^{\log}(x) = \{1/3\delta_{\bar{0}} + 1/3\delta_{\bar{1}} + 1/3\delta_{\bar{2}}\}$ and $V^{\log}(x) \cap V(x) = \emptyset$. We proceed as in the first example. First, we note that the asymptotic density of the set J_w of appearances of a block $w = w_1 w_2 \in \{0, 1, 2\}^2$ with $w_1 \neq w_2$ in x is zero. It follows that the support of every measure $\nu \in V(x) \cup V^{\log}(x)$ is contained in the set $\{\bar{0}, \bar{1}, \bar{2}\}$ consisting of three shift-invariant (fixed) points. Next, we compute (or conclude from [8, Lemma 2]) that

$$\delta(\{j \in \mathbb{N} : x_j = 0\}) = \delta(\{j \in \mathbb{N} : x_j = 1\}) = \delta(\{j \in \mathbb{N} : x_j = 2\}) = 1/3.$$

This shows that $V^{\log}(x) = \{1/3\delta_{\bar{0}} + 1/3\delta_{\bar{1}} + 1/3\delta_{\bar{2}}\}$.

Let $\nu \in V(x)$. It follows that there is a sequence (N_i) such that for every $k \geqslant 1$ and every block $w \in \{0, 1, 2\}^k$, we have

$$\lim_{i \to \infty} \frac{|\{0 \leqslant j < N_i : x_j = w_1, \ldots, x_{j+k-1} = w_k\}|}{N_i} = \nu(\{y : y_{[0,k)} = w\}).$$

Without loss of generality we can assume that there exists $p \in \{0, 1, 2\}$ such that $N_i \equiv p \bmod 3$ for every $i \in \mathbb{N}$. Let $q = p - 1 \bmod 3$. It is then easy to see that

$$\nu(\{y : y_0 = q\}) \leqslant \frac{3}{13} < \frac{1}{3},$$

which implies that $\nu \neq 1/3\delta_{\bar{0}} + 1/3\delta_{\bar{1}} + 1/3\delta_{\bar{2}}$.

Remark 12.14 With some more effort in can be seen that in the above example we have that

$$\overline{\mathrm{conv}}(V(x)) = \overline{\mathrm{conv}}\left(\left\{\frac{1}{13}\delta_{\bar{0}} + \frac{3}{13}\delta_{\bar{1}} + \frac{9}{13}\delta_{\bar{2}}, \frac{1}{13}\delta_{\bar{2}} + \frac{3}{13}\delta_{\bar{0}} + \frac{9}{13}\delta_{\bar{1}}, \right.\right.$$

$$\left.\left.\frac{1}{13}\delta_{\bar{1}} + \frac{3}{13}\delta_{\bar{2}} + \frac{9}{13}\delta_{\bar{0}}\right\}\right),$$

and $V(x)$ is the combinatorial boundary of that simplex, that is,

$$V(x) = \overline{\mathrm{conv}}\left(\left\{\frac{1}{13}\delta_{\bar{0}} + \frac{3}{13}\delta_{\bar{1}} + \frac{9}{13}\delta_{\bar{2}}, \frac{1}{13}\delta_{\bar{1}} + \frac{3}{13}\delta_{\bar{2}} + \frac{9}{13}\delta_{\bar{0}}\right\}\right)$$

$$\cup \overline{\mathrm{conv}}\left(\left\{\frac{1}{13}\delta_{\bar{2}} + \frac{3}{13}\delta_{\bar{0}} + \frac{9}{13}\delta_{\bar{1}}, \frac{1}{13}\delta_{\bar{1}} + \frac{3}{13}\delta_{\bar{2}} + \frac{9}{13}\delta_{\bar{0}}\right\}\right)$$

$$\cup \overline{\mathrm{conv}}\left(\left\{\frac{1}{13}\delta_{\bar{0}} + \frac{3}{13}\delta_{\bar{1}} + \frac{9}{13}\delta_{\bar{2}}, \frac{1}{13}\delta_{\bar{2}} + \frac{3}{13}\delta_{\bar{0}} + \frac{9}{13}\delta_{\bar{1}}\right\}\right).$$

Acknowledgements The research of the first author was supported by Narodowe Centrum Nauki grant 2014/13/B/ST1/03153. The research of the second author was supported by Narodowe Centrum Nauki grant UMO-2012/07/E/ST1/00185. The research of the third author was supported by Narodowe Centrum Nauki grant UMO-2014/15/B/ST1/03736.

References

1. M. Denker, C. Grillenberger, K. Sigmund, *Ergodic Theory on Compact Spaces*. Lecture Notes in Mathematics, vol. 527 (Springer, Berlin, 1976)
2. T. Downarowicz, *Entropy in Dynamical Systems* (Cambridge University Press, Cambridge, 2011)
3. A. Dymek, S. Kasjan, J. Kułaga-Przymus, M. Lemańczyk, $\mathscr{B}$-free sets and dynamics. Trans. Am. Math. Soc. (2015, to appear). arXiv:1509.08010
4. E.H. El Abdalaoui, J. Kułaga-Przymus, M. Lemańczyk, T. de la Rue, The Chowla and the Sarnak conjectures from ergodic theory point of view. Discrete Contin. Dyn. Syst. **37**(6), 2899–2944 (2017)
5. S. Ferenczi, J. Kułaga-Przymus, M. Lemańczyk, Sarnak's conjecture - what's new, in Ergodic Theory and Dynamical Systems in Their Interactions with Arithmetic and Combinatorics, ed. By S. Ferenczi et al. Lecture Notes in Mathematics, vol. 2213 (Springer, New York, 2018). arXiv:1710.04039v1
6. N. Frantzikinakis, Ergodicity of the Liouville system implies the Chowla conjecture. Discrete Anal. (2017, to appear). https://arxiv.org/abs/1611.09338
7. N. Frantzikinakis, B. Host, The logarithmic Sarnak conjecture for ergodic weights. Ann. Math. (2017, to appear). https://arxiv.org/abs/1708.00677
8. F. Luca, Š. Porubský, On asymptotic and logarithmic densities. Tatra Mt. Math. Publ. **31**, 75–86 (2005)
9. R.R. Phelps, *Lectures on Choquet's Theorem* (D. Van Nostrand Co., Inc., Princeton, 1966)
10. W. Rudin, *Functional Analysis*, 2nd edn. (McGrew-Hill Inc., New York, 1991)
11. P. Sarnak, Three lectures on the Möbius function, randomness and dynamics. http://publications.ias.edu/sarnak/
12. T. Tao, The logarithmically averaged Chowla and Elliot conjectures for two-point correlations. Forum Math. Pi **4**, e8–e36 (2016)
13. T. Tao, Equivalence of the logarithmically averaged Chowla and Sarnak conjectures, in *Number Theory – Diophantine Problems, Uniform Distribution and Applications: Festschrift in Honour of Robert F. Tichy's 60th Birthday*, ed. by C. Elsholtz, P. Grabner (Springer International Publishing, Cham, 2017), pp. 391–421

Chapter 13
On the Logarithmic Probability That a Random Integral Ideal Is $\mathcal{A}$-Free

Christian Huck

13.1 Introduction

Recently, the dynamical and spectral properties of so-called $\mathcal{A}$-free systems as given by the orbit closure of the square-free integers, visible lattice points and various number-theoretic generalisations have received increased attention; see [1, 2, 5, 6] and references therein. One reason is the connection of one-dimensional examples such as the square-free integers with Sarnak's conjecture [12] on the 'randomness' of the Möbius function, another the explicit computability of correlation functions as well as eigenfunctions for these systems together with intrinsic ergodicity properties. Here, we provide a very first step towards the study of a rather general notion of freeness for sets of integral ideals in an algebraic number field K.

A well known result by Benkoski [3] states that the probability that a randomly chosen m-tuple of integers is relatively l-free (the integers are not divisible by a common nontrivial lth power) is $1/\zeta(lm)$, where ζ is the Riemann zeta function. In a recent paper Sittinger [13] reproved that formula and gave an extension to arbitrary rings of algebraic integers in number fields K. Due to a lack of unique prime factorisation of integers in this general situation, one certainly passes to counting integral ideals as a whole and, with a natural notion of asymptotic density, the outcome is $1/\zeta_K(lm)$, where

$$\zeta_K(s) = \sum_{0 \neq \mathfrak{a} \subset O_K} \frac{1}{N(\mathfrak{a})^s}$$

C. Huck ($\boxtimes$)
Fakultät für Mathematik, Universität Bielefeld, Bielefeld, Germany
e-mail: huck@math.uni-bielefeld.de

© Springer International Publishing AG, part of Springer Nature 2018
S. Ferenczi et al. (eds.), *Ergodic Theory and Dynamical Systems in their Interactions with Arithmetics and Combinatorics*, Lecture Notes in Mathematics 2213, https://doi.org/10.1007/978-3-319-74908-2_13

is the Dedekind zeta function of K. This immediately leads to the question if the result allows for a further generalisation to more general notions of freeness, where one forbids common divisors from an arbitrary set $\mathcal{A}$ of non-zero integral ideals instead of considering merely the set consisting of all prime-powers of the form $\mathfrak{p}^l$ with $\mathfrak{p} \subset O_K$ prime. In the special case $K = \mathbb{Q}$ and $m = 1$, this was successfully done in a paper by Davenport and Erdös [7] from 1951. The goal of this short note is to provide a full generalisation of their result to arbitrary rings of algebraic integers. It turns out that, building on old and new results from analytic number theory, one can easily adjust their argument to the more general situation. In this generality, the case $m \geqslant 2$ remains open.

13.2   Preliminaries

Let K be a fixed algebraic number field of degree $d = [K : \mathbb{Q}] \in \mathbb{N}$. Let O_K denote the ring of integers of K and recall that O_K is a Dedekind domain [10]. Hence we have unique factorisation of non-zero ideals into prime ideals at our disposal, i.e. any non-zero integral ideal $\mathfrak{a} \subset O_K$ has a (up to rearrangement) unique representation of the form

$$\mathfrak{a} = \mathfrak{p}_1 \cdot \ldots \cdot \mathfrak{p}_l,$$

where the $\mathfrak{p}_i$ are prime ideals. Recall that the (absolute) norm $N(\mathfrak{a}) = [O_K : \mathfrak{a}]$ of a non-zero integral ideal $\mathfrak{a} \subset O_K$ is always finite. Moreover, the norm is completely multiplicative, i.e. one always has $N(\mathfrak{a}\mathfrak{b}) = N(\mathfrak{a})N(\mathfrak{b})$. A proof of the following fundamental result can be found in [9].

Proposition 13.1 *Let $H(x)$ be the number of non-zero integral ideals with norm less than or equal to x. Then*

$$H(x) = cx + O(x^{1-\frac{1}{d}})$$

for some positive constant c.

Corollary 13.2 *As $x \to \infty$, one has*

$$\sum_{N(\mathfrak{a}) \leqslant x} \frac{1}{N(\mathfrak{a})} \sim c \log x,$$

where c is the constant from Proposition 13.1.

Proof For $k \in \mathbb{N}$, let $h(k)$ denote the number of non-zero integral ideals with norm equal to k. Summation by parts yields

$$\sum_{N(\mathfrak{a}) \leqslant x} \frac{1}{N(\mathfrak{a})} = \sum_{k=1}^{\lfloor x \rfloor} \frac{h(k)}{k}$$

$$= \frac{H(\lfloor x \rfloor)}{\lfloor x \rfloor} + \sum_{k=1}^{\lfloor x \rfloor - 1} \frac{H(k)}{k(k+1)}$$

$$= c + O(x^{-\frac{1}{d}}) + c \sum_{k=1}^{\lfloor x \rfloor - 1} \frac{1}{k+1} + O\left(\sum_{k=1}^{\lfloor x \rfloor - 1} \frac{k^{-\frac{1}{d}}}{k+1} \right)$$

$$= c \sum_{k=1}^{\lfloor x \rfloor - 1} \frac{1}{k+1} + O(1)$$

$$\sim c \log x,$$

since $\sum_{k=1}^{\lfloor x \rfloor} \frac{1}{k} \sim \log x$ as $x \to \infty$.

The following generalisation of Mertens' third theorem to partial Euler products of the Dedekind zeta function $\zeta_K(s)$ of K at $s = 1$ was shown by Rosen. It will turn out to be crucial for our main result.

Theorem 13.3 ([11]) *There is a positive constant C such that*

$$\prod_{N(\mathfrak{p}) \leqslant x} \left(1 - \frac{1}{N(\mathfrak{p})} \right)^{-1} = C \log x + O(1),$$

where $\mathfrak{p}$ ranges over the prime ideals of O_K. In particular, $\prod_{N(\mathfrak{p}) \leqslant x} (1 - \frac{1}{N(\mathfrak{p})})^{-1} \sim C \log x$ as $x \to \infty$.

Remark 13.4 In fact, Rosen shows that the constant C above is given by $C = \alpha_K e^{\gamma}$, where α_K is the residue of $\zeta_K(s)$ at $s = 1$ and γ is the Euler-Mascheroni constant.

Let $\mathcal{A} = \{\mathfrak{a}_1, \mathfrak{a}_2, \dots\}$ be a fixed set of non-zero integral ideals $\mathfrak{a}_i \subset O_K$. We are interested in the set

$$\mathcal{M}_{\mathcal{A}} := \{\mathfrak{b} \neq 0 \mid \exists i \ \mathfrak{b} \subset \mathfrak{a}_i\}$$

of non-zero integral ideals that are multiples of some $\mathfrak{a}_i$ respectively its complement in the set of all non-zero integral ideals

$$\mathcal{V}_{\mathcal{A}} := \{\mathfrak{b} \mid \forall i \ \mathfrak{b} \not\subset \mathfrak{a}_i\}$$

of so-called $\mathcal{A}$-free (or $\mathcal{A}$-prime) integral ideals. More precisely, we ask if the natural asymptotic densities of these sets exist. In general, one defines densities of sets of non-zero integral ideals as follows.

Definition 13.5 Let S be a set of non-zero integral ideals $\mathfrak{b} \subset O_K$. And let $S(x)$ be the subset of those $\mathfrak{b}$ with $N(\mathfrak{b}) \leqslant x$.

(1) The *upper/lower (asymptotic) density* $D(S)/d(S)$ of S is defined as

$$\limsup_{x\to\infty}/\liminf_{x\to\infty} \frac{S(x)}{H(x)} \,.$$

If these numbers coincide, the common value is called the *(asymptotic) density* of S, denoted by $\mathrm{dens}(S)$.

(2) The *upper/lower (asymptotic) logarithmic density* $\Delta(S)/\delta(S)$ of S is defined as

$$\limsup_{x\to\infty}/\liminf_{x\to\infty} \frac{\displaystyle\sum_{\substack{\mathfrak{b}\in S \\ N(\mathfrak{b})\leqslant x}} \frac{1}{N(\mathfrak{b})}}{\displaystyle\sum_{\substack{0\neq\mathfrak{b}\subset O_K \\ N(\mathfrak{b})\leqslant x}} \frac{1}{N(\mathfrak{b})}} \,,$$

where one might substitute the denominator by $c \log x$ due to Corollary 13.2. Again, if these numbers coincide, the common value is called the *(asymptotic) logarithmic density* of S, denoted by $\mathrm{dens}_{\log}(S)$.

As in the well known special case of rational integers, the above lower and upper densities are related as follows.

Lemma 13.6 (Density Inequality) *For any set S of non-zero integral ideals of K, one has*

$$d(S) \leqslant \delta(S) \leqslant \Delta(S) \leqslant D(S) \,.$$

In particular, the existence of the density of S implies the existence of the logarithmic density of S.

Proof The assertion follows from summation by parts as follows. Let us first show that $\Delta(S) \leqslant D(S)$. To this end, let $\varepsilon > 0$ and choose $N \in \mathbb{N}$ such that $\frac{S(n)}{H(n)} \leqslant D(S) + \varepsilon$ for all $n \geqslant N$. For $k \in \mathbb{N}$, let $s(k)$ denote the number of non-zero integral ideals $\mathfrak{a} \in S$ with norm equal to k. Summation by parts yields for $n \geqslant N$

$$\sum_{k=1}^{n} \frac{s(k)}{k} = \frac{S(n)}{n} + \sum_{k=1}^{n-1} \frac{S(k)}{k(k+1)}$$

$$\leqslant \frac{H(n)}{n} + \sum_{k=1}^{N-1} \frac{S(k)}{k(k+1)} + \sum_{k=N}^{n-1} \frac{S(k)}{k(k+1)}$$

$$\leqslant \frac{H(n)}{n} + \sum_{k=1}^{N-1} \frac{S(k)}{k(k+1)} + (D(S)+\varepsilon) \sum_{k=N}^{n-1} \frac{H(k)}{k(k+1)} \,.$$

Since $\frac{H(n)}{n} \rightarrow c$ and $\sum_{k=N}^{n-1} \frac{H(k)}{k(k+1)} \sim c \log n$ as $n \rightarrow \infty$ (see the proof of Corollary 13.2), one obtains $\Delta(S) \leqslant D(S) + \varepsilon$. The assertion follows.

For the left inequality $d(S) \leqslant \delta(S)$, let $\varepsilon > 0$ and choose $N \in \mathbb{N}$ such that $\frac{S(n)}{H(n)} \geqslant d(S) - \varepsilon$ for all $n \geqslant N$. Again, summation by parts yields for $n \geqslant N$

$$\sum_{k=1}^{n} \frac{s(k)}{k} \geqslant \sum_{k=N}^{n-1} \frac{S(k)}{k(k+1)}$$

$$\geqslant (d(S) - \varepsilon) \sum_{k=N}^{n-1} \frac{H(k)}{k(k+1)},$$

which as above implies $\delta(S) \geqslant d(S) - \varepsilon$ and thus the assertion.

13.3 The Davenport-Erdős Theorem for Number Fields

Next, we shall study the densities of the set $\mathcal{M}_{\mathcal{A}}$. Let us start with the finite case. Note that, for a finite set $\mathcal{J}$ of integral ideals, their least common multiple is just the intersection $\bigcap \mathcal{J}$.

Proposition 13.7 *If $\mathcal{A}$ is finite, then the density of $\mathcal{M}_{\mathcal{A}}$ exists and is given by*

$$\mathrm{dens}(\mathcal{M}_{\mathcal{A}}) = \sum_{\varnothing \neq \mathcal{J} \subset \mathcal{A}} (-1)^{|\mathcal{J}|+1} \frac{1}{N(\bigcap \mathcal{J})}$$

Proof If $\mathfrak{b}$ is a non-zero integral ideal of norm $N(\mathfrak{b})$ and divisible by $\mathfrak{a}$, then there is a unique non-zero integral ideal $\mathfrak{a}'$ such that $\mathfrak{b} = \mathfrak{a}\mathfrak{a}'$. In particular, $N(\mathfrak{a}') = N(\mathfrak{b})/N(\mathfrak{a})$ by the multiplicativity of the norm. This provides a bijection from the set of multiples of $\mathfrak{a}$ of norm n to the set of non-zero integral ideals of norm $n/N(\mathfrak{a})$. Hence, by the inclusion-exclusion principle, one has

$$\frac{S(x)}{H(x)} = \sum_{\varnothing \neq \mathcal{J} \subset \mathcal{A}} (-1)^{|\mathcal{J}|+1} H\left(\frac{x}{N(\bigcap \mathcal{J})}\right) \Big/ H(x).$$

Application of Proposition 13.1 now yields the assertion. $\qquad \blacksquare$

Now let $\mathcal{A} = \{\mathfrak{a}_1, \mathfrak{a}_2, \dots\}$ be (countably) infinite. Since $\mathrm{dens}(\mathcal{M}_{\{\mathfrak{a}_1,\dots,\mathfrak{a}_r\}})$ is an increasing sequence with upper bound 1, we may define

$$A := \lim_{r \to \infty} \mathrm{dens}(\mathcal{M}_{\{\mathfrak{a}_1,\dots,\mathfrak{a}_r\}}).$$

It is then natural to ask if, in general, A is the density of $\mathcal{M}_{\mathcal{A}}$. Already in the special case $K = \mathbb{Q}$ the answer is negative in the sense that the natural lower and upper densities may differ; cf. [4].

Remark 13.8 Due to $\mathrm{dens}(\mathcal{M}_{\{\mathfrak{a}_1,\ldots,\mathfrak{a}_r\}}) \leqslant d(\mathcal{M}_{\mathcal{A}})$ for all $r \in \mathbb{N}$, one has $A \leqslant d(\mathcal{M}_{\mathcal{A}})$.

Proposition 13.9 *If the series $\sum_{\mathfrak{a} \in \mathcal{A}} \frac{1}{N(\mathfrak{a})}$ converges, then the density of $\mathcal{M}_{\mathcal{A}}$ exists and is equal to A.*

Proof For fixed $r \in \mathbb{N}$, the number of elements of $\mathcal{M}_{\mathcal{A}}$ up to norm n *not* divisible by any of $\mathfrak{a}_1, \ldots, \mathfrak{a}_r$ is at most $\sum_{i=r+1}^{\infty} H(\frac{n}{N(\mathfrak{a}_i)})$. Hence, the corresponding upper density is at most $\sum_{i=r+1}^{\infty} \frac{1}{N(\mathfrak{a}_i)}$ and this converges to 0 as $r \to \infty$. It follows that the upper density of $\mathcal{M}_{\mathcal{A}}$ is

$$\mathrm{dens}(\mathcal{M}_{\{\mathfrak{a}_1,\ldots,\mathfrak{a}_r\}}) + O\Big(\sum_{i=r+1}^{\infty} \frac{1}{N(\mathfrak{a}_i)} \Big),$$

which converges to A as $r \to \infty$. This yields $D(\mathcal{M}_{\mathcal{A}}) \leqslant A$ and thus the assertion by Remark 13.8.

Example 13.1 Recall that the Dedekind zeta function $\zeta_K(s)$ converges for all $s > 1$ and has the Euler product expansion

$$\zeta_K(s) = \sum_{\mathfrak{a} \neq 0} \frac{1}{N(\mathfrak{a})^s} = \prod_{\mathfrak{p}} \Big(1 - \frac{1}{N(\mathfrak{p})^s}\Big)^{-1}.$$

It follows that, for $l \geqslant 2$ fixed and $\mathcal{A} = \{\mathfrak{p}^l \mid \mathfrak{p} \text{ prime}\}$, the density of $\mathcal{M}_{\mathcal{A}}$ exists and is equal to

$$1 - \prod_{\mathfrak{p}} \Big(1 - \frac{1}{N(\mathfrak{p})^l}\Big) = 1 - \frac{1}{\zeta_K(l)}.$$

In other words, the density of $\mathcal{V}_{\mathcal{A}}$ exists and is equal to $\frac{1}{\zeta_K(l)}$, in accordance with [13, Thm. 4.1].

As a preparation of the proof below, we next introduce the so-called *multiplicative density* of $\mathcal{M}_{\mathcal{A}}$. Let $\{\mathfrak{p}_1, \mathfrak{p}_2, \ldots\}$ be the set of all prime ideals of O_K, with a numbering that corresponds to increasing order with respect to the norms, i.e. $i \leqslant j$ always implies $N(\mathfrak{p}_i) \leqslant N(\mathfrak{p}_j)$. For $k \in \mathbb{N}$ fixed, denote by $\mathfrak{n}'$ the general non-zero integral ideal composed entirely of the prime ideals $\mathfrak{p}_1, \ldots, \mathfrak{p}_k$ (a so-called $\mathfrak{p}_1, \ldots, \mathfrak{p}_k$-ideal). Then, one has the convergence

$$\sum_{\mathfrak{n}'} \frac{1}{N(\mathfrak{n}')} = \prod_{i=1}^{k} \Big(1 - \frac{1}{N(\mathfrak{p}_i)}\Big)^{-1} =: \Pi_k.$$

Further, denote by $\mathfrak{b}'$ those ideals from $\mathcal{M}_\mathcal{A}$ that are $\mathfrak{p}_1, \ldots, \mathfrak{p}_k$-ideals and let

$$B_k := \frac{\sum_{\mathfrak{b}'} \frac{1}{N(\mathfrak{b}')}}{\sum_{\mathfrak{n}'} \frac{1}{N(\mathfrak{n}')}} = \Pi_k^{-1} \sum_{\mathfrak{b}'} \frac{1}{N(\mathfrak{b}')}.$$

If the sequence B_k converges as $k \to \infty$, the limit is called the *multiplicative density* of $\mathcal{M}_\mathcal{A}$. Let $\mathcal{A}' := \{\mathfrak{a}'_1, \mathfrak{a}'_2, \ldots\}$ be the subset of $\mathcal{A}$ consisting of the $\mathfrak{p}_1, \ldots, \mathfrak{p}_k$-ideals only. Then the $\mathfrak{b}'$ from above are precisely those of the form $\mathfrak{a}'_i \mathfrak{n}'$. It follows from the inclusion-exclusion principle and Proposition 13.9 in conjunction with the convergence of $\sum_{\mathfrak{a}' \in \mathcal{A}'} \frac{1}{N(\mathfrak{a}')}$ that

$$\sum_{\mathfrak{b}'} \frac{1}{N(\mathfrak{b}')} = \sum_{\mathfrak{n}'} \frac{1}{N(\mathfrak{n}')} \sum_{\varnothing \neq \mathcal{J} \subset \mathcal{A}'} (-1)^{|\mathcal{J}|+1} \frac{1}{N(\bigcap \mathcal{J})}$$

$$= \Pi_k \, \mathrm{dens}(\mathcal{M}_{\mathcal{A}'}).$$

One obtains that $B_k = \mathrm{dens}(\mathcal{M}_{\mathcal{A}'})$ which shows that the B_k increase with k. Since the B_k are bounded above by 1, this proves that the B_k indeed converge, say $\lim_{k \to \infty} B_k =: B$.

Next, we shall show that $B = A$. Clearly, if k is sufficiently large in relation to r, then $\{\mathfrak{a}_1, \ldots, \mathfrak{a}_r\} \subset \mathcal{A}'$. Hence, one has

$$B \geqslant B_k = \mathrm{dens}(\mathcal{M}_{\mathcal{A}'}) \geqslant \mathrm{dens}(\mathcal{M}_{\{\mathfrak{a}_1, \ldots, \mathfrak{a}_r\}})$$

and therefore $B \geqslant A$. For the reverse inequality $A \geqslant B$, let k be fixed. The convergence of $\sum_{\mathfrak{a}' \, in \, \mathcal{A}'} \frac{1}{N(\mathfrak{a}')}$ implies that the density of $\mathcal{M}_{\mathcal{A}'}$ exists and satisfies (see the proof of Proposition 13.9)

$$\mathrm{dens}(\mathcal{M}_{\mathcal{A}'}) \leqslant \mathrm{dens}(\mathcal{M}_{\{\mathfrak{a}'_1, \ldots, \mathfrak{a}'_r\}}) + \sum_{i=r+1}^{\infty} \frac{1}{N(\mathfrak{a}'_i)}.$$

Now choose s large enough such that $\{\mathfrak{a}'_1, \ldots, \mathfrak{a}'_r\} \subset \{\mathfrak{a}_1, \ldots, \mathfrak{a}_s\}$. It follows that

$$\mathrm{dens}(\mathcal{M}_{\{\mathfrak{a}'_1 dots, \mathfrak{a}'_r\}}) \leqslant \mathrm{dens}(\mathcal{M}_{\{\mathfrak{a}_1, \ldots, \mathfrak{a}_s\}}) \leqslant A$$

and further, by letting $r \to \infty$, $\mathrm{dens}(\mathcal{M}_{\mathcal{A}'}) \leqslant A$, i.e. $B_k \leqslant A$. It follows that $B \leqslant A$. Altogether, this proves the claim $B = A$. We are now in a position to proof the main result of this short note.

Theorem 13.10 *The logarithmic density of $\mathcal{M}_\mathcal{A}$ exists and is equal to A. The number A also equals the lower density of $\mathcal{M}_\mathcal{A}$.*

Proof We have to show, for $S = M_{\mathcal{A}}$, the equality $d(S) = \delta(S) = \Delta(S) = A$, i.e. $\Delta(S) \leqslant A$ or, equivalently, $\Delta(S) \leqslant B$ since we have already seen above that $A = B$. Let $k \in \mathbb{N}$ be fixed. Divide the $\mathfrak{b}'$ from above of norm $\leqslant x$ into two classes, placing in the first class those from $M_{\mathcal{A}'}$ and in the second class the remaining ones. The $\mathfrak{b}'$ in the first class have density B_k (see above), hence the sum $\beta_1(x)$ corresponding to the $\mathfrak{b}'$ in the first class satisfies (the density inequality is an equality in this case)

$$\lim_{x \to \infty} \frac{\beta_1(x)}{c \log x} = B_k .$$

For the sum $\beta_2(x)$ corresponding to the $\mathfrak{b}'$ in the second class, let $\{\mathfrak{p}_1, \ldots, \mathfrak{p}_h\}$ be the set of all prime ideals with norm up to x. The $\mathfrak{b}'$ in the second class are $\mathfrak{p}_1, \ldots, \mathfrak{p}_h$-ideals, but are not in $M_{\mathcal{A}'}$. Denoting by $\mathfrak{b}^*$ the $\mathfrak{b}'$ of this kind (whether of norm $\leqslant x$ or not), one has

$$\beta_2(x) \leqslant \sum_{\mathfrak{b}^*} \frac{1}{N(\mathfrak{b}^*)} .$$

The $\mathfrak{b}^*$ are obtained by taking all $\mathfrak{p}_1, \ldots, \mathfrak{p}_h$-ideals $\mathfrak{b}''$, and removing from them all $\mathfrak{b}'\mathfrak{c}$, where $\mathfrak{b}'$ is a $\mathfrak{p}_1, \ldots, \mathfrak{p}_k$-ideal and $\mathfrak{c}$ is any $\mathfrak{p}_{k+1}, \ldots, \mathfrak{p}_h$-ideal. Hence

$$\sum_{\mathfrak{b}^*} \frac{1}{N(\mathfrak{b}^*)} = \sum_{\mathfrak{b}''} \frac{1}{N(\mathfrak{b}'')} - \sum_{\mathfrak{b}'} \frac{1}{N(\mathfrak{b}')} \sum_{\mathfrak{c}} \frac{1}{N(\mathfrak{c})} = \Pi_h B_h - \Pi_k B_k \sum_{\mathfrak{c}} \frac{1}{N(\mathfrak{c})} .$$

Since

$$\sum_{\mathfrak{c}} \frac{1}{N(\mathfrak{c})} = \prod_{i=k+1}^{h} \left(1 - \frac{1}{N(\mathfrak{p}_i)}\right)^{-1} = \Pi_h \Pi_k^{-1} ,$$

this shows that

$$\sum_{\mathfrak{b}^*} \frac{1}{N(\mathfrak{b}^*)} = \Pi_h (B_h - B_k) .$$

Finally, it follows from the Mertens type Theorem 13.3 by Rosen that

$$\beta_2(x) \leqslant \sum_{\mathfrak{b}^*} \frac{1}{N(\mathfrak{b}^*)} = \Pi_h (B_h - B_k) \leqslant C \log x (B_h - B_k)$$

and thus, with $\beta(x) := \beta_1(x) + \beta_2(x)$,

$$\limsup_{x \to \infty} \frac{\beta(x)}{c \log x} \leqslant B_k + \frac{C}{c}(B - B_k) ,$$

since $x \to \infty$ implies $h \to \infty$ which in turn implies $B_h \to B$. Letting $k \to \infty$ and thus $B_k \to B$, one obtains that $\Delta(M_{\mathcal{A}}) \leqslant B$.

Corollary 13.11 *The logarithmic density of $\mathcal{V}_{\mathcal{A}}$ exists and is equal to $1 - A$. This number also equals the upper density of $\mathcal{V}_{\mathcal{A}}$.*

Proof In general, one has $d(S) = 1 - D(S^c)$ and $\delta(S) = 1 - \Delta(S^c)$.

Remark 13.12 It is natural to ask for an extension of the above results to the case of m-tuples $(\mathfrak{b}_1, \ldots, \mathfrak{b}_m)$ of non-zero integral ideals, where one studies the set of those tuples that consist of simultaneous multiples of ideals from $\mathcal{A}$ respectively its complement consisting of the relatively $\mathcal{A}$-free tuples. This is work in progress.

Remark 13.13 There is a non-canonical possibility of defining upper and lower (asymptotic) densities of sets S of non-zero integral ideals $\mathfrak{b} \subset O_K$ by passing from S to the subset

$$\tilde{S} := \{a \in O_K \mid (a) \in S\}$$

of O_K and considering the image $\alpha(\tilde{S}) \subset \mathbb{Z}^d$ under any isomorphism $\alpha : O_K \to \mathbb{Z}^d$ of Abelian groups (recall that $d = [K : \mathbb{Q}]$). The set $\alpha(\tilde{S})$ then has natural upper and lower densities defined by counting points e.g. in centred balls (or cubes) of radius R in $\mathbb{R}^d$ divided by the volume and then considering the lim sup resp. lim inf as $R \to \infty$. Note that this also extends componentwise to the case of m-tuples mentioned in the last remark. In general, it is not clear if the outcome is independent of the embedding α or coincides with the corresponding densities introduced above. However, for the set of coprime m-tuples $(\mathfrak{b}_1, \ldots, \mathfrak{b}_m)$ of non-zero integral ideals (i.e. $\mathfrak{b}_1 + \ldots + \mathfrak{b}_m = O_K$) resp. the set of m-tuples $(a_1, \ldots, a_m) \in O_K^m$ with $(a_1) + \ldots + (a_m) = O_K$, even the (suitably defined) densities exist and all answers are affirmative (with both densities equal to $1/\zeta_K(m)$) as follows from [8, 13]. Another coincidence of the two ways of computing densities shows up (with both densities equal to $1/\zeta_K(l)$) in the case of l-free non-zero integral ideals (non-divisibility by any nontrivial lth power) resp. integers in O_K [5, 13]. Proving such a coincidence in our setting above for the lower density of $M_{\mathcal{A}}$ remains open, even for the case $m = 1$.

Acknowledgements It is a pleasure to thank Joanna Kułaga-Przymus and Jeanine Van Order for helpful discussions. This work was initiated during a research in pairs stay at CIRM (Luminy) in 2016, within the Jean Morlet Chair. It was supported by the German Research Council (DFG), within the CRC 701.

References

1. M. Baake, C. Huck, Ergodic properties of visible lattice points. Proc. Steklov Inst. Math. **288**, 165–188 (2015)
2. A. Bartnicka, S. Kasjan, J. Kułaga-Przymus, M. Lemańczyk, $\mathcal{B}$-free sets and dynamics. Trans. Am. Math. Soc. arXiv:1509.08010 (accepted)
3. S.J. Benkoski, The probability that k positive integers are relatively r-prime. J. Number Theory **8**, 218–223 (1976)
4. A.S. Besicovitch, On the density of certain sequences of integers. Math. Ann. **110**, 336–341 (1934)
5. F. Cellarosi, I. Vinogradov, Ergodic properties of k-free integers in number fields. J. Mod. Dyn. **7**, 461–488 (2013)
6. E.H. El Abdalaoui, M. Lemańczyk, T. de la Rue, A dynamical point of view on the set of $\mathcal{B}$-free integers. Int. Math. Res. Notices **2015**, 7258–7286 (2015)
7. P. Erdös, H. Davenport, On sequences of positive integers. J. Indian Math. Soc. **15**, 19–24 (1951)
8. A. Ferraguti, G. Micheli, On the Mertens-Cesàro theorem for number fields. Bull. Aust. Math. Soc. **93**, 199–210 (2016)
9. D.A. Marcus, *Number Fields*, 2nd edn. (Springer, New York, 1995)
10. J. Neukirch, *Algebraic Number Theory* (Springer, Berlin, 1999)
11. M. Rosen, A generalization of Mertens' theorem. J. Ramanujan Math. Soc. **14**, 1–19 (1999)
12. P. Sarnak, Three lectures on the Möbius function, randomness and dynamics (Lecture 1) (2010). http://publications.ias.edu/sarnak/
13. B.D. Sittinger, The probability that random algebraic integers are relatively r-prime. J. Number Theory **130**, 164–171 (2010)

Chapter 14
The Lagrange and Markov Spectra from the Dynamical Point of View

Carlos Matheus

14.1 Diophantine Approximations and Lagrange and Markov Spectra

14.1.1 Rational Approximations of Real Numbers

Given a real number $\alpha \in \mathbb{R}$, it is natural to compare the quality $|\alpha - p/q|$ of a rational approximation $p/q \in \mathbb{Q}$ and the size q of its denominator.

Since any real number lies between two consecutive integers, for every $\alpha \in \mathbb{R}$ and $q \in \mathbb{N}$, there exists $p \in \mathbb{Z}$ such that $|q\alpha - p| \leqslant 1/2$, i.e.

$$\left| \alpha - \frac{p}{q} \right| \leqslant \frac{1}{2q} \tag{14.1}$$

In 1842, Dirichlet [4] used his famous *pigeonhole principle* to improve (14.1).

Theorem 14.1 (Dirichlet) *For any $\alpha \in \mathbb{R} - \mathbb{Q}$, the inequality*

$$\left| \alpha - \frac{p}{q} \right| \leqslant \frac{1}{q^2}$$

has infinitely many rational solutions $p/q \in \mathbb{Q}$.

C. Matheus (✉)
Université Paris 13, Sorbonne Paris Cité, LAGA, CNRS (UMR 7539), Villetaneuse, France

© Springer International Publishing AG, part of Springer Nature 2018
S. Ferenczi et al. (eds.), *Ergodic Theory and Dynamical Systems in their Interactions with Arithmetics and Combinatorics*, Lecture Notes in Mathematics 2213, https://doi.org/10.1007/978-3-319-74908-2_14

Proof Given $Q \in \mathbb{N}$, we decompose the interval $[0, 1)$ into Q disjoint subintervals as follows:

$$[0, 1) = \bigcup_{j=0}^{Q-1} \left[\frac{j}{Q}, \frac{j+1}{Q} \right)$$

Next, we consider the $Q+1$ distinct[1] numbers $\{ i\alpha \}$, $i = 0, \ldots, Q$, where $\{x\}$ denotes the *fractional part*[2] of x. By the *pigeonhole principle*, some interval $\left[\frac{j}{Q}, \frac{j+1}{Q} \right)$ must contain two such numbers, say $\{n\alpha\}$ and $\{m\alpha\}$, $0 \leqslant n < m \leqslant Q$. It follows that

$$|\{m\alpha\} - \{n\alpha\}| < \frac{1}{Q},$$

i.e., $|q\alpha - p| < 1/Q$ where $0 < q := m - n \leqslant Q$ and $p := \lfloor m\alpha \rfloor - \lfloor n\alpha \rfloor$. Therefore,

$$\left| \alpha - \frac{p}{q} \right| < \frac{1}{qQ} \leqslant \frac{1}{q^2}$$

This completes the proof of the theorem.

In 1891, Hurwitz [12] showed that Dirichlet's theorem is essentially optimal:

Theorem 14.2 (Hurwitz) *For any $\alpha \in \mathbb{R} - \mathbb{Q}$, the inequality*

$$\left| \alpha - \frac{p}{q} \right| \leqslant \frac{1}{\sqrt{5}q^2}$$

has infinitely many rational solutions $p/q \in \mathbb{Q}$.

Moreover, for all $\varepsilon > 0$, the inequality

$$\left| \frac{1 + \sqrt{5}}{2} - \frac{p}{q} \right| \leqslant \frac{1}{(\sqrt{5} + \varepsilon)q^2}$$

has only finitely many rational solutions $p/q \in \mathbb{Q}$.

The first part of Hurwitz theorem is proved in Appendix 1, while the second part of Hurwitz theorem is left as an exercise to the reader:

---

[1] $\alpha \notin \mathbb{Q}$ is used here.

[2] $\{x\} := x - \lfloor x \rfloor$ and $\lfloor x \rfloor := \max\{n \in \mathbb{Z} : n \leqslant x\}$ is the integer part of x.

Exercise 14.3 Show the second part of Hurwitz theorem. (*Hint*: use the identity $p^2 - pq - q^2 = \left(q\frac{1+\sqrt{5}}{2} - p\right)\left(q\frac{1-\sqrt{5}}{2} - p\right)$ relating $\frac{1+\sqrt{5}}{2}$ and its Galois conjugate $\frac{1-\sqrt{5}}{2}$.)

Moreover, use your argument to give a bound on

$$\#\left\{\frac{p}{q} \in \mathbb{Q} : \left|\frac{1+\sqrt{5}}{2} - \frac{p}{q}\right| \leq \frac{1}{(\sqrt{5}+\varepsilon)q^2}\right\}$$

in terms of $\varepsilon > 0$.

Note that Hurwitz theorem does *not* forbid an improvement of "$\left|\alpha - \frac{p}{q}\right| \leq \frac{1}{\sqrt{5}q^2}$ has infinitely many rational solutions $p/q \in \mathbb{Q}$" for *certain* $\alpha \in \mathbb{R} - \mathbb{Q}$. This motivates the following definition:

Definition 14.4 The constant

$$\ell(\alpha) := \limsup_{p,q\to\infty} \frac{1}{|q(q\alpha - p)|}$$

is called the *best constant of Diophantine approximation* of α.

Intuitively, $\ell(\alpha)$ is the best constant ℓ such that $|\alpha - \frac{p}{q}| \leq \frac{1}{\ell q^2}$ has infinitely many rational solutions $p/q \in \mathbb{Q}$.

Remark 14.5 By Hurwitz theorem, $\ell(\alpha) \geq \sqrt{5}$ for all $\alpha \in \mathbb{R} - \mathbb{Q}$ and $\ell(\frac{1+\sqrt{5}}{2}) = \sqrt{5}$.

The collection of *finite* best constants of Diophantine approximations is the *Lagrange spectrum*:

Definition 14.6 The *Lagrange spectrum* is

$$L := \{\ell(\alpha) : \alpha \in \mathbb{R} - \mathbb{Q}, \ell(\alpha) < \infty\} \subset \mathbb{R}$$

Remark 14.7 Khinchin proved in 1926 a famous theorem implying that $\ell(\alpha) = \infty$ for Lebesgue almost every $\alpha \in \mathbb{R} - \mathbb{Q}$ (see, e.g., Khinchin's book [15] for more details).

14.1.2 Integral Values of Binary Quadratic Forms

Let $q(x, y) = ax^2 + bxy + cy^2$ be a *binary quadratic form* with real coefficients $a, b, c \in \mathbb{R}$. Suppose that q is *indefinite*[3] with positive *discriminant* $\Delta(q) := b^2 - 4ac$. What is the smallest value of $q(x, y)$ at non-trivial integral vectors $(x, y) \in \mathbb{Z}^2 - \{(0, 0)\}$?

Definition 14.8 The *Markov spectrum* is

$$M := \left\{ \frac{\sqrt{\Delta(q)}}{\displaystyle\inf_{(x,y)\in\mathbb{Z}^2-\{(0,0)\}} |q(x, y)|} \in \mathbb{R} : q \text{ is an indefinite binary quadratic form with } \Delta(q) > 0 \right\}$$

Remark 14.9 A similar Diophantine problem for *ternary* (and *n*-ary, $n \geqslant 3$) quadratic forms was proposed by Oppenheim in 1929. Oppenheim's conjecture was famously solved in 1987 by Margulis using *dynamics on homogeneous spaces*: the reader is invited to consult Witte Morris book [28] for more details about this beautiful portion of Mathematics.

In 1880, Markov [18] noticed a relationship between certain binary quadratic forms and rational approximations of certain irrational numbers. This allowed him to prove the following result:

Theorem 14.10 (Markov) $L \cap (-\infty, 3) = M \cap (-\infty, 3) = \{k_1 < k_2 < k_3 < k_4 < \dots\}$ *where* $k_1 = \sqrt{5}$, $k_2 = \sqrt{8}$, $k_3 = \frac{\sqrt{221}}{5}$, $k_4 = \frac{\sqrt{1517}}{13}$, $\dots$ *is an explicit increasing sequence of quadratic surds*[4] *accumulating at 3.*

In fact, $k_n = \sqrt{9 - \frac{4}{m_n^2}}$ *where* $m_n \in \mathbb{N}$ *is the n-th Markov number, and a Markov number is the largest coordinate of a Markov triple* (x, y, z), *i.e., an integral solution of* $x^2 + y^2 + z^2 = 3xyz$.

Remark 14.11 All Markov triples can be deduced from $(1, 1, 1)$ by applying the so-called *Vieta involutions* V_1, V_2, V_3 given by

$$V_1(x, y, z) = (x', y, z)$$

---

[3]I.e., q takes both positive and negative values.
[4]I.e., $k_n^2 \in \mathbb{Q}$ for all $n \in \mathbb{N}$.

where $x' = 3yz - x$ is the other solution of the second degree equation $X^2 - 3yzX + (y^2 + z^2) = 0$, etc. In other terms, all Markov triples appear in *Markov tree*[5]:

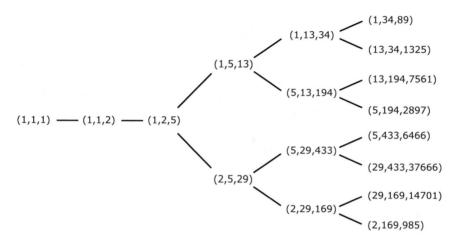

Remark 14.12 For more information on Markov numbers, the reader might consult Zagier's paper [30] on this subject. Among many conjectures and results mentioned in this paper, we have:

- Conjecturally, each Markov number z determines *uniquely* Markov triples (x, y, z) with $x \leqslant y \leqslant z$;
- If $M(x) := \#\{m \text{ Markov number} : m \leqslant x\}$, then $M(x) = c(\log x)^2 + O(\log x(\log \log x)^2)$ for an *explicit* constant $c \simeq 0.18071704711507\ldots$; conjecturally, $M(x) = c(\log(3x))^2 + o(\log x)$, i.e., if m_n is the n-th Markov number (counted with multiplicity), then $m_n \sim \frac{1}{3}A^{\sqrt{n}}$ with $A = e^{1/\sqrt{c}} \simeq 10.5101504\ldots$

14.1.3   Best Rational Approximations and Continued Fractions

The constant $\ell(\alpha)$ was defined in terms of rational approximations of $\alpha \in \mathbb{R} - \mathbb{Q}$. In particular,

$$\ell(\alpha) = \limsup_{n \to \infty} \frac{1}{|s_n(s_n\alpha - r_n)|}$$

---

[5]Namely, the tree where Markov triples (x, y, z) are displayed after applying permutations to put them in normalized form $x \leqslant y \leqslant z$, and two normalized Markov triples are connected if we can obtain one from the other by applying Vieta involutions.

where $(r_n/s_n)_{n \in \mathbb{N}}$ is the sequence of best rational approximations of α. Here, p/q is called a *best rational approximation*[6] whenever

$$\left| \alpha - \frac{p}{q} \right| < \frac{1}{2q^2}$$

The sequence $(r_n/s_n)_{n \in \mathbb{N}}$ of best rational approximations of α is produced by the so-called *continued fraction algorithm*.

Given $\alpha = \alpha_0 \notin \mathbb{Q}$, we define recursively $a_n = \lfloor \alpha_n \rfloor$ and $\alpha_{n+1} = \frac{1}{\alpha_n - a_n}$ for all $n \in \mathbb{N}$. We can write α as a *continued fraction*

$$\alpha = a_0 + \cfrac{1}{a_1 + \cfrac{1}{a_2 + \cfrac{1}{\ddots}}} =: [a_0; a_1, a_2, \ldots]$$

and we denote

$$\mathbb{Q} \ni \frac{p_n}{q_n} := a_0 + \cfrac{1}{a_1 + \cfrac{1}{\ddots + \frac{1}{a_n}}} := [a_0; a_1, \ldots, a_n]$$

Remark 14.13 Lévy's theorem [16] (from 1936) says that $\sqrt[n]{q_n} \to e^{\pi^2/12 \log 2} \simeq 3.27582291872\ldots$ for Lebesgue almost every $\alpha \in \mathbb{R}$. By elementary properties of continued fractions (recalled below), it follows from Lévy's theorem that $\sqrt[n]{|\alpha - \frac{p_n}{q_n}|} \to e^{-\pi^2/6 \log 2} \simeq 0.093187822954\ldots$ for Lebesgue almost every $\alpha \in \mathbb{R}$.

Proposition 14.14 p_n *and* q_n *are recursively given by*

$$\begin{cases} p_{n+2} = a_{n+2}p_{n+1} + p_n, \ p_{-1} = 1, p_{-2} = 0 \\ q_{n+2} = a_{n+2}q_{n+1} + q_n, \ q_{-1} = 0, q_{-2} = 1 \end{cases}$$

Proof Exercise.[7]

In other words, we have

$$[a_0; a_1, \ldots, a_{n-1}, z] = \frac{zp_{n-1} + p_{n-2}}{zq_{n-1} + q_{n-2}} \tag{14.2}$$

---

[6]This nomenclature will be justified later by Propositions 14.18 and 14.19 below.

[7]*Hint*: Use induction and the fact that $[t_0; t_1, \ldots, t_n, t_{n+1}] = [t_0; t_1, \ldots, t_n + \frac{1}{t_{n+1}}]$.

or, equivalently,

$$\begin{pmatrix} p_{n+1} & p_n \\ q_{n+1} & q_n \end{pmatrix} \cdot \begin{pmatrix} a_{n+2} & 1 \\ 1 & 0 \end{pmatrix} = \begin{pmatrix} p_{n+2} & p_{n+1} \\ q_{n+2} & q_{n+1} \end{pmatrix} \tag{14.3}$$

Corollary 14.15 $p_{n+1}q_n - p_n q_{n+1} = (-1)^n$ *for all* $n \geqslant 0$.

Proof This follows from (14.3) because the matrix $\begin{pmatrix} * & 1 \\ 1 & 0 \end{pmatrix}$ has determinant -1.

Corollary 14.16 $\alpha = \frac{\alpha_n p_{n-1} + p_{n-2}}{\alpha_n q_{n-1} + q_{n-2}}$ *and* $\alpha_n = \frac{p_{n-2} - q_{n-2}\alpha}{q_{n-1}\alpha - p_{n-1}}$.

Proof This is a consequence of (14.2) and the fact that $\alpha =: [a_0; a_1, \ldots, a_{n-1}, \alpha_n]$.

The relationship between $\frac{p_n}{q_n}$ and the sequence of best rational approximations is explained by the following two propositions:

Proposition 14.17 $\left| \alpha - \frac{p_n}{q_n} \right| \leqslant \frac{1}{q_n q_{n+1}} < \frac{1}{a_{n+1}q_n^2} \leqslant \frac{1}{q_n^2}$ *and, moreover, for all* $n \in \mathbb{N}$,

$$either \left| \alpha - \frac{p_n}{q_n} \right| < \frac{1}{2q_n^2} \ or \ \left| \alpha - \frac{p_{n+1}}{q_{n+1}} \right| < \frac{1}{2q_{n+1}^2}.$$

Proof Note that α belongs to the interval with extremities p_n/q_n and p_{n+1}/q_{n+1} (by Corollary 14.16). Since this interval has size

$$\left| \frac{p_{n+1}}{q_{n+1}} - \frac{p_n}{q_n} \right| = \left| \frac{p_{n+1}q_n - p_n q_{n+1}}{q_n q_{n+1}} \right| = \left| \frac{(-1)^n}{q_n q_{n+1}} \right| = \frac{1}{q_n q_{n+1}}$$

(by Corollary 14.15), we conclude that $\left| \alpha - \frac{p_n}{q_n} \right| \leqslant \frac{1}{q_n q_{n+1}}$.

Furthermore, $\frac{1}{q_n q_{n+1}} = \left| \frac{p_{n+1}}{q_{n+1}} - \alpha \right| + \left| \alpha - \frac{p_n}{q_n} \right|$. Thus, if

$$\left| \alpha - \frac{p_n}{q_n} \right| \geqslant \frac{1}{2q_n^2} \quad and \quad \left| \alpha - \frac{p_{n+1}}{q_{n+1}} \right| \geqslant \frac{1}{2q_{n+1}^2},$$

then

$$\frac{1}{q_n q_{n+1}} \geqslant \frac{1}{2q_n^2} + \frac{1}{2q_{n+1}^2},$$

i.e., $2q_n q_{n+1} \geqslant q_n^2 + q_{n+1}^2$, i.e., $q_n = q_{n+1}$, a contradiction.

In other terms, the sequence $(p_n/q_n)_{n \in \mathbb{N}}$ produced by the continued fraction algorithm contains best rational approximations with frequency at least $1/2$.

Conversely, the continued fraction algorithm detects *all* best rational approximations:

Proposition 14.18 *If* $|\alpha - \frac{p}{q}| < \frac{1}{2q^2}$, *then* $p/q = p_n/q_n$ *for some* $n \in \mathbb{N}$.

Proof Exercise.[8]

The terminology "best rational approximation" is motivated by the previous proposition and the following result:

Proposition 14.19 *For all* $q < q_n$, *we have* $|\alpha - \frac{p_n}{q_n}| < |\alpha - \frac{p}{q}|$.

Proof If $q < q_{n+1}$ and $p/q \neq p_n/q_n$, then

$$\left| \frac{p}{q} - \frac{p_n}{q_n} \right| \geqslant \frac{1}{qq_n} > \frac{1}{q_n q_{n+1}} = \left| \frac{p_{n+1}}{q_{n+1}} - \frac{p_n}{q_n} \right|$$

Hence, p/q does not belong to the interval with extremities p_n/q_n and p_{n+1}/q_{n+1}, and so

$$\left| \alpha - \frac{p_n}{q_n} \right| < \left| \alpha - \frac{p}{q} \right|$$

because α lies between p_n/q_n and p_{n+1}/q_{n+1}.

In fact, the approximations (p_n/q_n) of α are usually quite impressive:

Example 14.1 $\pi = [3; 7, 15, 1, 292, 1, 1, 1, 2, 1, 3, 1, 14, 2, 1, \dots]$ so that

$$\frac{p_0}{q_0} = 3, \quad \frac{p_1}{q_1} = \frac{22}{7}, \quad \frac{p_2}{q_2} = \frac{333}{106}, \quad \frac{p_3}{q_3} = \frac{355}{113}, \quad \cdots$$

The approximations p_1/q_1 and p_3/q_3 are called Yuelü and Milü (after Wikipedia) and they are somewhat spectacular:

$$\left| \pi - \frac{22}{7} \right| < \frac{1}{700} < \left| \pi - \frac{314}{100} \right| \text{ and } \left| \pi - \frac{355}{113} \right| < \frac{1}{3,000,000} < \left| \pi - \frac{3141592}{1,000,000} \right|$$

14.1.4   Perron's Characterization of Lagrange and Markov Spectra

In 1921, Perron interpreted $\ell(\alpha)$ in terms of Dynamical Systems as follows.

---

[8]*Hint*: Take $q_{n-1} < q \leqslant q_n$, suppose that $p/q \neq p_n/q_n$ and derive a contradiction in each case $q = q_n$, $q_n/2 \leqslant q < q_n$ and $q < q_n/2$ by analysing $|\alpha - \frac{p}{q}|$ and $|\frac{p}{q} - \frac{p_n}{q_n}|$ like in the proof of Proposition 14.19.

Proposition 14.20 $\alpha - \frac{p_n}{q_n} = \frac{(-1)^n}{(\alpha_{n+1}+\beta_{n+1})q_n^2}$ *where* $\beta_{n+1} := \frac{q_{n-1}}{q_n} =$ $[0; a_n, a_{n-1}, \ldots, a_1]$.

Proof Recall that $\alpha_{n+1} = \frac{p_{n-1}-q_{n-1}\alpha}{q_n\alpha-p_n}$ (cf. Corollary 14.16). Hence, $\alpha_{n+1} + \beta_{n+1} = \frac{p_{n-1}q_n-p_nq_{n-1}}{q_n(q_n\alpha-p_n)} = \frac{(-1)^n}{q_n(q_n\alpha-p_n)}$ (by Corollary 14.15). This proves the proposition.

Therefore, the proposition says that $\ell(\alpha) = \limsup_{n\to\infty}(\alpha_n+\beta_n)$. From the dynamical point of view, we consider the *symbolic space* $\Sigma = (\mathbb{N}^*)^{\mathbb{Z}} =: \Sigma^- \times \Sigma^+ = (\mathbb{N}^*)^{\mathbb{Z}^-} \times (\mathbb{N}^*)^{\mathbb{N}}$ equipped with the left *shift dynamics* $\sigma : \Sigma \to \Sigma$, $\sigma((a_n)_{n\in\mathbb{Z}}) := (a_{n+1})_{n\in\mathbb{Z}}$ and the *height function* $f : \Sigma \to \mathbb{R}$, $f((a_n)_{n\in\mathbb{Z}}) = [a_0; a_1, a_2, \ldots] + [0; a_{-1}, a_{-2}, \ldots]$. Then, the proposition above implies that

$$\ell(\alpha) = \limsup_{n\to+\infty} f(\sigma^n(\underline{\theta}))$$

where $\alpha = [a_0; a_1, a_2, \ldots]$ and $\underline{\theta} = (\ldots, a_{-1}, a_0, a_1, \ldots)$. In particular,

$$L = \{\ell(\underline{\theta}) : \underline{\theta} \in \Sigma, \ell(\underline{\theta}) < \infty\} \tag{14.4}$$

where $\ell(\underline{\theta}) := \limsup_{n\to+\infty} f(\sigma^n(\underline{\theta}))$.

Also, the Markov spectrum has a *similar description*:

$$M = \{m(\underline{\theta}) : \underline{\theta} \in \Sigma, m(\underline{\theta}) < \infty\} \tag{14.5}$$

where $m(\underline{\theta}) := \sup_{n\in\mathbb{Z}} f(\sigma^n(\underline{\theta}))$.

Remark 14.21 A *geometrical interpretation* of $\sigma : \Sigma \to \Sigma$ is provided by the so-called *Gauss map*[9]:

$$G(x) = \left\{\frac{1}{x}\right\} \tag{14.6}$$

for $0 < x \leqslant 1$.

---

[9]From Number Theory rather than Differential Geometry.

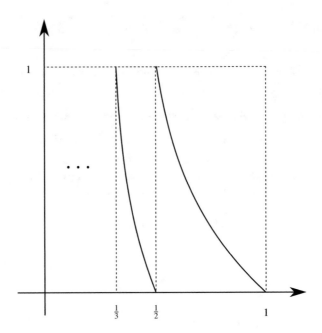

Indeed, $G([0; a_1, a_2, \dots]) = [0; a_2, \dots]$, so that $\sigma : \Sigma \to \Sigma$ is a symbolic version of the *natural extension* of G.

Furthermore, the identification $(\dots, a_{-1}, a_0, a_1, \dots) \simeq ([0; a_{-1}, a_{-2}, \dots], [a_0; a_1, a_2, \dots]) = (y, x)$ allows us to write the height function as $f((a_n)_{n \in \mathbb{Z}}) = x + y$.

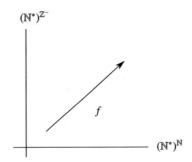

Perron's dynamical interpretation of the Lagrange and Markov spectra is the starting point of many results about L and M which are not so easy to guess from their definitions:

Exercise 14.22 Show that $L \subset M$ are closed subsets of $\mathbb{R}$.

Remark 14.23 $M - L \neq \emptyset$: for example, Freiman [6] proved in 1968 that

$$s = \overline{22122112211}\,\overline{1221122122} \in (\mathbb{N}^*)^{\mathbb{Z}}$$

has the property that $3.118120178 \simeq m(s) \in M - L$. (Here $\overline{\theta_1 \ldots \theta_n}$ means infinite repetition of the block $\theta_1 \ldots \theta_n$.)

Also, Freiman [7] showed in 1973 that $m(s_n) \in M - L$ and $m(s_n) \to m(s_\infty) \simeq 3.293044265 \in M - L$ where

$$s_n = \overline{2221121}\, \underbrace{22 \ldots 22}_{n \text{times}}\, 1211222121\overline{1122212}$$

for $n \geqslant 4$, and

$$s_\infty = \overline{2}1211222121\overline{1122212}$$

14.1.5 Digression: Lagrange Spectrum and Cusp Excursions on the Modular Surface

The Lagrange spectrum is related to the values of a certain height function H along the orbits of the geodesic flow g_t on the (unit cotangent bundle to) the modular surface: indeed, we will show that

$$L = \{\limsup_{t \to +\infty} H(g_t(x)) < \infty : x \text{ is a unit cotangent vector to the modular surface}\}$$

Remark 14.24 This fact is not surprising to experts: the Gauss map appears naturally by quotienting out the weak-stable manifolds of g_t as observed by Artin, Series, Arnoux, ... (see, e.g., [1]).

An *unimodular lattice* in $\mathbb{R}^2$ has the form $g(\mathbb{Z}^2)$, $g \in SL(2, \mathbb{Z})$, and the stabilizer in $SL(2, \mathbb{R})$ of the standard lattice $\mathbb{Z}^2$ is $SL(2, \mathbb{Z})$. In particular, the space of unimodular lattices in $\mathbb{R}^2$ is $SL(2, \mathbb{R})/SL(2, \mathbb{Z})$.

As it turns out, $SL(2, \mathbb{R})/SL(2, \mathbb{Z})$ is the unit cotangent bundle to the *modular surface* $\mathbb{H}/SL(2, \mathbb{Z})$ (where $\mathbb{H} = \{z \in \mathbb{C} : \text{Im}(z) > 0\}$ is the hyperbolic upper-half plane and $\begin{pmatrix} a & b \\ c & d \end{pmatrix} \in SL(2, \mathbb{R})$ acts on $z \in \mathbb{H}$ via $\begin{pmatrix} a & b \\ c & d \end{pmatrix} \cdot z = \frac{az+b}{cz+d}$).

The *geodesic flow* of the modular surface is the action of $g_t = \begin{pmatrix} e^t & 0 \\ 0 & e^{-t} \end{pmatrix}$ on $SL(2, \mathbb{R})/SL(2, \mathbb{Z})$. The *stable* and *unstable* manifolds of g_t are the orbits of the stable and unstable horocycle flows $h_s = \begin{pmatrix} 1 & 0 \\ s & 1 \end{pmatrix}$ and $u_s = \begin{pmatrix} 1 & s \\ 0 & 1 \end{pmatrix}$: indeed, this follows from the facts that $g_t h_s = h_{se^{-2t}} g_t$ and $g_t u_s = u_{se^t} g_t$.

The set of *holonomy* (or *primitive*) *vectors* of $\mathbb{Z}^2$ is

$$\text{Hol}(\mathbb{Z}^2) := \{(p, q) \in \mathbb{Z}^2 : \gcd(p, q) = 1\}$$

In general, the set $\mathrm{Hol}(X)$ of holonomy vectors of $X = g(\mathbb{Z}^2)$, $g \in SL(2, \mathbb{Z})$, is

$$\mathrm{Hol}(X) := g(\mathrm{Hol}(\mathbb{Z}^2)) \subset \mathbb{R}^2$$

The *systole* $\mathrm{sys}(X)$ of $X = g(\mathbb{Z}^2)$ is

$$\mathrm{sys}(X) := \min\{\|v\|_{\mathbb{R}^2} : v \in \mathrm{Hol}(X)\}$$

Remark 14.25 By Mahler's compactness criterion [17], $X \mapsto \frac{1}{\mathrm{sys}(X)}$ is a proper function on $SL(2, \mathbb{R})/SL(2, \mathbb{Z})$.

Remark 14.26 For later reference, we write $\mathrm{Area}(v) := |\mathrm{Re}(v)| \cdot |\mathrm{Im}(v)|$ for the area of the rectangle in $\mathbb{R}^2$ with diagonal $v = (\mathrm{Re}(v), \mathrm{Im}(v)) \in \mathbb{R}^2$.

Proposition 14.27 *The forward geodesic flow orbit of $X \in SL(2, \mathbb{R})/SL(2, \mathbb{Z})$ does not go straight to infinity (i.e., $\mathrm{sys}(g_t(X)) \to 0$ as $t \to +\infty$) if and only if there is no vertical vector in $\mathrm{Hol}(X)$. In this case, there are (unique) parameters $s, t, \alpha \in \mathbb{R}$ such that*

$$X = h_s g_t u_{-\alpha}(\mathbb{Z}^2)$$

Proof By unimodularity, any $X = g(\mathbb{Z}^2)$ has a single *short* holonomy vector. Since g_t contracts vertical vectors and expands horizontal vectors for $t > 0$, we have that $\mathrm{sys}(g_t(X)) \to 0$ as $t \to +\infty$ if and only if $\mathrm{Hol}(X)$ contains a vertical vector.

By Iwasawa decomposition, there are (unique) parameters $s, t, \theta \in \mathbb{R}$ such that $X = h_s g_t r_\theta$, where $r_\theta = \begin{pmatrix} \cos\theta & -\sin\theta \\ \sin\theta & \cos\theta \end{pmatrix}$. Since $\cos\theta \neq 0$ when $\mathrm{Hol}(X)$ contains no vertical vector and, in this situation,

$$r_\theta = h_{\tan\theta} g_{\log\cos\theta} u_{-\tan\theta},$$

we see that $X = h_{s+e^{-2t}\tan\theta} \cdot g_{t+\log\cos\theta} \cdot u_{-\tan\theta}(\mathbb{Z}^2)$ (because $h_s g_t r_\theta = h_s g_t h_{\tan\theta} g_{\log\cos\theta} u_{-\tan\theta} = h_{s+e^{-2t}\tan\theta} \cdot g_{t+\log\cos\theta} \cdot u_{-\tan\theta}$). This ends the proof of the proposition. $\qquad\blacksquare$

Proposition 14.28 *Let $X = h_s g_t u_{-\alpha}(\mathbb{Z}^2)$ be an unimodular lattice without vertical holonomy vectors. Then,*

$$\ell(\alpha) = \limsup_{\substack{|\mathrm{Im}(v)| \to \infty \\ v \in \mathrm{Hol}(X)}} \frac{1}{\mathrm{Area}(v)} = \limsup_{T \to +\infty} \frac{2}{\mathrm{sys}(g_T(X))^2}$$

Remark 14.29 This proposition says that the dynamical quantity $\displaystyle\limsup_{T \to +\infty} \frac{2}{\mathrm{sys}(g_T(X))^2}$ does *not* depend on the "weak-stable part" $h_s g_t$ (but only on α) and it can be computed *without* dynamics by simply studying almost vertical holonomy vectors in X.

Proof Note that $\text{Area}(g_t(v)) = \text{Area}(v)$ for all $t \in \mathbb{R}$ and $v \in \mathbb{R}^2$. Since $\text{Area}(v) = \frac{\|g_{t(v)}(v)\|^2}{2}$ for $t(v) := \frac{1}{2}\log\frac{|\text{Im}(v)|}{|\text{Re}(v)|}$, the equality $\limsup\limits_{\substack{|\text{Im}(v)|\to\infty \\ v\in\text{Hol}(X)}} \frac{1}{\text{Area}(v)} = \limsup\limits_{T\to+\infty} \frac{2}{\text{sys}(g_T(X))^2}$ follows.

The relation $g_T h_s = h_{se^{-2T}} g_T$ and the continuity of the systole function imply that $\limsup\limits_{T\to+\infty} \frac{2}{\text{sys}(g_T(X))^2}$ depends only on α. Because any $v \in \text{Hol}(u_{-\alpha}(\mathbb{Z}^2))$ has the form $v = (p - q\alpha, q) = u_{-\alpha}(p, q)$ with $(p, q) \in \text{Hol}(\mathbb{Z}^2)$, the equality $\limsup\limits_{\substack{|\text{Im}(v)|\to\infty \\ v\in\text{Hol}(X)}} \frac{1}{\text{Area}(v)} = \ell(\alpha)$.

In summary, the previous proposition says that the Lagrange spectrum L coincides with

$$\{\limsup_{T\to+\infty} H(g_T(x)) < \infty : x \in SL(2, \mathbb{R})/SL(2, \mathbb{Z})\}$$

where $H(y) = \frac{2}{\text{sys}(y)^2}$ is a (proper) height function and g_t is the geodesic flow on $SL(2, \mathbb{R})/SL(2, \mathbb{Z})$.

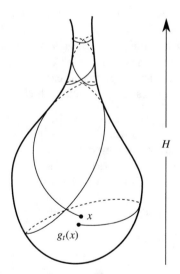

Remark 14.30 Several number-theoretical problems translate into dynamical questions on the modular surface: for example, Zagier [29] showed that the Riemann hypothesis is equivalent to a certain speed of equidistribution of u_s-orbits on $SL(2, \mathbb{R})/SL(2, \mathbb{Z})$.

14.1.6 Hall's Ray and Freiman's Constant

In 1947, Hall [9] proved that:

Theorem 14.31 (Hall) *The half-line* $[6, +\infty)$ *is contained in L.*

This result motivates the following nomenclature: the biggest half-line $[c_F, +\infty) \subset L(\subset M)$ is called *Hall's ray*.

In 1975, Freiman [8] determined Hall's ray:

Theorem 14.32 (Freiman) $c_F = 4 + \frac{253589820 + 283798\sqrt{462}}{491993569} \simeq 4.527829566\ldots$

The constant c_F is called *Freiman's constant*.

Let us sketch the proof of Hall's theorem based on the following lemma:

Lemma 14.33 (Hall) *Denote by* $C(4) := \{[0; a_1, a_2, \ldots] \in \mathbb{R} : a_i \in \{1, 2, 3, 4\} \ \forall i \in \mathbb{N}\}$. *Then,*

$$C(4) + C(4) := \{x + y \in \mathbb{R} : x, y \in C(4)\} = [\sqrt{2} - 1, 4(\sqrt{2} - 1)] = [0.414\ldots, 1.656\ldots]$$

Remark 14.34 The reader can find a proof of this lemma in Cusick-Flahive's book [3]. Interestingly enough, some of the techniques in the proof of Hall's lemma were rediscovered much later (in 1979) in the context of Dynamical Systems by Newhouse [26] (in the proof of his *gap lemma*).

Remark 14.35 $C(4)$ is a *dynamical Cantor set*[10] whose Hausdorff dimension is $> 1/2$ (see Remark 14.44 below). In particular, $C(4) \times C(4)$ is a planar Cantor set of Hausdorff dimension > 1 and Hall's lemma says that its image $f(C(4) \times C(4)) = C(4) + C(4)$ under the projection $f(x, y) = x + y$ contains an interval. Hence, Hall's lemma can be thought as a sort of "particular case" of *Marstrand's theorem* [19] (ensuring that typical projections of planar sets with Hausdorff dimension > 1 has positive Lebesgue measure).

For our purposes, the specific form $C(4) + C(4)$ is *not* important: the *key point* is that $C(4) + C(4)$ is an interval of length > 1.

Indeed, given $6 \leqslant \ell < \infty$, Hall's lemma guarantees the existence of $c_0 \in \mathbb{N}$, $5 \leqslant c_0 \leqslant \ell$ such that $\ell - c_0 \in C(4) + C(4)$. Thus,

$$\ell = c_0 + [0; a_1, a_2, \ldots] + [0; b_1, b_2, \ldots]$$

with $a_i, b_i \in \{1, 2, 3, 4\}$ for all $i \in \mathbb{N}$.

---

[10]See Sects. 14.2.2 and 14.2.3 below.

Define

$$\alpha := [0; \underbrace{b_1, c_0, a_1, \ldots}_{1^{st} \text{ block}}, \underbrace{b_n, \ldots, b_1, c_0, a_1, \ldots, a_n}_{n^{th} \text{ block}}, \ldots]$$

Since $c_0 \geqslant 5 > 4 \geqslant a_i, b_i$ for all $i \in \mathbb{N}$, Perron's characterization of $\ell(\alpha)$ implies that

$$L \ni \ell(\alpha) = \lim_{n \to \infty} (c_0 + [0; a_1, a_2, \ldots, a_n] + [0; b_1, b_2, \ldots, b_n]) = \ell$$

This proves Theorem 14.31.

14.1.7 Statement of Moreira's Theorem

Our discussion so far can be summarized as follows:

- $L \cap (-\infty, 3) = M \cap (-\infty, 3) = \{k_1 < k_2 < \cdots < k_n < \ldots\}$ is an *explicit* discrete set;
- $L \cap [c_F, \infty) = M \cap [c_F, \infty)$ is an *explicit* ray.

Moreira's theorem [21] says that the *intermediate parts* $L \cap [3, c_F]$ and $M \cap [3, c_F]$ of the Lagrange and Markov spectra have an intricate structure:

Theorem 14.36 (Moreira) *For each $t \in \mathbb{R}$, the sets $L \cap (-\infty, t)$ and $M \cap (-\infty, t)$ have the same Hausdorff dimension, say $d(t) \in [0, 1]$.*

Moreover, the function $t \mapsto d(t)$ is continuous, $d(3 + \varepsilon) > 0$ for all $\varepsilon > 0$ and $d(\sqrt{12}) = 1$ (even though $\sqrt{12} = 3.4641 \ldots < 4.5278 \ldots = c_F$).

Remark 14.37 Many results about L and M are *dynamical*.[11] In particular, it is not surprising that many facts about L and M have counterparts for *dynamical Lagrange and Markov spectra*[12]: for example, Hall ray or intervals in dynamical Lagrange spectra were found by Parkkonen-Paulin [27], Hubert-Marchese-Ulcigrai [11] and Moreira-Romaña [23], and the continuity result in Moreira's Theorem 14.36 was recently extended by Cerqueira, Moreira and the author in [2].

Before entering into the proof of Moreira's theorem, let us close this section by briefly recalling the notion of Hausdorff dimension.

---

[11]I.e., they involve Perron's characterization of L and M, the study of Gauss map and/or the geodesic flow on the modular surface, etc.
[12]I.e., the collections of "records" of height functions along orbits of dynamical systems.

14.1.8   Hausdorff Dimension

The *s-Hausdorff measure* $m_s(X)$ of a subset $X \subset \mathbb{R}^n$ is

$$m_s(X) := \lim_{\delta \to 0} \inf_{\substack{\bigcup_{i \in \mathbb{N}} U_i \supset X, \\ \mathrm{diam}(U_i) \leqslant \delta \ \forall i \in \mathbb{N}}} \sum_{i \in \mathbb{N}} \mathrm{diam}(U_i)^s$$

The *Hausdorff dimension* of X is

$$HD(X) := \sup\{s \in \mathbb{R} : m_s(X) = \infty\} = \inf\{s \in \mathbb{R} : m_s(X) = 0\}$$

Remark 14.38   There are many notions of dimension in the literature: for example, the *box-counting dimension* of X is $\lim_{\delta \to 0} \frac{\log N_X(\delta)}{\log(1/\delta)}$ where $N_X(\delta)$ is the smallest number of boxes of side lengths $\leqslant \delta$ needed to cover X. As an exercise, the reader is invited to show that the Hausdorff dimension is always smaller than or equal to the box-counting dimension.

The following exercise (whose solution can be found in Falconer's book [5]) describes several elementary properties of the Hausdorff dimension:

Exercise 14.39   Show that:

(a) if $X \subset Y$, then $HD(X) \leqslant HD(Y)$;
(b) $HD(\bigcup_{i \in \mathbb{N}} X_i) = \sup_{i \in \mathbb{N}} HD(X_i)$; in particular, $HD(X) = 0$ whenever X is a countable
      set (such as $X = \{p\}$ or $X = \mathbb{Q}^n$);
(c) if $f : X \to Y$ is α-Hölder continuous,[13] then $\alpha \cdot HD(f(X)) \leqslant HD(X)$;
(d) $HD(\mathbb{R}^n) = n$ and, more generally, $HD(X) = m$ when $X \subset \mathbb{R}^n$ is a smooth
      m-dimensional submanifold.

Example 14.2   Cantor's middle-third set $C = \{\sum_{i=1}^{\infty} \frac{a_i}{3^i} : a_i \in \{0, 2\} \ \forall i \in \mathbb{N}\}$ has

Hausdorff dimension $\frac{\log 2}{\log 3} \in (0, 1)$: see Falconer's book [5] for more details.

Using item (c) of Exercise 14.39 above, we have the following corollary of Moreira's Theorem 14.36:

Corollary 14.40 (Moreira)   *The function $t \mapsto HD(L \cap (-\infty, t))$ is not α-Hölder continuous for any $\alpha > 0$.*

---

[13]I.e., for some constant $C > 0$, one has $|f(x) - f(x')| \leqslant C|x - x'|^\alpha$ for all $x, x' \in X$.

Proof By Theorem 14.36, d maps $L \cap [3, 3+\varepsilon]$ to the non-trivial interval $[0, d(3+\varepsilon)]$ for any $\varepsilon > 0$. By item (c) of Exercise 14.39, if $t \mapsto d(t) = HD(L \cap (-\infty, t))$ were α-Hölder continuous for some $\alpha > 0$, then it would follow that

$$0 < \alpha = \alpha \cdot HD([0, d(3+\varepsilon)]) \leqslant HD(L \cap [3, 3+\varepsilon]) = d(3+\varepsilon)$$

for all $\varepsilon > 0$. On the other hand, Theorem 14.36 (and item (b) of Exercise 14.39) also says that

$$\lim_{\varepsilon \to 0} d(3+\varepsilon) = d(3) = HD(L \cap (-\infty, 3)) = 0$$

In summary, $0 < \alpha \leqslant \lim_{\varepsilon \to 0} d(3+\varepsilon) = 0$, a contradiction.

14.2 Proof of Moreira's Theorem

14.2.1 Strategy of Proof of Moreira's Theorem

Roughly speaking, the continuity of $d(t) = HD(L \cap (-\infty, t))$ is proved in four steps:

- if $0 < d(t) < 1$, then for all $\eta > 0$ there exists $\delta > 0$ such that $L \cap (-\infty, t - \delta)$ can be "*approximated from inside*" by $K + K' = f(K \times K')$ where K and K' are *Gauss-Cantor sets* with $HD(K) + HD(K') = HD(K \times K') > (1 - \eta)d(t)$ (and $f(x, y) = x + y$);
- by *Moreira's dimension formula* (derived from profound works of Moreira and Yoccoz on the geometry of Cantor sets), we have that

$$HD(f(K \times K')) = HD(K \times K')$$

- thus, if $0 < d(t) < 1$, then for all $\eta > 0$ there exists $\delta > 0$ such that

$$d(t - \delta) \geqslant HD(f(K \times K')) = HD(K \times K') \geqslant (1 - \eta)d(t);$$

hence, $d(t)$ is *lower semicontinuous*;
- finally, an elementary compactness argument shows the *upper semicontinuity* of $d(t)$.

Remark 14.41 This strategy is *purely dynamical* because the particular forms of the height function f and the Gauss map G are *not* used. Instead, we just need the *transversality* of the gradient of f to the stable and unstable manifolds (vertical and horizontal axis) and the *non-essential affinity* of Gauss-Cantor sets. (See [2] for more explanations.)

In the remainder of this section, we will implement (a version of) this strategy in order to deduce the continuity result in Theorem 14.36.

14.2.2 Dynamical Cantor Sets

A *dynamically defined Cantor set* $K \subset \mathbb{R}$ is

$$K = \bigcap_{n \in \mathbb{N}} \psi^{-n}(I_1 \cup \cdots \cup I_k)$$

where $I_1, \ldots, I_k$ are pairwise disjoint compact intervals, and $\psi : I_1 \cup \cdots \cup I_k \to I$ is a C^r-map from $I_1 \cup \cdots \cup I_k$ to its convex hull I such that:

- ψ is *uniformly expanding*: $|\psi'(x)| > 1$ for all $x \in I_1 \cup \cdots \cup I_k$;
- ψ is a (full) *Markov map*: $\psi(I_j) = I$ for all $1 \leqslant j \leqslant k$.

Remark 14.42 Dynamical Cantor sets are usually defined with a weaker Markov condition, but we stick to this definition for simplicity.

Example 14.3 Cantor's middle-third set $C = \{\sum_{i=1}^{\infty} \frac{a_i}{3^i} : a_i \in \{0, 2\} \ \forall i \in \mathbb{N}\}$ is

$$C = \bigcap_{n \in \mathbb{N}} \psi^{-n}([0, 1/3] \cup [2/3, 1])$$

where $\psi : [0, 1/3] \cup [2/3, 1] \to [0, 1]$ is given by

$$\psi(x) = \begin{cases} 3x, & \text{if } 0 \leqslant x \leqslant 1/3 \\ 3x - 2, & \text{if } 2/3 \leqslant x \leqslant 1 \end{cases}$$

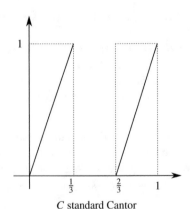

C standard Cantor

Remark 14.43 A dynamical Cantor set is called *affine* when $\psi|_{I_j}$ is affine for all j. In this language, Cantor's middle-third set is an *affine dynamical Cantor set*.

Example 14.4 Given $A \geqslant 2$, let $C(A) := \{[0; a_1, a_2, \ldots] : 1 \leqslant a_i \leqslant A \,\forall i \in \mathbb{N}\}$. This is a dynamical Cantor set associated to Gauss map: for example,

$$C(2) = \bigcap_{n \in \mathbb{N}} G^{-n}(I_1 \cup I_2)$$

where I_1 and I_2 are the intervals depicted below.

Remark 14.44 Hensley [10] showed that

$$HD(C(A)) = 1 - \frac{6}{\pi^2 A} - \frac{72 \log A}{\pi^4 A^2} + O(\frac{1}{A^2}) = 1 - \frac{1 + o(1)}{\zeta(2)A}$$

and Jenkinson-Pollicott [13, 14] used thermodynamical formalism methods to obtain that

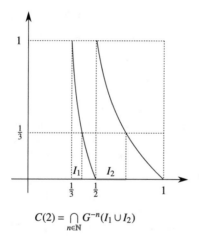

$$C(2) = \bigcap_{n \in \mathbb{N}} G^{-n}(I_1 \cup I_2)$$

$$HD(C(2)) = 0.531280506277205141624468647368471785493059109018 39\ldots,$$

$$HD(C(3)) \simeq 0.705\ldots, \quad HD(C(4)) \simeq 0.788\ldots$$

14.2.3   Gauss-Cantor Sets

The set $C(A)$ above is a particular case of *Gauss-Cantor set*:

Definition 14.45 Given $B = \{\beta_1, \ldots, \beta_l\}$, $l \geqslant 2$, a finite, primitive[14] alphabet of finite words $\beta_j \in (\mathbb{N}^*)^{r_j}$, the Gauss-Cantor set $K(B) \subset [0, 1]$ associated to B is

$$K(B) := \{[0; \gamma_1, \gamma_2, \ldots] : \gamma_i \in B \ \forall i\}$$

Example 14.5 $C(A) = K(\{1, \ldots, A\})$.

Exercise 14.46 Show that any Gauss-Cantor set $K(B)$ is dynamically defined.[15]

From the *symbolic* point of view, $B = \{\beta_1, \ldots, \beta_l\}$ as above induces a subshift

$$\Sigma(B) = \{(\gamma_i)_{i \in \mathbb{Z}} : \gamma_i \in B \ \forall i\} \subset \Sigma = (\mathbb{N}^*)^{\mathbb{Z}} = \Sigma^- \times \Sigma^+ := (\mathbb{N}^*)^{\mathbb{Z}^-} \times (\mathbb{N}^*)^{\mathbb{N}}$$

Also, the corresponding Gauss-Cantor is $K(B) = \{[0; \gamma] : \gamma \in \Sigma^+(B)\}$ where $\Sigma^+(B) = \pi^+(\Sigma(B))$ and $\pi^+ : \Sigma \to \Sigma^+$ is the natural projection (related to local unstable manifolds of the left shift map on Σ).

For later use, denote by $B^T = \{\beta^T : \beta \in B\}$ the *transpose* of B, where $\beta^T := (a_n, \ldots, a_1)$ for $\beta = (a_1, \ldots, a_n)$.

The following proposition (due to Euler) is proved in Appendix 2:

Proposition 14.47 (Euler) *If* $[0; \beta] = \frac{p_n}{q_n}$, *then* $[0; \beta^T] = \frac{r_n}{q_n}$.

A striking consequence of this proposition is:

Corollary 14.48 $HD(K(B)) = HD(K(B^T))$.

Proof (Sketch of Proof) The lengths of the intervals $I(\beta) = \{[0; \beta, a_1, \ldots] : a_i \in \mathbb{N} \ \forall i\}$ in the construction of $K(B)$ depend only on the denominators of the partial quotients of $[0; \beta]$. Therefore, we have from Proposition 14.47 that $K(B)$ and $K(B^T)$ are Cantor sets constructed from intervals with same lengths, and, *a fortiori*, they have the Hausdorff dimension.

Remark 14.49 This corollary is closely related to the existence of *area-preserving* natural extensions of Gauss map (see [1]) and the coincidence of stable and unstable dimensions of a horseshoe of an area-preserving surface diffeomorphism (see [20]).

14.2.4   Non-essentially Affine Cantor Sets

We say that

$$K = \bigcap_{n \in \mathbb{N}} \psi^{-n}(I_1 \cup \cdots \cup I_r)$$

---

[14]I.e., β_i doesn't begin by β_j for all $i \neq j$.
[15]*Hint*: For each word $\beta_j \in (\mathbb{N}^*)^{r_j}$, let $I(\beta_j) = \{[0; \beta_j, a_1, \ldots] : a_i \in \mathbb{N} \ \forall i\} = I_j$ and $\psi|_{I_j} := G^{r_j}$ where $G(x) = \{1/x\}$ is the Gauss map.

is *non-essentially affine* if there is *no* global conjugation $h \circ \psi \circ h^{-1}$ such that *all* branches

$$(h \circ \psi \circ h^{-1})|_{h(I_j)}, \quad j = 1, \ldots, r$$

are affine maps of the real line.

Equivalently, if $p \in K$ is a periodic point of ψ of period k and $h : I \to I$ is a diffeomorphism of the convex hull I of $I_1 \cup \cdots \cup I_r$ such that $h \circ \psi^k \circ h^{-1}$ is affine[16] on $h(J)$ where J is the connected component of the domain of ψ^k containing p, then K is non-essentially affine if and only if $(h \circ \psi \circ h^{-1})''(x) \neq 0$ for some $x \in h(K)$.

Proposition 14.50 *Gauss-Cantor sets are non-essentially affine.*

Proof The basic idea is to explore the fact that the second derivative of a non-affine Möbius transformation never vanishes.

More concretely, let $B = \{\beta_1, \ldots, \beta_m\}$, $\beta_j \in (\mathbb{N}^*)^{r_j}$, $1 \leqslant j \leqslant m$. For each β_j, let

$$x_j := [0; \beta_j, \beta_j, \ldots] \in I_j = I(\beta_j) \subset \{[0; \beta_j, \alpha] : \alpha \geqslant 1\}$$

be the fixed point of the branch $\psi|_{I_j} = G^{r_j}$ of the expanding map ψ naturally[17] defining the Gauss-Cantor set $K(B)$.

By Corollary 14.16, $\psi|_{I_j}(x) = \dfrac{q^{(j)}_{r_j-1} x - p^{(j)}_{r_j-1}}{p^{(j)}_{r_j} - q^{(j)}_{r_j} x}$ where $\dfrac{p^{(j)}_k}{q^{(j)}_k} = [0; b^{(j)}_1, \ldots, b^{(j)}_k]$ and $\beta_j = (b^{(j)}_1, \ldots, b^{(j)}_{r_j})$.

Note that the fixed point x_j of $\psi|_{I_j}$ is the positive solution of the second degree equation

$$q^{(j)}_{r_j} x^2 + (q^{(j)}_{r_j-1} - p^{(j)}_{r_j}) x - p^{(j)}_{r_j-1} = 0$$

In particular, x_j is a *quadratic surd*.

For each $1 \leqslant j \leqslant k$, the Möbius transformation $\psi|_{I_j}$ has a hyperbolic fixed point x_j. It follows (from Poincaré linearization theorem) that there exists a Möbius transformation

$$\alpha_j(x) = \frac{a_j x + b_j}{c_j x + d_j}$$

linearizing $\psi|_{I_j}$, i.e., $\alpha_j(x_j) = x_j$, $\alpha'(x_j) = 1$ and $\alpha_j \circ (\psi|_{I_j}) \circ \alpha_j^{-1}$ is an affine map.

Since non-affine Möbius transformations have non-vanishing second derivative, the proof of the proposition will be complete once we show that $\alpha_1 \circ (\psi|_{I_2}) \circ \alpha_1^{-1}$

---

[16]Such a diffeomorphism h linearizing *one* branch of ψ always exists by Poincaré's linearization theorem.

[17]Cf. Exercise 14.46.

is not affine. So, let us suppose by contradiction that $\alpha_1 \circ (\psi|_{I_2}) \circ \alpha_1^{-1}$ is affine. In this case, ∞ is a common fixed point of the (affine) maps $\alpha_1 \circ (\psi|_{I_2}) \circ \alpha_1^{-1}$ and $\alpha_1 \circ (\psi|_{I_1}) \circ \alpha_1^{-1}$, and, *a fortiori*, $\alpha_1^{-1}(\infty) = -d_1/c_1$ is a common fixed point of $\psi|_{I_1}$ and $\psi|_{I_2}$. Thus, the second degree equations

$$q_{r_1}^{(1)} x^2 + (q_{r_1-1}^{(1)} - p_{r_1}^{(1)})x - p_{r_1-1}^{(j)} = 0 \quad \text{and} \quad q_{r_2}^{(2)} x^2 + (q_{r_2-1}^{(2)} - p_{r_2}^{(2)})x - p_{r_2-1}^{(2)} = 0$$

would have a common root. This implies that these polynomials coincide (because they are polynomials in $\mathbb{Z}[x]$ which are irreducible[18]) and, hence, their other roots x_1, x_2 must coincide, a contradiction.

14.2.5 Moreira's Dimension Formula

The Hausdorff dimension of projections of products of non-essentially affine Cantor sets is given by the following formula:

Theorem 14.51 (Moreira) *Let K and K' be two C^2 dynamical Cantor sets. If K is non-essentially affine, then the projection $f(K \times K') = K + K'$ of $K \times K'$ under $f(x, y) = x + y$ has Hausdorff dimension*

$$HD(f(K + K')) = \min\{1, HD(K) + HD(K')\}$$

Remark 14.52 This statement is a *particular* case of Moreira's dimension formula (which is sufficient for our current purposes because Gauss-Cantor sets are non-essentially affine).

The proof of this result is out of the scope of these notes: indeed, it depends on the techniques introduced in two works (from 2001 and 2010) by Moreira and Yoccoz [24, 25] such as fine analysis of *limit geometries* and *renormalization operators*, "recurrence on scales", "compact recurrent sets of relative configurations", and *Marstrand's theorem*. We refer the reader to [22] for more details.

Remark 14.53 Moreira's dimension formula is coherent with Hall's Lemma 14.33: in fact, since $HD(C(4)) > 1/2$, it is natural that $HD(C(4) + C(4)) = 1$.

---

[18]Thanks to the fact that their roots $x_1, x_2 \notin \mathbb{Q}$.

14.2.6 First Step Towards Moreira's Theorem 14.36: Projections of Gauss-Cantor Sets

Let $\Sigma(B) \subset (\mathbb{N}^*)^{\mathbb{Z}}$ be a complete shift of finite type. Denote by $\ell(\Sigma(B))$, resp. $m(\Sigma(B))$, the pieces of the Lagrange, resp. Markov, spectrum generated by $\Sigma(B)$, i.e.,

$$\ell(\Sigma(B)) = \{\ell(\underline{\theta}) : \underline{\theta} \in \Sigma(B)\}, \quad \text{resp.} \quad m(\Sigma(B)) = \{m(\underline{\theta}) : \underline{\theta} \in \Sigma(B)\}$$

where $\ell(\underline{\theta}) = \lim\sup_{n\to\infty} f(\sigma^n(\underline{\theta}))$, $m(\underline{\theta}) = \sup_{n\in\mathbb{Z}} f(\sigma^n(\underline{\theta}))$, $f((\theta_i)_{i\in\mathbb{Z}}) = [\theta_0; \theta_1, \dots] + [0; \theta_{-1}, \dots]$ and $\sigma((\theta_i)_{i\in\mathbb{Z}}) = (\theta_{i+1})_{i\in\mathbb{Z}}$ is the shift map.

The following proposition relates the Hausdorff dimensions of the pieces of the Langrange and Markov spectra associated to $\Sigma(B)$ and the projection $f(K(B) \times K(B^T))$:

Proposition 14.54 *One has* $HD(\ell(\Sigma(B))) = HD(m(\Sigma(B))) = \min\{1, 2 \cdot HD(K(B))\}$.

Proof (Sketch of Proof) By definition,

$$\ell(\Sigma(B)) \subset m(\Sigma(B)) \subset \bigcup_{a=1}^{R} (a + K(B) + K(B^T))$$

where $R \in \mathbb{N}$ is the largest entry among all words of B.

Thus, $HD(\ell(\Sigma(B))) \leqslant HD(m(\Sigma(B))) \leqslant HD(K(B)) + HD(K(B^T))$. By Corollary 14.48, it follows that

$$HD(\ell(\Sigma(B))) \leqslant HD(m(\Sigma(B))) \leqslant \min\{1, 2 \cdot HD(K(B))\}$$

By Moreira's dimension formula (cf. Theorem 14.51), our task is now reduced to show that for all $\varepsilon > 0$, there are "replicas" K and K' of Gauss-Cantor sets such that

$$HD(K), HD(K') > HD(K(B)) - \varepsilon \quad \text{and} \quad f(K \times K') = K + K' \subset \ell(\Sigma(B))$$

In this direction, let us order B and B^T by declaring that $\gamma < \gamma'$ if and only if $[0; \gamma] < [0; \gamma']$.

Given $\varepsilon > 0$, we can replace if necessary B and/or B^T by $B^n = \{\gamma_1 \dots \gamma_n : \gamma_i \in B \,\forall\, i\}$ and/or $(B^T)^n$ for some large $n = n(\varepsilon) \in \mathbb{N}$ in such a way that

$$HD(K(B^*)), HD(K((B^T)^*)) > HD(K(B)) - \varepsilon$$

where $A^* := \{\min A, \max A\}$. Indeed, this holds because the Hausdorff dimension of a Gauss-Cantor set $K(A)$ associated to an alphabet A with a large number of words does not decrease too much after removing only two words from A.

We *expect* the values of ℓ on $((B^T)^*)^{\mathbb{Z}^-} \times (B^*)^{\mathbb{N}}$ to *decrease* because we removed the minimal and maximal elements of B and B^T (and, in general, $[a_0; a_1, a_2, \ldots] <$ $[b_0; b_1, b_2, \ldots]$ if and only if $(-1)^k(a_k - b_k) < 0$ where k is the smallest integer with $a_k \neq b_k$).

In particular, this gives *some* control on the values of ℓ on $((B^T)^*)^{\mathbb{Z}^-} \times (B^*)^{\mathbb{N}}$, but this does *not* mean that $K(B^*) + K((B^T)^*) \subset \ell(\Sigma(B))$.

We overcome this problem by studying *replicas* of $K(B^*)$ and $K((B^T)^*)$. More precisely, let $\widetilde{\theta} = (\ldots, \widetilde{\gamma}_0, \widetilde{\gamma}_1, \ldots) \in \Sigma(B)$, $\widetilde{\gamma}_i \in B$ for all $i \in \mathbb{Z}$, such that

$$m(\widetilde{\theta}) = \max m(\Sigma(B))$$

is attained at a position in the block $\widetilde{\gamma}_0$.

By compactness, there exists $\eta > 0$ and $m \in \mathbb{N}$ such that any

$$\theta = (\ldots, \gamma_{-m-2}, \gamma_{-m-1}, \widetilde{\gamma}_{-m}, \ldots, \widetilde{\gamma}_0, \ldots, \widetilde{\gamma}_m, \gamma_{m+1}, \gamma_{m+2}, \ldots)$$

with $\gamma_i \in B^*$ for all $i > m$ and $\gamma_i \in (B^T)^*$ for all $i < -m$ satisfies:

- $m(\theta)$ is attained in a position in the *central block* $(\widetilde{\gamma}_{-m}, \ldots, \widetilde{\gamma}_0, \ldots, \widetilde{\gamma}_m)$;
- $f(\sigma^n(\theta)) < m(\theta) - \eta$ for any *non-central position* n.

By exploring these properties, it is possible to enlarge the central block to get a word called $\tau^{\#} = (a_{-N_1}, \ldots, a_0, \ldots, a_{N_2})$ in Moreira's paper [21] such that the replicas

$$K = \{[a_0; a_1, \ldots, a_{N_2}, \gamma_1, \gamma_2, \ldots] : \gamma_i \in B^* \; \forall i > 0\}$$

and

$$K' = \{[0; a_{-1}, \ldots, a_{-N_1}, \gamma_{-1}, \gamma_{-2}, \ldots] : \gamma_i \in (B^T)^* \; \forall i < 0\}$$

of $K(B^*)$ and $K((B^T)^*)$ have the desired properties that

$$K + K' = f(K \times K') \subset \ell(\Sigma(B))$$

and

$$HD(K) = HD(K(B^*)) > HD(K) - \varepsilon, HD(K') = HD(K((B^T)^*)) > HD(K(B^T)) - \varepsilon$$

This completes our sketch of proof of the proposition.

14.2.7   Second Step Towards Moreira's Theorem 14.36: Upper Semi-continuity

Let $\Sigma_t := \{\theta \in (\mathbb{N}^*)^{\mathbb{Z}} : m(\theta) \leqslant t\}$ for $3 \leqslant t < 5$.

Our long term goal is to compare Σ_t with its projection $K_t^+ := \{[0; \gamma] : \gamma \in \pi^+(\Sigma_t)\}$ on the unstable part (where $\pi^+ : (\mathbb{N}^*)^{\mathbb{Z}} \to (\mathbb{N}^*)^{\mathbb{N}}$ is the natural projection).

Given $\alpha = (a_1, \ldots, a_n)$, its *unstable scale* $r^+(\alpha)$ is

$$r^+(\alpha) = \lfloor \log 1/(\text{length of } I^+(\alpha)) \rfloor$$

where $I^+(\alpha)$ is the interval with extremities $[0; a_1, \ldots, a_n]$ and $[0; a_1, \ldots, a_n + 1]$.

Denote by

$$P_r^+ := \{\alpha = (a_1, \ldots, a_n) : r^+(\alpha) \geqslant r, r^+(a_1, \ldots, a_{n-1}) < r\}$$

and

$$C^+(t, r) := \{\alpha \in P_r^+ : I^+(\alpha) \cap K_t^+ \neq \emptyset\}.$$

Remark 14.55 By symmetry (i.e., replacing γ's by γ^T's), we can define $K_t^-, r^-(\alpha)$, etc.

For later use, we observe that the unstable scales have the following behaviour under concatenations of words:

Exercise 14.56 Show that $r^+(\alpha\beta k) \geqslant r^+(\alpha) + r^+(\beta)$ for all α, β finite words and for all $k \in \{1, 2, 3, 4\}$.

In particular, since the family of intervals

$$\{I^+(\alpha\beta k) : \alpha \in C^+(t, r), \beta \in C^+(t, s), 1 \leqslant k \leqslant 4\}$$

covers K_t^+, it follows from Exercise 14.56 that

$$\#C^+(t, r + s) \leqslant 4\#C^+(t, r)\#C^+(t, s)$$

for all $r, s \in \mathbb{N}$ and, hence, the sequence $(4\#C^+(t, r))_{r \in \mathbb{N}}$ is *submultiplicative*.

So, the *box-counting dimension* (cf. Remark 14.38) $\Delta^+(t)$ of K_t^+ is

$$\Delta^+(t) = \inf_{m \in \mathbb{N}} \frac{1}{m} \log(4\#C^+(t, m)) = \lim_{m \to \infty} \frac{1}{m} \log \#C^+(t, m)$$

An elementary compactness argument shows that the upper-semicontinuity of $\Delta^+(t)$:

Proposition 14.57 *The function* $t \mapsto \Delta^+(t)$ *is upper-semicontinuous.*

Proof For the sake of contradiction, assume that there exist $\eta > 0$ and t_0 such that $\Delta^+(t) > \Delta^+(t_0) + \eta$ for all $t > t_0$.

By definition, this means that there exists $r_0 \in \mathbb{N}$ such that

$$\frac{1}{r} \log \#C^+(t, r) > \Delta^+(t_0) + \eta$$

for all $r \geqslant r_0$ and $t > t_0$.

On the other hand, $C^+(t, r) \subset C^+(s, r)$ for all $t \leqslant s$ and, by compactness, $C^+(t_0, r) = \bigcap_{t > t_0} C^+(t, r)$. Thus, if $r \to \infty$ and $t \to t_0$, the inequality of the previous paragraph would imply that

$$\Delta^+(t_0) > \Delta^+(t_0) + \eta,$$

a contradiction.

14.2.8 Third Step Towards Moreira's Theorem 14.36: Lower Semi-continuity

The main result of this subsection is the following theorem allowing us to "approximate from inside" Σ_t by Gauss-Cantor sets.

Theorem 14.58 *Given $\eta > 0$ and $3 \leqslant t < 5$ with $d(t) := HD(L \cap (-\infty, t)) > 0$, we can find $\delta > 0$ and a Gauss-Cantor set $K(B)$ associated to $\Sigma(B) \subset \{1, 2, 3, 4\}^{\mathbb{Z}}$ such that*

$$\Sigma(B) \subset \Sigma_{t-\delta} \quad and \quad HD(K(B)) \geqslant (1 - \eta)\Delta^+(t)$$

This theorem allows us to derive the continuity statement in Moreira's Theorem 14.36:

Corollary 14.59 $\Delta^-(t) = \Delta^+(t)$ *is a continuous function of t and $d(t) = \min\{1, 2 \cdot \Delta^+(t)\}$.*

Proof By Corollary 14.48 and Theorem 14.58, we have that

$$\Delta^-(t - \delta) \geqslant HD(K(B^T)) = HD(K(B)) \geqslant (1 - \eta)\Delta^+(t).$$

Also, a "symmetric" estimate holds after exchanging the roles of Δ^- and Δ^+. Hence, $\Delta^-(t) = \Delta^+(t)$. Moreover, the inequality above says that $\Delta^-(t) = \Delta^+(t)$ is a lower-semicontinuous function of t. Since we already know that $\Delta^+(t)$ is an upper-semicontinuous function of t thanks to Proposition 14.57, we conclude that $t \mapsto \Delta^-(t) = \Delta^+(t)$ is continuous. Finally, by Proposition 14.54, from

$\Sigma(B) \subset \Sigma_{t-\delta}$, we also have that

$$d(t - \delta) \geqslant HD(\ell(\Sigma(B))) = \min\{1, 2 \cdot HD(K(B))\} \geqslant (1 - \eta) \min\{1, 2\Delta^+(t)\}$$

Since $d(t) \leqslant \min\{1, \Delta^+(t) + \Delta^-(t)\}$ (because $\Sigma_t \subset \pi^-(\Sigma_t) \times \pi^+(\Sigma_t)$), the proof is complete.

Let us now sketch the construction of the Gauss-Cantor sets $K(B)$ approaching Σ_t from inside.

Proof (Sketch of Proof of Theorem 14.58) Fix $r_0 \in \mathbb{N}$ large enough so that

$$\left| \frac{\log \#C^+(t, r)}{r} - \Delta^+(t) \right| < \frac{\eta}{80} \Delta^+(t)$$

for all $r \geqslant r_0$.

Set $B_0 := C^+(t, r_0)$, $k = 8(\#B_0)^2 \lceil 80/\eta \rceil$ and

$$\widetilde{B} := \{\beta = (\beta_1, \ldots, \beta_k) : \beta_j \in B_0 \text{ and } I^+(\beta) \cap K_t^+ \neq \emptyset\} \subset B_0^k$$

It is not hard to show that $\widetilde{B}$ has a significant cardinality in the sense that

$$\#\widetilde{B} > 2(\#B_0)^{(1-\frac{\eta}{40})k}$$

In particular, one can use this information to prove that $HD(K(\widetilde{B}))$ is not far from $\Delta^+(t)$, i.e.

$$HD(K(\widetilde{B})) \geqslant (1 - \frac{\eta}{20})\Delta^+(t)$$

Unfortunately, since we have no control on the values of m on $\Sigma(\widetilde{B})$, there is no guarantee that $\Sigma(\widetilde{B}) \subset \Sigma_{t-\delta}$ for some $\delta > 0$.

We can overcome this issue with the aid of the notion of *left-good* and *right-good* positions. More concretely, we say that $1 \leqslant j \leqslant k$ is a right-good position of $\beta = (\beta_1, \ldots, \beta_k) \in \widetilde{B}$ whenever there are two elements $\beta^{(s)} = \beta_1 \ldots \beta_j \beta_{j+1}^{(s)} \ldots \beta_k^{(s)} \in \widetilde{B}$, $s \in \{1, 2\}$ such that

$$[0; \beta_j^{(1)}] < [0; \beta_j] < [0; \beta_j^{(2)}]$$

Similarly, $1 \leqslant j \leqslant k$ is a left-good position $\beta = (\beta_1, \ldots, \beta_k) \in \widetilde{B}$ whenever there are two elements $\beta^{(s)} = \beta_1 \ldots \beta_j \beta_{j+1}^{(s)} \ldots \beta_k^{(s)} \in \widetilde{B}$, $s \in \{3, 4\}$ such that

$$[0; (\beta_j^{(3)})^T] < [0; \beta_j^T] < [0; (\beta_j^{(2)})^T]$$

Furthermore, we say that $1 \leqslant j \leqslant k$ is a *good position* of $\beta = (\beta_1, \ldots, \beta_k) \in \widetilde{B}$ when it is both a left-good and a right-good position.

Since there are at most two choices of $\beta_j \in B_0$ when $\beta_1, \ldots, \beta_{j-1}$ are fixed and j is a right-good position, one has that the subset

$$\mathcal{E} := \{\beta \in \widetilde{B} : \beta \text{ has } 9k/10 \text{ good positions (at least)}\}$$

of *excellent* words in $\widetilde{B}$ has cardinality

$$\#\mathcal{E} > \frac{1}{2}\#\widetilde{B} > (\#B_0)^{(1-\frac{\eta}{40})k}$$

We *expect* the values of m on $\Sigma(\mathcal{E})$ to *decrease* because excellent words have many good positions. Also, the Hausdorff dimension of $K(\mathcal{E})$ is not far from $\Delta^+(t)$ thanks to the estimate above on the cardinality of $\mathcal{E}$. However, there is no reason for $\Sigma(\mathcal{E}) \subset \Sigma_{t-\delta}$ for some $\delta > 0$ because an *arbitrary* concatenation of words in $\mathcal{E}$ might not belong to Σ_t.

At this point, the idea is to build a complete shift $\Sigma(B) \subset \Sigma_{t-\delta}$ from $\mathcal{E}$ with the following combinatorial argument. Since $\beta = (\beta_1, \ldots, \beta_k) \in \mathcal{E}$ has $9k/10$ good positions, we can find good positions $1 \leqslant i_1 \leqslant i_2 \leqslant \cdots \leqslant i_{\lceil 2k/5 \rceil} \leqslant k - 1$ such that $i_s + 2 \leqslant i_{s+1}$ for all $1 \leqslant s \leqslant \lceil 2k/5 \rceil - 1$ and $i_s + 1$ are also good positions for all $1 \leqslant s \leqslant \lceil 2k/5 \rceil$. Because $k := 8(\#B_0)^2\lceil 80/\eta \rceil$, the pigeonhole principle reveals that we can choose positions $j_1 \leqslant \cdots \leqslant j_{3(\#B_0)^2}$ and words $\widehat{\beta}_{j_1}, \widehat{\beta}_{j_1+1}, \ldots, \widehat{\beta}_{j_{3(\#B_0)^2}}, \widehat{\beta}_{j_{3(\#B_0)^2}+1} \in B_0$ such that $j_s + 2\lceil 80/\eta \rceil \leqslant j_{s+1}$ for all $s < 3(\#B_0)^2$ and the subset

$$X = \{(\beta_1, \ldots, \beta_k) \in \mathcal{E} : j_s, j_s + 1 \text{ are good positions and}$$
$$\beta_{j_s} = \widehat{\beta}_{j_s}, \beta_{j_s+1} = \widehat{\beta}_{j_s+1} \ \forall \ s \leqslant 3(\#B_0)^2\}$$

of excellent words with prescribed subwords $\widehat{\beta}_{j_s}, \widehat{\beta}_{j_s+1}$ at the good positions j_s, j_s+1 has cardinality

$$\#X > (\#B_0)^{(1-\frac{\eta}{20})k}$$

Next, we convert X into the alphabet B of an appropriate complete shift with the help of the projections $\pi_{a,b} : X \to B_0^{j_b - j_a}$, $\pi_{a,b}(\beta_1, \ldots, \beta_k) = (\beta_{j_a+1}, \beta_{j_a+2}, \ldots, \beta_{j_b})$. More precisely, an elementary counting argument shows that we can take $1 \leqslant a < b \leqslant 3(\#B_0)^2$ such that $\widehat{\beta}_{j_a} = \widehat{\beta}_{j_b}$, $\widehat{\beta}_{j_a+1} = \widehat{\beta}_{j_b+1}$, and the image $\pi_{a,b}(X)$ of some projection $\pi_{a,b}$ has a significant cardinality

$$\#\pi_{a,b}(X) > (\#B_0)^{(1-\frac{\eta}{4})(j_b - j_a)}$$

From these properties, we get an alphabet $B = \pi_{a,b}(X)$ whose words concatenate in an appropriate way (because $\widehat{\beta}_{j_a} = \widehat{\beta}_{j_b}$, $\widehat{\beta}_{j_a+1} = \widehat{\beta}_{j_b+1}$), the Hausdorff dimension of $K(B)$ is $HD(K(B)) > (1-\eta)\Delta^+(t)$ (because $\#B > (\#B_0)^{(1-\frac{\eta}{4})(j_b-j_a)}$ and $j_b - j_a > 2\lceil\frac{80}{\eta}\rceil$), and $\Sigma(B) \subset \Sigma_{t-\delta}$ for some $\delta > 0$ (because the features of good positions forces the values of m on $\Sigma(B)$ to decrease). This completes our sketch of proof.

14.2.9   End of Proof of Moreira's Theorem 14.36

By Corollary 14.59, the function

$$t \mapsto d(t) = HD(L \cap (-\infty, t))$$

is continuous. Moreover, an inspection of the proof of Corollary 14.59 shows that we have also proved the equality $HD(M \cap (-\infty, t)) = HD(L \cap (-\infty, t))$.

Therefore, our task is reduced to prove that $d(3 + \varepsilon) > 0$ for all $\varepsilon > 0$ and $d(\sqrt{12}) = 1$.

The fact that $d(3 + \varepsilon) > 0$ for any ε uses explicit sequences $\theta_m \in \{1, 2\}^{\mathbb{Z}}$ such that $\lim_{m\to\infty} m(\theta_m) = 3$ in order to exhibit non-trivial Cantor sets in $M \cap (-\infty, 3 + \varepsilon)$.

More precisely, consider[19] the periodic sequences

$$\theta_m := \overline{2 \underbrace{1 \ldots 1}_{2m\text{times}} 2}$$

where $\overline{a_1 \ldots a_k} := \ldots a_1 \ldots a_k\, a_1 \ldots a_k \ldots$. Since the sequence $\theta_\infty = \overline{1}, 2, 2, \overline{1}$ has the property that $m(\theta_\infty) = [2; \overline{1}] + [0; 2, \overline{1}] = 3$, and $|[a_0; a_1, \ldots, a_n, b_1, \ldots] - [a_0; a_1, \ldots, a_n, c_1, \ldots]| < \frac{1}{2^{n-1}}$ in general,[20] we have that the alphabet B_m consisting of the two words $2 \underbrace{1 \ldots 1}_{2m\text{times}} 2$ and $2 \underbrace{1 \ldots 1}_{2m+2\text{times}} 2$ satisfies

$$\Sigma(B_m) \subset \Sigma_{3+\frac{1}{2^m}}$$

Thus, $d(3 + \frac{1}{2^m}) = HD(M \cap (-\infty, 3 + \frac{1}{2^m})) \geqslant HD(\Sigma(B_m)) = 2 \cdot HD(K(B_m)) > 0$ for all $m \in \mathbb{N}$.

Finally, the fact that $d(\sqrt{12}) = 1$ follows from Corollary 14.59 and Remark 14.44. Indeed, Perron showed that $m(\theta) \leqslant \sqrt{12}$ if and only if $\theta \in \{1, 2\}^{\mathbb{Z}}$ (see the proof of Lemma 7 in Chapter 1 of Cusick-Flahive book [3]). Thus,

---

[19]This choice of θ_m is motivated by the discussion in Chapter 1 of Cusick-Flahive book [3].
[20]See Lemma 2 in Chapter 1 of [3].

$K^+_{\sqrt{12}} = C(2)$. By Corollary 14.59, it follows that

$$d(\sqrt{12}) = \min\{1, 2 \cdot \Delta^+(\sqrt{12})\} = \min\{1, 2 \cdot HD(C(2))\}$$

Since Remark 14.44 tells us that $HD(C(2)) > 1/2$, we conclude that $d(\sqrt{12}) = 1$.

Appendix 1: Proof of Hurwitz Theorem

Given $\alpha \notin \mathbb{Q}$, we want to show that the inequality

$$\left| \alpha - \frac{p}{q} \right| \leqslant \frac{1}{\sqrt{5}q^2}$$

has infinitely many rational solutions.

In this direction, let $\alpha = [a_0; a_1, \dots]$ be the continued fraction expansion of α and denote by $[a_0; a_1, \dots, a_n] = p_n/q_n$. We affirm that, for every $\alpha \notin \mathbb{Q}$ and every $n \geqslant 1$, we have

$$\left| \alpha - \frac{p}{q} \right| < \frac{1}{\sqrt{5}q^2}$$

for some $\frac{p}{q} \in \{\frac{p_{n-1}}{q_{n-1}}, \frac{p_n}{q_n}, \frac{p_{n+1}}{q_{n+1}}\}$.

Remark 14.60 Of course, this last statement provides infinitely many solutions to the inequality $\left| \alpha - \frac{p}{q} \right| \leqslant \frac{1}{\sqrt{5}q^2}$. So, our task is reduced to prove the affirmation above.

The proof of the claim starts by recalling Perron's Proposition 14.20:

$$\alpha - \frac{p_n}{q_n} = \frac{(-1)^n}{(\alpha_{n+1} + \beta_{n+1})q_n^2}$$

where $\alpha_{n+1} := [a_{n+1}; a_{n+2}, \dots]$ and $\beta_{n+1} = \frac{q_{n-1}}{q_n} = [0; a_n, \dots, a_1]$.

For the sake of contradiction, suppose that the claim is false, i.e., there exists $k \geqslant 1$ such that

$$\max\{(\alpha_k + \beta_k), (\alpha_{k+1} + \beta_{k+1}), (\alpha_{k+2} + \beta_{k+2})\} \leqslant \sqrt{5} \tag{14.7}$$

Since $\sqrt{5} < 3$ and $a_m \leqslant \alpha_m + \beta_m$ for all $m \geqslant 1$, it follows from (14.7) that

$$\max\{a_k, a_{k+1}, a_{k+2}\} \leqslant 2 \tag{14.8}$$

If $a_m = 2$ for some $k \leqslant m \leqslant k+2$, then (14.8) would imply that $\alpha_m + \beta_m \geqslant 2 + [0; 2, 1] = 2 + \frac{1}{3} > \sqrt{5}$, a contradiction with our assumption (14.7). So, our hypothesis (14.7) forces

$$a_k = a_{k+1} = a_{k+2} = 1 \tag{14.9}$$

Denoting by $x = \frac{1}{\alpha_{k+2}}$ and $y = \beta_{k+1} = q_{k-1}/q_k \in \mathbb{Q}$, we have from (14.9) that

$$\alpha_{k+1} = 1 + x, \quad \alpha_k = 1 + \frac{1}{1+x}, \quad \beta_k = \frac{1}{y} - 1, \quad \beta_{k+2} = \frac{1}{1+y}$$

By plugging this into (14.7), we obtain

$$\max\left\{\frac{1}{1+x} + \frac{1}{y}, 1 + x + y, \frac{1}{x} + \frac{1}{1+y}\right\} \leqslant \sqrt{5} \tag{14.10}$$

On one hand, (14.10) implies that

$$\frac{1}{1+x} + \frac{1}{y} \leqslant \sqrt{5} \quad \text{and} \quad 1 + x \leqslant \sqrt{5} - y.$$

Thus,

$$\frac{\sqrt{5}}{y(\sqrt{5} - y)} = \frac{1}{\sqrt{5} - y} + \frac{1}{y} \leqslant \frac{1}{1+x} + \frac{1}{y} \leqslant \sqrt{5},$$

and, *a fortiori*, $y(\sqrt{5} - y) \geqslant 1$, i.e.,

$$\frac{\sqrt{5} - 1}{2} \leqslant y \leqslant \frac{\sqrt{5} + 1}{2} \tag{14.11}$$

On the other hand, (14.10) implies that

$$x \leqslant \sqrt{5} - 1 - y \quad \text{and} \quad \frac{1}{x} + \frac{1}{1+y} \leqslant \sqrt{5}.$$

Hence,

$$\frac{\sqrt{5}}{(1+y)(\sqrt{5} - 1 - y)} = \frac{1}{\sqrt{5} - 1 - y} + \frac{1}{1+y} \leqslant \frac{1}{x} + \frac{1}{1+y} \leqslant \sqrt{5},$$

and, *a fortiori*, $(1+y)(\sqrt{5} - 1 - y) \geqslant 1$, i.e.,

$$\frac{\sqrt{5} - 1}{2} \leqslant y \leqslant \frac{\sqrt{5} + 1}{2} \tag{14.12}$$

It follows from (14.11) and (14.12) that $y = (\sqrt{5} - 1)/2$, a contradiction because $y = \beta_{k+1} = q_{k-1}/q_k \in \mathbb{Q}$. This completes the argument.

Appendix 2: Proof of Euler's Remark

Denote by $[0; a_1, a_2, \ldots, a_n] = \frac{p(a_1,\ldots,a_n)}{q(a_1,\ldots,a_n)} = \frac{p_n}{q_n}$. It is not hard to see that

$$q(a_1) = a_1, \quad q(a_1, a_2) = a_1 a_2 + 1,$$

$$q(a_1, \ldots, a_n) = a_n q(a_1, \ldots, a_{n-1}) + q(a_1, \ldots, a_{n-2}) \quad \forall \, n \geqslant 3.$$

From this formula, we see that $q(a_1, \ldots, a_n)$ is a sum of the following products of elements of $\{a_1, \ldots, a_n\}$. First, we take the product $a_1 \ldots a_n$ of all a_i's. Secondly, we take all products obtained by removing any pair $a_i a_{i+1}$ of adjacent elements. Then, we iterate this procedure until no pairs can be omitted (with the convention that if n is even, then the empty product gives 1). This rule to describe $q(a_1, \ldots, a_n)$ was discovered by Euler.

It follows immediately from Euler's rule that $q(a_1, \ldots, a_n) = q(a_n, \ldots, a_1)$. This proves Proposition 14.47.

References

1. P. Arnoux, Le codage du flot géodésique sur la surface modulaire. Enseign. Math. (2) **40**(1–2), 29–48 (1994)
2. A. Cerqueira, C. Matheus, C.G. Moreira, Continuity of Hausdorff dimension across generic dynamical Lagrange and Markov spectra. Preprint (2016) available at arXiv:1602.04649
3. T. Cusick, M. Flahive, *The Markoff and Lagrange Spectra*. Mathematical Surveys and Monographs, vol. 30 (American Mathematical Society, Providence, RI, 1989), x+97 pp.
4. P.G. Dirichlet, Verallgemeinerung eines Satzes aus der Lehre von den Kettenbrüchen nebst einigen Anwendungen auf die Theorie der Zahlen. pp. 633–638 Bericht über die Verhandlungen der Königlich Preussischen Akademie der Wissenschaften. Jahrg. 1842, S. 93–95
5. K. Falconer, *The Geometry of Fractal Sets*. Cambridge Tracts in Mathematics, vol. 85 (Cambridge University Press, Cambridge, 1986), xiv+162 pp.
6. G. Freiman, Non-coincidence of the spectra of Markov and of Lagrange. Mat. Zametki **3**, 195–200 (1968)
7. G. Freiman, Non-coincidence of the spectra of Markov and of Lagrange, in *Number-Theoretic Studies in the Markov Spectrum and in the Structural Theory of Set Addition (Russian)* (Kalinin. Gos. Univ, Moscow, 1973), pp. 10–15, 121–125
8. G. Freiman, *Diophantine Approximations and the Geometry of Numbers (Markov's Problem)* (Kalininskii Gosudarstvennyi University, Kalinin, 1975), 144 pp.
9. M. Hall, On the sum and product of continued fractions. Ann. Math. (2) **48**, 966–993 (1947)
10. D. Hensley, Continued fraction Cantor sets, Hausdorff dimension, and functional analysis. J. Number Theory **40**(3), 336–358 (1992)

11. P. Hubert, L. Marchese, C. Ulcigrai, Lagrange spectra in Teichmüller dynamics via renormalization. Geom. Funct. Anal. **25**(1), 180–255 (2015)
12. A. Hurwitz, Ueber die angenäherte Darstellung der Irrationalzahlen durch rationale Brüche, Math. Ann. **39**(2), 279–284 (1891)
13. O. Jenkinson, M. Pollicott, Computing the dimension of dynamically defined sets: E2 and bounded continued fractions. Ergodic Theory Dyn. Syst. **21**(5), 1429–1445 (2001)
14. O. Jenkinson, M. Pollicott, Rigorous effective bounds on the Hausdorff dimension of continued fraction Cantor sets: a hundred decimal digits for the dimension of E_2. Preprint (2016) available at arXiv:1611.09276
15. A. Khinchin, *Continued Fractions* (The University of Chicago Press, Chicago, London, 1964), xi+95 pp.
16. P. Lévy, Sur le développement en fraction continue d'un nombre choisi au hasard. Compos. Math. **3**, 286–303 (1936)
17. K. Mahler, On lattice points in n-dimensional star bodies. I. Existence theorems. Proc. R. Soc. Lond. Ser. A **187**, 151–187 (1946)
18. A. Markoff, Sur les formes quadratiques binaires indéfinies. Math. Ann. **17**(3), 379–399 (1880)
19. J. Marstrand, Some fundamental geometrical properties of plane sets of fractional dimensions. Proc. Lond. Math. Soc. (3) **4**, 257–302 (1954)
20. H. McCluskey, A. Manning, Hausdorff dimension for horseshoes. Ergodic Theory Dyn. Syst. **3**(2), 251–260 (1983)
21. C.G. Moreira, Geometric properties of the Markov and Lagrange spectra. Preprint (2016) available at arXiv:1612.05782
22. C.G. Moreira, Geometric properties of images of cartesian products of regular Cantor sets by differentiable real maps. Preprint (2016) available at arXiv:1611.00933
23. C.G. Moreira, S. Romaña, On the Lagrange and Markov dynamical spectra, Ergodic Theory Dyn. Syst. 1–22 (2016). https://doi.org/10.1017/etds.2015.121
24. C.G. Moreira, J.-C. Yoccoz, Stable intersections of regular Cantor sets with large Hausdorff dimensions, Ann. Math. (2) **154**(1), 45–96 (2001)
25. C.G. Moreira, J.-C. Yoccoz, Tangences homoclines stables pour des ensembles hyperboliques de grande dimension fractale. Ann. Sci. Éc. Norm. Supér. (4) **43**(1), 1–68 (2010)
26. S. Newhouse, The abundance of wild hyperbolic sets and nonsmooth stable sets for diffeomorphisms. Inst. Hautes Études Sci. Publ. Math. **50**, 101–151 (1979)
27. J. Parkkonen, F. Paulin, Prescribing the behaviour of geodesics in negative curvature. Geom. Topol. **14**(1), 277–392 (2010)
28. D. Witte Morris, *Ratner's Theorems on Unipotent Flows*. Chicago Lectures in Mathematics (University of Chicago Press, Chicago, IL, 2005), xii+203 pp.
29. D. Zagier, Eisenstein series and the Riemann zeta function, in *Automorphic Forms, Representation Theory and Arithmetic (Bombay, 1979)*. Tata Institute of Fundamental Research Studies in Mathematics, vol. 10 (Tata Institute of Fundamental Research, Bombay, 1981), pp. 275–301
30. D. Zagier, On the number of Markoff numbers below a given bound. Math. Comput. **39**(160), 709–723 (1982)

Chapter 15
On the Missing Log Factor

Olivier Ramaré

15.1 Introduction

The Möbius function has attracted lots of attention in the last few years. As is classical in Analytic Number Theory, we are trying to estimate sums of the form $\sum_{n \leqslant x} \mu(n) g(x, n)$ for various and usually regular functions $g(x, n)$.

There are essentially three definitions of the Möbius function:

- It is the multiplicative fonction with $\mu(p) = -1$ and $\mu(p^k) = 0$ ($k \geqslant 2$),
- It is the convolution inverse of $\mathbb{1}$,
- It appears as the coefficients of the Dirichlet series of $1/\zeta(s)$.

All three are of course linked,[1] but this list enables a rough and empirical classification of proofs. In this talk, we concentrate on the *second definition*, and we shall often add an explicit angle to our looking glass. We will in particular see that this combinatorial definition leads to functional analysis problems.

---

[1] If only by the fact that they define the same function!

O. Ramaré (✉)
Aix Marseille Université, CNRS, Centrale Marseille, Institut de Mathématiques de Marseille, I2M – UMR 7373, Marseille, France
e-mail: olivier.ramare@univ-amu.fr

© Springer International Publishing AG, part of Springer Nature 2018   293
S. Ferenczi et al. (eds.), *Ergodic Theory and Dynamical Systems in their Interactions with Arithmetics and Combinatorics*, Lecture Notes in Mathematics 2213, https://doi.org/10.1007/978-3-319-74908-2_15

15.2 Meissel and Gram

Let us start our journey by an identity due the German mathematician Ernst Meissel[2] in 1854 which is equation (6) of [27]. Thanks to the DigiZeitschriften project hosted by the university of Göttingen, we can have access to this text online, though some knowledge of Latin is required. The classical reference book [10] on history of numbers of L.E. Dickson may serve as a first guide, and for instance, the paper [27] is mentioned in Chapter XIX of this series of three books. In modern notation, the identity in question reads

$$\sum_{n \leqslant x} \mu(n)[x/n] = 1 \tag{15.1}$$

where $[y]$ denotes the integer part of the real number y, while $\{y\}$ denotes its fractional part. This is established by noticing that $[y] = \sum_{1 \leqslant m \leqslant y} 1$ when y is non-negative and on using the property that $\sum_{mn=\ell} \mu(n) = \mathbb{1}_{\ell=1}$. Now let us replace $[y]$ by $y - \{y\}$ in the above; we get

$$\sum_{n \leqslant x} \mu(n)\{x/n\} = -1 + x \sum_{n \leqslant x} \frac{\mu(n)}{n}. \tag{15.2}$$

We stop to emphasize three surprising aspects of this equation:

1. *Error term treatment:* On the left-hand side, the summand $\mu(n)$ is contaminated by the error term $\{x/n\}$ while the contamination disappears on the right-hand side! The Prime Number Theorem thus implies that the left-hand side is indeed $o(x)$.
2. *Identity:* We have used an identity, and the question arises is naturally to know whether it is an accident or a feature.
3. *Log-factor:* When we bound $|\mu(n)|$ and $\{x/n\}$ by 1, we see that the trivial bound for the left-hand side is x, while the trivial bound for the right-hand side is $\dots O(x \log x)$! As a consequence, the Danish mathematician Jørgen Pedersen Gram showed in [17, pp. 196–197][3] that

$$\left| \sum_{n \leqslant x} \mu(n)/n \right| \leqslant 1 \tag{15.3}$$

---

[2]His full name is Daniel Friedrich Ernst Meissel. This student of Carl Gustav Jacob Jacobi and Johann Peter Gustav Lejeune Dirichlet was born in 1826 and passed away in 1895. His full biography can be found in [29].

[3]This reference has been kindly provided to us by M. Balazard. The reader is referred to the MacTutor archive maintained by the University of Saint Andrew, in Scotland for the biography of J.-P. Gram. We just mention here that Meissel travelled to Denmark in 1885 to meet the 23 years old Gram who had just won the Gold Medal of the Royal Danish Academy of Sciences for the memoir we refer to. The inequality we extract from this memoir is not its main matter but rather a pleasant side dish.

for every positive x. It is of course a consequence of the Prime Number Theorem that this sum goes to zero, but this partial result is striking.

The identity angle leads to more curious identities. Here is another one obtained much later by the Canadian mathematician Robert Allister MacLeod in [26]:

$$\sum_{n\leqslant x} \mu(n)\frac{\{x/n\}^2 - \{x/n\}}{x/n} = x\sum_{n\leqslant x}\frac{\mu(n)}{n} - \sum_{n\leqslant x}\mu(n) - 2 + \frac{2}{x}.$$

In fact, MacLeod exhibits a full family of similar identities, all valid for any $x \geqslant 1$. Yet again, the reader can see that the left-hand side is contaminated by an "error term"-like function, while this contamination is absent from the right-hand side.

I am showing you this identity to insist on the strange aspect these relations may take. Are these identities just curiosities or is a better understanding possible? Can we give some order to these facts?

15.3   Generalizing Meissel's Proof, I

While trying to shed some light on Meissel's identity I devised the next theorem that shows that, under rather general conditions, we *always* save a log factor. More refined version are possible, but this simplistic one captures the main power of Gram's statement. We first need two general lemmas.

Lemma 15.1 *When $Q \geqslant 0$, we have*

$$\sum_{p^\nu \leqslant Q} \nu^2 \log p \leqslant 3Q,$$

the sum being over every prime powers p^ν.

Proof We first use GP/Pari [44] to establish the claimed inequality when Q is below 10^6. Then we express our sum, say S, in the following manner:

$$S = \sum_{p^\nu \leqslant Q} \log p + \sum_{p^\nu \leqslant Q}(\nu^2 - 1)\log p \leqslant \psi(Q) + \sum_{p \leqslant \sqrt{Q}}\left(\frac{\log Q}{\log p}\right)^2 \log Q$$

$$\leqslant \psi(Q) + \pi(\sqrt{Q})\frac{\log^3 Q}{\log^2 2}$$

with the usual Tchebyshev function ψ and π. We recall that $\psi(x) \leqslant 1.04\,x$ for every $x > 0$ by Rosser and Schoenfeld [39, (3.35)] and that $\pi(Q) \leqslant 1.26x/\log x$ by Rosser and Schoenfeld [39, (3.6)]. A numerical application ends the proof of the lemma.

The next lemma follows the path initiated by Levin and Fainleib in [25], and trodden by several authors, like in [19].

Lemma 15.2 *Let h be a non-negative multiplicative function for which there exists a parameter H such that $|h(p^\nu)| \leqslant H\nu$ for every prime power p^ν. Then we have*

$$\sum_{n \leqslant x} h(n) \leqslant \frac{3Hx}{\log x} \sum_{n \leqslant x} \frac{h(n)}{n}.$$

Proof We start by

$$\sum_{n \leqslant x} h(n) \log x = \sum_{n \leqslant x} h(n) \log n + \sum_{n \leqslant x} h(n) \log \frac{x}{n}$$

$$\leqslant \sum_{n \leqslant x} h(n) \log n + x \sum_{n \leqslant x} \frac{h(n)}{n}.$$

Concerning the sum with $h(n) \log n$, we write

$$\log n = \sum_{p^\nu \| n} \log(p^\nu)$$

where the summation ranges over every prime power p^ν dividing n and such that $p^{\nu+1}$ does not divides n. In other words, ν is the proper power of p that divides n. We infer from this identity that:

$$\sum_{n \leqslant x} h(n) \log n = \sum_{p^\nu \leqslant x} \log(p^\nu) \sum_{p^\nu \| n \leqslant x} h(n) \leqslant \sum_{p^\nu \leqslant x} \log(p^\nu) h(p^\nu) \sum_{\substack{n \leqslant x/p^\nu, \\ (n,p)=1}} h(n)$$

$$\leqslant H \sum_{p^\nu \leqslant x} \nu \log(p^\nu) \sum_{n \leqslant x/p^\nu} h(n)$$

$$\leqslant H \sum_{n \leqslant x} h(n) \sum_{p^\nu \leqslant x/n} \nu \log(p^\nu).$$

To conclude, we use Lemma 15.1 above.

Theorem 15.3 *Let K be some real parameter and let g be a multiplicative function such that $|g(p^\nu)| \leqslant K$ for every prime power p^ν. Then we have*

$$\left| \sum_{n \leqslant x} \frac{g(n)}{n} \right| \leqslant \frac{9K}{\log x} \sum_{n \leqslant x} \frac{|g(n)| + |(\mathbb{1} \star g)(n)|}{n}.$$

On taking $g = \mu$, and $K = 1$, we recover the fact that the partial sum $\sum_{n \leqslant x} \mu(n)/n$ is bounded.

Proof We consider the sum $S = \sum_{n \leqslant x} (\mathbb{1} \star g)(n)$ which we write in the form

$$S = \sum_{m \leqslant x} g(m)[x/m] = x \sum_{m \leqslant x} \frac{g(m)}{m} - \sum_{m \leqslant x} g(m)\{x/m\}.$$

We deduce from the above that

$$\left| \sum_{n \leqslant x} \frac{g(n)}{n} \right| \leqslant \frac{1}{x} \sum_{n \leqslant x} (|g| + |\mathbb{1} \star g|)(n).$$

We next notice that both functions $|g|$ and $|\mathbb{1} \star g|$ are multiplicative and non-negative. Furthermore $\left| g(p^\nu) \right| \leqslant K \leqslant K\nu$ by hypothesis, while the reader will readily check that $\left| (\mathbb{1} \star g)(p^\nu) \right| \leqslant K(\nu+1) \leqslant 2K\nu$. We are thus in a position to apply Lemma 15.2 twice, namely to the two multiplicative functions $|g|$ and $|\mathbb{1} \star g|$. Completing the proof of the theorem is then straightforward.

15.3.1 An Intriguing Example

Seeing the appearance of $|g|$ and $|\mathbb{1} \star g|$, one may want to balance the effect of both factors; this almost happens when one selects $g(d) = \mu(d)/2^{\omega(d)}$. This case has in fact been considered long ago by Sigmund Selberg, a mathematician like his more famous brother Atle Selberg, in his 1954 paper [41] where he used Meissel's approach in a very careful manner to show the next theorem.

Theorem 15.4 (Selberg [41]) *We have, for every $x > 0$,*

$$0 \leqslant \sum_{n \leqslant x} \frac{\mu(n)}{2^{\omega(n)}n} \leqslant 1.$$

Proof Let us denote by f the function that associates $\mu(n)/2^{\omega(n)}$ to the integer n. The reader will readily check that $(\mathbb{1} \star f)(n) = 1/2^{\omega(n)}$. We thus get

$$\sum_{n \leqslant x} \frac{1}{2^{\omega(n)}} = \sum_{\ell \leqslant x} \frac{\mu(\ell)}{2^{\omega(\ell)}} \left[\frac{x}{\ell} \right] = x \sum_{\ell \leqslant x} \frac{\mu(\ell)}{\ell 2^{\omega(\ell)}} - \sum_{\ell \leqslant x} \frac{\mu(\ell)}{2^{\omega(\ell)}} \left\{ \frac{x}{\ell} \right\}$$

from which we deduce that

$$x \sum_{\ell \leqslant x} \frac{\mu(\ell)}{\ell 2^{\omega(\ell)}} = \sum_{n \leqslant x} \frac{1}{2^{\omega(n)}} \left(1 + \mu(n) \left\{ \frac{x}{n} \right\} \right).$$

This astounding equation immediately implies that the left hand side is non-negative and bounded above. To prove the more precise bound 1, we first notice that it is enough to prove it for positive integers x, in which case we first find that

$$\frac{1}{2^{\omega(1)}}\left(1+\mu(1)\left\{\frac{x}{1}\right\}\right)=1$$

and then that, as soon as $n\geqslant 2$, we have $2^{\omega(n)}\geqslant 2$, hence

$$\frac{1}{2^{\omega(n)}}\left(1+\mu(n)\left\{\frac{x}{n}\right\}\right)\leqslant\frac{2}{2}=1.$$

It is straightforward to conclude from these two inequalities.

A consequence of Theorem 15.3 is also that, when $x\geqslant 2$, we have

$$\sum_{n\leqslant x}\frac{\mu(n)}{2^{\omega(n)}n}\ll 1/\sqrt{\log x}. \tag{15.4}$$

This is for instance a consequence of the following theorem that we infer from the more precise [33, Theorem 21.1]. This theorem is in essence the one of Levin and Fainleib [25] we referred to above.

Theorem 15.5 *Let g be a non-negative multiplicative function. Let κ be a non-negative real parameter such that*

$$\begin{cases}\displaystyle\sum_{\substack{p\geqslant 2,v\geqslant 1\\p^{v}\leqslant Q}}g(p^{v})\log(p^{v})=\kappa\log Q+O(1)\quad(Q\geqslant 1),\\[2em]\displaystyle\sum_{p\geqslant 2}\sum_{v,k\geqslant 1}g(p^{k})g(p^{v})\log(p^{v})\ll 1.\end{cases}$$

Then, we have

$$\sum_{d\leqslant D}g(d)=\frac{(\log D)^{\kappa}}{\Gamma(\kappa+1)}\prod_{p\geqslant 2}\left\{\left(1-\frac{1}{p}\right)^{\kappa}\sum_{v\geqslant 0}g(p^{v})\right\}(1+O(1/\log D)).$$

To infer (15.4) from Theorem 15.3, we use Theorem 15.5 twice with $\kappa=1/2$. We leave the details to the reader.

We are thus in a position to prove elementarily and with no use of the Prime Number Theorem that the sum $\sum_{n\leqslant x}\frac{\mu(n)}{2^{\omega(n)}n}$ goes to 0! So why not try to reconstruct the Möbius function from this? This is easily achieved by employing Dirichlet's series. We first define the multiplicative function f_0 by $f_0(p^{v})=-(v-1)/2$ and then find that

$$\sum_{n \geq 1} \frac{\mu(n)}{n^s} = \left(\sum_{n \geq 1} \frac{\mu(n)}{2^{\omega(n)} n^s} \right)^2 \sum_{n \geq 1} \frac{f_0(n)}{n^s}. \tag{15.5}$$

The abscissa of absolute convergence of $D(f_0, s) = \sum_{n \geq 1} f_0(n)/n^s$ is $1/2$.

Proof All the implied functions being multiplicative, it is enough to check this identity on each local p-factor, i.e. that

$$1 - \frac{1}{p^s} = \left(1 - \frac{1}{2p^s} \right)^2 \left(1 - \sum_{v \geq 1} \frac{v-1}{2^v p^{vs}} \right).$$

This comes from the following formal identity, with $Y = X/2$:

$$\frac{1 - X}{(1 - \frac{X}{2})^2} = \frac{1}{1 - Y} - \frac{Y}{(1 - Y)^2} = \sum_{k \geq 0} Y^k - \sum_{k \geq 1} k Y^k.$$

To get the abscissa of absolute convergence we consider, with $\sigma = \Re s$,

$$\Delta = \sum_{p \geq 2} \left| \sum_{v \geq 2} \frac{v-1}{2^v p^{vs}} \right| \leq \sum_{p \geq 2} \frac{1}{4p^{2\sigma}} \left| \sum_{k \geq 0} \frac{k+1}{(2p^\sigma)^k} \right|$$

$$\leq \sum_{p \geq 2} \frac{1}{4p^{2\sigma}} \frac{1}{(1 - 1/(2p^\sigma))^2} \leq \sum_{p \geq 2} \frac{1}{p^{2\sigma}}.$$

This is bounded when $\sigma > 1/2$, showing that the product

$$\prod_{p \geq 2} \sum_{v \geq 0} \frac{f_0(p^v)}{p^{vs}}$$

is absolutely convergent when $\Re s > 1/2$. An immediate consequence is that the series is absolutely convergent in the same half-plane at least. The reader will readily see that the series of $|f_0(n)|/n^s$ diverges when $s = 1/2$, thus establishing that the half-plane $\Re s = 1/2$ is the actual half-plane of absolute convergence of $D(f_0, s)$.

The function $f_0(n)/n$ being much smaller than the function $\mu(n)2^{-\omega(n)}/n$, a first goal before finding bounds for $\sum_{n \leq x} \mu(n)/n$ from bounds on $\mu(n)2^{-\omega(n)}/n$ is to estimate the quantity

$$\sum_{\ell m \leq x} \frac{\mu(\ell)\mu(m)}{2^{\omega(\ell) + \omega(m)} \ell m}.$$

The Dirichlet hyperbola formula is made for that, i.e. we write

$$\sum_{\ell m \leqslant x} \frac{\mu(\ell)\mu(m)}{2^{\omega(\ell)+\omega(m)}\ell m} = 2 \sum_{\ell \leqslant \sqrt{x}} \frac{\mu(\ell)}{2^{\omega(\ell)}\ell} \sum_{m \leqslant x/\ell} \frac{\mu(m)}{2^{\omega(m)}m} - \left(\sum_{\ell \leqslant \sqrt{x}} \frac{\mu(\ell)}{2^{\omega(\ell)}\ell} \right)^2.$$

The second term is $O(1/\log x)$ while the first one is

$$\ll \sum_{\ell \leqslant \sqrt{x}} \frac{|\mu(\ell)|}{2^{\omega(\ell)}\ell} \frac{1}{\sqrt{\log x}} \ll 1$$

because we are losing the sign of the Möbius factor $\mu(\ell)$. The bound (15.4) fails to improve on (15.3)! The reader may want to use the non-negativity bound and distinguish as to whether ℓ has an even or an odd number of prime factors... And for instance aim at a lower estimate: when $\mu(\ell) = 1$, we use the fact that the summand is non-negative, and otherwise that it is $O(1/\sqrt{\log x})$. We have then to estimate

$$\sum_{\ell \leqslant \sqrt{x}} \frac{1 + \mu(\ell)}{2} \frac{|\mu(\ell)|}{2^{\omega(\ell)}\ell} = \frac{1}{2} \sum_{\ell \leqslant \sqrt{x}} \frac{|\mu(\ell)|}{2^{\omega(\ell)}\ell} + \frac{1}{2} \sum_{\ell \leqslant \sqrt{x}} \frac{\mu(\ell)}{2^{\omega(\ell)}\ell}$$

$$= \frac{1}{2} \sum_{\ell \leqslant \sqrt{x}} \frac{|\mu(\ell)|}{2^{\omega(\ell)}\ell} + O\left(\frac{1}{\sqrt{\log x}} \right)$$

and so, we have only saved a factor $1/2$.

Yet a third path opens before us: we may want to use the non-negativity of the sum $\sum_{n \leqslant x} \frac{\mu(n)}{2^{\omega(n)}n}$ in a stronger manner via Landau's Theorem on Mellin transform of non-negative functions, and maybe derive a stronger estimate! Indeed, the integral

$$\int_1^\infty \sum_{n \leqslant x} \frac{\mu(n)}{2^{\omega(n)}n} \frac{dx}{x^{s+1}}$$

represents the function $(1/s) \sum_{n \geqslant 1} \frac{\mu(n)}{2^{\omega(n)}n^{s+1}}$. Hence the abscissa of convergence of the integral should be a pole of the function represented. Can we show in this fashion that the integral converges for $\Re s > -1/2$ hence improving on (15.4)? This is tempting, but does not work: the series $\sum_{n \geqslant 1} \frac{\mu(n)}{2^{\omega(n)}n^{s+1}}$ behaves like $1/\sqrt{\zeta(s+1)}$, i.e. like $\sqrt{s}$ next to $s = 0$, and the innocent looking factor $(1/s)$ in front of the series above shows that the integral has a polar contribution at $s = 0$. In fact, S. Selberg already showed in [41] that $\sum_{n \leqslant x} \frac{\mu(n)}{2^{\omega(n)}n}$ is equivalent to $C/\sqrt{\log x}$, where C is some well-defined and non-zero constant.

The purpose of this digression was to show the reader that the results we are looking at are tight. Any improvement would have acute consequences.

15.4 The Axer-Landau Equivalence Theorem

We have studied the situation from the angle of general multiplicative functions; let us now restrict more closely our attention to the case of the Möbius function. Here is an enlightening result in this direction. We first recall how the van Mangoldt function Λ is defined:

$$\Lambda(n) = \begin{cases} \log p & \text{when } n = p^\nu \\ 0 & \text{else.} \end{cases} \tag{15.6}$$

Theorem 15.6 (Axer-Landau, 1899-1911) *The five following statements are equivalent:*

(S_1) *The number of primes up to x is asymptotic to $x/\log x$.*
(S_2) $M(x) = \sum_{n \leqslant x} \mu(n)$ *is* $o(x)$.
(S_3) $m(x) = \sum_{n \leqslant x} \mu(n)/n$ *is* $o(1)$.
(S_4) $\psi(x) = \sum_{n \leqslant x} \Lambda(n)$ *is asymptotic to x.*
(S_5) $\tilde{\psi}(x) = \sum_{n \leqslant x} \Lambda(n)/n$ *is* $\log x - \gamma + o(1)$.

In fact, proving that (S_3) implies (S_2) or that (S_5) implies (S_4) is a simple matter of summation by parts, as is the equivalence of (S_1) and (S_4). Edmund Landau in 1899 in [22] was the first to investigate this kind of result: he showed that (S_1) implies (S_3). The Viennese mathematician A. Axer continued in 1910 in [3] by establishing that (S_2) implies (S_3). Landau immediately applied Axer's method to prove that (S_4) and (S_5) are equivalent and concluded in [23] essentially by showing that (S_3) implies (S_4). See also [4] and [24].

Concerning our question, this theorem shows that we clearly need to save the logarithm factor over the trivial estimate for $m(x)$, as well as for $\tilde{\psi}$. A second aspect arises from this theorem: the call for an quantitative version of it. If one follows the proofs of Axer and Landau, the saving is essentially limited at $O(1/\sqrt{\log x})$, though some later authors, like the Swiss mathematician Alfred Kienast in [21], went further.

15.4.1 A Related Problem

In [34] and more fully in [31] with David Platt from Bristol, I investigated the problem of deriving quantitatively (near) optimal results on $\tilde{\psi}(x)$ once one supposes results for $\psi(x)$. This implication has been shown to be false in the general context of Beurling integers[4] by Harold Diamond and Wen-Bin Zhang in [9]. They even

---

[4]The Beurling integers are the multiplicative semi-group built on a family of "primes" to be chosen real numbers from $(1, \infty)$.

exhibit a Beurling system $\mathcal{B}$ where one has $\psi_{\mathcal{B}}(x) \sim x$ while $\tilde{\psi}_{\mathcal{B}}(x) - \log x \gg \log \log x$, with obvious notation. This means in particular that something special linked with the nature of the integers is required. It took us quite a while to understand what was happening, though I had essentially settled the problem in the q-aspect several years ago in [32]: instead of looking at primes, I was looking at primes in some arithmetic progression, say modulo some q; the error term has then a dependence in q and in x. In the mentioned paper, I resolved this question *provided* the question for $q = 1$ was solved! I thought that I had reduced the problem to a simpler one, but it is more correct to say that the x-aspect is the one that leads to real difficulties.

The first idea is of course to use a summation by parts, i.e. to write

$$\tilde{\psi}(x) - \log x = \frac{\psi(x) - x}{x} + 1 + \int_1^x \frac{\psi(t) - t}{t^2} dt. \qquad (15.7)$$

A careful look at this equation will in fact be enough to solve the question. We can understand on it the idea of Diamond and Zhang: they built a Beurling system where the integral above does not converge. A different approach from this same starting point leads to the next theorem we proved with D. Platt.

Theorem 15.7 (Platt and Ramaré [31]) *There exists $c > 0$ such that, when $x \geqslant 10$, we have:*

$$\tilde{\psi}(x) - \log x + \gamma \ll \max_{x \leqslant y \leqslant 2x} \frac{|\psi(y) - y|}{y} + \exp\left(-c \frac{\log x}{\log \log x}\right).$$

A similar statement for primes in arithmetic progressions holds true. This theorem is very efficient to compare $\tilde{\psi}$ together with ψ, and is in fact nearly optimal from a quantitative viewpoint. We are almost saving a power of x; a look at the proof discloses that the zero-free region for the Riemann-zeta function is used only up to the height $\log x$. This has the consequence that numerically, verifying the Riemann Hypothesis up to the height H gives control for x roughly up to e^H! And since X. Gourdon and P. Demichel [16] have checked this Riemann Hypothesis[5] up to height $2.445 \cdot 10^{12}$, we can assume the Riemann Hypothesis is available when $x \leqslant e^{10^{12}}$, which is enormous! Practically, this discussion shows that the factor $\exp(-c(\log x)/\log \log x)$ can be replaced by a very small quantity. We shall see below some very explicit consequences of this fact, but let us start by a rough explanation of the proof. This is not the way the proof appeared at first, but how I now understand it. We first note that

$$1 + \int_1^\infty \frac{\psi(t) - t}{t^2} dt = -\gamma. \qquad (15.8)$$

---

[5]This computation has not been the subject of any published paper. D. Platt in [30] has checked this hypothesis up to height 10^9 by with a very precise program using interval arithmetic.

This is highly non-obvious if seen like that. The Prime Number Theorem with a remainder term ensures that the integral converges, but the full proof requires the limited development $\zeta(s) = (s-1)^{-1} + \gamma + O(s-1)$ around $s = 1$ which implies that $-(\zeta'/\zeta)(s) = (s-1)^{-1} - \gamma + O(s-1)$. We leave the details to the reader. What is really important for us is that this quantity is indeed a constant so that we can rewrite (15.7) in

$$\tilde{\psi}(x) - \log x = \frac{\psi(x) - x}{x} - \gamma - \int_x^\infty \frac{\psi(t) - t}{t^2} dt. \tag{15.9}$$

This formula is not enough to conclude but a very small modification of it will suffice: let $F : [1, \infty) \to \mathbb{R}$ be a smooth function such that $F(y) = 1$ when $y \geqslant 2$. We have

$$\tilde{\psi}(x) - \log x = \frac{\psi(x) - x}{x} - \gamma - \int_x^{2x} (1 - F(t/x)) \frac{\psi(t) - t}{t^2} dt$$

$$+ \int_x^\infty F(t/x) \frac{\psi(t) - t}{t^2} dt. \tag{15.10}$$

The integral over $[x, 2x]$ can be controlled by $\max_{x < y < 2x} |\psi(y) - y|/x$, but what about the last integral? In short: we express it in terms of the zeros of the Riemann zeta-function and get in this manner a fast convergent sum. Why is that so? The reader may think it is because of the smoothing and the involvement of Mellin transforms... And would be right! A fact that had escaped my attention so long, and not only mine, is that this argument works for the point at infinity. Repeated integrations by parts for instance, when assuming F smooth enough, show that the corresponding Mellin transform decays vary rapidly in vertical strips.

Here are two very explicit consequences that we promised earlier.

Theorem 15.8 *We have*

$$\left| \sum_{n \leqslant x} \Lambda(n)/n - \log x + \gamma \right| \leqslant \frac{1}{149 \log x} \quad (x \geqslant 23).$$

The previous result is due to J. Rosser and L. Schoenfeld in [39] and had a 2 instead of 149.

Theorem 15.9 *We have*

$$\left| \sum_{n \leqslant x} \Lambda(n)/n - \log x + \gamma \right| \leqslant \frac{2}{(\log x)^2} \quad (x > 1).$$

This result has no ancestor that I know of. There are related work by P. Dusart [13], [12], L. Faber and H. Kadiri, H. Kadiri and A. Lumley [20] (and more to come), C. Axler [5], L. Panaitopol [28], R. Vanlalnagaia [45], etc.

The sketch I propose above is the manner I now explain the proof, but the two initial papers, [34] and [31] proceeded in a very different manner: the integral $\int_x^\infty (\psi(t) - t)dt/t^2$ was expressed in terms of the zeros of zeta and the relevant expression was compared with another one more convergent. The better understood scheme above will have an interesting consequence we shall see later.

15.4.2 The Horizon

It is time to set the horizon! Here are three conjectures.

Conjecture C There exists a constant $A > 0$ such that

$$m(x) \overset{?}{\ll} \max_{x/A < y \leqslant xA} |M(y)|/y + x^{-1/4}.$$

And since we would like to have control of $M(x)$ via[6] $\psi(x)$, I also believe the following.

Conjecture D There exists a constant $A > 0$ such that

$$m(x) \overset{?}{\ll} \max_{x/A < y \leqslant xA} |\psi(y) - y|/y + x^{-1/4}.$$

And we recall the conjecture of [34].

Conjecture A There exists a constant $A > 0$ such that

$$\tilde{\psi}(x) \overset{?}{\ll} \max_{x/A < y \leqslant xA} |\psi(y) - y|/y + x^{-1/4}.$$

These three conjectures are trivially true under the Riemann Hypothesis, even with the $x^{-1/4}$ replaced by $x^{-1/2+\varepsilon}$. This exponent $1/4$ is not particularly relevant, the saving of any power of x would be a true achievement. These three conjectures are obvious if we allow a factor $\log x$ in front of the maxima, simply by using integration by parts, but even if we allow a factor between 1 and $\log x$, like $\sqrt{\log x}$ for instance, the answer is not known.

---

[6]I formulated a more precise conjecture, say Conjecture B, in [35].

15.5   From M to m

The proof we presented of Theorem 15.7 allows one to dispense with the notion of zeros, though introducing them is numerically much more efficient. We can however express the function F in (15.10) in terms of its Mellin transform. This Mellin transform decreases fast in vertical strips[7] and this is enough to get the result! We provide a full proof in [38].

Theorem 15.10 *There exists $c > 0$ such that, when $x \geqslant 10$, we have:*

$$m(x) \ll \max_{x \leqslant y \leqslant 2x} \frac{|M(y)|}{y} + \exp\left(-c\,\frac{\log x}{\log \log x}\right).$$

The difference with the case of $\tilde{\psi}$ is that we do not have any efficient version of this theorem.

15.6   Generalizing Meissel's Proof, II

M. Balazard took another path to understand Meissel's formula. He rewrote this identity in the form:

$$\frac{1}{x}\int_1^x M(x/t)\frac{\{t\}}{t}\,dt = m(x) - \frac{M(x)}{x} - \frac{\log x}{x} \qquad (15.11)$$

and did the same for the MacLeod identity:

$$\frac{1}{x}\int_1^x M(x/t)\frac{(2\{t\}-1)t + \{t\} - \{t\}^2}{t^2}\,dt = m(x) - \frac{M(x)}{x} - \frac{2}{x} + \frac{2}{x^2}.$$

Some order emerges in this manner, but the question remains as to whether these identities are oldies to be thrown in the wastebasket or not. The situation has been further cleared by F. Daval[8] (2016, private communication) in the next theorem.

Theorem 15.11 (Daval, 2016) *Let $h : [0, 1] \to \mathbb{C}$ be a continuous function normalized by $\int_0^1 h(u)\,du = 1$. When $x \geqslant 1$, we have*

$$\frac{1}{x}\int_1^x M(x/t)\left(1 - \frac{1}{t}\sum_{n \leqslant t} h(n/t)\right)dt = m(x) - \frac{M(x)}{x} - \frac{1}{x}\int_{1/x}^1 \frac{h(y)}{y}\,dy.$$

---

[7] As already stated, we show that by classically repeated integrations by parts.

[8] F. Daval was at the time a PhD student of mine.

Like many identities, once it is written, it is not very difficult to establish. On selecting $h = 1$, we recover the Meissel identity, and on selecting $h(t) = 2t$, the MacLeod identity I showed is being recovered. We thus see that a "Riemann integral-remainder" appears; functional analysis is coming in! Among the natural questions, let us mention this one: given a function f over $[0, 1]$, can it be approximated by such a Riemann-remainder term? If not what is the best approximation? Before continuing, let us mention that there are some other identities in this area, and for instance, following J.-P. Gram [17], R. MacLeod [26] and M. Balazard [6], here is, in Balazard's form, a typical identity I obtained in [37, Lemma 3.2]:

$$\sum_{n \leqslant x} \frac{\mu(n)}{n} \log\left(\frac{x}{n}\right) - 1 = \frac{6 - 8\gamma}{3x} - \frac{5 - 4\gamma}{x^2} + \frac{6 - 4\gamma}{3x^4} - \frac{1}{x} \int_1^x M(x/t) h'(t) dt$$

where[9] the function h is differentiable except at integer points where it has left and right-derivative, and satisfies $0 \leqslant t^2 |h'(t)| \leqslant \frac{7}{4} - \gamma$. The function h is this time linked with the error term $\sum_{n \leqslant t} 1/n - \log t$. Similar identities have been proved with $\log^k(x/n)$ instead of $\log(x/n)$, for any positive integer k. The theory of F. Daval can most probably be adapted to these cases. One striking consequence is the next result.

Theorem 15.12 ([37, Theorem 1.5]) *When $x \geqslant 3155$, we have*

$$\left| \sum_{n \leqslant x} \frac{\mu(n)}{n} \log(x/n) - 1 \right| \leqslant \frac{1}{389 \log x}.$$

Note that one could try to derive such an estimate by writing

$$\sum_{n \leqslant x} \frac{\mu(n)}{n} \log(x/n) = (\log x) \sum_{n \leqslant x} \frac{\mu(n)}{n} - \sum_{n \leqslant x} \frac{\mu(n)}{n} \log n$$

and using estimates for both. But to attain the accuracy level of our theorem, one would need to prove at least that $\sum_{n \leqslant x} \mu(n)/n = O^*(1/(389 \log^2 x))$, and we are rather far from having this kind of results!

15.6.1 The Problem at Large

Let us try to formalize the problem. We start from a regular function $F : [1, \infty) \rightarrow \mathbb{C}$, for instance $F(t) = 1$ or $F(t) = \log t$. The question is to find two functions H and

---

[9] As a matter of fact, the mentioned lemma is slightly different, but a corrigendum is on its way.

G and a constant C such that

$$\sum_{n \leqslant x} \frac{\mu(n)}{n} F(x/n) - C \frac{M(x)}{x} = \frac{1}{x} \int_1^x M(x/t)G(t)dt + H(x).$$

To avoid trivial solutions, we assume that

$$\int_1^\infty |F(t)|dt/t = \infty, \qquad \int_1^\infty |G(t)|dt/t < \infty,$$

and that H is smooth and "small". This looks like a functional transform from F to G, but there is a lot of slack! Indeed, when $F = 1$ or when $F(t) = \log t$, there are several solutions.

15.6.2 Beginning of a Theory When $F = 1$

We start from Theorem 15.11 and, remembering the identities of MacLeod in Balazard's form, we aim at writing the integral with M in the form $\int M(x/t)f'(t)dt$. With this goal in sight we note that

$$\int_0^x \left(1 - \frac{1}{t}\sum_{n \leqslant t} h(n/t)\right)dt = \int_0^1 \{ux\}\frac{h(u)}{u}du.$$

So, given $f : [1, \infty) \to \mathbb{C}$, we want to solve $f(x) = \int_0^1 \{ux\}\frac{h(u)}{u}du$. The change of variable $y = 1/x$ leads to the problem: given $g : [0, 1] \to \mathbb{C}$, solve $g(y) = \int_0^1 \frac{\{u/y\}}{u/y}h(u)du$. We see another appearance of functional analysis! The operator T over the Hilbert space $L^2([0, 1])$ which associates $\int_0^1 \frac{\{u/y\}}{u/y}h(u)du$ to h is a Hilbert-Schmidt, compact and contracting operator. Indeed, we readily check that the kernel $(u, y) \mapsto \frac{\{u/y\}}{u/y}$ belongs to $L^2([0, 1]^2)$ and then, we for instance use [15] (around Eqs. (9.6)–(9.8)). Since

$$\int_0^1 \int_0^1 \left|\frac{\{u/y\}}{u/y}\right|^2 dudy = \int_0^1 \int_0^{1/y} \left|\frac{\{z\}}{z}\right|^2 dz\,ydy$$

$$\leqslant \int_0^1 \left(1 + \int_1^\infty \frac{\{z\}}{z^2}dz\right)ydy = 2(1 - \gamma) < 1$$

we readily see by invoking the Cauchy-Schwarz inequality that T is strictly contracting. The general theory tells us that there exist a sequence of complex numbers $(\lambda_n)_n$ and two orthonormal sequences of functions $(\psi_n)_n$ and $(\varphi_n)_n$ such that

$$\int_0^1 \{u/y\}\frac{h(u)}{u}du = \sum_{n \geqslant 1} \lambda_n \int_0^1 h(u)\overline{\psi_n(u)}du\varphi_n(y)$$

for every $y \in [0, 1]$. By Swann [42], this operator to be of Shatten class p for every $p > 1$, and I suspect it is *not* of trace class. The above decomposition is a consequence of the general theory of integral operator and a more specific study should be able to disclose arithmetical properties. For instance, the presence of the fractional part is not without recalling the Nyman-Beurling criteria. This is work in progress.

15.6.3 The Localization Problem, Case F = 1

We are here going back to what has been done rather than guessing what could be happening in the future! It is easier to first state a result and then describe the problem at hand from there. We start with a lemma.

Lemma 15.13 (F. Daval, 2016, private communication) *Let $h : [0, 1] \mapsto \mathbb{C}$ be a C^k-function for some $k \geqslant 2$, normalized with $\int_0^1 h(u)du = 1$. We further assume that*

- $h(0) = h'(0) = 0$,
- *When $3 \leqslant 2i + 1 \leqslant k - 1$, we have $h^{(2i+1)}(0) = 0$,*
- *When $0 \leqslant \ell \leqslant k - 2$, we have $h^{(\ell)}(1) = 0$.*

Then we have, for $t \geqslant 1$,

$$\left| 1 - \frac{1}{t} \sum_{n \leqslant t} h(n/t) \right| \ll 1/t^k.$$

Given an integer $k \geqslant 2$, let us call $\mathcal{H}_k$ the class of functions h described above. Then, for any $h \in \mathcal{H}_k$, there exists a constant $C_k(h)$ such that

$$\left| \int_1^x M(x/t) \left(1 - \frac{1}{t} \sum_{n \leqslant t} h(n/t) \right) dt \right| \leqslant \frac{C_k(h)}{x} \int_1^x M(t)(t/x)^{k-2} dt.$$

F. Daval (2016, private communication) has obtained the following table:

$k =$	3	4	5	6	7
$\min_h C_k(h) \leqslant$	1.05	1.44	2.52	5.9	13.2

This improves of earlier values of M. Balazard in [6]. It would be interesting to determine numerically these minima with more accuracy. The value for $k = 5$ has been obtained with the highly non-obvious choice $h_5(t) = 2t^2(1-t)^4(120t^2+52t+13)$. As a consequence, one can get for instance the following inequality:

$$\left| m(x) - \frac{M(x)}{x} \right| \leqslant \frac{33/13}{x^4} \int_1^x |M(t)| t^3 dt + \frac{19/7}{x}$$

and such an inequality should give improvements for many of the results I obtained in [37]. We call this problem the "localization problem" because a high power of t in the integral above means that the values of $M(t)$ for t close to x have more weight than the lower ones. We recall that conjecture C claims that one can use only the values of t that are a constant multiple of x.

All in all, a lot remains to be understood in this area. I for instance wonder whether functions like $\{t^2 + 1\}$ could appear in these identities rather than $\{t\}$… I thought at first that the answer should be no but I am not so sure anymore.

15.6.4 From Λ to μ/From ψ to M

Let us continue our journey around the Axer-Landau Equivalence Theorem. We first notice that Wen-Bin Zhang has exhibited in [46] a Beurling system of integers where one have $M_P(x) = o(x)$ without $\psi_P(x) \sim x$. Our final destination being numerical estimates, we are however more interested in the reverse implication, i.e. to derive bounds for M from bounds for the primes. This problem has been studied by A. Kienast in [21] and by L. Schoenfeld [40], and they proceeded as I later did in [35] by using some combinatorial identities. The family of identities I produced is simply more efficient. It is better to refer the reader to the cited paper but let us give the general flavour. The first interesting case reads

$$\sum_{\ell \leqslant x} \mu(\ell) \log^2 \ell = \sum_{d\ell \leqslant x} \mu(\ell)\big(\Lambda \star \Lambda(d) - \Lambda(d) \log d\big). \qquad (15.12)$$

It is worth mentioning that the Selberg[10] identity that is used for proving elementarily the Prime Number Theorem is $\Lambda \star \Lambda(d) + \Lambda(d) \log d = (\mu \star \log^2)(d)$ and that, assuming this Prime Number Theorem, both factors $\Lambda \star \Lambda(d)$ and $\Lambda(d) \log d$ contribute equally to the average. In particular, the function $\Lambda \star \Lambda(d) - \Lambda(d) \log d$ should be looked upon as a remainder term. We get information of its average order by using the Dirichlet hyperbola formula; it would most probably be better to use an explicit expression in terms of the zeros directly, but this involves the residues of $(\zeta'/\zeta)^2$ and there lacks a control of those, while the residues of ζ'/ζ are well understood. Some more thought discloses that we need essentially the L^1-norm of such residues, and since they are non-negative integers for ζ'/ζ, we may as well

---

[10]This one is Atle Selberg!

compute their simple average, which is readily achieved by a contour integration that has most of its path outside the critical strip. No such phenomenon is known to occur for $(\zeta'/\zeta)^2$! The reader may be wary of the Möbius factor that appears on the right-hand side of (15.12), but only one such factor appears. It is maybe more apparent in the next identity of this series:

$$\sum_{\ell \leqslant x} \mu(\ell) \log^3 \ell = \sum_{d\ell \leqslant x} \mu(\ell)\big(\Lambda \star \Lambda \star \Lambda(d) - 3\Lambda \star (\Lambda \log)(d) + \Lambda(d) \log^2 d\big).$$

When starting with the last identity with $k = 3$, one can expect to save a $\log^3 x$ on the trivial estimate x, but the presence of the Möbius factor on the right-hand side reduces that to a saving of one $\log x$ less, so $\log^2 x$. This is because the Dirichlet hyperbola method is not used, though one may employ a recursion process: indeed, L. Schoenfeld does that, followed by H. Cohen, F. Dress and M. El Marraki in [7], [11] and [14]. I did not introduce such a step as it is numerically costly, but a more careful treatment is here possible.

15.7   Generalizing Meissel's Proof, III

We now turn towards the third aspect of Meissel's identity, which is to provide a simple proof of $\sum_{n \leqslant x} \mu(n)/n \ll 1$. Here is a theorem I proved a long time back with Andrew Granville in [18, Lemma 10.2].

Theorem 15.14 *For $x \geqslant 1$ and $q \geqslant 1$, we have*

$$\left| \sum_{\substack{n \leqslant x, \\ \gcd(n,q)=1}} \frac{\mu(n)}{n} \right| \leqslant 1.$$

This result belongs to the family of the eternally-reproved lemmas! In fact, I discovered much later than it appeared already in an early paper of Harold Davenport as [8, Lemma 1]! The precise upper bound by 1 is not given, but the proof is already there. And Terence Tao reproved this result in [43], in a larger context, but the proof is again the same! We cannot even say that Davenport's paper or the one I co-authored are forgotten: they are simply cited for other reasons.

The main theme is the handling of the coprimality condition. Since we mentioned the investigations of Sigmund Selberg, it is worthwhile stating a surprising lemma that one finds in [41, Satz 4].

Theorem 15.15 *For $x \geqslant 1$ and $d, q \geqslant 1$, with $d|q$, we have*

$$0 \leqslant \sum_{\substack{n \leqslant x, \\ \gcd(n,d)=1}} \frac{\mu(n)}{2^{\omega(n)}n} \leqslant \sum_{\substack{n \leqslant x, \\ \gcd(n,q)=1}} \frac{\mu(n)}{2^{\omega(n)}n} \leqslant 1.$$

We should make a stop here; indeed the reader may think that removing the coprimality condition is an easy task. The standard manner goes by using the Möbius function and the identity:

$$\sum_{\substack{d|n,\\d|q}} \mu(d) = \sum_{d|\gcd(n,q)} \mu(d) = \begin{cases} 1 & \text{when } \gcd(n,q) = 1 \\ 0 & \text{else.} \end{cases} \quad (15.13)$$

However, here is what happens in our case:

$$\sum_{\substack{n\leqslant x,\\ \gcd(n,q)=1}} \frac{\mu(n)}{n} = \sum_{n\leqslant x} \sum_{d|\gcd(n,q)} \mu(d)\frac{\mu(n)}{n} = \sum_{d|q} \mu(d) \sum_{d|n\leqslant x} \frac{\mu(n)}{n}$$

$$= \sum_{d|q} \frac{\mu^2(d)}{d} \sum_{\substack{m\leqslant x/d,\\ \gcd(m,d)=1}} \frac{\mu(m)}{m} \quad (15.14)$$

and thus a coprimality condition comes back in play! E. Landau has devised a long time ago a manner to go around this problem: it consists in comparing the multiplicative function $f(n) = \mathbb{1}_{(n,q)=1}\mu(n)$ with the function μ, i.e. to find a function g such that $f = \mu \star g_q$, where $\star$ is the arithmetical convolution product. Determining g_q is an exercise resolved by comparing the Dirichlet series. Once the reader has found the expression for g_q, he or she will find that it is somewhat unwieldy. The foremost problem is that it has an infinite support and thus, when we write

$$\sum_{\substack{n\leqslant x,\\ \gcd(n,q)=1}} \frac{\mu(n)}{n} = \sum_{\ell\geqslant 1} g_q(\ell) \sum_{m\leqslant x/\ell} \mu(m)/m$$

one has to handle the case when ℓ is large, i.e. when x/ℓ is small. This leads to difficulties, for instance when one wants explicit estimates. But even if one aims only at theoretical results, difficulties appear: for instance, if one wants to bound $\sum_{\substack{n\leqslant x,\\ \gcd(n,q)=1}} \frac{\mu(n)}{n}$ from the estimate $|\sum_{n\leqslant x} \frac{\mu(n)}{n}| \leqslant 1$ and the function g_q, the resulting bound is $O(q/\phi(q))$, which can be infinitely larger than $O(1)$.

I devised in [36] and [37] a workaround to handle this question. The two remarks needed are first that the Liouville function[11] λ is rather close to the Möbius function, and second that the Liouville function being completely multiplicative, the proof above (leading to (15.14)) would this time succeed. This implies a process in three

---

[11]The Liouville function is the completely multiplicative function defined by $\lambda(n) = (-1)^{\Omega(n)}$, where $\Omega(n)$ is the number of prime factors of n, counted with multiplicity, so that $\Omega(12) = 3$.

steps:

1. Go from μ to λ.
2. Get rid of the coprimality with the Möbius function.
3. Study the resulting sum by comparing λ to μ and by using results on μ.

In the second paper, I noticed that it is possible to combine steps 1 and 3, hence gaining in efficiency. This process is however only half a cure: one indeed avoids short sums, and this is numerically important, but the factor $q/\phi(q)$ we talked about earlier still arises! Here is a typical result I obtained in this fashion.

Theorem 15.16 *When* $1 \leqslant q < x$, *we have*

$$\left| \sum_{\substack{d \leqslant x, \\ (d,q)=1}} \mu(d)/d \right| \leqslant \frac{4q/5}{\phi(q)\log(x/q)}.$$

Similar results with $\mu(d)\log(x/d)/d$ and $\mu(d)\log(x/d)^2/d$ are also presented. Let me end this section with a methodological remark: Theorem 15.14 does not contain in its statement a natural restriction of q with respect to x, and as such is hard to improve upon. Indeed q could be the product of all the primes below x, in which case the bound is optimal. The factor $1/\log(x/q)$ in Theorem 15.16 avoids this fact, which is why I believe it can be largely improved. The removal of the factor $q/\phi(q)$ would be a interesting step.

15.7.1 A Related Problem

Meissel's identity leads to an excellent handling of the coprimality condition, and we saw at the beginning of this section that it was not obvious to generalize. In another paper [1] with Akhilesh P. concerning the Selberg sieve density function, we encountered the problem of bounding the sum

$$\sum_{\substack{k>K, \\ \gcd(k,q)=1}} \frac{\mu(k)}{k\phi(k)} \tag{15.15}$$

uniformly in q. We were only able at the time to get a better than trivial estimate, but recently, together with Akhilesh P. in [2], we proved the next result by again employing the Liouville trick described above to which we added a sieving argument.

Theorem 15.17

$$\limsup_{K \to \infty} K \max_{q} \left| \sum_{\substack{k>K, \\ \gcd(k,q)=1}} \frac{\mu(k)}{k^2} \right| = 0.$$

Our result is more general and encompasses the sum (15.15). In essence, the proof runs as follows: when q has many prime factors, use a sieve bound; when q has few prime factors, remove the coprimality condition with Möbius. This time coprimality with $\prod_{p \leqslant K, p \nmid q} p$ comes into play. Both arguments take care of extremal ranges of q (i.e. when q as many or few prime factors). These ranges do not overlap: there is a middle zone where this time, the oscillation of μ comes into play, and it is where we use the λ-trick to get rid of the coprimality condition.

The rate of convergence is however unknown to us. Under the Riemann Hypothesis, our proof gives a rate of convergence in $1/(\log K)^{1/3-\varepsilon}$ for any $\varepsilon > 0$ but the best we have been able to prove concerning an Omega-result is that

$$\limsup_{K \to \infty} K \log K \max_q \left| \sum_{\substack{k > K, \\ (k,q)=1}} \frac{\mu(k)}{k^2} \right| \geqslant 1.$$

We have not even been able to improve on this last constant 1, which we got by considering $q = \prod_{p \leqslant K} p$. Our journey ends here!

References

1. P. Akhilesh, O. Ramaré, Explicit averages of non-negative multiplicative functions: going beyond the main term. Colloq. Math. **147**, 275–313 (2017)
2. P. Akhilesh, O. Ramaré, Tail of a moebius sum with coprimality conditions. Integers **18**, Paper No. A4, 6 (2018)
3. A. Axer, Beitrag zur Kenntnis der zahlentheoretischen Funktionen $\mu(n)$ und $\lambda(n)$. Prace Matematyczno-Fizyczne **21**(1), 65–95 (1910)
4. A. Axer, Über einige Grenzwertsätze. Wien. Ber. **120**, 1253–1298 (1911) (German)
5. C. Axler, New bounds for the prime counting function. Integers **16** (2016), Paper No. A22, 15.
6. M. Balazard, Elementary remarks on Möbius' function. Proc. Steklov Inst. Math. **276**, 33–39 (2012)
7. H. Cohen, F. Dress, M. El Marraki, Explicit estimates for summatory functions linked to the Möbius μ-function. Funct. Approx. Comment. Math. **37**(part 1), 51–63 (2007)
8. H. Davenport, On some infinite series involving arithmetical functions. Q. J. Math. Oxf. Ser. **8**, 8–13 (1937)
9. H.G. Diamond, W.B. Zhang, A PNT equivalence for Beurling numbers. Funct. Approx. Comment. Math. **46**(part 2), 225–234 (2012)
10. L.E. Dickson, *Theory of Numbers* (Chelsea Publishing Company, New York, 1971)
11. F. Dress, M. El Marraki, Fonction sommatoire de la fonction de Möbius 2. Majorations asymptotiques élémentaires. Exp. Math. **2**(2), 89–98 (1993)
12. P. Dusart, Autour de la fonction qui compte le nombre de nombres premiers. Ph.D. thesis, Limoges, 1998, 173 pp., http/string://www.unilim.fr/laco/theses/1998/T1998_01.pdf
13. P. Dusart, Estimates of some functions over primes without R. H. (2010), http://arxiv.org/abs/1002.0442
14. M. El Marraki, Fonction sommatoire de la fonction μ de Möbius, majorations asymptotiques effectives fortes. J. Théor. Nombres Bordx. **7**(2), 407–433 (1995)
15. I.C. Gohberg, M.G. Kreĭn, *Introduction to the Theory of Linear Nonselfadjoint Operators*. Translated from the Russian by A. Feinstein. Translations of Mathematical Monographs, vol. 18 (American Mathematical Society, Providence, 1969)

16. X. Gourdon, P. Demichel, The 10^{13} first zeros of the Riemann Zeta Function and zeros computations at very large height (2004), http://numbers.computation.free.fr/Constants/Miscellaneous/zetazeros1e13-1e24.pdf

17. J.P. Gram, Undersøgelser angaaende Maengden af Primtal under en given Graense. Résumé en français. Kjöbenhavn. Skrift. (6) II, 185–308 (1884)

18. A. Granville, O. Ramaré, Explicit bounds on exponential sums and the scarcity of squarefree binomial coefficients. Mathematika **43**(1), 73–107 (1996)

19. H. Halberstam, H.E. Richert, On a result of R. R. Hall. J. Number Theory **11**, 76–89 (1979)

20. H. Kadiri, A. Lumley, Short effective intervals containing primes. Integers **14** (2014), Paper No. A61, 18, arXiv:1407:7902

21. A. Kienast, Über die Äquivalenz zweier Ergebnisse der analytischen Zahlentheorie. Math. Ann. **95**, 427–445 (1926), https://doi.org/10.1007/BF01206619

22. E. Landau, Neuer Beweis der Gleichung $\sum_{k=1}^{\infty} \frac{\mu(k)}{k} = 0$, Berlin. 16 S. gr. 8° (1899)

23. E. Landau, Über die Äquivalenz zweier Hauptsätze der analytischen Zahlentheorie. Wien. Ber. **120**, 973–988 (1911) (German)

24. E. Landau, Über einige neuere Grenzwertsätze. Rendiconti del Circolo Matematico di Palermo (1884–1940) **34**, 121–131 (1912), https://doi.org/10.1007/BF03015010

25. B.V. Levin, A.S. Fainleib, Application of some integral equations to problems of number theory. Russ. Math. Surv. **22**, 119–204 (1967)

26. R.A. MacLeod, A curious identity for the Möbius function. Utilitas Math. **46**, 91–95 (1994)

27. E. Meissel, Observationes quaedam in theoria numerorum. J. Reine Angew. Math. **48**, 301–316 (1854) (Latin)

28. L. Panaitopol, A formula for $\pi(x)$ applied to a result of Koninck-Ivić. Nieuw Arch. Wiskd. (5) **1**(1), 55–56 (2000)

29. J. Peetre, Outline of a scientific biography of Ernst Meissel (1826–1895). Hist. Math. **22**(2), 154–178 (1995), https://doi.org/10.1006/hmat.1995.1015

30. D.J. Platt, Numerical computations concerning the GRH. Ph.D. thesis, 2013, http://arxiv.org/abs/1305.3087

31. D.J. Platt, O. Ramaré, Explicit estimates: from $\Lambda(n)$ in arithmetic progressions to $\Lambda(n)/n$. Exp. Math. **26**, 77–92 (2017)

32. O. Ramaré, Sur un théorème de Mertens. Manuscripta Math. **108**, 483–494 (2002)

33. O. Ramaré, *Arithmetical Aspects of the Large Sieve Inequality*. Harish-Chandra Research Institute Lecture Notes, vol. 1 (Hindustan Book Agency, New Delhi, 2009). With the collaboration of D.S. Ramana

34. O. Ramaré, Explicit estimates for the summatory function of $\Lambda(n)/n$ from the one of $\Lambda(n)$. Acta Arith. **159**(2), 113–122 (2013)

35. O. Ramaré, From explicit estimates for the primes to explicit estimates for the Moebius function. Acta Arith. **157**(4), 365–379 (2013)

36. O. Ramaré, Explicit estimates on the summatory functions of the Moebius function with coprimality restrictions. Acta Arith. **165**(1), 1–10 (2014)

37. O. Ramaré, Explicit estimates on several summatory functions involving the Moebius function. Math. Comp. **84**(293), 1359–1387 (2015)

38. O. Ramaré, Quantitative steps in Axer-Landau equivalence theorem (2017, submitted), 9pp.

39. J.B. Rosser, L. Schoenfeld, Approximate formulas for some functions of prime numbers. Ill. J. Math. **6**, 64–94 (1962)

40. L. Schoenfeld, An improved estimate for the summatory function of the Möbius function. Acta Arith. **15**, 223–233 (1969)

41. S. Selberg, Über die Summe $\sum_{n \leqslant x} \frac{\mu(n)}{nd(n)}$, in *Tolfte Skandinaviska Matematikerkongressen, Lund* (Lunds Universitets Matematiska Institution, Lund, 1954), pp. 264–272

42. D.W. Swann, Some new classes of kernels whose Fredholm determinants have order less than one. Trans. Am. Math. Soc. **160**, 427–435 (1971)

43. T. Tao, A remark on partial sums involving the Möbius function. Bull. Aust. Math. Soc. **81**(2), 343–349 (2010) (English)

44. The PARI Group, Bordeaux, PARI/GP, version 2.7.0 (2014), http://pari.math.u-bordeaux.fr/
45. R. Vanlalngaia, Fonctions de Hardy des séries L et sommes de Mertens explicites. Ph.D. thesis, Mathématique, Lille (2015), http://math.univ-lille1.fr/~ramare/Epsilons/theseRamdinmawiaVanlalngaia.pdf
46. W.B. Zhang, A generalization of Halász's theorem to Beurling's generalized integers and its application. Ill. J. Math. **31**(4), 645–664 (1987)

Chapter 16
Chowla's Conjecture: From the Liouville Function to the Moebius Function

Olivier Ramaré

16.1 Historical Setting

In Problem 57 of the book [1] (see equation (341) therein), S. Chowla formulated the following conjecture, where, by "integer polynomial", we understand "an element of $\mathbb{Z}[X]$".

Conjecture 16.1 (Chowla's Conjecture for the Liouville Function) For any integer polynomial $f(x)$ that is not of the form $cg(x)^2$ for some integer polynomial $g(x)$, one has

$$\sum_{n \leqslant x} \lambda(f(n)) = o(x)$$

where λ is the Liouville function.

This conjecture has then been stated and formulated in many different forms, very often by restricting f to be a product of linear factors as by T. Tao in [5], and even more often to be a product of monic linear factors. Furthermore, the Liouville function λ is sometimes replaced by the Möbius function μ as by P. Sarnak in [4]. The aim of this note is to establish some links between these conjectures.

O. Ramaré (✉)
Aix Marseille Université, CNRS, Centrale Marseille, Institut de Mathématiques de Marseille, I2M – UMR 7373, Marseille, France
e-mail: olivier.ramare@univ-amu.fr

© Springer International Publishing AG, part of Springer Nature 2018
S. Ferenczi et al. (eds.), *Ergodic Theory and Dynamical Systems in their Interactions with Arithmetics and Combinatorics*, Lecture Notes in Mathematics 2213, https://doi.org/10.1007/978-3-319-74908-2_16

We should also mention that some authors, like A. Hildebrand in [2] or K. Matomäki, M. Radziwiłł and T. Tao in [3], refer to another closely connected conjecture of Chowla that states that the sequence $(\lambda(n), \lambda(n+1), \cdots, \lambda(n+k))$ may take any value in $\{\pm 1\}^k$. This is indeed Problem 56 of [1].

Here are different forms that can be called "Chowla's conjecture".

Conjecture 16.2 For any finite tuple $((a_i, b_i))_{i \in I}$ of positive integers such that $a_i b_j - a_j b_i \neq 0$ as soon as $i \neq j$, we have

$$\sum_{n \leqslant x} \prod_{i \in I} \lambda(a_i n + b_i) = o(x).$$

Remark 16.1 Note that when the condition of the $((a_i, b_i))_{i \in I}$ is verified, it is also verified for the coefficients after the substitution $n \mapsto pn + q$.

Conjecture 16.3 For any positive integer a and any strictly increasing finite sequence $(b_i)_{i \in I}$ of non-negative integers, we have

$$\sum_{n \leqslant x} \prod_{i \in I} \lambda(an + b_i) = o(x).$$

Conjecture 16.4 For any finite tuple $((a_i, b_i))_{i \in I}$ of positive integers such that $a_i b_j - a_j b_i \neq 0$ as soon as $i \neq j$, we have

$$\sum_{n \leqslant x} \prod_{i \in I} \mu(a_i n + b_i) = o(x).$$

Conjecture 16.5 For any positive integer a and any strictly increasing finite sequence $(b_i)_{i \in I}$ of non-negative integers, we have

$$\sum_{n \leqslant x} \prod_{i \in I} \mu(an + b_i) = o(x).$$

Conjecture 16.6 For any finite tuple $((a_i, b_i))_{i \in I}$ of positive integers such that $a_i b_j - a_j b_i \neq 0$ as soon as $i \neq j$, and any additional finite tuple $((c_k, d_k))_{k \in K}$ we have

$$\sum_{n \leqslant x} \prod_{i \in I} \mu(a_i n + b_i) \prod_{k \in K} \mu^2(c_k n + d_k) = o(x)$$

provided the set I be non-empty.

Conjecture 16.7 For any positive integer a and any strictly increasing finite sequences $(b_i)_{i\in I}$ and $(d_k)_{k\in K}$ of non-negative integers, we have

$$\sum_{n\leqslant x}\prod_{i\in I}\mu(an+b_i)\prod_{k\in K}\mu^2(an+d_k) = o(x)$$

provided the set I be non-empty.

In the ergodic context, the hypothesis $\sum_{n\leqslant x}\prod_{i\in I}\mu(n+b_i)\prod_{k\in K}\mu^2(n+d_k) = o(x)$ is often seen, with some natural conditions on the parameters. Since it is applied to powers T^a of a same operator T, the proper statements are really Conjectures 16.2, 16.4 and 16.6.

The reader may wonder whether the non-negativity condition concerning the parameters b_i is restrictive or not. It is not and we can reduce to this case by a suitable shift of the variable n. This would implies discarding finitely many terms is the diverse sums we consider and increase the initial b_i by $b_i + (N_0 - 1)a_i$, assuming we replace n by $n+N_0-1$; there clearly exists an N_0 for which all the $b_i + (N_0 - 1)a_i$ are positive.

Theorem 16.8

- *Conjecture 16.2 implies Conjecture 16.4.*
- *Conjecture 16.4 implies Conjecture 16.6.*
- *Conjecture 16.3 implies Conjecture 16.5.*
- *Conjecture 16.5 implies Conjecture 16.7.*
- *Conjecture 16.6 implies Conjecture 16.2.*

Note that we have not been able to prove that Conjecture 16.7 implies Conjecture 16.3.

16.2 Lemmas

Lemma 16.9 *Let $(f(n))_n$ be a complex sequence such that $|f(n)| \leqslant 1$, let a and b be fixed and assume that, for every u and w, one has*

$$\sum_{n\leqslant y}\lambda(a(u^2n+w)+b)f(u^2n+w) = o_{u,w}(y). \tag{16.1}$$

Then $\sum_{n\leqslant x}\mu(an+b)f(n) = o(x)$.

Proof We readily check on the Dirichlet series that

$$\mu(m) = \sum_{u^2v=m}\mu(u)\lambda(v) \tag{16.2}$$

We define $X = ax + b$. We infer from this identity that, for any positive integer U, one has

$$\sum_{n \leqslant x} \mu(an + b)f(n) = \sum_{m \leqslant ax+b} \mu(m)f\left(\frac{m - b}{a}\right)\mathbb{1}_{m \equiv b[a]}$$

$$= \sum_{u \leqslant \sqrt{ax+b}} \mu(u) \sum_{v \leqslant (ax+b)/u^2} \lambda(v)f\left(\frac{u^2v - b}{a}\right)\mathbb{1}_{u^2v \equiv b[a]}$$

$$= \sum_{u \leqslant U} \mu(u) \sum_{v \leqslant (ax+b)/u^2} \lambda(v)f\left(\frac{u^2v - b}{a}\right)\mathbb{1}_{u^2v \equiv b[a]} + O^*\left(\sum_{u > U} X/u^2\right)$$

since

$$\left|\sum_{v \leqslant (ax+b)/u^2} \lambda(v)f\left(\frac{u^2v - b}{a}\right)\mathbb{1}_{u^2v \equiv b[a]}\right| \leqslant \sum_{v \leqslant X/u^2} 1 \leqslant X/u^2.$$

We next recall that a comparison to an integral ensures us that $\sum_{u > U} u^{-2} \leqslant 1/U$ (since U is an integer). Since we want the reader to follow as closely as possible the argument, we also use the notation $f = O^*(g)$ to mean that $|f| \leqslant g$. We have reached

$$\sum_{n \leqslant x} \mu(an + b)f(n) = \sum_{u \leqslant U} \mu(u) \sum_{v \leqslant (ax+b)/u^2} \lambda(u^2v)f\left(\frac{u^2v - b}{a}\right)\mathbb{1}_{u^2v \equiv b[a]} + O^*(X/U)$$

$$= \sum_{u \leqslant U} \mu(u) \sum_{\substack{m \leqslant x, \\ am+b \equiv 0[u^2]}} \lambda(am + b)f(m) + O^*(X/U)$$

(with $am = u^2v - b$) from which we infer that

$$\left|\sum_{n \leqslant x} \mu(an + b)f(n)\right| \leqslant \sum_{u \leqslant U} \left|\sum_{\substack{m \leqslant x, \\ am+b \equiv 0[u^2]}} \lambda(am + b)f(m)\right| + \frac{X}{U}.$$

The set $\{m/am + b \equiv 0[u^2]\}$ is a finite union of arithmetic progressions modulo u^2, say $\mathcal{W}$, hence

$$\sum_{\substack{m \leqslant x, \\ am+b \equiv 0[u^2]}} \lambda(am + b)f(m) = \sum_{\substack{0 \leqslant w < u^2, \\ w \in \mathcal{W}}} \sum_{k \leqslant \frac{ax+b-w}{u^2}} \lambda(a(u^2k + w) + b)f(u^2k + w).$$

Our hypothesis applies to the inner sum. The remainder of the proof is mechanical. First note that $X \leqslant (a+b)x$. Let $\varepsilon > 0$ be fixed. We select

$$U = \left[\frac{2}{(a+b)\varepsilon} \right] + 1 \leqslant \frac{3}{(a+b)\varepsilon}.$$

There exists $y_0(a, b, \varepsilon)$ such that, for every $u \leqslant U$, every $w \in \mathcal{W}$ and every $y \geqslant y_0(a, b, \varepsilon)$, one has

$$\left| \sum_{n \leqslant y} f(a^2 n) \right| \leqslant \tfrac{1}{6}\varepsilon^2 y.$$

We thus assume that $x \geqslant y_0(a, b, \varepsilon)/(a+b)$ and get, for such an x that

$$\left| \sum_{n \leqslant x} \mu(an+b)f(n) \right| \leqslant x\left(\frac{1}{6}\frac{3}{\varepsilon}\varepsilon^2 + \frac{\varepsilon}{2} \right) \leqslant \varepsilon x$$

(where we have bounded above $|\mathcal{W}|$ by u^2) as required.

Lemma 16.10 *Let $(f(n))_n$ be a sequence sur that $|f(n)| \leqslant 1$, let a and b be fixed and, assume that, for every u and w, one has*

$$\sum_{n \leqslant y} f(u^2 n + w) = o_{u,w}(y). \tag{16.3}$$

Then $\sum_{n \leqslant x} \mu^2(an+b)f(n) = o(x)$.

Proof We use the identity

$$\mu^2(n) = \sum_{d^2 \mid n} \mu(d).$$

After this initial step, the proof runs as the one of Lemma 16.9.

Lemma 16.11 *Let $(f(n))_n$ be a sequence sur that $|f(n)| \leqslant 1$, let a and b be fixed and, assume that, for every u and c such that $cu^2 \equiv b[a]$, one has*

$$\sum_{n \leqslant y} \mu(an+c)f\left(u^2 n + \frac{cu^2 - b}{a} \right) = o_{u,w}(y). \tag{16.4}$$

Then $\sum_{n \leqslant x} \lambda(an+b)f(n) = o(x)$.

Proof We readily check on the Dirichlet series that

$$\lambda(m) = \sum_{u^2v=m} \mu(v) \qquad (16.5)$$

We infer from this identity that, for any positive integer U, one has

$$\sum_{n \leqslant x} \lambda(an+b)f(n) = \sum_{m \leqslant ax+b} \lambda(m)f\left(\frac{m-b}{a}\right)\mathbb{1}_{m \equiv b[a]}$$

$$= \sum_{u \leqslant \sqrt{ax+b}} \sum_{v \leqslant (ax+b)/u^2} \mu(v)f\left(\frac{u^2v-b}{a}\right)\mathbb{1}_{u^2v \equiv b[a]}$$

$$= \sum_{u \leqslant U} \sum_{v \leqslant (ax+b)/u^2} \mu(v)f\left(\frac{u^2v-b}{a}\right)\mathbb{1}_{u^2v \equiv b[a]} + O^*(X/U)$$

again with $X = ax + b$. Once u is fixed, the set $\{v/u^2v \equiv b[a]\}$ is a finite union of arithmetic progressions modulo a, say $\mathcal{W}$, hence

$$\sum_{v \leqslant (ax+b)/u^2} \mu(v)f\left(\frac{u^2v-b}{a}\right)\mathbb{1}_{u^2v \equiv b[a]} = \sum_{\substack{0 \leqslant v_0 < a, \\ v_0 \in \mathcal{W}}} \sum_{k \leqslant \frac{ax+b}{au^2} - \frac{v_0}{a}} \mu(ak+v_0)f\left(u^2k + \frac{u^2v_0-b}{a}\right).$$

Our hypothesis applies to the inner sum. The remainder of the proof is mechanical.

16.3   Proof of the Theorem 16.8

Proof To prove that Conjecture 16.2 implies Conjecture 16.4, we prove that, assuming Conjecture 16.2 and following its notation and hypotheses, we have (with the shortcut $a_{i_0} = a$ and $b_{i_0} = b$)

$$\sum_{n \leqslant x} \mu(an+b) \prod_{i \in I'} \mu(a_in+b_i) \prod_{i \in I''\setminus\{i_0\}} \lambda(a_in+b_i) = o(x)$$

assuming that

$$\sum_{n \leqslant x} \lambda(an+b) \prod_{i \in I'} \mu(A_in+B_i) \prod_{i \in I''\setminus\{i_0\}} \lambda(A_in+B_i) = o(x)$$

where none of the vectors (A_i, B_i) and (a, b) are colinear. We use a recursion on the cardinality of I' to do so. Lemma 16.9 is tailored for this purpose, the only part that needs checking is that no two vectors of new set of parameters $(au^2, aw + b)$ and

$(a_i u^2, a_i w + b_i)$ are colinear, which is immediate:

$$\begin{vmatrix} au^2 & a_i u^2 \\ aw + b & a_i w + b_i \end{vmatrix} = au^2(a_i w + b_i) - a_i u^2(aw + b) = u^2 \begin{vmatrix} a & a_i \\ b & b_i \end{vmatrix} \neq 0.$$

The same proof shows that Conjecture 16.3 implies Conjecture 16.5: we simply have to note that the required coefficients a and a_i's remain the same.

The same proof again shows that Conjecture 16.4 implies Conjecture 16.6, we simply have to replace λ by 1, and Lemma 16.9 by Lemma 16.10, in its proof and similarly that Conjecture 16.5 implies Conjecture 16.7.

Let us finally turn towards the proof that Conjecture 16.4 implies Conjecture 16.2, a task for which we will use Lemma 16.11. We aim at proving that (again with the shortcut $a_{i_0} = a$ and $b_{i_0} = b$)

$$\sum_{n \leqslant x} \lambda(an + b) \prod_{i \in I'} \lambda(a_i n + b_i) \prod_{i \in I'' \setminus \{i_0\}} \mu(a_i n + b_i) = o(x)$$

assuming that

$$\sum_{n \leqslant x} \mu(an + b) \prod_{i \in I'} \lambda(A_i n + B_i) \prod_{i \in I'' \setminus \{i_0\}} \mu(A_i n + B_i) = o(x)$$

where none of the vectors (A_i, B_i) and (a, b) are colinear. We use a recursion on the cardinality of I' to do so. The only part that needs checking is the hypothesis in Lemma 16.11, namely that no two vectors of new set of parameters (a, c) and $(a_i u^2, a_i \frac{cu^2 - b}{a} + b_i)$ are colinear, which is immediate:

$$\begin{vmatrix} a & a_i u^2 \\ c & a_i \frac{cu^2 - b}{a} + b_i \end{vmatrix} = a\left(a_i \frac{cu^2 - b}{a} + b_i\right) - a_i u^2 c = \begin{vmatrix} a & a_i \\ b & b_i \end{vmatrix} \neq 0.$$

This concludes the proof of our Theorem.

References

1. S. Chowla, *The Riemann Hypothesis and Hilbert's Tenth Problem.* Mathematics and Its Applications, vol. 4 (Gordon and Breach Science Publishers, New York, 1965)
2. A. Hildebrand, On consecutive values of the Liouville function. Enseign. Math. (2) **32**(3–4), 219–226 (1986)
3. K. Matomäki, M. Radziwiłł, T. Tao, Sign patterns of the Liouville and Möbius functions. Forum Math. Sigma **4**, e14, 44 (2016)
4. P. Sarnak, Three lectures on the Möbius function randomness and dynamics. Technical report, Institute for Advanced Study (2011)
5. T. Tao, The logarithmically averaged Chowla and Elliott conjectures for two-point correlations. Forum Math. Pi. **4**, e8, 36 (2016)

Part III
Selected Topics in Dynamics

Chapter 17
Weak Mixing for Infinite Measure Invertible Transformations

Terrence Adams and Cesar E. Silva

17.1 Historical Perspective

Ergodic theory is a relatively modern branch of mathematics that studies statistical properties of time-varying systems. The term "ergodic" was coined by Ludwig Boltzmann while doing foundational work in statistical mechanics. The field of ergodic theory continues to see applications in areas as diverse as billiard dynamics, geodesic flows, smooth dynamics, group actions, combinatorial number theory and operator theory. At its core, ergodic theory focuses on properties of a single invertible transformation T that preserves an invariant measure μ. Statistical analysis for the transformation T entails a study of ergodic components (i.e., sets of positive measure X such that $\mu(TX \setminus X) = 0$). The case where X has finite measure, or equivalently can be scaled to a probability space, is the most tractable case, and enjoys many significant advances. In particular, many problems studied today from statistical learning theory can be modeled in the general setting of a process $f(x), f(Tx), f(T^2x), \ldots$ where T is an invertible measure-preserving transformation, and f maps into a compact metric space.[1]

---

[1] While the area of smooth dynamics is an important area for ergodic theory, we will not impose differentiability constraints in this survey. In practical terms, non-differentiable dynamical systems are seeing rapid applications as recurrent neural networks with programmed non-differentiable

T. Adams
U.S. Government, Ft. Meade, MD, USA
e-mail: terry@ganita.org

C. E. Silva (✉)
Department of Mathematics, Williams College, Williamstown, MA, USA
e-mail: csilva@williams.edu

© Springer International Publishing AG, part of Springer Nature 2018
S. Ferenczi et al. (eds.), *Ergodic Theory and Dynamical Systems in their Interactions with Arithmetics and Combinatorics*, Lecture Notes in Mathematics 2213, https://doi.org/10.1007/978-3-319-74908-2_17

The study of infinite measure-preserving transformations dates back to the early days of ergodic theory, as one can see mentions of it in Hopf's 1937 book [65], where an example of a transformation, later called Krickeberg mixing [4, 72], is constructed. Also, Hopf proved a ratio ergodic theorem for infinite measure-preserving transformations, extending the Birkhoff ergodic theorem.

The weak mixing condition is a notion in ergodic theory that lies between ergodicity and mixing. In the case of finite invariant measure, which is the main case that has been traditionally considered in ergodic theory, weak mixing has several equivalent formulations, many of which on a first look seem quite different; this is one of the reasons why the weak mixing property has had many applications. We know, since the work of Kakutani and Parry in 1963 [68], that in infinite invariant measure the situation is quite different and that many of these equivalences do not hold.

In this article we survey several weak mixing-type notions and discuss their equivalence, or counterexamples showing that they are different. We are mainly concerned with measure-preserving integer actions (i.e., invertible transformations), where we already can see different behavior in the finite and infinite cases; the questions have been studied for nonsingular actions of more general groups and we provide some references to that work. Our counterexamples are in the class of rank-one transformations, so we first discuss this class in some detail.

17.2 Preliminary Definitions

We will let $(X, \mathcal{B}, \mu)$ be a standard σ-finite measure space, i.e., $(X, \mathcal{B})$ is Borel isomorphic to the unit interval (or the real line) and μ is a σ-finite measure on $(X, \mathcal{B})$. Furthermore, we will assume μ is nonatomic, i.e., if $\mu(A) > 0$ then there exists a measurable set $B \subset A$ such that $0 < \mu(B) < \mu(A)$. By an invertible measurable transformation T on X we mean a map $T : X \rightarrow X$ such that A is measurable if and only if $T^{-1}(A)$ is measurable, and there exists a measure zero set N so that $T : X \setminus N \rightarrow X \setminus N$ is a bijection. It follows that for every measurable set A and all $n \in \mathbb{Z}$, the set $T^n A$ is measurable. All transformations will be assumed to be invertible measurable transformations. A transformations T is **nonsingular** if for all measurable sets A, $\mu(A) = 0$ if and only if $\mu(TA) = 0$, and it is **measure-preserving** if or all measurable sets A, $\mu(A) = \mu(TA)$. All transformations will be assumed measure-preserving unless we explicitly say they are nonsingular; they are **finite measure-preserving** when $\mu(X) < \infty$, in which case we will always assume $\mu(X) = 1$ and call them **probability-preserving**.

We say that transformation T is **ergodic** if whenever $\mu(A \triangle TA) = 0$ we have that $\mu(A) = 0$ or $\mu(A^c) = 0$ (where $A^c = X \setminus A$), and is **totally ergodic** if

---

properties such as rectified linear units and max pooling. Also, neurological spiking is naturally modeled as a discontinuous process.

T^n is ergodic for all $n \in \mathbb{Z} \setminus \{0\}$. A transformation T is **conservative** if for all sets A of positive measure there exists $n > 0$ such that $\mu(T^n A \cap A) > 0$. All finite measure-preserving transformations are conservative, a fact equivalent to the Poincaré recurrence theorem (see, e.g., [80, 87]), but there exist non-conservative transformations in infinite measure; for example $T(x) = x + 1$ on $\mathbb{R}$ with Lebesgue measure is measure-preserving but not conservative. On the other hand, as our measure is nonatomic, an ergodic invertible transformation is conservative. To see this suppose that $\mu(A \cap T^n A) = 0$ for all $n \neq 0$. There exists B such that $0 < \mu(B) < \mu(A)$. Then if we let $B^* = \cup_{n \in \mathbb{Z}} T^n B$, it follows that $\mu(B^*) > 0$ and $\mu(B^{*c}) > 0$, so T would not be ergodic, a contradiction. We also have that T is ergodic if and only if for all sets A and B of positive measure, there exists $n \in \mathbb{Z}$ such that $\mu(T^n A \cap B) > 0$. Then T is conservative and ergodic if and only if for all sets A and B of positive measure, there exists $n \in \mathbb{Z}, n > 0$, such that $\mu(T^n A \cap B) > 0$.

A transformation T is **partially rigid** if there exists a constant $c > 0$ such that for all sets A of finite measure

$$\limsup \mu(T^n A \cap A) \geq c\mu(A).$$

The transformation T is **rigid** if there exists an increasing sequence (n_i) such that $\lim_{i \to \infty} \mu(T^{n_i} A \triangle A) = 0$ for all sets A of finite measure; this is equivalent to partial rigidity with constant $c = 1$, see e.g., [23, 2.6].

Lemma 17.1 *If a transformation T is partially rigid, with partial rigidity constant $c > 0$, then for any $n \neq 0$, T^n is partially rigid with partial rigidity constant (at least) c^n. In particular, if T is rigid, then T^n is rigid.*

Proof The case $n = 2$ illustrates the main idea. (Of course, when $n = 0$ we obtain the identity, which is rigid.) Assume $c > 0$ is the partial rigidity constant for T. Let A be a set of finite measure. We know there is an increasing sequence (n_i) such that, for $i > 0$,

$$\mu(T^{n_i} A \cap A) \geq c\mu(A).$$

Write $n_i = 2k_i + r_i$, where $r_i \in \{0, 1\}$. If for infinitely may i we have that $r_i = 0$, then $\mu(T^{2k_i} A \cap A) \geq c\mu(A)$ and we are done. If for infinitely may i we have that $r_i = 1$, then there is a (say first) integer k_1 such that $\mu(T^{2k_1+1} A \cap A) \geq c\mu(A)$. Then using the partial rigidity of T with the set $T^{2k_1+1} A \cap A$, we have that there is an infinite sequence $(2m_i + q_i), q_i \in \{0, 1\}$, such that

$$\mu(T^{2m_i+q_i} A \cap T^{2k_1+1} A \cap A) \geq c\mu(T^{2k_1+1} A \cap A) \geq c^2 \mu(A).$$

If for infinitely many i we have that $q_i = 0$, then $\mu(T^{2m_i} A \cap A) \geq c^2 \mu(A)$, and we are done. If for infinitely many i we have that $q_i = 1$, then

$$\mu(T^{2m_i+1} A \cap T^{2k_1+1} A) \geq c^2 \mu(A),$$

so $\mu(T^{2(m_i-k_1)} A \cap A) \geq c^2 \mu(A)$, completing the proof.

A similar argument can be used to give a proof of the result of Halmos [63] that if T is conservative then T^n is conservative for all $n \in \mathbb{Z}$.

When $(X, \mathcal{B}, \mu)$ is a probability space we let $\mathcal{G}_1$ be the group of invertible measure-preserving transformations on $(X, \mathcal{B}, \mu)$ (where transformations are identified if they differ on a null set), and when $(X, \mathcal{B}, \mu)$ is an infinite, σ-finite measure space, we let $\mathcal{G}_\infty$ be the group of invertible measure-preserving transformations on $(X, \mathcal{B}, \mu)$. Then $\mathcal{G}_1$ and $\mathcal{G}_\infty$ can be endowed with a complete separable metric (see [61, 64] for the probability case and [3, 86] for the infinite case). Under this metric a sequence of transformations T_n converges to T if and only if

$$\mu(T_n A \,\triangle\, TA) \to 0 \text{ for all finite measure } A.$$

Also, under this topology, called the weak topology, the ergodic transformations are generic, or typical, (i.e., residual in the group of invertible measure-preserving transformations); this was shown by Halmos [61] in the finite case, and Sachdeva [86] in the infinite case. These results were extended to the nonsingular case in [27]. Genericity of the rigid transformations in the finite case is well known, see e.g., [23]; the infinite case is in [17].

17.3 Mixing Properties for Probability-Preserving Transformations

A probability-preserving transformation T is **strong mixing**[2] if for all measurable sets A and B,

$$\lim_{n \to \infty} \mu(A \cap T^n B) = \mu(A)\mu(B).$$

Probably the best known mixing transformations are Bernoulli shifts. Projection onto a single coordinate produces an independent identically distributed (i.i.d.) process. These are arguably the oldest studied mixing processes, dating back to "I Ching" (1150 B.C.) in terms of coin flips, and again to Cardano, Pascal, Fermat and Huygens for dice rolling.

However, zero entropy mixing transformations were not known until Ornstein constructed a rank-one mixing transformation using random spacers [78]. An explicit rank one transformation known as the staircase transformation, was conjectured by Smorodinsky to be mixing, and shown in [11] to be strong mixing.

While strong mixing has been observed for centuries, there is a sense in which it is not typical: the collection of strong mixing transformations is meager in this collection, see e.g., [82]. However, if the property of strong mixing is weakened

---

[2]Strong mixing is often referred to as mixing.

to the following characterization, then this property is typical [62, 64]: for all measurable sets A, B,

$$\lim_{n \to \infty} \frac{1}{n} \sum_{i=0}^{n-1} |\mu(A \cap T^i B) - \mu(A)\mu(B)| = 0. \tag{17.1}$$

This is known as **weak mixing**.

Another characterization of strong mixing uses the Koopman operator. Let $< \cdot, \cdot >$ denote the inner product in $L^2(X, \mu)$ given by $< f, g >= \int f\bar{g} \, d\mu$. Then T is strong mixing if and only if for all f, g in $L^2(X, \mu)$

$$\lim_{n \to \infty} < U^n f, g >=< f, 1 >< g, 1 > .$$

This can be shown equivalent to the fact that $U^n f$ converges weakly to 0 for all f in the orthogonal complement of the constants, $\mathbb{C}^\perp$, see e.g., [80].

17.3.1   Mixing Property Potpourri

Many other weaker variants of strong mixing were introduced during the study of probability-preserving transformations. These include topological mixing, mild mixing, light mixing and partial mixing. We will not go into details on these properties, but provide definitions below, and point out the relationship among these properties.

A transformation T is **topologically mixing**, if given open sets U and V, there exists $N \in \mathbb{N}$ such that for $n \geq N$,

$$\mu(U \cap T^n V) > 0. \tag{17.2}$$

A transformation T is **lightly mixing** if given sets A and B of positive measure,

$$\liminf_{n \to \infty} \mu(A \cap T^n B) > 0. \tag{17.3}$$

Given open sets have positive measure, then light mixing implies topological mixing. A transformation T is **partially mixing** if there exists $\alpha > 0$ such that for all sets A and B of positive measure,

$$\liminf_{n \to \infty} \mu(A \cap T^n B) \geq \alpha \mu(A)\mu(B). \tag{17.4}$$

Clearly, strong mixing implies partial mixing, and partial mixing implies light mixing, but one can have a countable product of partially mixing transformations that is not partially mixing but is lightly mixing [69]. Partial mixing does not imply strong mixing [49]. A transformation T is **mildly mixing** if for every set A of positive measure,

$$\liminf_{n \to \infty} \mu(A^c \cap T^n A) > 0. \tag{17.5}$$

Note $A^c = X \backslash A$ is the complement of A. So light mixing implies mild mixing, but the classic Chacón transformation [46] is mildly mixing but not lightly mixing [48] (but the two-subintervals at each stage Chacón transformation is lightly mixing [48]). Mild mixing implies weak mixing, but there are weakly mixing transformations that are not mildly mixing [51]. Furthermore, a probability-preserving transformation T is mildly mixing if and only if for all ergodic infinite measure-preserving transformations S, the transformation $T \times S$ is ergodic.

17.3.2 Weak Mixing Equivalences

In the probability-preserving scenario, there are several well known conditions that are equivalent to the weak mixing formulation given in (17.1). We now state those conditions, noting that they can be stated both when T is probability-preserving, or infinite measure-preserving. We remark, though, that while here we are interested in infinite measure-preserving transformations, these conditions can also be stated in the nonsingular case, and for actions of more general groups.

(WM) A transformation T is said to be **weakly mixing** if for all ergodic finite measure-preserving transformations S, the product transformation $T \times S$ is ergodic.

(DE) A transformation T is said to be **doubly ergodic**, or **Cartesian square ergodic**, if the product transformation $T \times T$ is ergodic.

(SWM) A transformation T is said to be **spectrally weak mixing**, if for all $f \in L^\infty$, whenever $f \circ T = zf$, for some $z \in \mathbb{C}$, then f is constant a.e.

(WDE) A transformation T is **weakly doubly ergodic** if for all sets A, B of positive measure there exists an integer $n > 0$ such that $\mu(T^n A \cap A) > 0$ and $\mu(T^n A \cap B) > 0$.

(EIC) A transformation T is **ergodic with isometric coefficients**, or **EIC**, if whenever $\phi : X \to Z$ is a factor map where Z is a metric space and the factor transformation is an isometry for the metric, then ϕ is constant, where the factor map is considered in the nonsingular sense.

(EUC) A transformation T is **ergodic with unitary coefficients**, or **EUC**, if whenever $\phi : X \to H$ is a factor map where H is a separable Hilbert space and the factor transformation is a unitary operator, then ϕ is constant, where the factor map is considered in the nonsingular sense.

(k-EI) A transformation is said to have **ergodic index** k, for some positive integer k, if the k-fold Cartesian product of T with itself, denoted by $T^{(k)}$, is ergodic but the $k + 1$-fold product is not.

(IEI) $T^{(k)}$ is ergodic for all $k > 0$; such a transformation is said to be of **infinite ergodic index**.

(PWM)   A transformation T is **power weakly mixing** if for all nonzero $k_1, \ldots, k_k$, the transformation $T^{k_1} \times \cdots \times T^{k_r}$ is ergodic.

Here we discuss the equivalence of these notions only in the probability-preserving case. The infinite measure-preserving case, as well as additional definitions, are discussed in Sect. 17.6. The equivalence of condition (17.1) with WM and SWM is standard and can be found in e.g., [64, 80]. WM clearly implies DE as one can take $S = T$, and thus it also implies IEI and PWM. It is clear that DE implies SWM; the converse usually uses the Spectral Theorem, [80]. The equivalence of WM with WDE is in [50], though the WDE property is not given a name. EIC and EUC are in [41], and discussed in [53], and references therein, for nonsingular actions of more general groups. For other equivalences in the probability-preserving case, and other group actions, see [22].

17.4   The Terrain of Infinite Rank-One

We describe an important class of transformations that has been used as a source of examples and counterexamples. There are many equivalent definitions of this notion and we refer the reader to [45] for a survey of rank-one transformations in finite measure. Here we will mainly be concerned with infinite measure rank-one transformations. As before, $(X, \mathcal{B}, \mu)$ will be a standard nonatomic, σ-finite measure space. The reader could think of X as the unit interval or the positive real line, and μ Lebesgue measure.

Many of the examples in this survey are rank-one transformations. We start with the abstract definition, which is similar to the finite case. We use the notion of a **column**, or **Rohlin column**, or **Rohlin tower**, which consists of a finite sequence disjoint measurable sets $B, T(B), \ldots, T^{h-1}(B)$ of the same measure; here h is said to be the **height** of the column. We say that T is **rank-one** if there exists a sequence of columns $C_n = \{B_n, \ldots, T^{h_n-1}(B_n)\}$ such that for any measurable set $A \subset X$ of finite measure and $\varepsilon > 0$, there exists an integer N such that for all $n \geq N$, we have that

$$\mu(A \triangle C_n(A)) < \varepsilon,$$

where $C_n(A)$ is a union of levels of C_n with minimal $\mu(A \triangle C_n(A))$. It can be shown that we can assume the base of $\{C_n\}$ is a union of elements of C_{n+1} and $B_{n+1} \subset B_n$; see [20, Lemma 9] for finite measure, and [25] for a verification that essentially the same proof works in infinite measure. Under the weak topology, the infinite measure-preserving rank-one transformations are generic [25].

There is also a constructive definition of rank-one transformations; see [25] for the infinite measure case. A column in this case consists of a finite sequence of intervals of the same length, sometimes also called levels, that we usually take left-closed and right-open; its height is the number of intervals in the column. A column C defines a column map T_C so that each interval is sent to the interval right above it by translation; so T_C is defined on every interval of the column except the top. It is clear that on the intervals where it is defined, T_C is measurable and measure-preserving with respect to Lebesgue measure. To specify a rank-one construction we are given a sequence of positive integers $(r_n)_{n \geqslant 0}$, $r_n \geqslant 2$, and a doubly indexed sequence of nonnegative integers $(s_{n,i})$, $n \geqslant 0$, $i = 0, \ldots, r_n - 1$. We start with column C_0 consisting of a single finite interval, so its height is $h_0 = 1$. In all of our infinite measure-preserving constructions we will take $C_0 = \{[0, 1)\}$. For the inductive step, suppose we are given column C_n of height h_n. Then cut each interval of column C_n into r_n subintervals of equal length to obtain r_n subcolums of equal measure. For each $i = 0, \ldots, r_n - 1$, place $s_{n,i}$ new intervals, called **spacers**, above the ith subcolumn. Then C_{n+1} is constructed by stacking each subcolumn, right over left, with its associated spacers under the next subcolumn. The example below shows this in more detail. Thus C_{n+1} will consist of r_n copies of C_n possibly separated by spacers. From the construction is it clear that $T_{C_{n+1}}$ agrees with T_{C_n} wherever T_{C_n} is defined. We let X be the union of the levels in the columns and $T(x) = \lim_{n \to \infty} T_{C_n}$. The height of $T_{C_{n+1}}$ is given by

$$h_{n+1} = r_n h_n + \sum_{i=0}^{r_n-1} s_{n,i}.$$

As an example of an infinite measure-preserving rank-one construction we give the Hajian-Kakutani transformation [59], an early example of a rank-one infinite measure-preserving transformation; see also [40]. We start with a column C_0 consisting of the unit interval $[0, 1)$. For the inductive step, given a column C_n of h_n intervals or levels, of the same length, cut each interval in half to obtain two subcolumns $C_{n,0}$ and $C_{n,1}$. Create a new column S_n consisting of $2h_n$ new intervals, each half the length of the intervals in C_n, so each of the same length as the intervals in $C_{n,0}$ and $C_{n,1}$. The intervals in S_n are chosen disjoint from the intervals in C_n. To form column C_{n+1} place $C_{n,1}$ on top of $C_{n,0}$, and S_n on top of $C_{n,1}$. So the top interval in $C_{n,0}$ is sent to the bottom interval of $C_{n,1}$, and the top interval in $C_{n,1}$ is sent to the bottom interval of S_n. In this case $r_n = 2$ and $s_{n,0} = 0$, $s_{n,1} = 2h_n$ for all $n \geqslant 0$. It follows that $h_{n+1} = 4h_n$. Figure 17.1 illustrates this step of the construction. The resulting transformation T is defined on $[0, \infty)$.

All rank-one transformations invertible measure-preserving, and ergodic; for the finite case see [45] for this and other properties, and for the infinite case see [88], and where other examples can also be found. It follows that the Hajian-Kakutani transformation is ergodic and infinite measure-preserving (it is conservative as well since it is invertible). If one takes a set E consisting of the even levels of every column, we find a set that is invariant for T^2, so T^2 is not ergodic, so the Hajian-Kakutani transformation is not totally ergodic.

Fig. 17.1 Construction of the
Hajian-Kakutani transformation

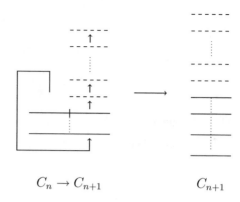

$$C_n \to C_{n+1} \qquad\qquad C_{n+1}$$

17.5   Swimming in Infinite Measure

17.5.1   Weakly Wandering Sets and Obstacles to Strong Mixing

We state a special case of the ergodic theorem, which is all that we need, see e.g.,
[80, 87].

Theorem 17.2 (Birkhoff   Ergodic   Theorem–Infinite   Measure   Case) *Let*
$(X, \mathcal{B}, \mu)$ *be a σ-finite measure space with $\mu(X) = \infty$ and let $T : X \to X$ be*
an ergodic measure-preserving transformation. If A is a measurable set of finite
measure, then there exists a null set N so that for all $x \in X \setminus N$,

$$\lim_{n\to\infty} \frac{1}{n} \sum_{i=0}^{n-1} \mathbb{I}_A(T^i x) = 0.$$

Then we have the following interesting consequence.

Corollary 17.3 *Let $(X, \mathcal{B}, \mu)$ be a σ-finite infinite measure space and $T : X \to X$*
be an ergodic measure-preserving transformation. If A, B are measurable sets of
finite measure, then

$$\lim_{n\to\infty} \frac{1}{n} \sum_{i=0}^{n-1} \mu(T^i A \cap B) = 0. \tag{17.6}$$

Proof By Theorem 17.2, there exists a null set $N \subset X$ so that for all x outside N,

$$\lim_{n\to\infty} \frac{1}{n} \sum_{i=0}^{n-1} \mathbb{I}_B(T^i x) = 0.$$

After multiplying both sides by $\mathbb{I}_A$ and integrating, the reader can verify that we obtain

$$\lim_{n\to\infty} \frac{1}{n} \sum_{i=0}^{n-1} \mu(T^{-i}B \cap A) = 0,$$

which is equivalent to (17.6).

It follows from (17.6) that there exists a sequence (n_k) so that

$$\lim_{k\to\infty} \mu(T^{n_k}A \cap B) = 0.$$

A consequence is that the standard definition of mixing behavior does not extend to infinite measure. Also, an ergodic infinite measure-preserving transformation cannot satisfy the light mixing condition in (17.3). We will, instead, explore conditions related to the weak mixing notion. As our measure spaces have a countable subset that approximates the measurable sets, one can in fact choose the sequence to work for all finite measure sets A, B.

Some authors have looked for conditions that hold in infinite measure and resemble mixing properties in finite measure. One such condition is already suggested by our discussion and is called Koopman mixing, and was investigated further by Krengel and Sucheston [71] and is mentioned below. Another condition was considered by Friedman [47], but because of the existence of weakly wandering sets, that condition is required to hold only for a dense family of sets, as it cannot hold for all measurable sets. Both of these conditions do not imply ergodicity, for example. A different definition has been proposed in [73].

Another important consequence of (17.6) is the following. For any set A of finite positive measure, one can further choose a strictly increasing sequence (n_i), $n_0 = 0$, so that for $0 \leqslant j < i$,

$$\mu(T^{n_i - n_j}A \cap A) < \frac{\mu(A)}{2^{i+j+2}}.$$

It follows that if we set

$$W = A \setminus \bigcup_{i \neq j} (T^{n_i}A \cap T^{n_j}A),$$

then

$$\mu(W) > 0 \text{ and } T^{n_i}W \cap T^{n_j}W = \emptyset, \text{ for } i \neq j.$$

Such a set W is called a **weakly wandering** set of positive measure, and they were proved to exist for every ergodic infinite measure-preserving transformation by Hajian and Kakutani [58]. The existence of weakly wandering sets of positive measure causes some obstructions when trying to generalize notions from the finite to the infinite case. See [44] for a discussion of extensions and related properties.

In the non-invertible case there are some notions, such as complete mixing, that do not hold for invertible transformations and the reader is referred to [3].

The notion of rank-one has been extended to other group actions; see [32] and references therein. Also, while the infinite measure-preserving case is already significantly different from the finite case, many of the properties we consider have been studied in the nonsingular case, and for actions of more general groups; the reader may refer to [30, 32, 37, 53, 54].

17.6   The Many Islands of Weak Mixing Notions

From now on we only consider invertible infinite measure-preserving transformations.

17.6.1   Conditions Weaker then Cartesian Square Ergodic

17.6.1.1   Weak Mixing and Spectral Weak Mixing

The equivalence of WM and SWM was proved in [9] (see below). WM transformations that are not DE were first constructed in [9] using Markov shifts; rank-one examples were constructed in [16]. SWM is also equivalent to no nonzero T invariant finite dimensional subspaces of L^∞.

If T is SWM, then T^n, $n \neq 0$ is SMW [77]; so if T is WM, then it must be totally ergodic. A totally ergodic infinite measure-preserving transformation that is not WM can be obtained by taking $T \times R$, where T is infinite measure-preserving WM and R is a probability-preserving irrational rotation. Then nontrivial eigenvalues of R give L^∞ eigenvalues for $T \times R$. As T^n, $n \neq 0$ is WM, so $T^n \times R^n$ is ergodic for all $n \neq 0$. The Hajian-Kakutani transformation gives an example of an ergodic, as it is rank-one, non-rigid, that is not totally ergodic, so not WM. A rank-one example that is rigid and not totally ergodic, so not WM, can be obtained as follows. Let $r_n = 2^n$, $s_{n,0} = \cdots = s_{n,r_n-2} = 0$ and $s_{n,r_n-1} = 2^n h_n$; then the rigidity sequence is (h_n) and T^2 is not ergodic.

If z satisfies $f \circ T = zf$ for some nonnull $f \in L^\infty$ we say it is an L^∞ **eigenvalue** of T and denote by $e(T)$ the set of all L^∞ eigenvalues of T. This set is a multiplicative Borel subgroup of the unit circle that has Lebesgue measure zero but can be uncountable, see e.g. [3]. The following theorem is very useful in understanding the SWM property; the analogue for probability-preserving transformations is well known, [83].

Theorem 17.4 (Keane, See [3]) *Let T be a conservative, ergodic infinite measure-preserving transformation and let S be an ergodic probability-preserving transformation. Then $T \times S$ is ergodic if and only if $\sigma_0(e(T)) = 0$, where σ_0 is the reduced maximal spectral type of S.*

A direct consequence is the following equivalence in infinite measure, first shown in [9].

Corollary 17.5 *T is WM if and only if it is SWM.*

The equivalence of WM with EUC is in [53].

17.6.1.2 Weakly Doubly Ergodic

The WDE property implies higher version of it as the following lemma shows.

Lemma 17.6 ([24]) *If T is weakly doubly ergodic, then for all sets A_i, B_i, $i = 1, \ldots, k$ of positive measure, there exists an integer $n > 0$ such that $\mu(T^n A_i \cap A_i) > 0$ and $\mu(T^n A_i \cap B_i) > 0$ for all $i \in \{1, \ldots, k\}$.*

If T is WDE, then so is T^k for all $k \neq 0$ [24]. If $T \times T$ is ergodic it is easy to see that T is WDE, and if T is WDE, then T is WM [24]. The WDE property for infinite measure (and nonsingular transformations) was defined in [24], where it was called *doubly ergodic*. We have already mentioned that this property, in the probability-preserving case and without a specific name, already appeared in [50], where it was shown equivalent to weak mixing. The property of having $T \times T$ ergodic has also been called doubly ergodic (see [53]), so to differentiate it, the property of [24] has been called weak doubly ergodic since [74]. A topological version of weak doubly ergodic was considered in [81]. A version of weak doubly ergodic for pairs of nonsingular transformations was considered in [42].

A modification of the Hajian-Kakutani transformation, called the HK+1 transformation, was constructed and shown to be SWM in [16]; it is clearly partially rigid, so it has infinite conservative index, but is it not WDE as shown in [24]. The construction of this transformation is similar to the Hajian-Kakutani transformation, except that at the nth stage the column of spacers S_n is chosen of height $2h_n + 1$ instead of height $2h_n$. A rigid SWM transformation that is not WDE is constructed in [21].

Infinite staircase transformations that are WDE but with $T \times T$ not conservative, hence not ergodic, are constructed in [24]. In [74], WDE rigid transformations (so with $T \times T$ conservative) but not DE are constructed. Weak double ergodicity for powers is studied in [56].

17.6.1.3 Ergodic with Isometric Coefficients

In the definition of EIC, the metric space can be assumed to be separable [53]. In [53], it is shown that EIC implies WM, and in [74] it is shown that WDE implies EIC. We do not know about the converses of these implications. We already know that WDE does not imply DE.

Thus far we have shown the following. For illustration purposes we write $p \implies q$ when p implies q and we know the converse does not hold, and write $p \to q$ when p implies q and the converse is open.

$$\text{DE} \Rightarrow \text{WDE} \to \text{EIC} \to \text{WM} \Leftrightarrow \text{SWM} \Leftrightarrow \text{EUC} \Rightarrow \text{TE} \Rightarrow \text{E}$$

17.6.2 Conditions Stronger than Cartesian Square Ergodic

17.6.2.1 Ergodic Index k

The fact that weak mixing conditions in infinite measure are different from the analogue conditions in finite measure was first shown by Kakutani and Parry [68] when they constructed infinite measure-preserving transformations T such that $T \times T$ is ergodic but $T \times T \times T$ is not conservative, hence not ergodic. The examples of Kakutani and Parry [68] were infinite measure-preserving Markov shifts. They also constructed Markov shifts T such that $T^{(k)}$ is ergodic but $T^{(k+1)}$ is not conservative, hence not ergodic, and such that $T^{(k)}$ is ergodic for all $k > 0$.

Rank-one transformations that are rigid and of ergodic index k, for each fixed $k \in \mathbb{N} \cup \{\infty\}$, were constructed in [14].

17.6.2.2 Infinite Ergodic Index

In [16], an infinite version of the classic Chacón transformation is constructed, and now called the infinite Chacón transformation; it is a rank-one construction that has $r_n = 3$ and $s_{n,0} = 0, s_{n,1} = 1, s_{n,2} = 3h_n + 1$ (the classic Chacón transformation is probability-preserving and has $r_n = 3$ and $s_{n,0} = 0, s_{n,1} = 1, s_{n,2} = 0$).

The infinite Chacón transformation has IEI and is partially rigid [16]. A transformation T is said to be of **infinite conservative index**, **ICI**, if $T^{(k)}$ is conservative for all k, similarly one defines **conservative index** k. If T is partially rigid, in particular if it is rigid, then it has infinite conservative index [16]. Clearly, IEI implies ICI. More generally, a transformation of positive type has infinite conservative index [7]. It is clear that a partially rigid transformation is of positive type.

In [56], it is shown that the infinite Chacón transformation is not PWM, and not multiply recurrent (see definition below), so it cannot be rigid; another proof that it is not rigid is given in [21], where there is also an example of a rigid transformation that is WM but not WDE.

17.6.2.3 Power Weakly Mixing

A rank-one transformation that is power weakly mixing was constructed in [38]; it is partially rigid, so not Koopman mixing (see definition below). In [13], the authors constructed a rank-one transformation T that has infinite ergodic index but such that $T \times T^2$ is not conservative, hence not ergodic. Thus IEI does not imply PWM. Bergelson asked if there exists a transformation that has IEI but such that $T \times T^{-1}$ is not ergodic. In [28], a rank-one transformation T is constructed such that $T \times T$ is ergodic but $T \times T^{-1}$ is not ergodic, but Bergelson's question remains open, see also [33]. It is also shown in [28] that when T is rank-one, the product $T \times T^{-1}$ is always conservative. There exist rank-one transformations such that $T \times T$ is not conservative [16]; and there exist Markov shifts such that $T \times T^{-1}$ is not conservative [28, 33]. A transformation is **power product conservative**, **PPC**, if all finite Cartesian products of all its powers are conservative. The infinite Chacón transformation is not PPC [56]. There exist transformations that are IEI and PPC but not PWM [31].

17.6.2.4 R-set Weak Mixing

T is **R-set weak mixing**, **R-sWM**, if for some nonempty $R \subset \mathbb{Q} \cap (0, 1)$, $T^p \times T^q$ is ergodic if and only if $p/q \in R$. It is shown in [14], that R-set WM implies WM, and that if $R_1 \subset R_2 \subset \mathbb{Q} \cap (0, 1)$, then there exists an infinite Lebesgue measure-preserving transformation T such that $T^p \times T^q$ is conservative ergodic for $\frac{p}{q} \in R_1$, $T^p \times T^q$ is conservative, but not ergodic for $\frac{p}{q} \in R_2 \setminus R_1$, and $T^p \times T^q$ is not conservative for $\frac{p}{q} \in \mathbb{Q}_0^1 \setminus R_2$. This definition can be applied to transformations such that $T \times T$ is ergodic. For non-rational directions see [34].

We have the following implications, with the same convention as above.

$$\text{PWM} \Rightarrow \infty\text{-EI} \Rightarrow k+1\text{-EI} \Rightarrow k\text{-EI} \Rightarrow \text{DE}$$

17.6.3 Conditions Independent of Cartesian Square Ergodic

17.6.3.1 Koopman Mixing

A transformation T is **Koopman mixing** if $U^n(f) \to 0$ weakly for all f in L^2 that are orthogonal to the constants. As we have mentioned, in the probability-preserving case this is equivalent to strong mixing, [80]. This condition is a spectral property; we note that Parry [79] showed that in infinite measure, ergodicity is not a spectral property, while it is known to be a spectral property in finite measure [80].

In the infinite measure-preserving case, as the only constant in L^2 is 0, Koopman mixing is equivalent to

$$\lim_{n\to\infty} \mu(T^n A \cap B) = 0, \tag{17.7}$$

for all A, B of finite measure.

Condition (17.7) is equivalent to

$$\lim_{n\to\infty} \mu(T^n A \cap A) = 0, \tag{17.8}$$

for all A of finite measure. (This follows by applying (17.8) to the set $A \cup B$ to obtain (17.7).) Condition (17.8) was defined by Hajian and Kakutani and called **zero type**. It was shown by Hajian and Kakutani, see [43], that if T is conservative, ergodic then it is either of zero type (Koopman mixing), or it is of **positive type**:

$$\limsup \mu(T^n A \cap A) > 0$$

for all A of positive finite measure.

One could argue that Koopman mixing is a possible notion of mixing in infinite measure, and in fact it is called mixing in [71], where it is also shown equivalent to another condition that turns out to be mixing when interpreted in finite measure. On the other hand, Koopman mixing does not even imply ergodic (just consider two disjoint copies of a transformation that is Koopman mixing). It is clear from the definition that if T is Koopman mixing, then for every S, the product $T \times S$ is Koopman mixing. If T is partially rigid, it cannot be Koopman mixing, and there are rigid, infinite ergodic index rank-one transformations [14]; in fact, rigid, PWM and rank-one are generic [17, 25]. There are Koopman mixing, PWM rank-one transformations [35]. If T is Koopman mixing and WM and R is a probability-preserving irrational rotation, then $T \times R$ is ergodic and Koopman mixing but not WM; rank-one examples are in [74]. The PWM example in [38] is partially rigid, so not Koopman mixing.

17.6.3.2 K-automorphisms

A transformation T is a K-**automorphism** if there exists a σ-finite subsigma-algebra $\mathcal{F}$ with $T^{-1}\mathcal{F} \subset \mathcal{F}$ and such that $\bigcap_{n\geq 0} T^{-n}\mathcal{F} = \emptyset$ and $\bigvee_{n\geq 0} T^n\mathcal{F} = \mathcal{B}$. Parry [79] showed that if a K-automorphism is conservative, then it is ergodic, and that there are dissipative K-automorphisms. A conservative K-automorphism is Koopman mixing [71]. Examples of K-automorphisms are given by irreducible, recurrent aperiodic Markov shifts [3]. A conservative K-automorphism is WM [3, 89], and there are examples of K-automorphisms with nonergodic Cartesian square [68]. Rank-one transformations are not K-automorphisms as K-automorphisms have positive Krengel entropy [70].

17.6.3.3 Rational Weak Mixing

If T is infinite measure-preserving and ergodic, for all sets A, B of finite measure, the ergodic theorem implies that

$$\lim_{n \to \infty} \frac{1}{n} \sum_{i=0}^{n-1} \mu(T^i A \cap B) = 0,$$

thus the ergodic average does not give us quantitative information depending on the measure of A and B.

A definition that gives a quantitative estimate for some classes of sets was introduced by Aaronson in [1]. A transformation T is *weak rationally ergodic*, **WRE** [1], if it is ergodic (and conservative) and there exists a distinguished set F of finite positive measure, such that for all measurable A, B in F, we have

$$\frac{1}{a_n(F)} \sum_{i=0}^{n-1} \mu(T^i A \cap B) \to \mu(A)\mu(B) \tag{17.9}$$

where

$$a_n(F) = \sum_{i=0}^{n-1} \frac{\mu(T^i F \cap F)}{\mu(F)}.$$

Note that when the measure of X is 1, letting $F = X$ gives $a_n(F) = n$, and the limit in (17.9) is the usual ergodic limit.

Many examples have been shown to be WRE, [3, 15]. In [29], it is shown rank-one transformations with bounded cuts are boundedly rationally ergodic [2], a condition that implies WRE. So for example the PWM transformation of Sect. 17.6.2.3 is WRE. Extensions have appeared in [8, 25].

There is a notion of subsequence weak rational ergodicity. It was shown in [25, 29] that rank-one transformations are subsequence weakly rationally ergodic. A consequence of this was that a class of transformations called Maharam transformations are not rank-one.

More recently, a stronger version of weak rational ergodicity was defined by Aaronson in [4]. A transformation T is **rational weak mixing**, **RWM**, if it is conservative, ergodic and there exists a set of finite measure F, such that for all $A, B \subset F$,

$$\frac{1}{a_n(F)} \sum_{i=0}^{n-1} \left| \mu(T^i A \cap B) - \mu(A)\mu(B) \cdot \frac{\mu(T^i F \cap F)}{\mu(F)} \right| \to 0.$$

When T is probability-preserving, letting $F = X$ one can check that this condition is equivalent to weak mixing. If T is RWM, then it is WRE and WM [4]. Aaronson asked if WRE and WM implies RWM; it was shown independently in [5] and [29] that this is not the case. In [29], it was shown that RWM implies WDE. It would be very interesting to know what condition is needed, in addition to WRE, to imply RWM. It is open whether WRE and PWM imply RWM.

The notions of RWM and Koopman mixing are independent. That is, there exist transformations that are rationally weakly mixing but not Koopman mixing, and vice-versa [29]. All early examples of RWM transformations were of product form with a Markov shift, until one was constructed in rank-one [29].

There are subsequence versions of RWM that we have not discussed [5], as well as power versions of them [5, 12].

17.6.3.4 Multiple Recurrence

As is well known, in the mid 1970s Furstenberg used ergodic theory to give a new proof of Szemerédi's theorem on arithmetic progression, starting a fruitful area of applications of ergodic theory to Ramsey theory. He did this by formulating a correspondence that associates, to a set of positive upper density set in the positive integers, a probability-preserving transformation, so that arithmetic progressions in the number set can be related to a property on the measurable set that he called multiple recurrence, see [50]. Multiple recurrence is a generalization of Poincaré recurrence. A transformation T is said to be d-**recurrent** if for all sets A of positive measure there exists a positive integer n so that

$$\mu(A \cap T^n A \cap T^{2n} A \cap \cdots \cap T^{dn} A) > 0.$$

The transformation is **multiply recurrent**, or **MR**, if it is d-recurrent for all d. Furstenberg then proved that all probability-preserving transformations are multiply recurrent, and used this to obtain Szemeredi's theorem, see [50]. It was natural to then ask if multiple recurrence still holds in infinite measure. Eigen, Hajian and Halverson showed that there exist infinite transformations that are not 2-recurrent, see [44]. These transformations, however, are not WM. As is well-known, the proof of multiple recurrence for weakly mixing transformations is different and easier than the proof for the general case, so it is of interest to know what happens when one assumes a property such as WM or PWM. In [7], Aaronson and Nakada showed that for a Markov shift T, T is d-recurrent if and only if the d-fold Cartesian product $T^{(d)}$ is conservative. In Gruher et al. [56], it was shown that in infinite measure, even PWM does not imply multiple recurrence. Multiple recurrence in infinite measure remains mysterious. Other examples are in [36]; in particular, while it is clear that rigid implies MR, there are rigid that are not polynomially multiply recurrent [36].

17.6.3.5   Measurable Sensitivity

Sensitive dependence on initial conditions is now a central notion in the standard characterizations of chaos, [19, 39, 52, 57, 85]. It is a topological notion but it has also been considered in a measurable context by several authors, see e.g., [10, 26, 66] for finite measure, and [67, 76] for the nonsingular case. Sensitive dependence essentially says that every point of the space has another point arbitrarily close to it such that at some future time the two points are significantly far apart. It needs a metric to make the notion precise.

More precisely, a transformation T on a metric space (X, d) is said to exhibit **sensitive dependence** with respect to d if there exists a number $\delta > 0$ such that for all $\varepsilon > 0$ and $x \in X$, there exists $n \in \mathbb{N}$ and $y \in B_\varepsilon(x)$ such that $d(T^n(x), T^n(y)) > \delta$.

Given a measure space $(X, \mathcal{B}, \mu)$, consider metrics d defined on X such that the balls defined by the metric are measurable and have positive μ-measure.

It was shown in [67] that if $(X, S(X), \mu, T)$ is a WDE nonsingular transformation, then one has a measurable version of sensitive dependence for all such metrics d: there exists a $\delta > 0$ such that for a.e. $x \in X$ and all $\varepsilon > 0$ there exists an n such that

$$\mu\{y \in B_\varepsilon(x) : d(T^n x, T^n y) > \delta\} > 0,$$

i.e., there is a set of positive measure of points close to x that are δ-apart at some future time.

It does make a difference to require that sensitivity hold for all such metrics. For example, while irrational rotations are far from chaotic in any reasonable definition of chaos, they have a metric for which they are sensitive. In fact, it can be shown every ergodic T has a (compatible) metric for which it is sensitive [60].

There is a stronger notion of sensitivity. That is, there exists $\delta > 0$ such that for all $x \in X$ and all $\varepsilon > 0$ there exists $N \in \mathbb{N}$ such that for all integers $n \geqslant N$ it is the case that $\mu\{y \in B_\varepsilon(x) : d(T^n(x), T^n(y)) > \delta\} > 0$. When T is probability-preserving, light mixing implies this stronger condition [67]; however it is interesting that in infinite measure if T is ergodic, conservative, and infinite measure-preserving, then there is a compatible metric d such there is a ball $B(x, \varepsilon)$ of positive measure and a sequence $\{n_k\}$ so that the diameter of $T^{n_k}B(x, \varepsilon)$ tends to 0, so points in this ball cannot stay separated, [67].

Again this kind of strong measurable sensitivity can happen for a particular measure, for example it is shown in [67] that the Hajian-Kakutani ergodic infinite measure-preserving transformation, which is not even totally ergodic, admits a metric with this strong version of sensitivity. On the other hand, it is not clear if WM in infinite measure implies basic measurable sensitivity. Grigoriev et al. showed in [55] that for conservative ergodic nonsingular transformations there is a dichotomy so that a transformation is either measurably sensitive or is isomorphic to a minimal uniformly rigid isometry. Various notions of sensitivity for dynamical systems have been studied extensively and we end with a stronger notion studied in [18]. In the standard definition there is no restriction on the time n when points have to be separated. A definition is introduced in [18] where one puts a quantitative, asymptotic bound on the sensitive time, restricting the first sensitive time of a point x to be at most asymptotically logarithmic in the measure of the ε-ball where candidate points are chosen. It is shown that in the probability-preserving case this is related to the entropy of the transformation, and it would be very interesting to explore this further for infinite measure.

We end with some implications.

RWM $\rightarrow$ WDE, RWM $\nRightarrow$ DE, RWM $\Rightarrow$ WRE and WM
PWM $\nRightarrow$ MR

17.7 Open Questions

Several questions have been mentioned earlier, and there are some implications whose converses remain open. In [9], the authors ask for a condition that implies the following property for T: if whenever $T \times S$ is conservative (for a conservative, ergodic S), then $T \times S$ is ergodic. Partial results are in [9, 89].

The following question is from [53] (see [6, 84] for related questions). Suppose that T has no non-trivial factor, in the nonsingular sense, that has discrete spectrum and is measure-preserving. Is it WM? The converse is clear.

Notions that have been proposed for mixing, such as Koopman mixing, do not imply ergodic; in rank-one (which is ergodic), Koopman mixing does not imply WM. It would be interesting to have a condition that implied PWM.

In finite measure, when $T \times T$ is ergodic, then $T \times T$ is WM. Is this true in infinite measure? (By Aaronson et al. [9], if $T \times T \times T$ is ergodic, then $T \times T$ is WM.) Other questions related to ergodicity and conservative of products are in [33]; Question 1 from [33, Section 4] has been answered in [75].

Acknowledgements We would like to thank the referee and Isaac Loh for suggestions and corrections.

References

1. J. Aaronson, Rational ergodicity and a metric invariant for Markov shifts. Israel J. Math. **27**(2), 93–123 (1977)
2. J. Aaronson, Rational ergodicity bounded rational ergodicity and some continuous measures on the circle. Israel J. Math. **33**(3–4), 181–197 (1979). https://doi.org/10.1007/BF02762160.
3. J. Aaronson, *An Introduction to Infinite Ergodic Theory.* Mathematical Surveys and Monographs, vol. 50 (American Mathematical Society, Providence, RI, 1997)
4. J. Aaronson, Rational weak mixing in infinite measure spaces. Ergod. Theory Dyn. Syst. **33**(6), 1611–1643 (2013). https://doi.org/10.1017/etds.2012.102
5. J. Aaronson, Conditions for rational weak mixing. Stoch. Dyn. **16**(2), 1660004, 12 (2016). https://doi.org/10.1142/S0219493716600042

6. J. Aaronson, M. Nadkarni, L_∞ eigenvalues and L_2 spectra of nonsingular transformations. Proc. Lond. Math. Soc. (3) **55**(3), 538–570 (1987). https://doi.org/10.1112/plms/s3-55.3.538

7. J. Aaronson, H. Nakada, Multiple recurrence of Markov shifts and other infinite measure preserving transformations. Israel J. Math. **117**, 285–310 (2000). https://doi.org/10.1007/BF02773574

8. J. Aaronson, B. Weiss, Symmetric Birkhoff sums in infinite ergodic theory (2013), http://arxiv.org/abs/1307.7490

9. J. Aaronson, M. Lin, B. Weiss, Mixing properties of Markov operators and ergodic transformations, and ergodicity of Cartesian products. Israel J. Math. **33**(3–4), 198–224 (1979). https://doi.org/10.1007/BF02762161

10. C. Abraham, G. Biau, B. Cadre, Chaotic properties of mappings on a probability space. J. Math. Anal. Appl. **266**(2), 420–431 (2002). https://doi.org/10.1006/jmaa.2001.7754

11. T.M. Adams, Smorodinsky's conjecture on rank-one mixing. Proc. Am. Math. Soc. **126**(3), 739–744 (1998). https://doi.org/10.1090/S0002-9939-98-04082-9

12. T. Adams, Rigidity sequences of power rationally weakly mixing transformations (2015), https://arxiv.org/abs/1503.05806

13. T. Adams, C.E. Silva, Rank-one power weakly mixing non-singular transformations. Ergod. Theory Dyn. Syst. **21**(5), 1321–1332 (2001). https://doi.org/10.1017/S0143385701001626

14. T. Adams, C.E. Silva, On infinite transformations with maximal control of ergodic two-fold product powers. Israel J. Math. **209**(2), 929–948 (2015). https://doi.org/10.1007/s11856-015-1241-1

15. T. Adams, C.E. Silva, Weak rational ergodicity does not imply rational ergodicity. Israel J. Math. **214**(1), 491–506 (2016). https://doi.org/10.1007/s11856-016-1371-0

16. T. Adams, N. Friedman, C.E. Silva, Rank-one weak mixing for non-singular transformations. Israel J. Math. **102**, 269–281 (1997). https://doi.org/10.1007/BF02773802

17. O.N. Ageev, C.E. Silva, Genericity of rigid and multiply recurrent innite measure-preserving and nonsingular transformations, in *Proceedings of the 16th Summer Conference on General Topology and its Applications* (New York), vol. 26, pp. 357–365 (2001/2002)

18. D. Aiello, H. Diao, Z. Fan, D.O. King, J. Lin, C.E. Silva, Measurable time-restricted sensitivity. Nonlinearity **25**(12), 3313–3325 (2012). https://doi.org/10.1088/0951-7715/25/12/3313

19. J. Auslander, J.A. Yorke, Interval maps, factors of maps, and chaos. Tôhoku Math. J. (2) **32**(2), 177–188 (1980). https://doi.org/10.2748/tmj/1178229634

20. J.R. Baxter, A class of ergodic transformations having simple spectrum. Proc. Am. Math. Soc. **27**, 275–279 (1971). https://doi.org/10.2307/2036306

21. R.L. Bayless, K.B. Yancey, Weakly mixing and rigid rank-one transformations preserving an infinite measure. New York J. Math. **21**, 615–636 (2015)

22. V. Bergelson, A. Gorodnik, *Weakly Mixing Group Actions: A Brief Survey and an Example*. Modern Dynamical Systems and Applications (Cambridge University Press, Cambridge, 2004), pp. 3–25

23. V. Bergelson, A. del Junco, M. Lemańczyk, J. Rosenblatt, Rigidity and non-recurrence along sequences. Ergod. Theory Dyn. Syst. **34**(5), 1464–1502 (2014). https://doi.org/10.1017/etds.2013.5

24. A. Bowles, L. Fidkowski, A.E. Marinello, C.E. Silva, Double ergodicity of nonsingular transformations and infinite measure-preserving staircase transformations. Ill. J. Math. **45**(3), 999–1019 (2001)

25. F. Bozgan, A. Sanchez, C.E. Silva, D. Stevens, J. Wang, Sub-sequence bounded rational ergodicity of rank-one transformations. Dyn. Syst. **30**(1), 70–84 (2015). https://doi.org/10.1080/14689367.2014.970518

26. B. Cadre, P. Jacob, On pairwise sensitivity. J. Math. Anal. Appl. **309**(1), 375–382 (2005). https://doi.org/10.1016/j.jmaa.2005.01.061

27. J.R. Choksi, S. Kakutani, Residuality of ergodic measurable transformations and of ergodic transformations which preserve an infinite measure. Indiana Univ. Math. J. **28**(3), 453–469 (1979). https://doi.org/10.1512/iumj.1979.28.28032

28. J. Clancy, R. Friedberg, I. Kasmalkar, I. Loh, T. Pădurariu, C.E. Silva, S. Vasudevan, Ergodicity and conservativity of products of infinite transformations and their inverses. Colloq. Math. **143** (2), 271–291 (2016)

29. I. Dai, X. Garcia, T. Pădurariu, C.E. Silva, On rationally ergodic and rationally weakly mixing rank-one transformations. Ergod. Theory Dyn. Syst. **35** (4), 1141–1164 (2015). https://doi.org/ 10.1017/etds.2013.96

30. A.I. Danilenko, Funny rank-one weak mixing for nonsingular abelian actions. Israel J. Math. **121**, 29–54 (2001). https://doi.org/10.1007/BF02802494

31. A.I. Danilenko, Infinite rank one actions and nonsingular Chacon transformations. Ill. J. Math. **48**(3), 769–786 (2004)

32. A.I. Danilenko, (C, F)-Actions in Ergodic Theory. Geometry and Dynamics of Groups and Spaces (Springer Science and Business Media, Berlin, 2008), pp. 325–351

33. A.I. Danilenko, Finite ergodic index and asymmetry for infinite measure preserving actions. Proc. Am. Math. Soc. **144**(6), 2521–2532 (2016). https://doi.org/10.1090/proc/12906

34. A.I. Danilenko, Directional recurrence and directional rigidity for infinite measure preserving actions of nilpotent lattices. Ergod. Theory Dyn. Syst. **37** (6), 1841–1861 (2017). https://doi. org/10.1017/etds.2015.127

35. A.I. Danilenko, V.V. Ryzhikov, Mixing constructions with infinite invariant measure and spectral multiplicities. Ergod. Theory Dyn. Syst. **31** (3), 853–873 (2011). https://doi.org/10. 1017/S0143385710000052

36. A.I. Danilenko, C.E. Silva, Multiple and polynomial recurrence for abelian actions in infinite measure. J. Lond. Math. Soc. (2) **69** (1), 183–200 (2004). https://doi.org/10.1112/ S0024610703004885

37. A.I. Danilenko, C.E. Silva, *Ergodic Theory: Non-Singular Transformations* (Springer, New York, 2012)

38. S.L. Day, B.R. Grivna, E.P. McCartney, C.E. Silva, Power weakly mixing infinite transformations. New York J. Math. **5**, 17–24 (1999) (electronic)

39. R.L. Devaney, *An Introduction to Chaotic Dynamical Systems*. Addison-Wesley Studies in Nonlinearity, 2nd edn. (Addison-Wesley Publishing Company Advanced Book Program, Redwood City, CA, 1989)

40. Y.N. Dowker, P. Erdős, Some examples in ergodic theory. Proc. Lond. Math. Soc. (3) **9**, 227–241 (1959). https://doi.org/10.1112/plms/s3-9.2.227

41. H.A. Dye, On the ergodic mixing theorem. Trans. Am. Math. Soc. **118**, 123–130 (1965). https:// doi.org/10.2307/1993948

42. S.J. Eigen, Ergodic cartesian products à la triangle sets, in *Measure Theory Oberwolfach 1983 (Oberwolfach, 1983)* (1984), pp. 263–270. https://doi.org/10.1007/BFb0072620

43. S. Eigen, A. Hajian, K. Halverson, Multiple recurrence and infinite measure preserving odometers. Israel J. Math. **108**, 37–44 (1998). https://doi.org/10.1007/BF02783041

44. S. Eigen, A. Hajian, Y. Ito, V. Prasad, *Weakly Wandering Sequences in Ergodic Theory*. Springer Monographs in Mathematics (Springer, Tokyo, 2014)

45. S. Ferenczi, Systems of finite rank. Colloq. Math. **73** (1), 35–65 (1997)

46. N.A. Friedman, *Introduction to Ergodic Theory*. Van Nostrand Reinhold Mathematical Studies, No. 29 (Van Nostrand Reinhold Co., New York, 1970)

47. N.A. Friedman, *Mixing Transformations in an Infinite Measure Space*. Studies in Probability and Ergodic Theory (Academic Press, New York, 1978), pp. 167–184

48. N.A. Friedman, J.L. King, Rank one lightly mixing. Israel J. Math. **73** (3), 281–288 (1991). https://doi.org/10.1007/BF02773841

49. N.A. Friedman, D.S. Ornstein, On partially mixing transformations. Indiana Univ. Math. J. **20**, 767–775 (1970/1971). https://doi.org/10.1512/iumj.1971.20.20061

50. H. Furstenberg, *Recurrence in Ergodic Theory and Combinatorial Number Theory* (Princeton University Press, Princeton, 1981). M. B. Porter Lectures

51. H. Furstenberg, B. Weiss, The finite multipliers of infinite ergodic transformations, in *The Structure of Attractors in Dynamical Systems (Proc. Conf., North Dakota State Univ., Fargo, ND,1977)* (1978), pp. 127–132

52. E. Glasner, B. Weiss, Sensitive dependence on initial conditions. Nonlinearity **6** (6), 1067–1075 (1993)
53. E. Glasner, B. Weiss, Weak mixing properties for non-singular actions. Ergod. Theory Dyn. Syst. **36** (7), 2203–2217 (2016). https://doi.org/10.1017/etds.2015.16
54. A. Glücksam, Ergodic multiplier properties. Ergod. Theory Dyn. Syst. **36** (3), 794–815 (2016). https://doi.org/10.1017/etds.2014.79
55. I. Grigoriev, M.C. Iordan, A. Lubin, N. Ince, C.E. Silva, On μ-compatible metrics and measurable sensitivity. Colloq. Math. **126** (1), 53–72 (2012). https://doi.org/10.4064/cm126-1-3
56. K. Gruher, F. Hines, D. Patel, C.E. Silva, R. Waelder, Power weak mixing does not imply multiple recurrence in infinite measure and other counterexamples. New York J. Math. **9**, 1–22 (2003) (electronic)
57. J. Guckenheimer, Sensitive dependence to initial conditions for one-dimensional maps. Comm. Math. Phys. **70**(2), 133–160 (1979)
58. A.B. Hajian, S. Kakutani, Weakly wandering sets and invariant measures. Trans. Am. Math. Soc. **110**, 136–151 (1964)
59. A.B. Hajian, S. Kakutani, Example of an ergodic measure preserving transformation on an infinite measure space, in *Contributions to Ergodic Theory and Probability (Proc. Conf., Ohio State Univ., Columbus, OH, 1970)* (1970), pp. 45–52
60. J. Hallett, L. Manuelli, C.E. Silva, On Li-Yorke measurable sensitivity. Proc. Am. Math. Soc. **143**(6), 2411–2426 (2015). https://doi.org/10.1090/S0002-9939-2015-12430-6
61. P.R. Halmos, Approximation theories for measure preserving transformations. Trans. Am. Math. Soc. **55**, 1–18 (1944). https://doi.org/10.2307/1990137
62. P.R. Halmos, In general a measure preserving transformation is mixing. Ann. Math (2) **45**, 786–792 (1944). https://doi.org/10.2307/1969304
63. P.R. Halmos, Invariant measures. Ann. of Math.(2) **48**, 735–754 (1947). https://doi.org/10.2307/1969138
64. P.R. Halmos, *Lectures on Ergodic Theory* (Chelsea Publishing Co., New York, 1960)
65. E. Hopf, *Ergodentheorie*. Ergebnisse der Mathematik und ihrer Grenzgebiete. 3, 5 (Springer, Berlin, 1937)
66. W. Huang, P. Lu, X. Ye, Measure-theoretical sensitivity and equicontinuity. Israel J. Math. **183**, 233–283 (2011). https://doi.org/10.1007/s11856-011-0049-x
67. J. James, T. Koberda, K. Lindsey, C.E. Silva, P. Speh, Measurable sensitivity. Proc. Am. Math. Soc. **136**(10), 3549–3559 (2008). https://doi.org/10.1090/S0002-9939-08-09294-0
68. S. Kakutani, W. Parry, Infinite measure preserving transformations with "mixing". Bull. Am. Math. Soc. **69**, 752–756 (1963)
69. J.L. King, Joining-rank and the structure of finite rank mixing transformations. J. Anal. Math. **51**, 182–227 (1988). https://doi.org/10.1007/BF02791123
70. U. Krengel, Entropy of conservative transformations. Z. Wahrscheinlichkeitstheorie und Verw Gebiete **7**, 161–181 (1967). https://doi.org/10.1007/BF00532635
71. U. Krengel, L. Sucheston, On mixing in infinite measure spaces. Z. Wahrscheinlichkeitstheorie und Verw Gebiete **13**, 150–164 (1969). https://doi.org/10.1007/BF00537021
72. K. Krickeberg, Strong mixing properties of Markov chains with infinite invariant measure, in *Proceedings of Fifth Berkeley Symposium on Mathematical Statistics and Probability (Berkeley, CA, 1965/66), Vol. II: Contributions to Probability Theory Part 2* (1967), pp. 431–446
73. M. Lenci, On infinite-volume mixing. Comm. Math. Phys. **298**(2), 485–514 (2010). https://doi.org/10.1007/s00220-010-1043-6
74. I. Loh, C.E. Silva, Strict doubly ergodic infinite transformations. Dyn. Syst. 1–25 (2017, to appear). https://doi.org/10.1080/14689367.2017.1280665
75. I. Loh, C.E. Silva, B. Athiwaratkun, Infinite symmetric ergodic index and related examples in infinite measure. Stud. Math. (to appear). https://arxiv.org/abs/1702.01455
    Isaac Loh, Cesar E. Silva, and Ben Athiwaratkun, Innite symmetric ergodic index and related examples in innite measure, Studia Mathematica, to appear.

76. C.A. Morales, Partition sensitivity for measurable maps. Math. Bohem. **138**(2), 133–148 (2013)
77. E.J. Muehlegger, A.S. Raich, C.E. Silva, M.P. Touloumtzis, B. Narasimhan, W. Zhao, Infinite ergodic index $\mathbf{Z}^d$-actions in infinite measure. Colloq. Math. **82**(2), 167–190 (1999)
78. D.S. Ornstein, *On the Root Problem in Ergodic Theory* (University California Press, Berkeley, 1972)
79. W. Parry, Ergodic and spectral analysis of certain infinite measure preserving transformations. Proc. Am. Math. Soc. **16**, 960–966 (1965)
80. W. Parry, *Topics in Ergodic Theory*. Cambridge Tracts in Mathematics, vol. 75 (Cambridge University Press, Cambridge, 2004). Reprint of the 1981 original
81. K.E. Petersen, Disjointness and weak mixing of minimal sets. Proc. Am. Math. Soc. **24**, 278–280 (1970). https://doi.org/10.2307/2036347
82. V. Rohlin, A "general" measure-preserving transformation is not mixing. Doklady Akad. Nauk SSSR (N.S.) **60**, 349–351 (1948)
83. D.J. Rudolph, *Fundamentals of Measurable Dynamics. Ergodic Theory on Lebesgue Spaces* (Oxford Science Publications, The Clarendon Press, Oxford University Press, New York, 1990)
84. D.J. Rudolph, C.E. Silva, Minimal self-joinings for nonsingular transformations. Ergod. Theory Dyn. Syst. **9**(4), 759–800 (1989). https://doi.org/10.1017/S0143385700005320
85. D. Ruelle, Dynamical systems with turbulent behavior, in *Mathematical Problems in Theoretical Physics (Proc. Internat. Conf., Univ. Rome, Rome, 1977)* (1978), pp. 341–360
86. U. Sachdeva, On category of mixing in infinite measure spaces. Math. Syst. Theory **5**, 319–330 (1971)
87. C.E. Silva, *Invitation to Ergodic Theory*. Student Mathematical Library, vol. 42 (American Mathematical Society, Providence, 2008)
88. C.E. Silva, On Mixing-Like Notions in Infinite Measure. Am. Math. Mon. **124**(9), 807–825 (2017)
89. C.E. Silva, P. Thieullen, A skew product entropy for nonsingular transformations. J. Lond. Math. Soc. (2) **52**(3), 497–516 (1995)

Chapter 18
More on Tame Dynamical Systems

Eli Glasner and Michael Megrelishvili

18.1 Introduction

Tame dynamical systems were introduced by Köhler [51] in 1995 and their theory
developed during the last decade in a series of works by several authors (see e.g.
[23, 24, 27, 29, 41, 50, 67]). Recently, connections to other areas of mathematics like:
Banach spaces, circularly ordered systems, substitutions and tilings, quasicrystals,
cut and project schemes and even model theory and logic were established (see e.g.
[3, 11, 42] and the survey [32] for more details).

Recall that for any topological group G and any dynamical G-system X (defined
by a continuous homomorphism $j : G \to H(X)$ into the group of homeomorphisms
of the compact space X) the corresponding enveloping semigroup $E(X)$ was defined
by Robert Ellis as the pointwise closure of the subgroup $j(G)$ of $H(X)$ in the
product space X^X. $E(X)$ is a compact right topological semigroup whose algebraic
and topological structure often reflects properties of (G, X) like almost periodicity
(AP), weak almost periodicity (WAP), distality, hereditary nonsensitivity (HNS)
and tameness, to mention a few. In the domain of symbolic dynamics WAP (and
even HNS) systems are necessarily countable, and in these classes minimal tame
subshifts are necessarily finite. In contrast there are many interesting symbolic (both
minimal and non-minimal) tame systems which are not HNS. Sturmian subshifts is
an important class of such systems.

E. Glasner (✉)
Department of Mathematics, Tel-Aviv University, Ramat Aviv, Israel
e-mail: glasner@math.tau.ac.il

M. Megrelishvili
Department of Mathematics, Bar-Ilan University, Ramat-Gan, Israel
e-mail: megereli@math.biu.ac.il

© Springer International Publishing AG, part of Springer Nature 2018
S. Ferenczi et al. (eds.), *Ergodic Theory and Dynamical Systems in their
Interactions with Arithmetics and Combinatorics*, Lecture Notes
in Mathematics 2213, https://doi.org/10.1007/978-3-319-74908-2_18

A metric dynamical G-system X is tame if and only if every element $p \in E(X)$ of the enveloping semigroup $E(X)$ is a limit of a sequence of elements from G, [27, 40], if and only if its enveloping semigroup $E(X)$ has cardinality at most $2^{\aleph_0}$ [27, 31]. For example, the enveloping semigroup of a Sturmian system has the form $E(X) = \mathbb{T}_\mathbb{T} \cup \mathbb{Z}$, the union of the "double-circle" $\mathbb{T}_\mathbb{T}$ and $\mathbb{Z}$. Thus its cardinality is $2^{\aleph_0}$. Another interesting property of a Sturmian system X is that both X and $E(X)$ are circularly ordered dynamical systems. As it turns out all circularly ordered systems are tame [34].

Another characterization of tameness of combinatorial nature, via the notion of independence tuples, is due to Kerr and Li [50]. Finally, the metrizable tame systems are exactly those systems which admit a representation on a separable Rosenthal Banach space [29] (a Banach space is called *Rosenthal* if it does not contain an isomorphic copy of l_1). As a by-product of the latter characterization we were able in [29] to show that, e.g. for a Sturmian system, the corresponding representation must take place on a separable Rosenthal space V with a non-separable dual, thereby proving the existence of such a Banach space. The question whether such Banach spaces exist was a major open problem until the mid of 70s (the first counterexamples were constructed independently by James, and Lindenstrauss and Stegall). For a survey of Banach representation theory for dynamical systems we refer the reader to [32].

In Sects. 18.2 and 18.3 we review and amplify some basic results concerning tame systems, fragmentability, and independent families of functions. In Sect. 18.4 we provide a new characterization of tame symbolic dynamical systems, Theorem 18.22, and a combinatorial characterization of tame subsets $D \subset \mathbb{Z}$ (i.e., subsets D such that the associated subshift $X_D \subset \{0, 1\}^\mathbb{Z}$ is tame), Theorem 18.25. In Sect. 18.5 we briefly review results relating tameness to independence and entropy.

In Sect. 18.6 we study coding functions that yield tame dynamical systems. A closely related task is to produce invariant families of real valued functions which do not contain independent infinite sequences (*tame families*). Theorem 18.38 gives some useful sufficient conditions for the tameness of families. For instance, as a corollary of this theorem, we show that if X is a compact metric equicontinuous G-system and $f : X \to \mathbb{R}$ is a bounded function with countably many discontinuities, then fG is a tame family. We also describe some old and new interesting examples of symbolic tame systems. E.g. in Theorem 18.44 we present a Sturmian-like extension of a rotation on $\mathbb{T}^d$ whose enveloping semigroup has the form $E(X) = \mathbb{Z} \cup (\mathbb{T}^d \times \mathcal{F})$, where $\mathcal{F}$ is the collection of ordered orthonormal bases for $\mathbb{R}^d$.

In Sect. 18.7 we consider dynamical properties which are related to order preservation. If X is a compact space equipped with some kind of order $\leqslant$, then subgroups of $H_+(X, \leqslant)$, the group of order preserving homeomorphisms of X, often have some special properties. E.g. $H_+(\mathbb{T})$, the group of orientation preserving homeomorphisms of the circle $\mathbb{T}$, is Rosenthal representable, Theorem 18.55, and we observe in Theorem 18.53 that it is Roelcke precompact. The recipe described in Theorem 18.58 yields many tame coding functions for subgroups of $H(\mathbb{T})$. Considering $\mathbb{Z}^k$ or $\mathrm{PSL}_2(\mathbb{R})$ as subgroups of $H(\mathbb{T})$ we obtain in this way tame Sturmian like $\mathbb{Z}^k$ and $\mathrm{PSL}_2(\mathbb{Z})$ dynamical systems.

Every topological group G has a universal minimal system $M(G)$, a universal minimal tame system $M_t(G)$, which is the largest tame G-factor of $M(G)$, and also a universal irreducible affine G-system $IA(G)$. In the final Sect. 18.8 we discuss some examples, where $M(G)$ and $IA(G)$ are tame. When $M(G)$ is tame, so that $M(G) = M_t(G)$, we say that G is intrinsically tame. Of course every extremely amenable group (i.e. a group with trivial $M(G)$) is intrinsically tame, and in Theorem 18.63 we show that the Polish groups $G = H_+(\mathbb{T})$ as well as the groups $Aut\,(\mathbf{S}(2))$ and $Aut\,(\mathbf{S}(3))$, of automorphisms of the circular directed graphs $\mathbf{S}(2)$ and $\mathbf{S}(3)$ respectively, are all intrinsically tame (but have nontrivial $M(G)$). When the universal system $IA(G)$ is tame we say that the group G is convexly intrinsically tame. Trivially every amenable group is convexly intrinsically tame, and every intrinsically tame group is convexly intrinsically tame. We show here that the group $H_+(\mathbb{T})$ is a nonamenable convexly intrinsically tame topological group. Also, every semisimple Lie group G with finite center and no compact factors (e.g., $SL_n(\mathbb{R})$) is convexly intrinsically tame (but not intrinsically tame).[1]

18.2 Preliminaries

By a topological space we mean a Tychonoff (completely regular Hausdorff) space. The closure operator in topological spaces will be denoted by cls. A function $f : X \to Y$ is *Baire class 1 function* if the inverse image $f^{-1}(O)$ of every open set $O \subset Y$ is F_σ in X [46]. For a pair of topological spaces X and Y, $C(X, Y)$ is the set of continuous functions from X into Y. We denote by $C(X)$ the Banach algebra of *bounded* continuous real valued functions even when X is not necessarily compact.

All semigroups S are assumed to be monoids, i.e., semigroups with a neutral element which will be denoted by e. A (left) *action* of S on a space X is a map $\pi : S \times X \to X$ such that $\pi(e, x) = x$ and $\pi(st, x) = \pi(s, \pi(t, x))$ for every $s, t \in S$ and $x \in X$. We usually write simply sx for $\pi(s, x)$.

An *S-space* is a topological space X equipped with a continuous action $\pi : S \times X \to X$ of a topological semigroup S on the space X. A compact S-space X is called a *dynamical S-system* and is denoted by (S, X). Note that in [29] and [31] we deal with the more general case of separately continuous actions. We reserve the symbol G for the case where S is a topological group. As usual, a continuous map $\alpha : X \to Y$ between two S-systems is called an *S-map* or a *homomorphism* when $\alpha(sx) = s\alpha(x)$ for every $(s, x) \in S \times X$. For every function $f : X \to \mathbb{R}$ and $s \in S$ denote by fs the composition $f \circ \tilde{s}$. That is, $(fs)(x) := f(sx)$.

For every S-system X we have a monoid homomorphism $j : S \to C(X, X)$, $j(s) = \tilde{s}$, where $\tilde{s} : X \to X, x \mapsto sx = \pi(s, x)$ is the *s-translation* ($s \in S$). The action is said to be *effective* (*topologically effective*) if j is an injection (respectively, a topological embedding).

---

[1] An earlier version of this work is posted on the Arxiv (arXiv:1405.2588).

The *enveloping semigroup* $E(S, X)$ (or just $E(X)$) for a compact S-system X is defined as the pointwise closure cls $(j(S))$ of $\tilde{S} = j(S)$ in X^X. Then $E(S, X)$ is a right topological compact monoid; i.e. for each $p \in E(X)$ right multiplication by p is a continuous map.

By a *cascade* on a compact space X we mean a $\mathbb{Z}$-action $\mathbb{Z} \times X \to X$. When dealing with cascades we usually write (T, X) instead of $(\mathbb{Z}, X)$, where T is the s-translation $X \to X$ corresponding to $s = 1 \in \mathbb{Z}$ (0 acts as the identity).

18.2.1 Background on Fragmentability and Tame Families

The following definitions provide natural generalizations of the fragmentability concept [43].

Definition 18.1 Let (X, τ) be a topological space and (Y, μ) a uniform space.

1. [44, 54] X is (τ, μ)-*fragmented* by a (typically, not continuous) function $f : X \to Y$ if for every nonempty subset A of X and every $\varepsilon \in \mu$ there exists an open subset O of X such that $O \cap A$ is nonempty and the set $f(O \cap A)$ is ε-small in Y. We also say in that case that the function f is *fragmented*. Notation: $f \in \mathcal{F}(X, Y)$, whenever the uniformity μ is understood. If $Y = \mathbb{R}$ then we write simply $\mathcal{F}(X)$.
2. [27] We say that a *family of functions* $F = \{f : (X, \tau) \to (Y, \mu)\}$ is *fragmented* if condition (1) holds simultaneously for all $f \in F$. That is, $f(O \cap A)$ is ε-small for every $f \in F$.
3. [29] We say that F is an *eventually fragmented family* if every infinite subfamily $C \subset F$ contains an infinite fragmented subfamily $K \subset C$.

In Definition 18.1(1) when $Y = X, f = id_X$ and μ is a metric uniformity, we retrieve the usual definition of fragmentability (more precisely, (τ, μ)-fragmentability) in the sense of Jayne and Rogers [43]. Implicitly it already appears in a paper of Namioka and Phelps [60].

Lemma 18.2 ([27, 29])

1. *It is enough to check the conditions of Definition 18.1 only for $\varepsilon \in \gamma$ from a subbase γ of μ and for closed nonempty subsets $A \subset X$.*
2. *If $f : (X, \tau) \to (Y, \mu)$ has a point of continuity property PCP (i.e., for every closed nonempty $A \subset X$ the restriction $f_{|A} : A \to Y$ has a continuity point) then it is fragmented. If (X, τ) is hereditarily Baire (e.g., compact, or Polish) and (Y, μ) is a pseudometrizable uniform space then f is fragmented if and only if f has PCP. So, in particular, for compact X, the set $\mathcal{F}(X)$ is exactly $B'_r(X)$ in the notation of [71].*
3. *If X is Polish and Y is a separable metric space then $f : X \to Y$ is fragmented iff f is a Baire class 1 function (i.e., the inverse image of every open set is F_σ).*
4. *Let (X, τ) be a separable metrizable space and (Y, ρ) a pseudometric space. Suppose that $f : X \to Y$ is a fragmented onto map. Then Y is separable.*

For other properties of fragmented maps and fragmented families we refer to [27, 29, 33, 44, 48, 54, 55, 59]. Basic properties and applications of fragmentability in topological dynamics can be found in [29, 32, 33].

18.2.2 Independent Sequences of Functions

Let $\{f_n : X \to \mathbb{R}\}_{n \in \mathbb{N}}$ be a uniformly bounded sequence of functions on a *set* X. Following Rosenthal [68] we say that this sequence is an l_1-*sequence* on X if there exists a real constant $a > 0$ such that for all $n \in \mathbb{N}$ and choices of real scalars $c_1, \ldots, c_n$ we have

$$a \cdot \sum_{i=1}^{n} |c_i| \leqslant \left\| \sum_{i=1}^{n} c_i f_i \right\|_\infty.$$

A Banach space V is said to be *Rosenthal* if it does not contain an isomorphic copy of l_1, or equivalently, if V does not contain a sequence which is equivalent to an l_1-sequence.

A Banach space V is an *Asplund* space if the dual of every separable Banach subspace is separable. Every Asplund space is Rosenthal and every reflexive space is Asplund.

A sequence f_n of real valued functions on a set X is said to be *independent* (see [68, 71]) if there exist real numbers $a < b$ such that

$$\bigcap_{n \in P} f_n^{-1}(-\infty, a) \cap \bigcap_{n \in M} f_n^{-1}(b, \infty) \neq \emptyset$$

for all finite disjoint subsets P, M of $\mathbb{N}$.

Definition 18.3 We say that a bounded family F of real valued (not necessarily, continuous) functions on a set X is *tame* if F does not contain an independent sequence.

Every bounded independent sequence is an l_1-sequence [68]. The sequence of projections on the Cantor cube

$$\{\pi_n : \{0, 1\}^{\mathbb{N}} \to \{0, 1\}\}_{n \in \mathbb{N}}$$

and the sequence of Rademacher functions

$$\{r_n : [0, 1] \to \mathbb{R}\}_{n \in \mathbb{N}}, \quad r_n(x) := \operatorname{sgn}(\sin(2^n \pi x))$$

both are independent (hence, nontame).

The following useful theorem synthesizes some known results. It mainly is based on results of Rosenthal and Talagrand. The equivalence of (1), (3) and (4) is a part of [71, Theorem 14.1.7] For the case (1) ⇔ (2) note that every bounded independent sequence $\{f_n : X \to \mathbb{R}\}_{n\in\mathbb{N}}$ is an l_1-sequence (in the sup-norm), [68, Prop. 4]. On the other hand, as the proof of [68, Theorem 1] shows, if $\{f_n\}_{n\in\mathbb{N}}$ has no independent subsequence then it has a pointwise convergent subsequence. Bounded pointwise-Cauchy sequences in $C(X)$ (for compact X) are weak-Cauchy as it follows by Lebesgue's theorem. Now Rosenthal's dichotomy theorem [68, Main Theorem] asserts that $\{f_n\}$ has no l_1-sequence. In [29, Sect. 4] we show why eventual fragmentability of F can be included in this list (item (5)).

Theorem 18.4 *Let X be a compact space and $F \subset C(X)$ a bounded subset. The following conditions are equivalent:*

1. *F does not contain an l_1-sequence.*
2. *F is a tame family (does not contain an independent sequence).*
3. *Each sequence in F has a pointwise convergent subsequence in $\mathbb{R}^X$.*
4. *The pointwise closure $cls(F)$ of F in $\mathbb{R}^X$ consists of fragmented maps, that is, $cls(F) \subset \mathcal{F}(X)$.*
5. *F is an eventually fragmented family.*

Let X be a topological space and $F \subset l_\infty(X)$ be a norm bounded family. Recall that F has Grothendieck's *Double Limit Property* (DLP) on X if for every sequence $\{f_n\} \subset F$ and every sequence $\{x_m\} \subset X$ the limits

$$\lim_n \lim_m f_n(x_m) \quad \text{and} \quad \lim_m \lim_n f_n(x_m)$$

are equal whenever they both exist.

The following examples are mostly reformulations of known results; the details can be found in [29, 33, 57].

Example 18.5

1. A Banach space V is Rosenthal iff every bounded subset $F \subset V$ is tame (as a family of functions) on every bounded subset $Y \subset V^*$ of the dual space V^*, iff F is eventually fragmented on Y.
2. A Banach space V is Asplund iff every bounded subset $F \subset V$ is a fragmented family of functions on every bounded subset $Y \subset V^*$.
3. A Banach space is reflexive iff every bounded subset $F \subset V$ has DLP on every bounded subset $X \subset V^*$.
4. ((DLP) ⇒ Tame) Let F be a bounded family of real valued (not necessarily continuous) functions on a set X such that F has DLP. Then F is tame.
5. The family Homeo $[0, 1]$, of all self homeomorphisms of $[0, 1]$, is tame (but does not have the DLP on $[0, 1]$).
6. Let X be a circularly (e.g., linearly) ordered set. Then any bounded family F of real functions with bounded total variation is tame.

Note that in (1)–(3) the converse statements are true; as it follows from results of [33] every bounded tame (DLP) family F on X can be represented on a Rosenthal (Asplund, reflexive) Banach space. Recall that a representation of F on a Banach space V consists of a pair (v, α) of bounded maps $v : F \to V$, $\alpha : X \to V^*$ (with weak-star continuous α) such that

$$f(x) = \langle v(f), \alpha(x) \rangle \quad \forall f \in F, \quad \forall x \in X.$$

In other words, the following diagram commutes

$$
\begin{array}{ccc}
F \times X & \longrightarrow & \mathbb{R} \\
{\scriptstyle v}\downarrow \quad {\scriptstyle \alpha}\downarrow & & \downarrow {\scriptstyle id} \\
V \times V^* & \longrightarrow & \mathbb{R}
\end{array}
$$

18.2.3   More Properties of Fragmented Families

Here we discuss a general principle: the fragmentability of a family of continuous maps defined on a compact space is "countably-determined". The following theorem is inspired by results of Namioka and can be deduced, after some reformulations, from [59, Theorems 3.4 and 3.6]. See also [9, Theorem 2.1].

Theorem 18.6 *Let $F = \{f_i : X \to Y\}_{i \in I}$ be a bounded family of **continuous** maps from a compact (not necessarily metrizable) space (X, τ) into a pseudometric space (Y, d). The following conditions are equivalent:*

1. *F is a fragmented family of functions on X.*
2. *Every countable subfamily K of F is fragmented.*
3. *For every countable subfamily K of F the pseudometric space $(X, \rho_{K,d})$ is separable, where $\rho_{K,d}(x_1, x_2) := \sup_{f \in K} d(f(x_1), f(x_2))$.*

Proof (1) $\Rightarrow$ (2) is trivial.

(2) $\Rightarrow$ (3): Let K be a countable subfamily of F. Consider the natural map

$$\pi : X \to Y^K, \pi(x)(f) := f(x).$$

By (2), K is a fragmented family. This means (see [27, Def. 6.8]) that the map π is (τ, μ_K)-fragmented, where μ_K is the uniformity of d-uniform convergence on $Y^K := \{f : K \to (Y, d)\}$. Then the map π is also (τ, d_K)-fragmented, where d_K is the pseudometric on Y^K defined by

$$d_K(z_1, z_2) := \sup_{f \in K} d(z_1(f), z_2(f)).$$

Since d is bounded, $d_K(z_1, z_2)$ is finite and d_K is well-defined. Denote by (X_K, τ_p) the subspace $\pi(X) \subset Y^K$ in pointwise topology. Since $K \subset C(X)$, the induced map $\pi_0 : X \to X_K$ is a continuous map onto the compact space (X_K, τ_p). Denote by $i : (X_K, \tau_p) \to (Y^K, d_K)$ the inclusion map. So, $\pi = i \circ \pi_0$, where the map π is (τ, d_K)-fragmented. This easily implies (see [29, Lemma 2.3.5]) that i is (τ_p, d_K)-fragmented. It immediately follows that the identity map $id : (X_K, \tau_p) \to (X_K, d_K)$ is (τ_p, d_K)-fragmented.

Since K is countable, $(X_K, \tau_p) \subset Y^K$ is metrizable. Therefore, (X_K, τ_p) is second countable (being a metrizable compactum). Now, since d_K is a pseudometric on Y^K, and $id : (X_K, \tau_p) \to (X_K, d_K)$ is (τ_p, d_K)-fragmented, we can apply Lemma 18.2(4). It directly implies that the set X_K is a separable subset of (Y^K, d_K). This means that $(X, \rho_{K,d})$ is separable.

(3) $\Rightarrow$ (1) : Suppose that F is not fragmented. Thus, there exists a non-empty closed subset $A \subset X$ and an $\varepsilon > 0$ such that for each non-empty open subset $O \subset X$ with $O \cap A \neq \emptyset$ there is some $f \in F$ such that $f(O \cap A)$ is not ε-small in (Y, d). Let V_1 be an arbitrary non-empty relatively open subset in A. There are $a, b \in V_1$ and $f_1 \in F$ such that $d(f_1(a), f_1(b)) > \varepsilon$. Since f_1 is continuous we can choose relatively open subsets V_2, V_3 in A with $\mathrm{cls}\,(V_2 \cup V_3) \subset V_1$ such that $d(f_1(x), f_1(y)) > \varepsilon$ for every $(x, y) \in V_2 \times V_3$.

By induction we can construct a sequence $\{V_n\}_{n \in \mathbb{N}}$ of non-empty relatively open subsets in A and a sequence $K := \{f_n\}_{n \in \mathbb{N}}$ in F such that:

(i) $V_{2n} \cup V_{2n+1} \subset V_n$ for each $n \in \mathbb{N}$;
(ii) $d(f_n(x), f_n(y)) > \varepsilon$ for every $(x, y) \in V_{2n} \times V_{2n+1}$.

We claim that $(X, \rho_{K,d})$ is not separable, where

$$\rho_{K,d}(x_1, x_2) := \sup_{f \in K} d(f(x_1), f(x_2)).$$

In fact, for each *branch*

$$\alpha := V_1 \supset V_{n_1} \supset V_{n_2} \supset \cdots$$

where for each i, $n_{i+1} = 2n_i$ or $2n_i + 1$, by compactness of X one can choose an element

$$x_\alpha \in \bigcap_{i \in \mathbb{N}} \mathrm{cls}\,(V_{n_i}).$$

If $x = x_\alpha$ and $y = x_\beta$ come from different branches, then there is an $n \in \mathbb{N}$ such that $x \in \mathrm{cls}\,(V_{2n})$ and $y \in \mathrm{cls}\,(V_{2n+1})$ (or vice versa). In any case it follows from (ii) and the continuity of f_n that $d(f_n(x), f_n(y)) \geqslant \varepsilon$, hence $\rho_{K,d}(x, y) \geqslant \varepsilon$. Since there are uncountably many branches we conclude that A and hence also X are not $\rho_{K,d}$-separable.

Definition 18.7 ([16, 55]) Let X be a compact space and $F \subset C(X)$ a norm bounded family of continuous real valued functions on X. Then F is said to be an *Asplund family for* X if for every countable subfamily K of F the pseudometric space $(X, \rho_{K,d})$ is separable, where

$$\rho_{K,d}(x_1, x_2) := \sup_{f \in K} |f(x_1) - f(x_2)|.$$

Corollary 18.8 *Let X be a compact space and $F \subset C(X)$ a norm bounded family of continuous real valued functions on X. Then F is fragmented if and only if F is an Asplund family for X.*

Theorem 18.9 *Let $F = \{f_i : X \to Y\}_{i \in I}$ be a family of continuous maps from a compact (not necessarily metrizable) space (X, τ) into a uniform space (Y, μ). Then F is fragmented if and only if every countable subfamily $A \subset F$ is fragmented.*

Proof The proof can be reduced to Theorem 18.6. Every uniform space can be uniformly approximated by pseudometric spaces. Using Lemma 18.2(1) we can assume that (Y, μ) is pseudometrizable; i.e. there exists a pseudometric d such that $\text{unif}(d) = \mu$. Moreover, replacing d by the uniformly equivalent pseudometric $\frac{d}{1+d}$ we can assume that $d \leqslant 1$.

18.3 Classes of Dynamical Systems

Definition 18.10 A compact dynamical S-system X is said to be *tame* if one of the following equivalent conditions are satisfied:

1. for every $f \in C(X)$ the family $fS := \{fs : s \in S\}$ has no independent subsequence.
2. every $p \in E(X)$ is a *fragmented map* $X \to X$.
3. for every $p \in E(X)$ and every $f \in C(X)$ the composition $fp : X \to \mathbb{R}$ has PCP.

The following principal result is a dynamical analog of the Bourgain-Fremlin-Talagrand dichotomy [8, 72].

Theorem 18.11 ([27] A Dynamical Version of BFT Dichotomy) *Let X be a compact metric dynamical S-system and let $E = E(X)$ be its enveloping semigroup. Either*

1. *E is a separable Rosenthal compact space (hence E is Fréchet and card $E \leqslant 2^{\aleph_0}$); or*
2. *the compact space E contains a homeomorphic copy of $\beta \mathbb{N}$ (hence card $E = 2^{2^{\aleph_0}}$).*

The first possibility holds iff X is a tame S-system.

Thus, a metrizable dynamical system is tame iff card$(E(X)) \leqslant 2^{\aleph_0}$ iff $E(X)$ is a Rosenthal compactum (or a Fréchet space). Moreover, by Glasner et al. [40] a metric S-system is tame iff every $p \in E(X)$ is a Baire class 1 map $p : X \to X$.

The class of tame dynamical systems is quite large. It is closed under subsystems, products and factors. Recall that an S-dynamical system X is weakly almost periodic (WAP) if and only if every $p \in E(X)$ is a continuous map. As every continuous map $X \to X$ is fragmented it follows that every WAP system is tame. A metrizable S-system X is WAP iff (S, X) is representable on a reflexive Banach space. The class of hereditarily nonsensitive systems (HNS) is an intermediate class of systems [32]. The property HNS admits a reformulation in terms of enveloping semigroup: (S, X) is HNS iff $E(S, X)$ (equivalently, $\tilde{S}$) is a fragmented family. Of course, this implies that every $p \in E(X)$ is fragmented. So, indeed, WAP $\subset$ HNS $\subset$ Tame. A metrizable S-system X is HNS iff $E(X)$ is metrizable iff (S, X) is Asplund representable (RN, in another terminology) [27, 40].

18.3.1   Some Classes of Functions

A *compactification* of X is a continuous map $\gamma : X \to Y$ with a dense range where Y is compact. When X and Y are S-spaces and γ is an S-map we say that γ is an S-*compactification*.

A function $f \in C(X)$ on an S-space X is said to be Right Uniformly Continuous if the induced right action $C(X) \times S \to C(X)$ is continuous at the points (f, s), where $s \in S$. Notation: $f \in \text{RUC}(X)$. If X is a compact S-space then $\text{RUC}(X) = C(X)$. Note that $f \in \text{RUC}(X)$ if and only if there exists an S-compactification $\gamma : X \to Y$ such that $f = \tilde{f} \circ \gamma$ for some $\tilde{f} \in C(Y)$. In this case we say that f *comes* from the S-compactification $\gamma : X \to Y$.

The function f is said to be: (a) *WAP*; (b) *Asplund*; (c) *tame* if f comes from an S-compactification $\gamma : X \to Y$ such that (S, Y) is: WAP, HNS or tame respectively. For the corresponding classes of functions we use the notation: WAP(X), Asp(X), Tame(X), respectively. Each of these is a norm closed S-invariant subalgebra of the S-invariant algebra RUC(X) and WAP$(X) \subset$ Asp$(X) \subset$ Tame(X). For more details see [31, 32]. As a particular case we have defined the algebras WAP(S), Asp(S), Tame(S) corresponding to the left action of S on itself.

The S-invariant subalgebra Tame(S) of RUC(S) induces an S-compactification of S which we denote by $S \to S^{\text{Tame}}$. Recall that it is a semigroup compactification of S and that S^{Tame} is a compact right topological semigroup [31]. Similarly, one defines the compactifications S^{AP}, S^{WAP}, S^{Asp}. Here AP means *almost periodic*. AP compact G-systems (for groups $S := G$) are just equicontinuous systems.

18.3.2   Cyclic S-compactifications

Let X be an S-space. For every $f \in \mathrm{RUC}(X)$ define the following pointwise continuous natural S-map

$$\delta_f : X \to \mathrm{RUC}(S), \quad \delta_f(x)(g) := f(gx).$$

It induces an S-compactification $\delta_f : X \to X_f$, where X_f is the pointwise closure of $\delta_f(X)$ in $\mathrm{RUC}(S)$. Denote by $\mathcal{A}_f := \langle fS \rangle$ the smallest S-invariant unital Banach subalgebra of $\mathrm{RUC}(X)$ which contains f. The corresponding Gelfand S-compactification is equivalent to $\delta_f : X \to X_f$. Let $\tilde{f} := \hat{e}|_{X_f}$, where $\hat{e}$ is the evaluation at e functional on $\mathrm{RUC}(S)$. Then f comes from the S-system X_f. Moreover, $\tilde{f}S$ separates points of X_f.

We call $\delta_f : X \to X_f$ the *cyclic compactification* of X (induced by f) [5, 27, 31].

Definition 18.12 ([27, 31]) We say that a compact dynamical S-system X is *cyclic* if there exists $f \in C(X)$ such that (S, X) is topologically S-isomorphic to the Gelfand space X_f of the S-invariant unital subalgebra $\mathcal{A}_f \subset C(X)$ generated by the orbit fS.

Lemma 18.13 *Let $\gamma : X \to Y$ be an S-compactification and $f \in C(X)$.*

1. *f comes from γ (i.e., $f = \bar{f} \circ \gamma$ for some $\bar{f} \in C(Y)$) if and only if there exists a continuous onto S-map $q : Y \to X_f$ such that $\bar{f} = \tilde{f} \circ q$ and the following diagram is commutative*

2. *$q : Y \to X_f$ in (1) is an isomorphism of S-compactifications if and only if $\bar{f}S$ separates points of Y (where, as before, $\bar{f} = \tilde{f} \circ q$).*

Proof Use Gelfand's description of compactifications in terms of the corresponding algebras and the Stone-Weierstrass Theorem.

Remark 18.14 Let X be a (not necessarily compact) S-space and $f \in \mathrm{RUC}(X)$. Then, as was shown in [31], there exist a cyclic S-system X_f, a continuous S-compactification $\pi_f : X \to X_f$, and a continuous function $\tilde{f} : X_f \to \mathbb{R}$ such that $f = \tilde{f} \circ \pi_f$; that is, f comes from the S-compactification $\pi_f : X \to X_f$. The collection of functions $\tilde{f}S$ separates points of X_f.

Theorem 18.15 *Let X be a compact S-space and $f \in C(X)$. The following conditions are equivalent:*

1. *$f \in \mathrm{Tame}(X)$ (i.e. f comes from a tame dynamical system).*
2. *fS is a tame family.*

3. $cls_p(fS) \subset \mathcal{F}(X)$.
4. fS is an eventually fragmented family.
5. For every countable infinite subset $A \subset S$ there exists a countable infinite subset $A' \subset A$ such that the corresponding pseudometric

$$\rho_{f,A'}(x, y) := \sup\{|f(gx) - f(gy)| : \ g \in A'\}$$

on X is separable.
6. The cyclic S-space X_f is Rosenthal representable (i.e., WRN).

Proof The equivalence of (1)–(4) follows from Theorem 18.4. For (4) $\Leftrightarrow$ (5) use Theorem 18.6. For (4) $\Leftrightarrow$ (6) we refer to [29, 33].

18.4 A Characterization of Tame Symbolic Systems

18.4.1 Symbolic Systems and Coding Functions

The binary *Bernoulli shift system* is defined as the cascade (Ω, σ), where $\Omega :=$ $\{0, 1\}^{\mathbb{Z}}$. We have the natural $\mathbb{Z}$-action on the compact metric space Ω induced by the σ-shift:

$$\mathbb{Z} \times \Omega \to \Omega, \quad \sigma^m(\omega_i)_{i \in \mathbb{Z}} = (\omega_{i+m})_{i \in \mathbb{Z}} \ \forall(\omega_i)_{i \in \mathbb{Z}} \in \Omega, \ \forall m \in \mathbb{Z}.$$

More generally, for a discrete monoid S and a finite alphabet $A := \{0, 1, \ldots, n\}$ the compact space $\Omega := A^S$ is an S-space under the action

$$S \times \Omega \to \Omega, \quad (s\omega)(t) = \omega(ts), \ \omega \in A^S, \ s, t \in S.$$

A closed S-invariant subset $X \subset A^S$ defines a subsystem (S, X). Such systems are called *subshifts* or *symbolic dynamical systems*.

Definition 18.16

1. Let $S \times X \to X$ be an action on a (not necessarily compact) space $X, f : X \to \mathbb{R}$ a bounded (not necessarily continuous) function, and $z \in X$. Define a *coding function* as follows:

$$\varphi := m(f, z) : S \to \mathbb{R}, \ s \mapsto f(sz).$$

2. When $f(X) \subseteq \{0, 1, \ldots, d\}$ every such code generates a point transitive subshift S_φ of A^S, where $A = \{0, 1, \ldots, d\}$ and

$$S_\varphi := cls_p\{g\varphi : g \in S\} \subset A^S \quad (\text{where } g\varphi(t) = \varphi(tg))$$

is the pointwise closure of the left S-orbit $S\varphi$ in the space $\{0, 1, \cdots, d\}^S$.

When $S = \mathbb{Z}^k$ we say that f is a (k, d)-code. In the particular case of the characteristic function $\chi_D : X \to \{0, 1\}$ for a subset $D \subset X$ and $S = \mathbb{Z}$ we get a $(1, 1)$-code, i.e. a binary function $m(D, z) : \mathbb{Z} \to \{0, 1\}$ which generates a $\mathbb{Z}$-subshift of the Bernoulli shift on $\{0, 1\}^{\mathbb{Z}}$.

Regarding some dynamical and combinatorial aspects of coding functions see [7, 17].

Among others we will study the following question

Question 18.17 When is a coding φ function tame? Equivalently, when is the associated transitive subshift system $S_\varphi \subset \{0, 1\}^{\mathbb{Z}}$ with $\varphi = m(D, z)$ tame?

Some restrictions on D are really necessary because *every* binary bisequence $\varphi : \mathbb{Z} \to \{0, 1\}$ can be encoded as $\varphi = m(D, z)$.

It follows from results in [29] that a coding bisequence $c : \mathbb{Z} \to \mathbb{R}$ is tame iff it can be represented as a generalized matrix coefficient of a Rosenthal Banach space representation. That is, iff there exist: a Rosenthal Banach space V, a linear isometry $\sigma \in \mathrm{Iso}(V)$ and two vectors $v \in V$, $\varphi \in V^*$ such that

$$c_n = \langle \sigma^n(v), \varphi \rangle = \varphi(\sigma^n(v)) \quad \forall n \in \mathbb{Z}.$$

Let, as above, A^S be the full symbolic shift S-system. For a nonempty $L \subseteq S$ define the natural projection

$$\pi_L : A^S \to A^L.$$

The compact zero-dimensional space A^S is metrizable iff S is countable (and, in this case, A^S is homeomorphic to the Cantor set).

It is easy to see that the full shift system $\Omega = A^S$ (hence also every subshift) is *uniformly expansive*. This means that there exists an entourage $\varepsilon_0 \in \mu$ in the natural uniform structure of A^S such that for every distinct $\omega_1 \neq \omega_2$ in Ω one can find $s \in S$ with $(s\omega_1, s\omega_2) \notin \varepsilon_0$. Indeed, take

$$\varepsilon_0 := \{(u, v) \in \Omega \times \Omega : u(e) = v(e)\},$$

where e, as usual, is the neutral element of S.

Lemma 18.18 *Every symbolic dynamical S-system $X \subset \Omega = A^S$ is cyclic (Definition 18.12).*

Proof It suffices to find $f \in C(X)$ such that the orbit fS separates the points of X since then, by the Stone-Weierstrass theorem, (S, X) is isomorphic to its cyclic S-factor (S, X_f). The family

$$\{\pi_s : X \to A = \{0, 1, \ldots, n\} \subset \mathbb{R}\}_{s \in S}$$

of basic projections clearly separates points on X and we let $f := \pi_e : X \to \mathbb{R}$. Now observe that $fS = \{\pi_s\}_{s \in S}$.

A topological space (X, τ) is *scattered* (i.e., every nonempty subspace has an isolated point) iff X is (τ, ξ)-fragmented, for arbitrary uniform structure ξ on the set X.

Proposition 18.19 ([55, Prop. 7.15]) *Every scattered compact jointly continuous S-space X is RN (that is, Asplund representable).*

Proof A compactum X is scattered iff $C(X)$ is Asplund [60]. Now use the canonical S-representation of (S, X) on the Asplund space $V := C(X)$.

The following result recovers and extends [27, Sect. 10] and [55, Sect. 7].

Theorem 18.20 *For a discrete monoid S and a finite alphabet A let $X \subset A^S$ be a subshift. The following conditions are equivalent:*

1. *(S, X) is Asplund representable (that is, RN).*
2. *(S, X) is HNS.*
3. *X is scattered.*
   If, in addition, X is metrizable (e.g., if S is countable) then each of the conditions above is equivalent also to:
4. *X is countable.*

Proof (1) $\Rightarrow$ (2): It was proved in [27, Lemma 9.8].

(2) $\Rightarrow$ (3): Let μ be the natural uniformity on X and μ_S the (finer) uniformity of uniform convergence on $X \subset X^S$ (we can treat X as a subset of X^S under the assignment $x \mapsto \widehat{x}$, where $\widehat{x}(s) = sx$). If X is HNS then the family $\widetilde{S}$ is fragmented. This means that X is μ_S-fragmented. As we already mentioned, every subshift X is uniformly S-expansive. Therefore, μ_S coincides with the discrete uniformity μ_Δ on X (the largest possible uniformity on the *set* X). Hence, X is also μ_Δ-fragmented. This means that X is a scattered compactum.

(3) $\Rightarrow$ (1): Use Proposition 18.19.

If X is metrizable then

(4) $\Leftrightarrow$ (3): A scattered compactum is metrizable iff it is countable.

Every zero-dimensional compact $\mathbb{Z}$-system X can be embedded into a product $\prod X_f$ of (cyclic) subshifts X_f (where, one may consider only continuous functions $f : X \to \{0, 1\}$) of the Bernoulli system $\{0, 1\}^{\mathbb{Z}}$.

For more information about countable (HNS and WAP) subshifts see [2, 10, 70].

Problem 18.21 Find a nice characterization for WAP (necessarily, countable) $\mathbb{Z}$-subshifts.

Next we consider tame subshifts.

Theorem 18.22 *Let X be a subshift of $\Omega = A^S$. The following conditions are equivalent:*

1. *(S, X) is a tame system.*

2. *For every infinite subset $L \subseteq S$ there exists an infinite subset $K \subseteq L$ and a countable subset $Y \subseteq X$ such that*

$$\pi_K(X) = \pi_K(Y).$$

That is,

$$\forall x = (x_s)_{s \in S} \in X, \ \exists y = (y_s)_{s \in S} \in Y \quad \text{with} \quad x_k = y_k \ \forall k \in K.$$

3. *For every infinite subset $L \subseteq S$ there exists an infinite subset $K \subseteq L$ such that $\pi_K(X)$ is a countable subset of A^K.*
4. *(S, X) is Rosenthal representable (that is, WRN).*

Proof (1) $\Leftrightarrow$ (2): As in the proof of Lemma 18.18 define $f := \pi_e \in C(X)$. Then X is isomorphic to the cyclic S-space X_f. (S, X) is a tame system iff $C(X) = \text{Tame}(X)$. By Lemma 18.18, $C(X) = \mathcal{A}_f$, so we have only to show that $f \in \text{Tame}(X)$.

By Theorem 18.15, $f := \pi_e : X \to \mathbb{R}$ is a tame function iff for every infinite subset $L \subset S$ there exists a countable infinite subset $K \subset L$ such that the corresponding pseudometric

$$\rho_{f,K}(x, y) := \sup_{k \in K}\{|(\pi_e)(kx) - (\pi_e)(ky)|\} = \sup_{k \in K}\{|x_k - y_k|\}$$

on X is separable. The latter assertion means that there exists a countable subset Y which is $\rho_{f,K}$-dense in X. Thus for every $x \in X$ there is a point $y \in Y$ with $\rho_{f,K}(x, y) < 1/2$. As the values of the function $f = \pi_0$ are in the set A, we conclude that $\pi_K(x) = \pi_K(y)$, whence

$$\pi_K(X) = \pi_K(Y).$$

The equivalence of (2) and (3) is obvious.

(1) $\Rightarrow$ (4): (S, X) is Rosenthal-approximable (Theorem 18.15(1)). On the other hand, (S, X) is cyclic (Lemma 18.18). By Theorem 18.15(7) we can conclude that (S, X) is WRN.

(4) $\Rightarrow$ (1): Follows directly by Theorem 18.15(1). $\qed$

Remark 18.23 From Theorem 18.22 we can deduce the following peculiar fact. If X is a tame subshift of $\Omega = \{0, 1\}^{\mathbb{Z}}$ and $L \subset \mathbb{Z}$ an infinite set, then there exist an infinite subset $K \subset L, k \geqslant 1$, and $a \in \{0, 1\}^{2k+1}$ such that $X \cap [a] \neq \emptyset$ and $\forall x, x' \in X \cap [a]$ we have $x|_K = x'|_K$. Here $[a] = \{z \in \{0, 1\}^{\mathbb{Z}} : z(j) = a(j), \ \forall |j| \leqslant k\}$. In fact, since $\pi_K(X)$ is a countable closed set it contains an isolated point, say w, and then the open set $\pi_K^{-1}(w)$ contains a subset $[a] \cap X$ as required.

18.4.2   Tame and HNS Subsets of $\mathbb{Z}$

We say that a subset $D \subset \mathbb{Z}$ is *tame* if the characteristic function $\chi_D : \mathbb{Z} \to \mathbb{R}$ is a tame function on the group $\mathbb{Z}$. That is, when this function *comes* from a pointed compact tame $\mathbb{Z}$-system (X, x_0). Analogously, we say that D is *HNS* (or *Asplund*), *WAP*, or *Hilbert* if $\chi_D : \mathbb{Z} \to \mathbb{R}$ is an Asplund, WAP or Hilbert function on $\mathbb{Z}$, respectively. By basic properties of the *cyclic system* $X_D := \text{cls}\{\chi_D \circ T^n : n \in \mathbb{Z}\} \subset \{0, 1\}^{\mathbb{Z}}$ (see Remark 18.14), the subset $D \subset \mathbb{Z}$ is tame (Asplund, WAP) iff the associated subshift X_D is tame (Asplund, WAP).

Surprisingly it is not known whether $X_f := \text{cls}\{f \circ T^n : n \in \mathbb{Z}\} \subset \mathbb{R}^{\mathbb{Z}}$ is a Hilbert system when $f : \mathbb{Z} \to \mathbb{R}$ is a Hilbert function (see [38]). The following closely related question from [56] is also open: Is it true that Hilbert representable compact metric $\mathbb{Z}$-spaces are closed under factors?

Remark 18.24   The definition of WAP sets was introduced by Ruppert [69]. He has the following characterisation [69, Theorem 4]:

$D \subset \mathbb{Z}$ is a WAP subset if and only if every infinite subset $B \subset \mathbb{Z}$ contains a finite subset $F \subset B$ such that the set

$$\bigcap_{b \in F} (b + D) \setminus \bigcap_{b \in B \setminus F} (b + D)$$

is finite. See also [25].

Theorem 18.25   *Let D be a subset of $\mathbb{Z}$. The following conditions are equivalent:*

1. *D is a tame subset (i.e., the associated subshift $X_D \subset \{0, 1\}^{\mathbb{Z}}$ is tame).*
2. *For every infinite subset $L \subseteq \mathbb{Z}$ there exists an infinite subset $K \subseteq L$ and a countable subset $Y \subseteq \beta\mathbb{Z}$ such that for every $x \in \beta\mathbb{Z}$ there exists $y \in Y$ such that*

$$n + D \in x \Longleftrightarrow n + D \in y \quad \forall n \in K$$

(treating x and y as ultrafilters on the set $\mathbb{Z}$).

Proof   By the universality of the greatest ambit $(\mathbb{Z}, \beta\mathbb{Z})$ it suffices to check when the function

$$f = \chi_{\overline{D}} : \beta\mathbb{Z} \to \{0, 1\}, \ f(x) = 1 \Leftrightarrow x \in \overline{D},$$

the natural extension function of $\chi_D : \mathbb{Z} \to \{0, 1\}$, is tame (in the usual sense, as a function on the compact cascade $\beta\mathbb{Z}$), where we denote by $\overline{D}$ the closure of D in $\beta\mathbb{Z}$ (a clopen subset). Applying Theorem 18.15 to f we see that the following condition is both necessary and sufficient: For every infinite subset $L \subseteq \mathbb{Z}$ there exists an infinite subset $K \subseteq L$ and a countable subset $Y \subseteq \beta\mathbb{Z}$ which is dense in the pseudometric space $(\beta\mathbb{Z}, \rho_{f,K})$. Now saying that Y is dense is the same as the requiring that Y be ε-dense for every $0 < \varepsilon < 1$. However, as f has values in $\{0, 1\}$ and $0 < \varepsilon < 1$ we conclude that for every $x \in \beta\mathbb{Z}$ there is $y \in Y$ with

$$x \in n + \overline{D} \Longleftrightarrow y \in n + \overline{D} \quad \forall n \in K,$$

and the latter is equivalent to

$$n + D \in x \Longleftrightarrow n + D \in y \quad \forall n \in K.$$

Theorem 18.26 *Let D be a subset of $\mathbb{Z}$. The following conditions are equivalent:*

1. *D is an Asplund subset (i.e., the associated subshift $X_D \subset \{0, 1\}^{\mathbb{Z}}$ is Asplund).*
2. *There exists a countable subset $Y \subseteq \beta\mathbb{Z}$ such that for every $x \in \beta\mathbb{Z}$ there exists $y \in Y$ such that*

$$n + D \in x \Longleftrightarrow n + D \in y \quad \forall n \in \mathbb{Z}.$$

Proof (Sketch) One can modify the proof of Theorem 18.25. Namely, if in assertion (4) of Theorem 18.15 eventual fragmentability of F is replaced by fragmentability then this characterization of Asplund functions, [27] follows.

Example 18.27 $\mathbb{N}$ is an Asplund subset of $\mathbb{Z}$ which is not a WAP subset. In fact, let $X_{\mathbb{N}}$ be the corresponding subshift. Clearly $X_{\mathbb{N}}$ is homeomorphic to the two-point compactification of $\mathbb{Z}$, with $\{0\}$ and $\{1\}$ as minimal subsets. Since a transitive WAP system admits a unique minimal set, we conclude that $X_{\mathbb{N}}$ is not WAP (see e.g. [22]). On the other hand, since $X_{\mathbb{N}}$ is countable we can apply Theorem 18.20 to show that it is HNS. Alternatively, using Theorem 18.26, we can take Y to be $\mathbb{Z} \cup \{p, q\}$, where we choose p and q to be any two non-principal ultrafilters such that p contains $\mathbb{N}$ and q contains $-\mathbb{N}$.

18.5   Entropy and Null Systems

We begin by recalling the basic definitions of topological (sequence) entropy. Let (X, T) be a cascade, i.e., a $\mathbb{Z}$-dynamical system, and $A = \{a_0 < a_1 < \ldots\}$ a sequence of integers. Given an open cover $\mathcal{U}$ define

$$h_{top}^A(T, \mathcal{U}) = \limsup_{n \to \infty} \frac{1}{n} N\left(\bigvee_{i=0}^{n-1} T^{-a_i}(\mathcal{U}) \right)$$

The *topological entropy along the sequence A* is then defined by

$$h_{top}^A(T) = \sup\{h_{top}^A(T, \mathcal{U}) : \mathcal{U} \text{ an open cover of } X\}.$$

When the phase space X is zero-dimensional, one can replace open covers by clopen partitions. We recall that a dynamical system (T, X) is called *null* if $h_{top}^A(T) = 0$ for every infinite $A \subset \mathbb{Z}$. With $A = \mathbb{N}$ one retrieves the usual definition of topological entropy. Finally when $Y \subset \{0, 1\}^{\mathbb{Z}}$, and $A \subset \mathbb{Z}$ is a given subset of $\mathbb{Z}$, we say that Y *is free on A* or that *A is an interpolation set for Y*, if $\{y|_A : y \in Y\} = \{0, 1\}^A$.

By theorems of Kerr and Li [49, 50] every null $\mathbb{Z}$-system is tame, and every tame system has zero topological entropy. From results of Glasner-Weiss [35] (for (1)) and Kerr-Li [50] (for (2) and (3)), the following results can be easily deduced. (See Propositions 3.9.2, 6.4.2 and 5.4.2 of [50] for the positive topological entropy, the untame, and the nonnull claims, respectively.)

Theorem 18.28

1. *A subshift $X \subset \{0, 1\}^{\mathbb{Z}}$ has positive topological entropy iff there is a subset $A \subset \mathbb{Z}$ of positive density such that X is free on A.*
2. *A subshift $X \subset \{0, 1\}^{\mathbb{Z}}$ is not tame iff there is an infinite subset $A \subset \mathbb{Z}$ such that X is free on A.*
3. *A subshift $X \subset \{0, 1\}^{\mathbb{Z}}$ is not null iff for every $n \in \mathbb{N}$ there is a finite subset $A_n \subset \mathbb{Z}$ with $|A_n| \geqslant n$ such that X is free on A_n.*

Proof We consider the second claim; the other claims are similar.

Certainly if there is an infinite $A \subset \mathbb{Z}$ on which X is free then X is not tame (e.g. use Theorem 18.22). Conversely, if X is not tame then, by Propositions 6.4.2 of [50], there exists a non diagonal IT pair (x, y). As x and y are distinct there is an n with, say, $x(n) = 0, y(n) = 1$. Since $T^n(x, y)$ is also an IT pair we can assume that $n = 0$. Thus $x \in U_0$ and $y \in U_1$, where these are the cylinder sets $U_i = \{z \in X : z(0) = i\}, i = 0, 1$. Now by the definition of an IT pair there is an infinite set $A \subset \mathbb{Z}$ such that the pair (U_0, U_1) has A as an independence set. This is exactly the claim that X is free on A.

The following theorem was proved (independently) by Huang [41], Kerr and Li [50], and Glasner [24]. See [26] for a recent generalization of this result.

Theorem 18.29 (A Structure Theorem for Minimal Tame Dynamical Systems)
Let (G, X) be a tame minimal metrizable dynamical system with G an abelian group. Then:

1. *(G, X) is an almost one to one extension $\pi : X \to Y$ of a minimal equicontinuous system (G, Y).*
2. *(G, X) is uniquely ergodic and the factor map π is, measure theoretically, an isomorphism of the corresponding measure preserving system on X with the Haar measure on the equicontinuous factor Y.*

Example 18.30

1. According to Theorem 18.29 the Morse minimal system, which is uniquely ergodic and has zero entropy, is nevertheless not tame as it fails to be an almost 1-1 extension of its adding machine factor. We can therefore deduce that, a fortiori, it is not null.

2. Let $L = IP\{10^t\}_{t=1}^{\infty} \subset \mathbb{N}$ be the IP-sequence generated by the powers of ten, i.e.

$$L = \{10^{a_1} + 10^{a_2} + \cdots + 10^{a_k} : 1 \leqslant a_1 < a_2 < \cdots < a_k\}.$$

Let $f = 1_L$ and let $X = \bar{O}_\sigma(f) \subset \{0, 1\}^{\mathbb{Z}}$, where σ is the shift on $\Omega = \{0, 1\}^{\mathbb{Z}}$. The subshift (σ, X) is not tame. In fact it can be shown that L is an interpolation set for X.

3. Take u_n to be the concatenation of the words $a_{n,i}0^n$, where $a_{n,i}$, $i = 1, 2, 3, \ldots, 2^n$ runs over $\{0, 1\}^n$. Let $v_n = 0^{|u_n|}$, $w_n = u_n v_n$ and w_∞ the infinite concatenation $\{0, 1\}^{\mathbb{N}} \ni w_\infty = w_1 w_2 w_3 \cdots$. Finally define $w \in \{0, 1\}^{\mathbb{Z}}$ by $w(n) = 0$ for $n \leqslant 0$ and $w(n) = w_\infty(n)$. Then $X = \bar{O}_\sigma(w) \subset \{0, 1\}^{\mathbb{Z}}$ is a countable subshift, hence HNS and a fortiori tame, but for an appropriately chosen sequence the sequence entropy of X is $\log 2$. Hence, X is not null. Another example of a countable nonnull subshift can be found in [41, Example 5.12].

4. In [50, Section 11] Kerr and Li construct a Toeplitz subshift (= a minimal almost one-to-one extension of an adding machine) which is tame but not null.

5. In [39, Theorem 13.9] the authors show that for interval maps being tame is the same as being null.

Remark 18.31 Let $T : [0, 1] \to [0, 1]$ be a continuous self-map on the closed interval. In an unpublished paper [52] the authors show that the enveloping semigroup $E(X)$ of the cascade (an $\mathbb{N} \cup \{0\}$-system) $X = [0, 1]$ is either metrizable or it contains a topological copy of $\beta\mathbb{N}$. The metrizable enveloping semigroup case occurs exactly when the system is HNS. This was proved in [40] for group actions but it remains true for semigroup actions [31]. The other case occurs iff σ is Li-Yorke chaotic. Combining this result with Example 18.30(5) one gets: HNS = null = tame, for any cascade $(T, [0, 1])$.

18.6   Some Examples of Tame Functions and Systems

In this section we give some methods for constructing tame systems and functions. It is closely related to the question whether given family of real (not necessarily, continuous) functions is tame.

Recall (see for example [7]) that a bisequence $\mathbb{Z} \to \{0, 1\}$ is *Sturmian* if it is recurrent and has the minimal complexity $p(n) = n + 1$.

Example 18.32

1. (See [27]) Consider an irrational rotation $(R_\alpha, \mathbb{T})$. Choose $x_0 \in \mathbb{T}$ and split each point of the orbit $x_n = x_0 + n\alpha$ into two points $x_n^{\pm}$. This procedure results is a *Sturmian* (symbolic) dynamical system (σ, X) which is a minimal almost 1-1 extension of $(R_\alpha, \mathbb{T})$. Then $E(X, \sigma) \setminus \{\sigma^n\}_{n \in \mathbb{Z}}$ is homeomorphic to the two arrows space, a basic example of a non-metrizable Rosenthal compactum. It follows that $E(\sigma, X)$ is also a Rosenthal compactum. Hence, (σ, X) is tame but not HNS.

2. Let P_0 be the set $[0, c)$ and P_1 the set $[c, 1)$; let z be a point in $[0, 1)$ (identified with $\mathbb{T}$) via the rotation R_α we get the binary bisequence $u_n, n \in \mathbb{Z}$ defined by $u_n = 0$ when $R_\alpha^n(z) \in P_0, u_n = 1$ otherwise. These are called *Sturmian like codings*. With $c = 1 - \alpha$ we retrieve the previous example. For example, when $\alpha := \frac{\sqrt{5}-1}{2}$ the corresponding sequence, computed at $z = 0$, is called the *Fibonacci bisequence*.

Example 18.33

1. In his paper [15] Ellis, following Furstenberg's classical work [18], investigates the projective action of $GL(n, \mathbb{R})$ on the projective space $\mathbb{P}^{n-1}$. It follows from his results that the corresponding enveloping semigroup is not first countable. However, in a later work [1], Akin studies the action of $G = GL(n, \mathbb{R})$ on the sphere $\mathbb{S}^{n-1}$ and shows that here the enveloping semigroup is first countable (but not metrizable). It follows that the dynamical systems $D_1 = (G, \mathbb{P}^{n-1})$ and $D_2 = (G, \mathbb{S}^{n-1})$ are tame but not HNS. Note that $E(D_1)$ is Fréchet, being a quotient of a first countable compact space, namely $E(D_2)$.
2. (Huang [41]) An almost 1-1 extension $\pi : X \to Y$ of an equicontinuous metric $\mathbb{Z}$-system Y with $X \setminus X_0$ countable, where $X_0 = \{x \in X : |\pi^{-1}\pi(x)| = 1\}$, is tame.

We will see that many coding functions are tame, including some multidimensional analogues of Sturmian sequences. The latter are defined on the groups $\mathbb{Z}^k$ and instead of the characteristic function $f := \chi_D$ (with $D = [0, c)$) one may consider coloring of the space leading to shifts with finite alphabet. We give a precise definition which (at least in some partial cases) was examined in several papers. Regarding some dynamical and combinatorial aspects of coding functions see for example [6, 17, 65], and the survey paper [7].

Definition 18.34 Consider an arbitrary finite partition

$$\mathbb{T} = \cup_{i=0}^d [c_i, c_{i+1})$$

of $\mathbb{T}$ by the ordered d-tuple of points $c_0 = 0, c_1, \ldots, c_d, c_{d+1} = 1$ and any coloring map

$$f : \mathbb{T} \to A := \{0, \ldots, d\}.$$

Now for a given k-tuple $(\alpha_1, \ldots, \alpha_k) \in \mathbb{T}^k$ and a given point $z \in \mathbb{T}$ consider the corresponding coding function

$$m(f, z) : \mathbb{Z}^k \to \{0, \ldots, d\} \quad (n_1, \ldots, n_k) \mapsto f(z + n_1\alpha_1 + \cdots + n_k\alpha_k).$$

We call such a sequence a *multidimensional (k, d)-Sturmian like sequence*.

Lemma 18.35

1. Let $q : X_1 \to X_2$ be a map between sets and $\{f_n : X_2 \to \mathbb{R}\}_{n \in \mathbb{N}}$ a bounded sequence of functions (with no continuity assumptions on q and f_n). If $\{f_n \circ q\}$ is an independent sequence on X_1 then $\{f_n\}$ is an independent sequence on X_2.
2. If q is onto then the converse is also true. That is $\{f_n \circ q\}$ is independent if and only if $\{f_n\}$ is independent.
3. Let $\{f_n\}$ be a bounded sequence of continuous functions on a topological space X. Let Y be a dense subset of X. Then $\{f_n\}$ is an independent sequence on X if and only if the sequence of restrictions $\{f_n|_Y\}$ is an independent sequence on Y.

Proof Claims (1) and (2) are straightforward.

(3) Since $\{f_n\}$ is an independent sequence for every pair of finite disjoint sets $P, M \subset \mathbb{N}$, the set

$$\bigcap_{n \in P} f_n^{-1}(-\infty, a) \cap \bigcap_{n \in M} f_n^{-1}(b, \infty)$$

is non-empty. This set is open because every f_n is continuous. Hence, each of them meets the dense set Y. As $f_n^{-1}(-\infty, a) \cap Y = f_n|_Y^{-1}(-\infty, a)$ and $f_n^{-1}(b, \infty) \cap Y = f_n|_Y^{-1}(b, \infty)$, this implies that $\{f_n|_Y\}$ is an independent sequence on Y.

Conversely if $\{f_n|_Y\}$ is an independent sequence on a subset $Y \subset X$ then by (1) (where q is the embedding $Y \hookrightarrow X$), $\{f_n\}$ is an independent sequence on X.

Below we will sometimes deal with (not necessarily continuous) functions $f : X \to \mathbb{R}$ such that the orbit fS of f in $\mathbb{R}^X$ is a tame family (Definition 18.3). An example of such Baire 1 function (which is not tame, being discontinuous), is the characteristic function χ_D of an arc $D = [a, a + s) \subset \mathbb{T}$ defined on the system $(R_\alpha, \mathbb{T})$, where R_α is an irrational rotation of the circle $\mathbb{T}$. See Theorem 18.40.

Lemma 18.36 *Let S be a semigroup, X a (not necessarily compact) S-space and $f : X \to \mathbb{R}$ a bounded (not necessarily continuous) function.*

1. Let $f \in \mathrm{RUC}(X)$; then $f \in \mathrm{Tame}(X)$ if and only if fS is a tame family. Moreover, there exists an S-compactification $v : X \to Y$ where the action $S \times Y \to Y$ is continuous, Y is a tame system and $f = \tilde{f} \circ v$ for some $\tilde{f} \in C(Y)$.
2. Let G be a topological group and $f \in \mathrm{RUC}(G)$. Then $f \in \mathrm{Tame}(G)$ if and only if fG is a tame family.
3. Let L be a discrete semigroup and $f : L \to \mathbb{R}$ a bounded function. Then $f \in \mathrm{Tame}(L)$ if and only if fL is a tame family.
4. Let $h : L \to S$ be a homomorphism of semigroups, $S \times Y \to Y$ be an action (without any continuity assumptions) on a set Y and $f : Y \to \mathbb{R}$ be a bounded function such that fL is a tame family. Then for every point $y \in Y$ the corresponding coding function $m(f, y) : L \to \mathbb{R}$ is tame on the discrete semigroup (L, τ_{discr}).

Proof For (1) consider the cyclic S-compactification $v : X \to Y = X_f$ (see Definition 18.12). Since $f \in \mathrm{RUC}(X)$ the action $S \times X_f \to X_f$ is jointly continuous

(Remark 18.14). By the basic property of the cyclic compactification there exists a continuous function $\tilde{f} : X_f \to \mathbb{R}$ such that $f = \tilde{f} \circ \nu$. The family fS has no independent sequence. By Lemma 18.35(3) we conclude that also $\tilde{f}S$ has no independent sequence. This means, by Theorem 18.15, that $\tilde{f}$ is tame. Hence (by Definition) so is f. The converse follows from Lemma 18.35(1).

(2) and (3) follow easily from (1) (with $X = G = L$) taking into account that on a discrete semigroup L every bounded function $L \to \mathbb{R}$ is in RUC(L).

(4) By (3) it is enough to show for the coding function $f_0 := m(f, y)$ that the family f_0L has no independent subsequence. Define $q : L \to Y, s \mapsto h(s)y$. Then $f_0t = (ft) \circ q$ for every $t \in L$. If f_0t_n is an independent sequence for some sequence $t_n \in L$ then Lemma 18.35(1) implies that the sequence of functions ft_n on Y is independent. This contradicts the assumption that fL has no independent subsequence.

Let $f : X \to \mathbb{R}$ be a real function on a topological space X. We denote by $cont(f)$ and $disc(f)$ the sets of points of continuity and discontinuity for f respectively.

Definition 18.37 Let F be a family of functions on X and $Y \subset X$. We say that F is:

1. *Strongly almost continuous* on Y if for every $x \in Y$ we have $x \in cont(f)$ for almost all $f \in F$ (i.e. with the exception of at most a finite set of elements which may depend on x).
2. *Almost continuous* if for every infinite (countable) subset $F_1 \subset F$ there exists an infinite subset $F_2 \subset F_1$ and a countable subset $C \subset X$ such that F_2 is strongly almost continuous on the complement $X \setminus C$.

For example, if $\bigcup\{disc(f) : f \in F\}$ is countable then F is almost continuous.

Theorem 18.38 *Let X be a compact metric space and F a bounded family of real valued functions on X such that F is almost continuous. Further assume that:*

(∗) *for every sequence $\{f_n\}_{n\in\mathbb{N}}$ in F there exists a subsequence $\{f_{n_m}\}_{m\in\mathbb{N}}$ and a countable subset $C \subset X$ such that $\{f_{n_m}\}_{m\in\mathbb{N}}$ pointwise converges on $X \setminus C$ to a function $\phi : X \setminus C \to \mathbb{R}$ where $\phi \in \mathcal{B}_1(X \setminus C)$.*

Then F is a tame family.

Proof Assuming the contrary let $\{f_n\}$ be an independent sequence in F. Then, by assumption, there exists a countable subset $C \subset X$ and a subsequence $\{f_{n_m}\}$ such that $\{f_{n_m} : X \setminus C \to \mathbb{R}\}$ pointwise converges on $X \setminus C$ to a function $\phi : X \setminus C \to \mathbb{R}$ such that $\phi \in \mathcal{B}_1(X \setminus C)$.

Independence is preserved by subsequences so this subsequence $\{f_{n_m}\}$ remains independent. For simplicity of notation assume that $\{f_n\}$ itself has the properties of $\{f_{n_m}\}$. Moreover we can suppose in addition, by Definition 18.37 with the same $C \subset X$, that $\{f_n\}$ is strongly almost continuous. That is, for every $x \in X \setminus C$ we have $x \in cont(f_n)$ for almost all f_n.

By the definition of independence, there exist $a < b$ such that for every pair of disjoint finite sets $P, M \subset \mathbb{N}$ we have

$$\bigcap_{n \in P} A_n \cap \bigcap_{n \in M} B_n \neq \emptyset,$$

where $A_n := f_n^{-1}(-\infty, a)$ and $B_n := f_n^{-1}(b, \infty)$. Now define a tree of nested sets as follows:

$$\Omega_1 := X$$
$$\Omega_2 := \Omega_1 \cap A_1 = A_1 \qquad \Omega_3 := \Omega_1 \cap B_1 = B_1$$
$$\Omega_4 := \Omega_2 \cap A_2 \qquad \Omega_5 := \Omega_2 \cap B_2 \qquad \Omega_6 := \Omega_3 \cap A_2 \qquad \Omega_7 := \Omega_3 \cap B_2,$$

and so on. In general,

$$\Omega_{2^{n+1}+2k} := \Omega_{2^n+k} \cap A_{n+1}, \qquad \Omega_{2^{n+1}+2k+1} := \Omega_{2^n+k} \cap B_{n+1}$$

for every $0 \leqslant k < 2^n$ and every $n \in \mathbb{N}$.

We obtain a system $\{\Omega_n\}_{n \in \mathbb{N}}$ which satisfies:

$$\Omega_{2n} \cup \Omega_{2n+1} \subset \Omega_n \text{ and } \Omega_{2n} \cap \Omega_{2n+1} = \emptyset \text{ for each } n \in \mathbb{N}.$$

Since $\{(A_n, B_n)\}_{n \in \mathbb{N}}$ is independent (in the sense of [68]), every Ω_n is nonempty.

For every binary sequence $u = (u_1, u_2, \dots) \in \{0, 1\}^{\mathbb{N}}$ we have the corresponding uniquely defined *branch*

$$\alpha_u := \Omega_1 \supset \Omega_{n_1} \supset \Omega_{n_2} \supset \cdots$$

where for each $i \in \mathbb{N}$ with $2^{i-1} \leqslant n_i < 2^i$ we have

$$n_{i+1} = 2n_i \text{ iff } u_i = 0 \text{ and } n_{i+1} = 2n_i + 1 \text{ iff } u_i = 1.$$

Let us say that $u, v \in \{0, 1\}^{\mathbb{N}}$ are *essentially distinct* if they have infinitely many different coordinates. Equivalently, if u and v are in different cosets of the Cantor group $\{0, 1\}^{\mathbb{N}}$ with respect to the subgroup H consisting of the binary sequences with finite support. Since H is countable there are uncountably many pairwise essentially distinct elements in the Cantor group. We choose a subset $T \subset \{0, 1\}^{\mathbb{N}}$ which intersects each coset in exactly one point. Clearly, $card(T) = 2^\omega$. Now for every branch α_u where $u \in T$ choose one element

$$x_u \in \bigcap_{i \in \mathbb{N}} cl(\Omega_{n_i}).$$

Here we use the compactness of X which guarantees that $\bigcap_{i \in \mathbb{N}} cl(\Omega_{n_i}) \neq \emptyset$. We obtain a set $X_T := \{x_u : u \in T\} \subset X$ and an onto function $T \to X_T, \ u \mapsto x_u$.

Define also $T_0 := \{u \in T : x_u \in X_T \cap C\}$ which is at most countable (possibly empty).

Claim

1. The restricted function $T \setminus T_0 \to X_T \setminus C$, $u \mapsto x_u$ is injective. In particular, $X_T \setminus C$ is uncountable.
2. $|\phi(x_u) - \phi(x_v)| \geq \varepsilon := b - a$ for every distinct $x_u, x_v \in X_T \setminus C$.

Proof of the Claim (1) Let $u = (u_i)$ and $v = (v_i)$ are distinct elements in $T \setminus T_0$. Denote by $\alpha_u := \{\Omega_{n_i}\}_{i \in \mathbb{N}}$ and $\alpha_v := \{\Omega_{m_i}\}_{i \in \mathbb{N}}$ the corresponding branches. Then, by the definition of X_T, we have the uniquely defined points $x_u \in \bigcap_{i \in \mathbb{N}} cl(\Omega_{n_i})$ and $x_v \in \bigcap_{i \in \mathbb{N}} cl(\Omega_{m_i})$ in $X_T \setminus C$. Since $u, v \in T \setminus T_0$ are essentially distinct they have infinitely many different indices. As $\{f_n\}$ is strongly almost continuous on $X \setminus C$ there exists a sufficiently large $t_0 \in \mathbb{N}$ such that the points x_u and x_v are both points of continuity of f_n for every $n \geq t_0$.

Now note that if $u_i \neq v_i$ then the sets Ω_{n_i+1} and Ω_{m_i+1} are contained (respectively) in a pair of disjoint sets $A_k := f_k^{-1}(-\infty, a)$ and $B_k := f_k^{-1}(b, \infty)$ with $k \geq 2^i$. Since u and v are essentially distinct we can assume that i is sufficiently large in order to ensure that $k \geq t_0$. We necessarily have exactly one of the cases:

$$(a) \quad \Omega_{n_i+1} \subset A_k, \quad \Omega_{m_i+1} \subset B_k$$

or

$$(b) \quad \Omega_{n_i+1} \subset B_k, \quad \Omega_{m_i+1} \subset A_k.$$

For simplicity we only check the first case (a). For (a) we have

$$x_u \in \mathrm{cls}\,(\Omega_{n_i+1}) \subset \mathrm{cls}\,(f_k^{-1}(-\infty, a)) \quad \text{and} \quad x_v \in \mathrm{cls}\,(\Omega_{n_i+1}) \subset \mathrm{cls}\,(f_k^{-1}(b, \infty)).$$

Since $\{x_u, x_v\} \subset cont(f_n)$ are continuity points for every $n \geq t_0$ and since $k \geq t_0$ by our choice, we obtain $f_k(x_u) \leq a$ and $f_k(x_v) \geq b$. So, we can conclude that $|f_k(x_u) - f_k(x_v)| \geq \varepsilon := b - a$ for every $k \geq t_0$. In particular, x_u and x_v are distinct. This proves (1).

(2) Furthermore, for our distinct $x_u, x_v \in X_T \setminus C$ by (*) we have $\lim f_k(x_u) = \phi(x_u)$ and $\lim f_k(x_v) = \phi(x_v)$. It follows that $|\phi(x_u) - \phi(x_v)| \geq \varepsilon$ and the condition (2) of our claim is also proved.

Define

$$Q := \{x \in X_T \setminus C : \text{there exists a countable open nbd } O_x \text{ of } x \text{ in the space } X_T \setminus C\}.$$

Observe that $Q = \bigcup\{O_x : x \in Q\}$. Since Q is second countable, by Lindelof property there exists a countable subcover. Hence, Q is at most a countable subset of $X_T \setminus C$ and any point $y \in Y := (X_T \setminus C) \setminus Q$ is a condensation point.

Now, it follows by assertion (2) of the Claim that for every open subset U in X with $U \cap Y \neq \emptyset$ we have $\mathrm{diam}(\phi(U \cap Y)) \geq \varepsilon$. This means that $\phi : X \setminus C \to \mathbb{R}$ is not fragmented. Since C is countable and X is compact metrizable the subset $X \setminus C$ is Polish. On Polish spaces fragmentability and Baire 1 property are the same for real valued functions (Lemma 18.2(3)). So, we obtain that $\phi : X \setminus C \to \mathbb{R}$ is not Baire 1. This contradicts the assumption that $\phi \in \mathcal{B}_1(X \setminus C)$.

Theorem 18.39 *Let X be a compact metric tame G-system and $f : X \to \mathbb{R}$ be a bounded function such that:*

$$p^{-1}(disc(f)) \text{ is countable for every } p \in E(X).$$

Then fG is a tame family.

Proof Assuming the contrary let $\{g_n\}_{n \in \mathbb{N}}$ be a sequence in G such that $F :=$ $\{fg_n\}_{n \in \mathbb{N}}$ is independent. Since X is a tame metric system there exists a subsequence of $\{g_n\}_{n \in \mathbb{N}}$ which converges to some $p \in E(X)$. For simplicity of the notation we assume that $\{fg_n\}_{n \in \mathbb{N}}$ is independent and also p is the pointwise limit of $\{g_n\}_{n \in \mathbb{N}}$ (in fact, of $\{j(g_n)\}_{n \in \mathbb{N}}$), where $j : G \to E(X) \subset X^X$ is the Ellis compactification).

$$C := \bigcup \{t^{-1}(disc(f)) : t \in \{p, g_n\}_{n \in \mathbb{N}}\} \text{ is countable}$$

In particular, since $g_n^{-1} cont(f) = cont(fg_n)$, this implies that F is almost continuous. Indeed, there exists the pointwise limit of the sequence $\{fg_n\}_{n \in \mathbb{N}}$ on $X \setminus C$ and it equals to fp. Since every fg_n is continuous on the Polish space $X \setminus C$ we obtain that $fp \in \mathcal{B}_1(X)$. Now Theorem 18.38 applies.

Corollary 18.40 *Let X be a compact metric equicontinuous G-system and $f : X \to \mathbb{R}$ be a bounded function with countably many discontinuities. Then fG is a tame family.*

Proof Since $disc(f)$ is countable and every $p \in E(X)$ is reversible (since X is a distal G-system) $p^{-1}(disc(f))$ remains countable.

Remark 18.41 As a consequence of previous results (Theorems 18.39, 18.40 and Lemma 18.36) coding functions $m(f, z) : G_0 \to \mathbb{R}$ are tame on every subgroup G_0 of G where G_0 is endowed with the discrete topology (for every given point $z \in X$). So, the corresponding cyclic system (G_0, X_f) is tame. Moreover, if $m(f, z)(G_0) = A$ is finite we get a tame symbolic system on the alphabet A (see Definition 18.16).

Example 18.42

1. For every irrational rotation α of the circle $\mathbb{T}$ and an arc $D := [a, b) \subset \mathbb{T}$ the following classical coding function is tame.

$$\varphi_D := \mathbb{Z} \to \mathbb{R}, \quad n \mapsto \chi_D(n\alpha)$$

2. More generally, the multidimensional Sturmian (k, d)-sequences $\mathbb{Z}^k \to \{0, 1, \ldots, d\}$ (Definition 18.34) are tame.

Proof In terms of Definition 18.34 consider the homomorphism

$$h : \mathbb{Z}^k \to \mathbb{T}, \quad (n_1, \ldots, n_k) \mapsto n_1\alpha_1 + \cdots + n_k\alpha_k.$$

Then any coloring function $f : \mathbb{T} \to A := \{0, \ldots, d\}$, has only finitely many discontinuities. Now Corollary 18.40 guarantees that the corresponding (k, d)-coding function $m(f, z) : \mathbb{Z}^k \to A \subset \mathbb{R}$ is tame for every $z \in \mathbb{T}$.

Note that the tameness of the functions on $\mathbb{Z}^k$ from Example 18.42 follows also by results from [34]. In fact, such functions come from circularly ordered metric dynamical $\mathbb{Z}^k$-systems. See also Theorem 18.58 below.

18.6.1 A Special Class of Generalized Sturmian Systems

Let R_α be an irrational rotation of the torus $\mathbb{T}^d$. In many cases a reasonably chosen subset $D \subset \mathbb{T}^d$ will yield a generalized Sturmian system.

Example 18.43 Let $\alpha = (\alpha_1, \ldots, \alpha_d)$ be a vector in $\mathbb{R}^d$, $d \geq 2$ with $1, \alpha_1, \ldots, \alpha_d$ independent over $\mathbb{Q}$. Consider the minimal equicontinuous dynamical system (R_α, Y), where $Y = \mathbb{T}^d = \mathbb{R}^d/\mathbb{Z}^d$ (the d-torus) and $R_\alpha y = y + \alpha$. Let D be a small closed d-dimensional ball in $\mathbb{T}^d$ and let $C = \partial D$ be its boundary, a $(d - 1)$-sphere. Fix $y_0 \in \text{int}D$ and let $X = X(D, y_0)$ be the symbolic system generated by the function

$$x_0 \in \{0, 1\}^{\mathbb{Z}} \text{ defined by } x_0(n) = \chi_D(R_\alpha^n y_0), \qquad X = \overline{O_\sigma x_0} \subset \{0, 1\}^{\mathbb{Z}},$$

where σ denotes the shift transformation. This is a well known construction and it is not hard to check that the system (σ, X) is minimal and admits (R_α, Y) as an almost 1-1 factor:

$$\pi : (\sigma, X) \to (R_\alpha, Y).$$

Theorem 18.44 *There exists a ball $D \subset \mathbb{T}^d$ as above such that the corresponding symbolic dynamical system (σ, X) is tame. For such D we then have a precise description of $E(\sigma, X) \setminus \mathbb{Z}$ as the product set $\mathbb{T}^d \times \mathcal{F}$, where $\mathcal{F}$ is the collection of ordered orthonormal bases for $\mathbb{R}^d$.*

Proof

1. First we show that a sphere $C \subset [0, 1)^d \cong \mathbb{T}^d$ can be chosen so that for every $y \in \mathbb{T}^d$ the set $(y + \{n\alpha : n \in \mathbb{Z}\}) \cap C$ is finite. We thank Benjamin Weiss for providing the following proof of this fact.

(a) For the case $d = 2$ the argument is easy. If A is any countable subset of the square $[0, 1) \times [0, 1)$ there are only a countable number of circles that contain three points of A. These circles have some countable collection of radii. Take any circle with a radius which is different from all of them and no translate of it will contain more than two points from the set A. Taking $A = \{n\alpha : n \in \mathbb{Z}\}$ we obtain the required circle.

(b) We next consider the case $d = 3$, which easily generalizes to the general case $d \geqslant 3$. What we have to show is that there can not be infinitely many points in

$$A = \{(n\alpha_1 - [n\alpha_1], \alpha_2 - [n\alpha_2], \alpha_3 - [n\alpha_3]) : n \in \mathbb{Z}\}$$

that lie on a plane. For if that is the case, we consider all 4-tuples of elements from the set A that do not lie on a plane to get a countable set of radii for the spheres that they determine. Then taking a sphere with radius different from that collection we obtain our required sphere. In fact, if a sphere contains infinitely many points of A and no 4-tuple from A determines it then they all lie on a single plane.

So suppose that there are infinitely many points in A whose inner product with a vector $v = (z, x, y)$ is always equal to 1. This means that there are infinitely many equations of the form:

$$z\alpha_1 + x\alpha_2 + y\alpha_3 = 1/n + z[n\alpha_1]/n + x[n\alpha_2]/n + y[n\alpha_3]/n. \qquad (*)$$

Subtract two such equations with the second using m much bigger than n so that the coefficient of y cannot vanish. We can express $y = rz + sx + t$ with r, s and t rational. This means that we can replace $(*)$ by

$$z\alpha_1 + x\alpha_2 + y\alpha_3 = 1/n + t[n\alpha_3]/n + z([n\alpha_1]/n + r[n\alpha_3]/n)$$
$$+ x([n\alpha_2]/n + s[n\alpha_3]/n). \qquad (**)$$

Now r, s and t have some fixed denominators and (having infinitely many choices) we can take another equation like $(**)$ where n (and the corresponding r, s, t) is replaced by some much bigger k, then subtract again to obtain an equation of the form $x = pz + q$ with p and q rational. Finally one more step will show that z itself is rational. However, in view of $(*)$, this contradicts the independence of $1, \alpha_1, \alpha_2, \alpha_3$ over $\mathbb{Q}$ and our proof is complete.

2. Next we show that for C as above

for every converging sequence $n_i\alpha$, say $n_i\alpha \to \beta \in \mathbb{T}^d \cong E(R_\alpha, \mathbb{T}^d)$, there exists a subsequence $\{n_{i_j}\}$ such that for every $y \in \mathbb{T}^d$, $y + n_{i_j}\alpha$ is either eventually in the interior of D or eventually in its exterior.

Clearly we only need to consider points $y \in C - \beta$. Renaming we can now assume that $n_i\alpha \to 0$ and that $y \in C$. Passing to a subsequence if necessary we can further assume that the sequence of unit vectors $\frac{n_i\alpha}{\|n_i\alpha\|}$ converges,

$$\frac{n_i\alpha}{\|n_i\alpha\|} \to v_0 \in \mathbb{S}^{d-1}.$$

In order to simplify the notation we now assume that C is centered at the origin. For every point $y \in C$ where $\langle y, v_0 \rangle \neq 0$ we have that $y + n_i\alpha$ is either eventually in the interior of D or eventually in its exterior. On the other hand, for the points $y \in C$ with $\langle y, v_0 \rangle = 0$ this is not necessarily the case. In order to deal with these points we need a more detailed information on the convergence of $n_i\alpha$ to β. At this stage we consider the sequence of orthogonal projections of the vectors $n_i\alpha$ onto the subspace $V_1 = \{u \in \mathbb{R}^d : \langle u, v_0 \rangle = 0\}$, say $u_i = \mathrm{proj}_{v_0}(n_i\alpha) \to u = \mathrm{proj}_{v_0}(\beta)$. If it happens that eventually $u_i = 0$, this means that all but a finite number of the $n_i\alpha$'s are on the line defined by v_0 and our required property is certainly satisfied.[2] Otherwise we choose a subsequence (again using the same index) so that

$$\frac{u_i}{\|u_i\|} \to v_1 \in \mathbb{S}^{d-2}.$$

Again (as we will soon explain) it is not hard to see that for points $y \in C \cap V_1$ with $\langle y, v_1 \rangle \neq 0$ we have that $y + n_i\alpha$ is either eventually in the interior of D or eventually in its exterior. For points $y \in C \cap V_1$ with $\langle y, v \rangle = 0$ we have to repeat this procedure. Considering the subspace $V_2 = \{u \in V_1 : \langle u, v_1 \rangle = 0\}$, we define the sequence of projections $u_i' = \mathrm{proj}_{v_1}(u_i) \in V_2$ and pass to a further subsequence which converges to a vector v_2

$$\frac{u_i'}{\|u_i'\|} \to v_2 \in \mathbb{S}^{d-3}.$$

Inductively this procedure will produce an **ordered orthonormal basis** $\{v_0, v_1, \ldots, v_{d-1}\}$ for $\mathbb{R}^d$ and a final subsequence (which for simplicity we still denote as n_i) such that

for each $y \in \mathbb{T}^d$, $y + n_i\alpha$ is either eventually in the interior of D or it is eventually in its exterior.

This is clear for points $y \in \mathbb{T}^d$ such that $y + \beta \notin C$. Now suppose we are given a point y with $y + \beta \in C$. We let k be the first index with $\langle y + \beta, v_k \rangle \neq 0$. As $\{v_0, v_1, v_2, \ldots, v_{d-1}\}$ is a basis for $\mathbb{R}^d$ such k exists. We claim that the sequence $y + n_i\alpha$ is either eventually in the interior of D or it is eventually in its exterior. To see this consider the affine hyperplane which is tangent to C at $y + \beta$ (which

---

[2] Actually this possibility can not occur, as is shown in the first step of the proof.

contains the vectors $\{v_0, \ldots, v_{k-1}\}$). Our assumption implies that the sequence $y + n_i \alpha$ is either eventually on the opposite side of this hyperplane from the sphere, in which case it certainly lies in the exterior of D, or it eventually lies on the same side as the sphere. However in this latter case it can not be squeezed in between the sphere and the tangent hyperplane, as this would imply $\langle y+\beta, v_k \rangle = 0$, contradicting our assumption. Thus it follows that in this case the sequence $y + n_i \alpha$ is eventually in the interior of D.

3. Let now p be an element of $E(\sigma, X)$. We choose a **net** $\{n_\nu\} \subset \mathbb{Z}$ with $\sigma^{n_\nu} \to p$. It defines uniquely an element $\beta \in E(Y) \cong \mathbb{T}^d$ so that $\pi(px) = \pi(x) + \beta$ for every $x \in X$. Taking a subnet if necessary we can assume that the net $\frac{\beta - n_\nu \alpha}{\|\beta - n_\nu \alpha\|}$ converges to some $v_0 \in S^{d-1}$. And, as above, proceeding by induction we assume likewise that all the corresponding limits $\{v_0, \ldots, v_{k-1}\}$ exist.

   Next we choose a **sequence** $\{n_i\}$ such that $n_i \alpha \to \beta$, $\frac{\beta - n_i \alpha}{\|\beta - n_i \alpha\|} \to v_0$ etc. We conclude that $\sigma^{n_i} \to p$. Thus every element of $E(\sigma, X)$ is obtained as a limit of a sequence in $\mathbb{Z}$ and is therefore of Baire class 1.

4. From the proof we see that the elements of $E(\sigma, X) \setminus \mathbb{Z}$ can be parametrized by the set $\mathbb{T}^d \times \mathcal{F}$, where $\mathcal{F}$ is the collection of ordered orthonormal bases for $\mathbb{R}^d$, $p \mapsto (\beta, \{v_0, \ldots, v_{d-1}\})$.

18.6.2   Strong Almost 1-1 Equivalence and Tameness

Recall that a G-factor $\pi : X \to Y$ is said to be an *almost one-to-one extension* if

$$X_0 := \{x \in X : \ |\pi^{-1}(\pi(x))| = 1\}$$

is a residual subset of X. We will say that $\pi : X \to Y$ is a *strongly almost 1-1 extension* if $X \setminus X_0$ is at most countable.

We say that compact dynamical G-systems X, Y are *strongly almost 1-1 equivalent* if there exist a continuous G-map $\pi : X \to Y$ and two countable subsets $X_1 \subset X, Y_1 \subset Y$ such that the restriction $\pi : X \setminus X_1 \to Y \setminus Y_1$ is bijective. One may show that a surjective strongly almost 1-1 equivalence $\pi : X \to Y$ is exactly a strongly almost 1-1 extension.

Remark 18.45 In [45, p. 30] Jolivet calls a strong almost 1-1 equivalence "semi-conjugation". However, the name semi-conjugation is often used as a synonym to factor map; so we use "strong almost 1-1 equivalence" instead.

The next lemma is well known; for completeness we provide the short proof.

Lemma 18.46 *Let $\pi : X \to Y$ be a continuous onto G-map of compact metric G-systems. Set*

$$X_0 := \{x \in X : \ |\pi^{-1}(\pi(x))| = 1\}.$$

Then the restriction map $\pi : X_0 \to Y_0$ is a topological homeomorphism of G-subspaces, where $Y_0 := \pi(X_0)$.

Proof Since G is a group X_0 and Y_0 are G-invariant. The map $\pi : X_0 \to Y_0$ is a continuous bijection. For every converging sequence $y_n \to y$, where $y_n, y \in Y_0$ the preimage $\pi^{-1}(\{y\} \cup \{y_n\}_{n\in\mathbb{N}})$ is a compact subset of X. On the other hand, $\pi^{-1}(\{y\} \cup \{y_n\}) \subset X_0$ by the definition of X_0. It follows that the restriction of π to $\pi^{-1}(\{y\} \cup \{y_n\})$ is a homeomorphism and $\pi^{-1}(y_n)$ converges to $\pi^{-1}(y)$.

As a corollary of Theorem 18.38 one can derive the following result which generalizes the above mentioned result of Huang [41] from Example 18.33(2).

Theorem 18.47 ([33]) *Let $\pi : X \to Y$ be a strongly almost 1-1 extension of compact metric G-systems. Assume that the dynamical system (G, Y) is tame and that the set $p^{-1}(y)$ is (at most) countable for every $p \in E(Y)$ and $y \in Y$,[3] then (G, X) is also tame.*

Proof We have to show that every $f \in C(X)$ is tame. Assuming the contrary, suppose fG contains an independent sequence fs_n. Since Y is metrizable and tame, one can assume (by Theorem 18.11) that the sequence s_n converges pointwise to some element p of $E(G, Y)$. Consider the set $Y_0 \cap p^{-1}Y_0$, where $Y_0 = \pi(X_0)$. Since $p^{-1}(y)$ is countable for every $y \in Y \setminus Y_0$ it follows that $Y \setminus (Y_0 \cap p^{-1}Y_0)$ is countable. Therefore, by the definition of X_0 and the countability of $X \setminus X_0$, we see that $X \setminus \pi^{-1}(Y_0 \cap p^{-1}Y_0)$ is also countable. Now observe that the sequence $(fs_n)(x)$ converges for every $x \in \pi^{-1}(Y_0 \cap p^{-1}Y_0)$. Indeed if we denote $y = \pi(x)$ then $s_n y$ converges to py in Y. In fact we have $py \in Y_0$ (by the choice of x) and $s_n y \in Y_0$. By Lemma 18.46, $\pi : X_0 \to Y_0$ is a G-homeomorphism. So we obtain that $s_n x$ converges to $\pi^{-1}(py)$ in X_0. Since $f : X \to \mathbb{R}$ is continuous, $(fs_n)(x)$ converges to $f(\pi^{-1}(py))$ in $\mathbb{R}$. Each fs_n is a continuous function, hence so is also its restriction to $\pi^{-1}(Y_0 \cap p^{-1}Y_0)$. Therefore the limit function $\phi : \pi^{-1}(Y_0 \cap p^{-1}Y_0) \to \mathbb{R}$ is Baire 1. Since $C := X \setminus \pi^{-1}(Y_0 \cap p^{-1}Y_0)$ is countable and fs_n is an independent sequence and Theorem 18.38 provides the sought-after contradiction.

Corollary 18.48 (Huang [41] for Cascades) *Let $\pi : X \to Y$ be a strong almost 1-1 equivalence of compact metric G-systems, where Y is equicontinuous. Then X is tame.*

Proof Consider the induced factor $X \to f(X) \subset Y$ and apply Theorem 18.47.

Recall the following:

Problem 18.49 (A Version of Pisot Conjecture [45, page 31]) Is it true that every (unimodular) irreducible Pisot substitution dynamical system is semi-conjugate to a toral translation?

---

[3]E.g., this latter condition is always satisfied when Y is distal. Another example of such (non-distal) system is the Sturmian system.

By Corollary 18.48 a related question is:

Problem 18.50 (A Weaker Form of Pisot Conjecture) Is it true that every substitutional symbolic dynamical system with Pisot conditions above is always tame?

Remark 18.51 Jolivet [45, Theorem 3.1.1], in the context of Pisot conjecture, discusses some (substitution) dynamical systems which semi-conjugate to a translation on the two-dimensional torus. In particular:

(a) (Rouzy) Tribonacci 3-letter substitution.
(b) Arnoux-Rouzy substitutions.
(c) Brun substitution.
(d) Jacobi-Perron substitution.

By Corollary 18.48 all these systems are tame.

18.7   Remarks About Order Preserving Systems

18.7.1   Order Preserving Action on the Unit Interval

Recall that for the group $G = H_+[0, 1]$ comprising the orientation preserving self-homeomorphisms of the unit interval, the G-system $X = [0, 1]$ with the obvious G-action is tame [31]. One way to see this is to observe that the enveloping semigroup of this dynamical system naturally embeds into the Helly compact space (and hence is a Rosenthal compact space). By Theorem 18.11, (G, X) is tame. We list here some other properties of $H_+[0, 1]$.

Remark 18.52 Let $G := H_+[0, 1]$. Then

1. *(Pestov [63])* G is extremely amenable.
2. [28] WAP(G) = Asp(G) = SUC(G) = {*constants*} and every Asplund representation of G is trivial.
3. [31] G is representable on a (separable) Rosenthal space.
4. *(Uspenskij [74, Example 4.4])* G is Roelcke precompact.
5. UC$(G) \subset$ Tame(G), that is, the Roelcke compactification of G is tame.
6. Tame$(G) \neq$ UC(G).
7. Tame$(G) \neq$ RUC(G), that is, G admits a transitive dynamical system which is not tame.
8. [58] $H_+[0, 1]$ and $H_+(\mathbb{T})$ are minimal topological groups.

In properties (5) and (6) we answer two questions of Ibarlucia which are related to [42]. For the details see [33].

Theorem 18.53 *The Polish group $G = H_+(\mathbb{T})$ is Roelcke precompact.*

Proof First a general fact: if a topological group G can be represented as $G = KH$, where K is a compact subset and H a Roelcke-precompact subgroup, then G is also Roelcke-precompact. This is easy to verify either directly or by applying [66, Prop. 9.17]. As was mentioned in Theorem 18.52(4), $H_+[0, 1]$ is Roelcke precompact. Now, observe that in our case $G = KH$, where $H := St(1) \cong H_+[0, 1]$ is the stability group of $1 \in \mathbb{T}$ and $K \cong \mathbb{T}$ is the subgroup of G consisting of the rotations of the circle. Indeed, the coset space G/H is homeomorphic to $\mathbb{T}$ and there exists a natural continuous section $s : \mathbb{T} \to K \subset G$.

18.7.2   Circularly Ordered Systems

In [34] we introduce the class of circularly ordered (c-ordered) dynamical systems which naturally generalizes the class of linearly ordered systems. A compact S-system X is said to be c-ordered (notation $(S, X) \in \text{CODS}$) if the topological space X is c-ordered and every s-translation $X \to X$ is c-order preserving.

Example 18.54

1. With every c-ordered compact space there is the associated topological group $H_+(X)$ of c-order preserving homeomorphisms. Certainly, X is a c-ordered $H_+(X)$-system. Every linearly ordered G-system is c-ordered.
2. The Sturmian like $\mathbb{Z}^k$-subshifts from Example 18.42 admit a circular order. Moreover, their enveloping semigroups also are c-ordered systems [34].
3. Every element g of the projective group $\text{PGL}(2, \mathbb{R})$ defines a homeomorphism on the circle $\mathbb{T} \to \mathbb{T}$ which is either c-order preserving or c-order reversing.

Theorem 18.55 ([34])

1. *Every c-ordered compact, not necessarily metrizable, S-space X is Rosenthal representable (that is, WRN), hence, in particular, tame. So, $\text{CODS} \subset \text{WRN} \subset$ Tame.*
2. *The topological group $H_+(X)$ (with compact open topology) is Rosenthal representable for every c-ordered compact space X. For example, this is the case for $H_+(\mathbb{T})$.*

The Ellis compactification $j : G \to E(G, \mathbb{T})$ of the group $G = H_+(\mathbb{T})$ is a topological embedding. In fact, observe that the compact open topology on $j(G) \subset C_+(\mathbb{T}, \mathbb{T})$ coincides with the pointwise topology. This observation implies, by [32, Remark 4.14], that $\text{Tame}(G)$ separates points and closed subsets. For any group G having sufficiently many tame functions the universal tame semigroup compactification $G \to G^{\text{Tame}}$ is a topological embedding.

Remark 18.56

1. Regarding Theorem 18.52(2) we note that recently Ben-Yaacov and Tsankov [4] found some other Polish groups G for which WAP$(G) = \{constants\}$ (and which are therefore also reflexively trivial).
2. Although $G = H_+(\mathbb{T})$ is representable on a (separable) Rosenthal Banach space, the group $H_+(\mathbb{T})$ is Asplund-trivial. Indeed, it is algebraically simple [19, Theorem 4.3] and contains a copy of $H_+[0, 1] = St(z)$ (a stabilizer group of some point $z \in \mathbb{T}$) which is Asplund-trivial [28]. Now, as in [28, Lemma 10.2], use an observation of Pestov, which implies that any continuous Asplund representation of $H_+(\mathbb{T})$ is trivial.

Question 18.57 Is it true that the universal tame compactification $u_t : G \to G^{\text{Tame}}$ is an embedding for every Polish group G?

The universal Polish group $G = H[0, 1]^{\mathbb{N}}$ is a natural candidate for a counterexample.

18.7.3   Noncommutative Sturmian Like Symbolic Systems

The following construction yields many tame coding functions for subgroups of $H_+(\mathbb{T})$, and via any abstract homomorphism $h : G \to H_+(\mathbb{T})$, we obtain coding functions on G.

Theorem 18.58 *Let $h : G \to H(\mathbb{T})$ be a group homomorphism and let*

$$f : \mathbb{T} \to A := \{0, \ldots, d\}$$

be a finite coloring map induced by a finite partition of the circle $\mathbb{T}$ comprising disjoint arcs.

Then, for any given point $z \in \mathbb{T}$ we have:

1. *the coding function $\varphi = m(f, z) : G \to \{0, \ldots, d\}$ is tame on the discrete copy of G.*
2. *the corresponding symbolic G-system $G_\varphi \subset \{0, 1, \cdots, d\}^G$ is tame.*
3. *if the action of G on $\mathbb{T}$ is minimal then, in many cases, the G-system G_φ is minimal and circularly ordered.*

Proof (A Sketch) The coloring map $f : \mathbb{T} \to A := \{0, \ldots, d\}$ has bounded variation. Every circle homeomorphism $g \in H(\mathbb{T})$ is either circular order preserving or reversing. This implies that the orbit $fG = \{fg : g \in G\}$, as a bounded family of real (discontinuous) functions on $\mathbb{T}$, has bounded total variation. As we know by Megrelishvili [57] and Glasner and Megrelishvili [34] any such family on $\mathbb{T}$ (or, on any other circularly ordered set) is tame. From Lemma 18.36(4) we conclude that $\varphi = m(f, z)$ is a tame function. This yields (1) and (2). For (3) we use some results from [34].

Remark 18.59 Some particular cases of this construction (for a suitable h) are as follows:

1. Sturmian and Sturmian like multidimensional symbolic $\mathbb{Z}^k$-systems (Example 18.42).
2. Consider a subgroup G of $PSL_2(\mathbb{R})$, isomorphic to F_2, which is generated by two Möbius transformations as in [30], say an irrational rotation and a parabolic transformation. When $d = 1$ (two colors) we get the corresponding minimal tame subshift $X \subset \{0, 1\}^{F_2}$.
3. One can consider coding functions on any Fuchsian group G. E.g., for the noncommutative modular group $G = PSL_2(\mathbb{Z}) \simeq \mathbb{Z}_2 * \mathbb{Z}_3$.
4. More generally, at least in the assertions (1) and (2) of Theorem 18.58, one may replace the circle $\mathbb{T}$ by any circularly ordered set.

18.8   Tame Minimal Systems and Topological Groups

Recall that for every topological group G there exists a unique universal minimal G-system $M(G)$. Frequently $M(G)$ is nonmetrizable. For example, this is the case for every locally compact noncompact G. On the other hand, many interesting massive Polish groups are extremely amenable that is, having trivial $M(G)$. See for example [62–64, 74]. The first example of a nontrivial yet metrizable $M(G)$ was found by Pestov. In [63] he shows that for $G := H_+(\mathbb{T})$ the universal minimal system $M(G)$ can be identified with the natural action of G on the circle $\mathbb{T}$. Glasner and Weiss [36, 37] gave an explicit description of $M(G)$ for the symmetric group S_∞ and for $H(C)$ (the Polish group of homeomorphisms of the Cantor set C). Using model theory Kechris, Pestov and Todorčević gave in [47] many new examples of various subgroups of S_∞ with metrizable (and computable) $M(G)$.

Note that the universal almost periodic factor $M_{AP}(G)$ of $M(G)$ is the Bohr compactification $b(G)$ of G. When the induced homomorphism $G \to \text{Homeo } b(G)$ is injective (trivial) the topological group G is called maximally (resp., minimally) almost periodic. Every topological group G has a universal minimal tame system $M_t(G)$ which is the largest tame G-factor of $M(G)$. It is not necessarily AP (in contrast to the HNS and WAP cases). There are (even discrete) minimally almost periodic groups which however admit effective minimal tame systems, or in other words, groups for which the corresponding homomorphism $G \to \text{Homeo } (M_t(G))$ is injective. For example, the countable group $PSL_2(\mathbb{Q})$ is minimally almost periodic (von Neumann and Wigner) even in its discrete topology. It embeds densely into the group $PSL_2(\mathbb{R})$ which acts effectively and transitively on the circle. Thus, the circle provides a topologically effective minimal action for every dense subgroup G of $PSL_2(\mathbb{R})$. In particular, it is effective (though not topologically effective) for the discrete copy of $PSL_2(\mathbb{Q})$.

Question 18.60 Which Polish groups (e.g., discrete countable groups) G have effective tame minimal actions? Equivalently, when is the homomorphism $G \to$ Homeo $(M_t(G))$ injective?

Next we will discuss in more details the question "when is $M(G)$ tame?".

Definition 18.61 We say that a topological group G is *intrinsically tame* if one of the following equivalent conditions is satisfied:

1. every continuous action of G on a compact space X admits a G-subsystem $Y \subset X$ which is tame.
2. any minimal compact G-system is tame.
3. the universal minimal G-system $M(G)$ is tame.
4. the natural projection $M(G) \to M_t(G)$ is an isomorphism.

The G-space $M_t(G)$ can also be described as a minimal left ideal in the universal tame G-system G^{Tame}. Recall that $G \to G^{\text{Tame}}$ is a semigroup G-compactification determined by the algebra Tame(G). The latter is isomorphic to its own enveloping semigroup and thus has a structure of a compact right topological semigroup. Moreover, any two minimal left ideals there, are isomorphic as dynamical systems.

In [27] we defined, for a topological group G and a dynamical property P, the notion of P-fpp (P fixed point property). Namely G has the P-fpp if every G-system which has the property P admits a G fixed point. Clearly this is the same as demanding that every minimal G-system with the property P be trivial. Thus for $P =$ Tame a group G has the tame-fpp iff $M_t(G)$ is trivial.

We will need the following theorem which extends a result in [24].

Theorem 18.62 *Let (G, X) be a metrizable minimal tame dynamical system and suppose it admits an invariant probability measure. Then (G, X) is point distal. If moreover, with respect to μ the system (G, μ, X) is weakly mixing then it is a trivial one point system.*

Proof With notations as in [24] we observe that for any minimal idempotent $v \in E(G, X)$ the set C_v of continuity points of v restricted to the set $\overline{vX}$, is a dense G_δ subset of $\overline{vX}$ and moreover $C_v \subset vX$ [24, Lemma 4.2.(ii)]. Also, by [24, Proposition 4.3] we have $\mu(vX) = 1$, and it follows that $\overline{vX} = X$. The proof of the claim that (G, X) is point distal is now finished as in [24, Proposition 4.4].

Finally, if the measure preserving system (G, μ, X) is weakly mixing it follows that it is also topologically weakly mixing. By the Veech-Ellis structure theorem for point distal systems [14, 75], if (G, X) is nontrivial it admits a nontrivial equicontinuous factor, say (G, Y). However (G, Y), being a factor of (G, X), is at the same time also topologically weakly mixing which is a contradiction.

Theorem 18.63

1. *Every extremely amenable group is intrinsically tame.*
2. *The Polish group* $H_+(\mathbb{T})$ *of orientation preserving homeomorphisms of the circle is intrinsically tame.*

3. *The Polish groups* Aut (**S**(2)) *and* Aut (**S**(3)), *of automorphisms of the circular directed graphs* **S**(2) *and* **S**(3), *are intrinsically tame.*
4. *A discrete group which is intrinsically tame is finite.*
5. *For an abelian infinite countable discrete group G, its universal minimal tame system $M_t(G)$ is a highly proximal extension of its Bohr compactification G^{AP} (see e.g. [24]).*
6. *The Polish group* H(C), *of homeomorphisms of the Cantor set, is not intrinsically tame.*
7. *The Polish group $G = S_\infty$, of permutations of the natural numbers, is not intrinsically tame. In fact $M_t(G)$ is trivial; i.e. G has the tame-fpp.*

Proof

(1) Is trivial.
(2) Follows from Pestov's theorem [63], which identifies $(G, M(G))$ for $G = H_+(\mathbb{T})$ as the tautological action of G on $\mathbb{T}$, and from Theorem 18.55 which asserts that this system is tame (being c-ordered).
(3) The universal minimal G-systems for the groups Aut (**S**(2)) and Aut (**S**(3)) are computed in [61]. In both cases it is easy to check that every element of the enveloping semigroup $E(M(G))$ is an order preserving map. As there are only $2^{\aleph_0}$ order preserving maps, it follows that the cardinality of $E(M(G))$ is $2^{\aleph_0}$, whence, in both cases, the dynamical system $(G, M(G))$ is tame.

   In order to prove Claim (4) we assume, to the contrary, that G is infinite and apply a result of Weiss [76], to obtain a minimal model, say (G, X, μ), of the Bernoulli probability measure preserving system $(G, \{0, 1\}^G, (\frac{1}{2}(\delta_0 + \delta_1))^G)$. Now (G, X, μ) is metrizable, minimal and tame, and it carries a G-invariant probability measure with respect to which the system is weakly mixing. Applying Theorem 18.62 we conclude that X is trivial. This contradiction finishes the proof.[4]
(5) In [41, 50] and [24] it is shown that a metric minimal tame G-system is an almost one-to-one extension of an equicontinuous system. Now tameness is preserved under sub-products, and because our group G is countable, it follows that $M_t(G)$ is a minimal sub-product of all the minimal tame metrizable systems. In turn this implies that $M_t(G)$ is a (non-metrizable) highly proximal extension of the Bohr compactification G^{AP} of G.
(6) To see that $G = H(C)$ is not intrinsically tame it suffices to show that the tautological action (G, C), which is a factor of $M(G)$, is not tame. To that end note that the shift transformation σ on $X = \{0, 1\}^{\mathbb{Z}}$ is a homeomorphism of the Cantor set. Now the enveloping semigroup $E(\sigma, X)$ of the cascade (σ, X), a subset of $E(G, X)$, is homeomorphic to $\beta\mathbb{N}$.

---

[4]Modulo an extension of Weiss' theorem, which does not yet exist, a similar idea would work for any locally compact group. The more general statement would be: A locally compact group which is intrinsically tame is compact.

(7) To see that $G = S_\infty$ is not intrinsically tame we recall first that, by Glasner and Weiss [35], the universal minimal dynamical system for this group can be identified with the natural action of G on the compact metric space $X = LO(\mathbb{N})$ of linear orders on $\mathbb{N}$. Also, it follows from the analysis of this dynamical system that for any minimal idempotent $u \in E(G, X)$ the image of u contains exactly two points, say $uX = \{x_1, x_2\}$. A final fact that we will need concerning the system (G, X) is that it carries a G-invariant probability measure μ of full support [35]. Now to finish the proof, suppose that (G, X) is tame. Then there is a **sequence** $g_n \in G$ such that $g_n \to u$ in $E(G, X)$. If $f \in C(X)$ is any continuous real valued function, then we have, for each $x \in X$,

$$\lim_{n \to \infty} f(g_n x) = f(ux) \in \{f(x_1), f(x_2)\}.$$

But then, choosing a function $f \in C(X)$ which vanishes at the points x_1 and x_2 and with $\int f \, d\mu = 1$, we get, by Lebesgue's theorem,

$$1 = \int f \, d\mu = \lim_{n \to \infty} \int f(g_n x) \, d\mu = \int f(ux) \, d\mu = 0.$$

Finally, the property of supporting an invariant measure, as well as the fact that the cardinality of the range of minimal idempotents is $\leqslant 2$, are inherited by factors and thus the same argument shows that $M(G)$ admits no nontrivial tame factor. Thus $M_t(G)$ is trivial.

We will say that G is *intrinsically c-ordered* if the G-system $M(G)$ is circularly ordered. Using this terminology Theorem 18.63 says that the Polish groups $G = H_+(\mathbb{T})$, $Aut(S(2))$ and $Aut(S(3))$ are intrinsically c-ordered. Note that for $G = H_+(\mathbb{T})$ every compact G-space X contains a copy of $\mathbb{T}$ as G-subspace or a G-fixed point.

The (nonamenable) group $G = H_+(\mathbb{T})$ has one more remarkable property. Besides $M(G)$, one can also effectively compute the affine analogue of $M(G)$. Namely, the *universal irreducible affine system* of G (we denote it by $IA(G)$) which was defined and studied in [20, 21]. It is uniquely determined up to affine isomorphisms. For any topological group G the corresponding affine compactification $G \to IA(G)$ coincides with the affine compactification $G \to P(M_{sp}(G))$, where, $M_{sp}(G)$ is the *universal strongly proximal minimal system* of G and $P(M_{sp}(G))$ is the space of probability measures on the compact space $M_{sp}(G)$. For more information regarding affine compactifications of dynamical systems we refer to [31].

Definition 18.64 We say that G is *convexly intrinsically tame* if one of the following equivalent conditions is satisfied:

1. every compact affine dynamical system (G, Q) admits an affine tame G-subsystem.
2. every compact affine dynamical system (G, Q) admits a tame G-subsystem.
3. every irreducible affine G-system is tame.
4. the universal irreducible affine G-system $IA(G)$ is tame.

Note that the G-system $P(X)$ is affinely universal for a G-system X; also, $P(X)$ is tame whenever X is [29, 31, Theorem 6.11]. In particular, it follows that any intrinsically tame group is convexly intrinsically tame.

It is well known that a topological group G is amenable iff $M_{sp}(G)$ is trivial (see [21]). Thus G is amenable iff $IA(G)$ is trivial and it follows that every amenable group is trivially convexly intrinsically tame.

Thus we have the following diagram which emphasizes the analogy between the two pairs of properties:

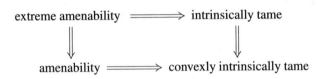

Remark 18.65 Given a class P of compact G-systems which is stable under subdirect products, one can define the notions of an intrinsically P group and a convexly intrinsically P group in a manner analogous to the one we adopted for $P = $ Tame. We then note that in this terminology a group is convexly intrinsically HNS (and, hence, also conv-int-WAP) iff it is amenable. This follows easily from the fact that the algebra $Asp(G)$ is left amenable [30]. This "collapsing effect" together with the special role of tameness in the dynamical BFT dichotomy 18.11 suggest that the notion of convex intrinsic tameness is a natural analogue of amenability.

At least for discrete groups, if G is intrinsically HNS then it is finite. In fact, for any group, an HNS minimal system is equicontinuous (see [27]), so that for a group G which is intrinsically HNS the universal minimal system $M(G)$ coincides with its Bohr compactification G^{AP}. Now for a discrete group, it is not hard to show that an infinite minimal equicontinuous system admits a nontrivial almost one to one (hence proximal) extension which is still minimal. Thus $M(G)$ must be finite. However, by a theorem of Ellis [13], for discrete groups the group G acts freely on $M(G)$, so that G must be finite as claimed. Probably similar arguments will show that a locally compact intrinsically HNS group is necessarily compact.

Theorem 18.66

1. S_∞ *is amenable (hence convexly intrinsically tame) but not intrinsically tame.*
2. $H(C)$ *is not convexly intrinsically tame.*
3. $H([0, 1]^{\mathbb{N}})$ *is not convexly intrinsically tame.*
4. $H_+(\mathbb{T})$ *is a (convexly) intrinsically tame nonamenable topological group.*
5. $SL_n(\mathbb{R})$, $n > 1$ *(more generally, any semisimple Lie group G with finite center and no compact factors) is convexly intrinsically tame nonamenable topological group.*

Proof

(1) S_∞ is amenable [12]. It is not intrinsically tame by Theorem 18.63(7).

(2) Natural action of H(C) on the Cantor set C is minimal and strongly proximal, but this action is not tame; it contains, as a subaction, a copy of the full shift $(\mathbb{Z}, C) \cong (\sigma, \{0, 1\}^{\mathbb{Z}})$.

(3) The group $H([0, 1]^{\mathbb{N}})$ is a universal Polish group (see Uspenskij [73]). It is not convexly intrinsically tame. This can be established by observing that the action of this group on the Hilbert cube is minimal, strongly proximal and not tame. The strong proximality of this action can be easily checked. The action is not tame because it is a *universal action* (see [53]) for all Polish groups on compact metrizable spaces.

(4) The (universal) minimal G-system $\mathbb{T}$ for $G = H_+(\mathbb{T})$ is strongly proximal. Hence, $IA(G)$ in this case is easily computable and it is exactly $P(\mathbb{T})$ which, as a G-system, is tame (by Theorem 18.63(4)). Thus, $H_+(\mathbb{T})$ is a (convexly) intrinsically tame.

(5) By Furstenberg's result [18] the universal minimal strongly proximal system $M_{sp}(G)$ is the homogeneous space $X = G/P$, where P is a minimal parabolic subgroup (see [21]). Results of Ellis [15] and Akin [1] (Example 18.33(1)) show that the enveloping semigroup $E(G, X)$ in this case is a Rosenthal compact space, whence the system (G, X) is tame by the dynamical BFT dichotomy (Theorem 18.11).

In particular, for $G = SL_2(\mathbb{R})$ note that in any compact *affine* G-space we can find either a 1-dimensional real projective G-space (a copy of the circle) or a fixed point. For general $SL_n(\mathbb{R})$, $n \geqslant 2$—flag manifolds and their G-quotients.

Acknowledgements This research was supported by a grant of Israel Science Foundation (ISF 668/13). The first named author thanks the Hausdorff Institute at Bonn for the opportunity to participate in the Program "Universality and Homogeneity" where part of this work was written, November 2013.

References

1. E. Akin, Enveloping linear maps, in *Topological Dynamics and Applications*. Contemporary Mathematics, vol. 215 (a volume in honor of R. Ellis) (American Mathematical Society, Providence, 1998), pp. 121–131

2. E. Akin, E. Glasner, WAP systems and labeled subshifts. Mem. AMS (2014, to appear). arXiv:1410.4753

3. J.B. Aujogue, Ellis enveloping semigroup for almost canonical model sets of an Euclidean space. Algebr. Geom. Topol. **15**(4), 2195–2237 (2015)

4. I. Ben Yaacov, T. Tsankov, Weakly almost periodic functions, model-theoretic stability, and minimality of topological groups. Trans. Am. Math. Soc. **368**, 8267–8294 (2016)

5. J.F. Berglund, H.D. Junghenn, P. Milnes, *Analysis on Semigroups* (Wiley, New York, 1989)

6. V. Berthe, L. Vuillon, Palindromes and two-dimensional Sturmian sequences. J. Autom. Lang. Combin. **6**, 121–138 (2001)

7. V. Berthe, S. Ferenczi, L.Q. Zamboni, Interactions between dynamics, arithmetics and combinatorics: the good, the bad, and the ugly. Contemp. Math. **385**, 333–364 (2005)

8. J. Bourgain, D.H. Fremlin, M. Talagrand, Pointwise compact sets in Baire-measurable functions. Am. J. Math. **100**(4), 845–886 (1977)
9. B. Cascales, I. Namioka, J. Orihuela, The Lindelof property in Banach spaces. Stud. Math. **154**, 165–192 (2003)
10. D. Cenzer, A. Dashti, F. Toska, S. Wyman, Computability of countable subshifts in one dimension. Theory Comput. Syst. **51**(3), 352–371 (2012)
11. A. Chernikov, P. Simon, Definably amenable NIP groups (2015). arXiv:1502.04365v1
12. P. de la Harpe, Moyennabilité de quelques groupes topologiques de dimension infinie. C.R. Acad. Sci. Paris Sér. A **277**, 1037–1040 (1973)
13. R. Ellis, Universal minimal sets. Proc. Am. Math. Soc. **11**, 540–543 (1960)
14. R. Ellis, The Veech structure theorem. Trans. Am. Math. Soc. **186**, 203–218 (1973)
15. R. Ellis, The enveloping semigroup of projective flows. Ergod. Theory Dyn. Syst. **13**, 635–660 (1993)
16. M. Fabian, *Gateaux Differentiability of Convex Functions and Topology. Weak Asplund Spaces*. Canadian Mathematical Society Series of Monographs and Advanced Texts (Wiley, New York, 1997)
17. T. Fernique, Multi-dimensional Sturmian sequences and generalized substitutions. Int. J. Found. Comput. Sci. **17** 575–600 (2006)
18. H. Furstenberg, A poisson formula for semi-simple Lie groups. Ann. Math. **77**, 335–386 (1963)
19. E. Ghys, Groups acting on the circle. Enseign. Math. (2) **47**, 329–407 (2001)
20. S. Glasner, Compressibility properties in topological dynamics. Am. J. Math. **97**, 148–171 (1975)
21. E. Glasner, *Proximal Flows*. Lecture Notes, vol. 517 (Springer, Berlin, 1976)
22. E. Glasner, *Ergodic Theory via Joinings*. Mathematical Surveys and Monographs, vol. 101 (American Mathematical Society, Providence, 2003)
23. E. Glasner, On tame dynamical systems. Colloq. Math. **105**, 283–295 (2006)
24. E. Glasner, The structure of tame minimal dynamical systems. Ergod. Theory Dyn. Syst. **27**, 1819–1837 (2007)
25. E. Glasner, Translation-finite sets (2011). ArXiv: 1111.0510
26. E. Glasner, The structure of tame mini systems for general groups. Invent. Math. **211**, 213–244 (2018). https://doi.org/10.1007/s00222-017-0747-z
27. E. Glasner, M. Megrelishvili, Linear representations of hereditarily non-sensitive dynamical systems. Colloq. Math. **104**(2), 223–283 (2006)
28. E. Glasner, M. Megrelishvili, New algebras of functions on topological groups arising from G-spaces. Fundam. Math. **201**, 1–51 (2008)
29. E. Glasner, M. Megrelishvili, Representations of dynamical systems on Banach spaces not containing l_1. Trans. Am. Math. Soc. **364**, 6395–6424 (2012). ArXiv e-print: 0803.2320
30. E. Glasner, M. Megrelishvili, On fixed point theorems and nonsensitivity. Isr. J. Math. **190**, 289–305 (2012). ArXiv e-print: 1007.5303
31. E. Glasner, M. Megrelishvili, Banach representations and affine compactifications of dynamical systems, in *Fields Institute Proceedings Dedicated to the 2010 Thematic Program on Asymptotic Geometric Analysis*, ed. by M. Ludwig, V.D. Milman, V. Pestov, N. Tomczak-Jaegermann (Springer, New York, 2013). ArXiv version: 1204.0432
32. E. Glasner, M. Megrelishvili, Representations of dynamical systems on Banach spaces, in *Recent Progress in General Topology III*, ed. by K.P. Hart, J. van Mill, P. Simon (Springer/Atlantis Press, New York/Pairs, 2014), pp. 399–470
33. E. Glasner, M. Megrelishvili, Eventual nonsensitivity and tame dynamical systems (2014). arXiv:1405.2588
34. E. Glasner, M. Megrelishvili, Circularly ordered dynamical systems. Monatsh. Math. (2016, to appear). arXiv:1608.05091
35. E. Glasner, B. Weiss, Quasifactors of zero-entropy systems. J. Am. Math. Soc. **8**, 665–686 (1995)
36. E. Glasner, B. Weiss, Minimal actions of the group $S(\mathbb{Z})$ of permutations of the integers. Geom. Funct. Anal. **12**, 964–988 (2002)

37. E. Glasner, B. Weiss, The universal minimal system for the group of homeomorphisms of the Cantor set. Fundam. Math. **176**, 277–289 (2003)
38. E. Glasner, B. Weiss, On Hilbert dynamical systems. Ergod. Theory Dyn. Syst. **32**(2), 629–642 (2012)
39. E. Glasner, X. Ye, Local entropy theory. Ergod. Theory Dyn. Syst. **29**, 321–356 (2009)
40. E. Glasner, M. Megrelishvili, V.V. Uspenskij, On metrizable enveloping semigroups. Isr. J. Math. **164**, 317–332 (2008)
41. W. Huang, Tame systems and scrambled pairs under an abelian group action. Ergod. Theory Dyn. Syst. **26**, 1549–1567 (2006)
42. T. Ibarlucia, The dynamical hierarchy for Roelcke precompact Polish groups. Isr. J. Math. **215**, 965–1009 (2016)
43. J.E. Jayne, C.A. Rogers, Borel selectors for upper semicontinuous set-valued maps. Acta Math. **155**, 41–79 (1985)
44. J.E. Jayne, J. Orihuela, A.J. Pallares, G. Vera, σ-Fragmentability of multivalued maps and selection theorems. J. Funct. Anal. **117**(2), 243–273 (1993)
45. T. Jolivet, Combinatorics of Pisot Substitutions, TUCS Dissertations No 164 (2013)
46. A.S. Kechris, *Classical Descriptive Set Theory*. Graduate Texts in Mathematics, vol. 156 (Springer, Berlin, 1991)
47. A.S. Kechris, V.G. Pestov, S. Todorčević, Fraïssé limits, Ramsey theory, and topological dynamics of automorphism groups. Geom. Funct. Anal. **15**(1), 106–189 (2005)
48. P.S. Kenderov, W.B. Moors, Fragmentability of groups and metric-valued function spaces. Top. Appl. **159**, 183–193 (2012)
49. D. Kerr, H. Li, Dynamical entropy in Banach spaces. Invent. Math. **162**, 649–686 (2005)
50. D. Kerr, H. Li, Independence in topological and C^*-dynamics. Math. Ann. **338**, 869–926 (2007)
51. A. Köhler, Enveloping semigroups for flows. Proc. R. Ir. Acad. **95A**, 179–191 (1995)
52. A. Komisarski, H. Michalewski, P. Milewski, Bourgain-Fremlin-Talagrand dichotomy and dynamical systems. Preprint (2004)
53. M. Megrelishvili, Free topological G-groups. N. Z. J. Math. **25**, 59–72 (1996)
54. M. Megrelishvili, Fragmentability and continuity of semigroup actions. Semigroup Forum **57**, 101–126 (1998)
55. M. Megrelishvili, Fragmentability and representations of flows. Topol. Proc. **27**(2), 497–544 (2003). See also: www.math.biu.ac.il/~megereli
56. M. Megrelishvili, Topological transformation groups: selected topics, in *Open Problems in Topology II*, ed. by E. Pearl (Elsevier Science, Amsterdam, 2007), pp. 423–438
57. M. Megrelishvili, A note on tameness of families having bounded variation. Topol. Appl. **217**, 20–30 (2017)
58. M. Megrelishvili, L. Polev, Order and minimality of some topological groups. Topol. Appl. **201**, 131–144 (2016)
59. I. Namioka, Radon-Nikodým compact spaces and fragmentability. Mathematika **34**, 258–281 (1987)
60. I. Namioka, R.R. Phelps, Banach spaces which are Asplund spaces. Duke Math. J. **42**, 735–750 (1975)
61. L. Nguyen van Thé, More on the Kechris-Pestov-Todorcevic correspondence: precompact expansions. Fund. Math. **222**(1), 19–47 (2013)
62. L. Nguyen van Thé, Fixed points in compactifications and combinatorial counterparts (2017). arXiv:1701.04257
63. V.G. Pestov, On free actions, minimal flows, and a problem by Ellis. Trans. Am. Math. Soc. **350**, 4149–4165 (1998)
64. V.G. Pestov, *Dynamics of Infinite-Dimensional Groups. The Ramsey-Dvoretzky-Milman Phenomenon*. University Lecture Series, vol. 40 (American Mathematical Society, Providence, 2006)
65. R. Pikula, Enveloping semigroups of affine skew products and sturmian-like systems, Dissertation, The Ohio State University, 2009

66. W. Roelcke, S. Dierolf, *Uniform Structures on Topological Groups and Their Quotients* (McGraw-Hill, New York, 1981)
67. A.V. Romanov, Ergodic properties of discrete dynamical systems and enveloping semigroups. Ergod. Theory Dyn. Syst. **36**, 198–214 (2016)
68. H.P. Rosenthal, A characterization of Banach spaces containing l_1. Proc. Natl. Acad. Sci. U.S.A. **71**, 2411–2413 (1974)
69. W. Ruppert, On weakly almost periodic sets. Semigroup Forum **32**, 267–281 (1985)
70. S.A. Shapovalov, A new solution of one Birkhoff problem. J. Dyn. Control Syst. **6**(3), 331–339 (2000)
71. M. Talagrand, Pettis integral and measure theory. Mem. AMS **51** (1984)
72. S. Todorčević, *Topics in Topology*. Lecture Notes in Mathematics, vol. 1652 (Springer, Berlin, 1997)
73. V.V. Uspenskij, A universal topological group with countable base. Funct. Anal. Appl. **20**, 160–161 (1986)
74. V.V. Uspenskij, Compactifications of topological groups, in *Proceedings of the Ninth Prague Topological Symposium (Prague, August 19–25, 2001)*, ed. by P. Simon. Published April 2002 by Topology Atlas (electronic publication), pp. 331–346. ArXiv:math.GN/0204144
75. W.A. Veech, Point-distal flows. Am. J. Math. **92**, 205–242 (1970)
76. B. Weiss, Minimal models for free actions, in *Dynamical Systems and Group Actions*. Contemporary Mathematics, vol. 567 (American Mathematical Society, Providence, 2012), pp. 249–264

Chapter 19
A Piecewise Rotation of the Circle, IPR Maps and Their Connection with Translation Surfaces

Kae Inoue and Hitoshi Nakada

19.1 Introduction

In the authors' paper [3], we extended Cruz and da Rocha's idea [2] and gave a continuum version of castles arising from a piecewise rotation map of the circle and defined a map, which gives an inverse of the invertible Rauzy-Veech induction [7, 9], on a suitable set of castles. Moreover we showed that from every interval exchange map one can construct a piecewise rotation which produces the same Rauzy class arising from the given interval exchange map. This Rauzy class depends on the choice of the discontinuous point of the piecewise rotation map of the circle. However, in [3], it was not clear that which Rauzy class we can get when we choose a different discontinuous point of the same piecewise rotation map. Indeed we left this problem as an open question. Also there was no discussion of translation surfaces from piecewise rotation maps. In this paper, we show that we can get any Rauzy class in the same extended Rauzy class by choosing an appropriate discontinuous point. In other words, we see that each Rauzy class from the same extended Rauzy class appears by the choice of the discontinuous point. This result is shown by the construction of the translation surface with fixed of the singularities. Here the construction gives a dynamic point of view for the structure of translation

K. Inoue
Faculty of Pharmacy, Keio University, Tokyo, Japan
e-mail: inoue-ke@pha.keio.ac.jp

H. Nakada (✉)
Department of Mathematics, Keio University, Yokohama, Japan
e-mail: nakada@math.keio.ac.jp

© Springer International Publishing AG, part of Springer Nature 2018
S. Ferenczi et al. (eds.), *Ergodic Theory and Dynamical Systems in their Interactions with Arithmetics and Combinatorics*, Lecture Notes in Mathematics 2213, https://doi.org/10.1007/978-3-319-74908-2_19

surfaces. We recall [1, 6, 9] that

$$\sum_{j=1}^{s} d_j = 2g - 2 \qquad (19.1)$$

is the only restriction for translation surfaces where $(d_1, \ldots, d_s)$'s components are the orders of singularities with the translation surface.

We start with the definition of interval exchange maps and the Rauzy induction. To make it simpler, we use Keane's definition (see [4, 5, 8, 9]) rather than Viana [10] and Yoccoz [11].

We fix an integer $d > 1$ and π is a permutation of $\{1, \ldots, d\}$ such that $\pi\{1, \ldots, k\} = \{1, \ldots, k\}$ implies $k = d$. We call such π an irreducible permutation. We consider $I = [0, 1)$ and its partition $\{I_1, \ldots, I_d\}$ where $I_j = [\beta_{j-1}, \beta_j)$ with $0 = \beta_0 < \beta_1 < \cdots < \beta_d$. We put $\lambda_j = |I_j| = \beta_j - \beta_{j-1}$ and define $\beta_j^{\pi} = \sum_{k \leqslant j} \lambda_{\pi^{-1}(k)}$ for $1 \leqslant j \leqslant d$. Then the interval exchange T_{π} is defined by

$$T_{\pi}(x) = x - \beta_{j-1} + \beta_{\pi(j)-1}^{\pi}$$

for $x \in I_j$, $1 \leqslant j \leqslant d$. We call $(\lambda_1, \ldots, \lambda_d)$ the length data and π the combinatorial data of T.

The Rauzy-Veech induction $\mathcal{R}$ (or Rauzy induction) is defined as follows: there are two cases.

(Case i) $\lambda_d > \lambda_{\pi^{-1}(d)}$
(Case ii) $\lambda_d < \lambda_{\pi^{-1}(d)}$

We consider the induced map of T to $[0, \beta_{d-1})$ or $[0, \beta_{d-1}^{\pi})$, denoted by $T_{[0,\beta_{d-1})}$ and $T_{[0,\beta_{d-1}^{\pi})}$, respectively, and define $\mathcal{R}T$ by

$$\mathcal{R}T(x) = \begin{cases} T_{[0,\beta_{d-1}^{\pi})}(\beta_{d-1}^{\pi}x) & \text{if Case i} \\ T_{[0,\beta_{d-1})}(\beta_{d-1}x) & \text{if Case ii} \end{cases}$$

for $x \in I$. Here we note the these two induced maps are also d-interval exchange maps. We exclude the case $\lambda_d = \lambda_{\pi^{-1}(d)}$ for the discussion of $\mathcal{R}$.

Since $\mathcal{R}T$ is also an interval exchange, there exists an irreducible permutation π' associated to $\mathcal{R}T$. The Rauzy class of π is the set of irreducible permutations associated to $\mathcal{R}^n T_{\pi}$ for all $n \geqslant 0$ with all possible choices of $\beta_1, \ldots, \beta_{d-1}$.

Next we consider a permutation σ on $\{0, 1, \ldots, d\}$ defined as follows,

$$\sigma_{\pi}(j) = \begin{cases} \pi^{-1}(1) - 1 & j = 0 \\ d & j = \pi^{-1}(d) \\ \pi^{-1}(\pi(j) + 1) - 1 & \text{otherwise.} \end{cases}$$

It turns out later that this gives the order of a singularity of a translation surface arising from π. To construct a translation surface, we consider $\mathbf{h}$, $\mathbf{a} \in \mathbb{R}^d$ such that

$$h_j - a_j = h_{\sigma_\pi(j)+1} - a_{\sigma_\pi(j)} \tag{19.2}$$

for $0 \leqslant j \leqslant d$, where $a_0 = h_0 = h_{d+1} = 0$. We put some additional condition on $\mathbf{h}$ and $\mathbf{a}$, see [9]:

$$\begin{cases} h_j \geqslant 0 & 1 \leqslant j \leqslant d \\ a_j \geqslant 0 & 1 \leqslant j \leqslant d-1 \\ -h_{\pi^{-1}(d)} \leqslant a_d \leqslant h_d \\ a_{\pi^{-1}(d)} \leqslant h_{\pi^{-1}(d)+1} \\ a_j \leqslant \min(h_j, h_{j+1}) & 0 \leqslant j < d, j \neq \pi^{-1}(d). \end{cases} \tag{19.3}$$

Then we can construct "zippered rectangles" [9]. However, here, we directly construct a translation surface. The definition of the zippered rectangles will be given later in Sect. 19.5.

For each $1 \leqslant j \leqslant d$, we define

$$\begin{cases} \xi_j = \beta_j + a_j i \\ \xi_j^* = \beta_j^\pi - i(h_{\pi^{-1}(j)} - a_{\pi^{-1}(j)-1}) \end{cases} \tag{19.4}$$

Here we also put $\xi_0 = \xi_0^* = 0$. We note that $\xi_d = \xi_d^*$ holds.

Then, under an additional condition, we can construct a polygon by the set of vertices $\{\xi_j, \xi_j^* : 0 \leqslant j \leqslant d\}$ and sides are determined by (ξ_{j-1}, ξ_j) and also (ξ_{j-1}^*, ξ_j^*), $1 \leqslant j \leqslant d$. For every $1 \leqslant j \leqslant d$, two sides $(\xi_{\pi^{-1}(j)-1}, \xi_{\pi^{-1}(j)})$ and (ξ_{j-1}^*, ξ_j^*) are parallel from (19.2) and have the same length.

Hence we can identify $(\xi_{\pi^{-1}(j)-1}, \xi_{\pi^{-1}(j)})$ and (ξ_{j-1}^*, ξ_j^*), we get a Riemann surface which is called a translation surface. Here two sides (ξ_{d-1}, ξ_d) and (ξ_{d-1}^*, ξ_d^*) may intersect. In this case we can reform the figure given by $\{\xi_j, \xi_j^* : 0 \leqslant j \leqslant d\}$ to get a polygon with $2d$ sides (see Viana [10] for example).

The map T to $\mathcal{R}T$ is extended to a map $\widehat{\mathcal{R}}$ of a set of $(\mathbf{h}, \mathbf{a})$ satisfying (19.2) and (19.3) to itself by $(\mathbf{h}, \mathbf{a})$ to $(\mathbf{h}', \mathbf{a}')$:

(Case i) $\lambda_d > \lambda_{\pi^{-1}(d)}$

$$h_j' = \begin{cases} h_{\pi^{-1}(d)} + h_d & \text{if } j = \pi^{-1}(d) \\ h_j & \text{otherwise}, \end{cases}$$

$$a_j' = \begin{cases} -(h_{\pi^{-1}(d)} - a_{\pi^{-1}(d)-1}) & \text{if } j = d \\ a_j & \text{otherwise}. \end{cases}$$

(Case ii) $\lambda_d < \lambda_{\pi^{-1}(d)}$

$$h'_j = \begin{cases} h_j & \text{if } 1 \leqslant j \leqslant \pi^{-1}(d) \\ h_{\pi^{-1}(d)} + h_d & \text{if } j = \pi^{-1}(d) + 1 \\ h_{j-1} & \text{otherwise,} \end{cases}$$

$$a'_j = \begin{cases} a_j & \text{if } 1 \leqslant j < \pi^{-1}(d) \\ h_{\pi^{-1}(d)} + a_{d-1} & \text{if } j = \pi^{-1}(d) \\ a_{j-1} & \text{otherwise.} \end{cases}$$

We see that $\hat{\mathcal{R}}$ is bijective. We note that even if two sides (ξ_{d-1}, ξ_d) and (ξ^*_{d-1}, ξ^*_d) intersect, we always have a polygon (i.e. no intersection) by iterations of $\hat{\mathcal{R}}$.

Now we consider the orders of singularities of the translation surface given in the above. For j and j', $0 \leqslant j, j' \leqslant d$, if there exists $n \in \mathbb{Z}$ such that $\sigma^n_\pi(j) = j'$, then ξ_j and $\xi_{j'}$, denote the same branch point. Then $\#\{j' : 1 \leqslant j' \leqslant d - 1, j' = \sigma^n_\pi(j), n \in \mathbb{Z}\} - 1$ is the order of a singularity of branch point with respect to j. Let $s \geqslant 1$ be the number of branch points and $d_1, \ldots, d_s$ be their orders of singularities. For a branch point arising from $0(= 0 + ia_0)$ (the σ_π-equivalence class of ξ_j, ξ^*_j, $0 \leqslant j \leqslant d$, including ξ_0) is said to be the marked singularity. For $(d_1, \ldots, d_s)$ satisfies (19.1), g is the genus of the translation surface. For any j, $1 \leqslant j \leqslant s$, there exists an interval exchange map which produces a translation surface with the vector $(d_1, \ldots d_s)$ which consists of a given order of a singularity and the marked singularity of order d_j. The set of irreducible d-permutations which appears in the combinatorial data of the interval exchange map associated to $(d_1, \ldots, d_s)$ (satisfying (19.1)) is called the extended Rauzy class and the number of Rauzy classes in the extended Rauzy class is the number of different values of $(d_1, \ldots d_s)$, see [1].

In this paper, we show that for any $(d_1, \ldots, d_s)$ satisfying (19.1), there exists a piecewise rotation map S of the circle with $d - 1 = \sum_{j=1}^s d_j + s - 1$ discontinuous points such that the induced equivalence relation of the discontinuous points induces m disjoint classes and given by $(d_1, \ldots, d_s)$. Moreover we construct translation surfaces with its vector $(d_1, \ldots d_s)$ and a given marked singularity order d_j, $1 \leqslant j \leqslant s$.

The main purpose of this paper is to show that a construction of translation surfaces from a piecewise rotation of the circle based on the idea by Cruz and da Rocha [2] is equivalent to the construction of translation surfaces stated in the above. Moreover, we give the map of a translation surface to a translation surface (a castle to a castle) which is gives the inverse of $\hat{\mathcal{R}}$. Concerning this map, we give a sufficient condition on our construction being a translation surface. This process naturally induces the inverse of the Zorich map (e.g. [10, 12]). Most of results extend the authors' previous paper [3] in terms of translation surfaces. The introduction of the equivalence relation to the set of discontinuous points of a piecewise rotation map (in Sect. 19.2) is crucial in this paper.

In Sect. 19.2, we give the explicit definition of the piecewise rotation map S_α and introduce an equivalence relation on the set of discontinuous points of S_α. Then we show that for any positive integer vector $(d_1, d_2, \ldots, d_s)$ such that $\sum_{j=1}^{s} d_j$ is even, there exists a piecewise rotation map S_α with $\sum_{j=1}^{s} d_j + s - 1$ discontinuous points such that the equivalence relation of the set of discontinuous points defined by S_α consists of s classes Λ_j, $1 \leqslant j \leqslant s$, such that Λ_j has d_j points. In Sect. 19.3, we recall the notions of the critical iterates, IPR maps (Induced Piecewise Rotation maps) induced in [3], and (discrete) castles following Cruz and da Rocha [2]. As an application, we construct a translation surface from a castle arising from a castle defined by an critical iterate. In Sect. 19.4, we recall the notion of (continuum) castles associated to IPR maps. This is a simple generalization of castles defined in Sect. 19.3. Then we construct a translation surface from a castle and show that there is a natural correspondence between this translation surface and a translation surface defined by (19.4) in the above. We also show that the equivalence relation introduced $\hat{\sigma}_\pi$ introduced in Sect. 19.2 corresponds to the equivalence relation σ_π in the above as the singularities of the translation surface. Finally, we give a brief explanation between castles and zippered rectangles and between the invertible extension of Rauzy-Veech induction and a map defined on castles reforming [3]. We also remark one simple condition (inducing a jump of the transformation) on which a castle gives a translation surface. This condition leads us the inverse of the Zorich map as stated in the above.

19.2 Piecewise Rotation of the Circle

We consider $\mathbb{T} = [0, 1)$ $(= \mathbb{R}/\mathbb{Z})$ as the unit circle. For a given positive integer $d \geqslant 4$, we give

$$0 = \delta_1 < \delta_2 < \cdots < \delta_{d-1}$$

and a permutation p of $\{0, 1, \ldots, d-1\}$. We consider the partition $\{\mathcal{I}_1, \ldots, \mathcal{I}_{d-1}\}$, which consists of intervals (arcs), generated by δ_j, $1 \leqslant j \leqslant d - 1$, i.e.

$$\begin{cases} \mathcal{I}_1 = [\delta_1, \delta_2) \\ \mathcal{I}_2 = [\delta_2, \delta_3) \\ \vdots \\ \mathcal{I}_{d-1} = [\delta_{d-1}, \delta_1) \end{cases}$$

and we put

$$
\begin{cases}
\ell_1 = \delta_2 - \delta_1 \\
\ell_2 = \delta_3 - \delta_2 \\
\vdots \\
\ell_{d-1} = 1 - \delta_{d-1}.
\end{cases}
$$

For $x \in \mathbb{T}$, we define

$$
S(x) = x - \delta_j + \sum_{k:p(k)<p(j)} \ell_k \tag{19.5}
$$

for $x \in \mathcal{I}_j$, $1 \leqslant i < d$ and

$$
S_\alpha(x) = S(x) + \alpha \qquad (\text{mod } 1) \tag{19.6}
$$

for $0 \leqslant \alpha < 1$. We assume that $\overline{\{S_\alpha^n(x)\}} = \mathbb{T}$ for any $x \in \mathbb{T}$ and that each δ_j, $1 \leqslant j \leqslant d-1$, is a discontinuous point.

From the permutation p of $\{1, 2, \ldots, d-1\}$, we define a map $\hat{\sigma}_p$ of $\{1, 2, \ldots, d-1\}$ onto itself by

$$
\hat{\sigma}_p(j) =
\begin{cases}
1 & \text{if } 2 \leqslant p(j) \text{ and } p^{-1}(p(j)-1) = d-1 \\
1 & \text{if } p(j) = 1 \text{ and } p^{-1}(d-1) = d-1 \\
p^{-1}(d-1) + 1 & \text{if } p(j) = 1 \text{ and } p^{-1}(d-1) \neq d-1 \\
p^{-1}(p(j)-1) + 1 & \text{if } 2 \leqslant p(j) \text{ and } p^{-1}(p(j)-1) \neq d-1.
\end{cases}
\tag{19.7}
$$

We consider the equivalence relation "$\sim$" of $\{\delta_j : 1 \leqslant j \leqslant d-1\}$ defined by $\hat{\sigma}_p$, i.e. $\delta_j \sim \delta_{j'}$ if and only if there exists $n \in \mathbb{Z}$ such that $\hat{\sigma}_p^n(j) = j'$. We denote by Λ_k, $1 \leqslant k \leqslant s$, its equivalence class i.e. δ_j and $\delta_{j'}$ are in the same Λ_k if and only if $\delta_j \sim \delta_{j'}$. The number of classes s is determined by p.

Lemma 19.1 *For any j, $1 \leqslant j \leqslant d-1$,*

$$
\hat{\sigma}_p(j) = j'
$$

if and only if

$$
\lim_{\varepsilon \searrow 0} S_\alpha(\delta_{j'} - \varepsilon) = S_\alpha(\delta_j)
$$

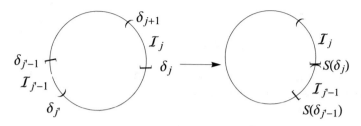

Fig. 19.1 $S(\delta_j)$ and $S(\delta_{j'-1})$

Proof We show the assertion of this lemma. Because $x \in I_{j'-1}$, from (19.6) and (19.5) (Fig. 19.1),

$$\lim_{\varepsilon \searrow 0} S_\alpha(\delta_{j'} - \varepsilon) = \lim_{\varepsilon \searrow 0} S_\alpha(\delta_{\hat{\sigma}_p(j)} - \varepsilon)$$

$$= \lim_{\varepsilon \searrow 0} S(\delta_{\hat{\sigma}_p(j)} - \varepsilon) - \delta_{\hat{\sigma}_p(j)-1} + \sum_{k:p(k)<p(\hat{\sigma}_p(j)-1)} \ell_k + \alpha.$$

As $\hat{\sigma}_p(j) = p^{-1}(p(j)-1) + 1$, where we read $1 - 1 = d - 1$ and $d - 1 + 1 = 1$ if necessary,

$$p(\hat{\sigma}_p(j) - 1) = p(p^{-1}(p(j) - 1) + 1 - 1)$$
$$= p(j) - 1.$$

Then,

$$\lim_{\varepsilon \searrow 0} S(\delta_{\hat{\sigma}_p(j)} - \varepsilon) - \delta_{\hat{\sigma}_p(j)-1} + \sum_{k:p(k)<p(\hat{\sigma}_p(j)-1)} \ell_k + \alpha$$

$$= \lim_{\varepsilon \searrow 0} S(\delta_{\hat{\sigma}_p(j)} - \varepsilon) - \delta_{\hat{\sigma}_p(j)-1} + \sum_{k:p(k)<p(j)-1} \ell_k + \alpha$$

$$= \sum_{k:p(k)<p(j)} \ell_k + \alpha$$

$$= S_\alpha(\delta_j).$$

$\square$

We consider the partition $\{\Lambda_1, \ldots, \Lambda_s\}$ of $\{1, \ldots, d - 1\}$ given by periodic orbits of $\hat{\sigma}_p$.

We define the order of Λ_k by

$$\mathrm{ord}_{\hat{\sigma}_p}(\Lambda_k) = |\Lambda_k| - 1 \tag{19.8}$$

for $1 \leqslant k \leqslant s$, where $|\Lambda_k|$ denotes the cardinality of Λ_k. We note that if $\Lambda_k = 1$, then there exists a δ_j such that $\hat{\sigma}_p(\delta_j) = \delta_j$ which means the map S_α is continuous at δ_j. Since we assumed that δ_j, $1 \leqslant j \leqslant d - 1$ are discontinuous points, this is not possible. Consequently we have $\mathrm{ord}_{\hat{\sigma}_p}(\Lambda_k) > 0$.

From this lemma, one can show that $j \sim j'$ if and only if there exist $j = k_1, \ldots, k_n = j' \in \{1, \ldots, d - 1\}$ such that

$$S_\alpha(\delta_{k_t}) = \lim_{\varepsilon \searrow 0} S_\alpha(\delta_{k_{t+1}} - \varepsilon) \tag{19.9}$$

for $1 \leqslant t < n$.

Suppose that $\{d_1, \ldots, d_s\}$, $d_j > 0$ such that $\sum_{j=1}^{s} d_j$ is even. We will show that there exists a piecewise rotation map S_α such that the partition generated by $\hat{\sigma}_p$ consists of s elements $\{\Lambda_1, \ldots, \lambda_s\}$ and $\mathrm{ord}_{\hat{\sigma}_p}(\lambda_k) = d_k$ for every $1 \leqslant k \leqslant s$.

Theorem 19.2 *For any positive integer valued vector* $(d_1, \ldots, d_s)$ *such that* $\sum_{j=1}^{s} d_j$ *is even, there exists a piecewise rotation map* S_α *such that the equivalence relation of the set of discontinuous points arising from the associated permutation* p *consists of* s *classes* $\Lambda_1, \ldots, \Lambda_s$ *and each* Λ_j *consists of* $d_j + 1$ *discontinuous points,* $1 \leqslant j \leqslant s$. *Moreover* $\overline{\{S_\alpha^n(x) : n \geqslant 0\}} = \mathbb{T}$ *for any* $x \in \mathbb{T}$.

We show this theorem in the following steps:

Step 1 $d_1, \ldots, d_s$ are all odd (and consequently s is even.)
Step 2 $d_1, \ldots, d_s$ are all even.
Step 3 $d_1, \ldots, d_s, d_{s+1}, \ldots, d_{s+t}$: $d_1, \ldots, d_s$ are odd with even s and $d_{s+1}, \ldots, d_{s+t}$ are even.

Idea of the Proof We construct an S_α for the case of Step 1 (Proposition 19.3). Then we use Propositions 19.6 and 19.8 to construct an S_α for the case of Step 2. Finally we use Proposition 19.8 again to combine Steps 1 and 2 for the case of Step 3.

Proposition 19.3 *(Step 1) Suppose* $d_1, \ldots, d_{2s}$ *are all positive and odd. Then there exists* S_α *such that the equivalence relation of the set of discontinuous points arising from the associated permutation* p *consists of* $2s$ *classes* $\Lambda_1, \ldots, \Lambda_{2s}$ *and each* Λ_j *consists of* $d_j + 1$ *discontinuous points,* $1 \leqslant j \leqslant 2s$.

Proof We show this proposition by constructing such an S_α. We fix positive real numbers $\ell_{11}, \ell_{12}, \ldots, \ell_{1d_1+1}, \ell_{21}, \ell_{22}, \ldots, \ell_{2d_2+1}, \ell_{31}, \ldots, \ell_{2s1}, \ldots, \ell_{2s\,d_{2s}+1}$ and α such that $\sum_{1 \leqslant j \leqslant 2s\, 1 \leqslant k \leqslant d_j+1} \ell_{jk} = 1$ and $\ell_{11}, \ldots, \ell_{2s\,d_{2s}+1}$ and α are independent over rational numbers. The latter assures that S_α constructed below satisfies the condition $\overline{\{S_\alpha^n(x) : n \geqslant 0\}} = \mathbb{T}$.

We divide $\mathbb{T}$ into $\sum_{j=1}^{2s} d_j + 2s$ arcs $\mathcal{I}_{jk}$ of length ℓ_{jk}, $1 \leqslant j \leqslant 2s$, $1 \leqslant k \leqslant d_j + 1$, respectively so that the right side of $\mathcal{I}_{jk}$ is

$$
\begin{cases}
\mathcal{I}_{jk+1} & \text{if } 1 \leqslant j \leqslant 2s \text{ and } 1 \leqslant k < d_j \\
\mathcal{I}_{j+11} & \text{if } 1 \leqslant j < 2s \text{ and } k = d_j \\
\mathcal{I}_{1\,d_1+1} & \text{if } j = 2s \text{ and } k = d_{2s} \\
\mathcal{I}_{j+1\,d_{j+1}+1} & \text{if } 1 \leqslant j < 2s \text{ and } k = d_j + 1 \\
\mathcal{I}_{11} & \text{if } j = 2s \text{ and } k = d_{2s} + 1.
\end{cases}
\tag{19.10}
$$

We denotes by δ_{jk} the left end point of $\mathcal{I}_{jk}$. Then the right end point of $\mathcal{I}_{jk}$ is determined by the above order of the arcs. We assume that $\delta_{11} = 0$.

We put $S(\delta_{11}) = S(0) = 0$ and define S by permutation of arcs $\mathcal{I}_{jk}$ such that

$$
\begin{aligned}
S(\delta_{11}) &< S(\delta_{13}) < \cdots < S(\delta_{1\,d_1}) < S(\delta_{22}) < S(\delta_{24}) < \cdots < S(\delta_{2\,d_2+1}) \\
&< S(\delta_{31}) < S(\delta_{33}) < \cdots < S(\delta_{3\,d_3}) < S(\delta_{42}) < S(\delta_{44}) < \cdots \\
&< S(\delta_{2s\,d_{2s}+1}) < S(\delta_{12}) < S(\delta_{14}) < \cdots < S(\delta_{1\,d_1+1}) < S(\delta_{21}) \\
&< S(\delta_{23}) < \cdots < S(\delta_{32}) < S(\delta_{34}) < \cdots < S(\delta_{2s\,1}) < \cdots \\
&< S(\delta_{2s\,d_{2s}}) < 1.
\end{aligned}
$$

With this definition of S_α, (19.7) can be rewritten in terms of (j, k) by using Lemma 19.1. Indeed, we see the following. We use $\mathrm{bwd}(j, k)$ and $\mathrm{fwd}(j, k)$ instead of -1 and $+1$.

By (19.10), we have

$$
p^{-1}(\mathrm{bwd}(p(j, k))) =
\begin{cases}
\begin{cases}
(2s, d_{2s}) & j = 1 \\
(j-1, d_{j-1}+1) & \text{otherwise}
\end{cases} & k = 1 \\[2em]
\begin{cases}
(2s, d_{2s}+1) & j = 1 \\
(j-1, d_{j-1}) & \text{otherwise}
\end{cases} & k = 2 \\[2em]
(j, k-2) & \text{otherwise,}
\end{cases}
$$

then,

$$
\hat{\sigma}_p(j, k) = \mathrm{fwd}(p^{-1}(\mathrm{bwd}(p(j, k))))
$$

$$
= \begin{cases} \begin{cases} (1\,d_1 + 1) & j = 1 \\ (j,\, d_j + 1) & \text{otherwise} \end{cases} & k = 1 \\[2mm] \begin{cases} (1,\, 1) & j = 1 \\ (j,\, 1) & \text{otherwise} \end{cases} & k = 2 \\[2mm] (j,\, k - 1) & \text{otherwise.} \end{cases}
$$

Hence, $\hat{\sigma}_p(j, \cdot)$ is the cyclic permutation on $\{(j, 1), (j, 2), \ldots, (j, d_j + 1)\}$:

$$
\hat{\sigma}_p(j,\, k) = \begin{cases} (j,\, k - 1) & \text{if } 2 \leqslant k \leqslant d_j + 1 \\ (j,\, d_j + 1) & \text{if } k = 1 \end{cases}
$$

and thus we conclude that

$$
\Lambda_j = \{\delta_{j1}, \ldots, \delta_{j\,d_j+1}\}
$$

is an equivalence class of arising from $\hat{\sigma}_p$ for $1 \leqslant j \leqslant 2s$. This shows the assertion of this proposition since the translation by α does not change $\hat{\sigma}_p$. □

Remark 19.4 To adjust the index notation, we can put as follows. If j is odd,

$$
p(j) = \begin{cases} \dfrac{j+1}{2} & \text{if } 1 \leqslant j \leqslant \sum_{k=1}^{2} d_k - 1 \\[3mm] \dfrac{j + 2k' - 1}{2} & \text{if } \sum_{k=1}^{2k'} d_k + 1 \leqslant j \leqslant \sum_{k=1}^{2k'+2} d_k - 1, \\ & \qquad\qquad 1 \leqslant k' \leqslant s - 1 \\[3mm] \dfrac{\sum_{k=1}^{2s} d_k + 2s + \sum_{k=1,odd}^{2k'-1} d_k + 2k' - 1}{2} & \\ & \text{if } j = \sum_{k=1}^{2s} d_k + 2k' - 1,\ 1 \leqslant k' \leqslant s. \end{cases}
$$

If j is even,

$$
p(j) = \begin{cases} \dfrac{\sum_{k=1}^{2s} d_k + 2sj}{2} & \text{if } 1 < j < d_1 \\[3mm] \dfrac{\sum_{k=1}^{2s} d_k + 2s + k'}{2} & \text{if } \sum_{k=1}^{2k'-1} d_k + 1 \leqslant j < \sum_{k=1}^{2k'+1} d_k - 1, \\ & \qquad\qquad 1 \leqslant k' \leqslant s - 1 \\[3mm] \dfrac{\sum_{k=1}^{2k'} d_k}{2} + k' & \text{if } j = \sum_{k=1}^{2s} d_k + 2k',\ 1 \leqslant k' \leqslant s, \end{cases}
$$

and

$$S_\alpha(x) = S(x) + \alpha \qquad \text{mod } 1 \qquad \text{for } x \in \mathbb{T}.$$

Example 19.5

(1) Suppose that $s = 2$, $d_1 = d_2 = 1$ and we put

$$p : \begin{pmatrix} 1\ 2\ 3\ 4 \\ 1\ 3\ 4\ 2 \end{pmatrix}.$$

Then we have

$$\hat{\sigma}_p : \begin{pmatrix} 1\ 2\ 3\ 4 \\ 2\ 1\ 4\ 3 \end{pmatrix}$$

and choose suitable $0 \leqslant \delta_1 < \delta_2 < \delta_3 < \delta_4 < 1$ and α. In this case, we have $\Lambda_1 = \{\delta_1, \delta_2\}$ and $\Lambda_2 = \{\delta_3, \delta_4\}$ (Fig. 19.2).

(2) Suppose that $s = 2$, $d_1 = 3$, $d_2 = 5$ and we put

$$p : \begin{pmatrix} 1\ 2\ 3\ 4\ 5\ 6\ 7\ 8\ 9\ 10 \\ 1\ 6\ 2\ 8\ 3\ 9\ 4\ 10\ 7\ 5 \end{pmatrix}.$$

Then we have

$$\hat{\sigma}_p : \begin{pmatrix} 1\ 2\ 3\ 4\ 5\ 6\ 7\ 8\ 9\ 10 \\ 9\ 1\ 2\ 10\ 4\ 5\ 6\ 7\ 3\ 8 \end{pmatrix}.$$

We have $\Lambda_1 = \{\delta_1, \delta_2, \delta_3, \delta_9\}$ and $\Lambda_2 = \{\delta_4, \delta_5, \delta_6, \delta_7, \delta_8, \delta_{10}\}$ (Fig. 19.3).

To show Step 2, we start with following.

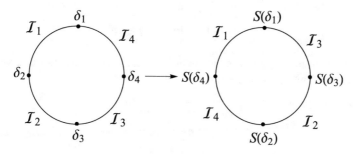

Fig. 19.2 $s = 2$, $d_1 = d_2 = 1$

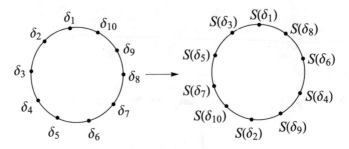

Fig. 19.3 $s = 2, d_1 = 3, d_2 = 5$

Proposition 19.6 *Suppose that d_1 is positive and even. Then there exists S_α such that the equivalence of the set of discontinuous points arising from the associated permutation p consists of only one class which is the set of discontinuous points $\{d_1, \ldots, d_{d_1+1}\}$.*

Proof As the proof of the previous proposition, we fix positive real numbers $\ell_1, \ldots, \ell_{d_1+1}$ and α such that $\sum_{j=1}^{d_1+1} \ell_j = 1$ and $\ell_1, \ldots, \ell_{d_1+1}$ and α arc independent over rational numbers. We divide $\mathbb{T}$ into $d_1 + 1$ arcs $\mathcal{I}_j$ of length ℓ_j, $1 \leqslant j \leqslant d_1 + 1$ such that the right side of $\mathcal{I}_j$ is $\mathcal{I}_{j+1}$ if $1 \leqslant j \leqslant d_1$ and $\mathcal{I}_1$ if $j = d_1 + 1$. We denote by δ_j the left end point of $\mathcal{I}_j$ and assume $\delta_1 = 0$. We put $S(\delta_1) = S(0) = 0$ and define S by a permutation of arcs $\mathcal{I}_j$ so that

$$0 = S(\delta_2) < S(\delta_4) < \cdots < S(\delta_{d_1}) < S(\delta_1) < S(\delta_3) < S(\delta_{d_1+1}) < 1$$

and define S_α as

$$S_\alpha(x) = S(x) + \alpha \qquad \text{mod } 1 \qquad \text{for } x \in \mathbb{T}.$$

Then we see

$$\hat{\sigma}_p(j) = \begin{cases} d_1 + 1 & \text{for } j = 1 \\ j - 1 & \text{for } 2 \leqslant j \leqslant d_1 + 1. \end{cases}$$

Thus it turns out that $\hat{\sigma}_p$ induces the cyclic permutation on the set of discontinuous points, which shows the assertion of this proposition. $\qquad\square$

Example 19.7 Suppose that $s = 1, d_1 = 4$ and we put

$$p : \begin{pmatrix} 1\,2\,3\,4\,5 \\ 3\,1\,4\,2\,5 \end{pmatrix}.$$

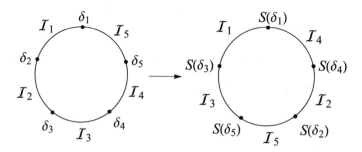

Fig. 19.4 $s = 1, d_1 = 4$

Then we have

$$\hat{\sigma}_p : \begin{pmatrix} 1\ 2\ 3\ 4\ 5 \\ 5\ 1\ 2\ 3\ 4 \end{pmatrix},$$

We have $\Lambda_1 = \{\delta_1,\ \delta_2,\ \delta_3,\ \delta_4,\ \delta_5\}$ (Fig. 19.4).

Proposition 19.8 *Suppose that S_α induces the equivalence classes $\Lambda_1, \ldots, \Lambda_s$ of the set of discontinuous points $\{\delta_1, \ldots, \delta_k\}$. Then for any odd integer $2t + 1$, there exists a piecewise rotation of the circle $\hat{S}_\alpha$ with $k + 2t + 1$ discontinuous points such that the associated equivalence classes are $\Lambda_1, \ldots, \Lambda_s, \Lambda_{s+1}$ with $\Lambda_{s+1} = \{\delta_{k+1}, \ldots, \delta_{k+2t+1}\}$.*

Proof We assume that $0 = \delta_1 < \delta_2 < \cdots < \delta_k < 1$. We put $\delta_{k+1}, \ldots, \delta_{k+2t+1}$ as

$$\delta_k < \delta_{k+1} < \cdots < \delta_{k+2t+1} < 1$$

keeping independence over rational numbers for length of arcs and α. Then $\hat{S}_\alpha$ can be defined as follows,

- $\hat{S}_\alpha(x) = S_\alpha(x)$ for $x \in \mathbb{T} \setminus [\delta_k, 1)$
- for $x \in [\delta_k, 1)$, $\hat{S}_\alpha(x)$ is given by the permutation of arcs $[\delta_k, \delta_{k+1}), [\delta_{k+1}, \delta_{k+2}), \cdots, [\delta_{k+2t}, \delta_{k+2t+1}), [\delta_{k+2t+1}, 1)$ determined by

$$\hat{S}_\alpha(\delta_k) < \hat{S}_\alpha(\delta_{k+2}) < \hat{S}_\alpha(\delta_{k+4}) < \cdots < \hat{S}_\alpha(\delta_{k+2t}) < \hat{S}_\alpha(\delta_{k+1})$$
$$< \hat{S}_\alpha(\delta_{k+3}) < \cdots < \hat{S}_\alpha(\delta_{k+2t+1}).$$

We see that this $\hat{S}_\alpha$ gives the desired map.   □

As mentioned before, we get the assertion of Theorem 19.2 from Propositions 19.3, 19.6, and 19.8.

Example 19.9 We start with S_α constructed in Example 19.5 (2) i.e. the equivalence classes of discontinuous points are

$$\Lambda_1 = \{\delta_1, \, \delta_2, \, \delta_3, \, \delta_9\} \text{ and } \Lambda_2 = \{\delta_4, \, \delta_5, \, \delta_6, \, \delta_7, \, \delta_{10}\}.$$

We construct $\hat{S}_\alpha$ by adding five discontinuous points

$$\{\delta_{11}, \, \delta_{12}, \, \delta_{13}, \, \delta_{14}, \, \delta_{15}\}.$$

so that

$$\begin{cases} \Lambda_1 = \{\delta_1, \, \delta_2, \, \delta_3, \, \delta_9\} \\ \Lambda_2 = \{\delta_4, \, \delta_5, \, \delta_6, \, \delta_8, \, \delta_9, \, \delta_{10}\} \\ \Lambda_3 = \{\delta_{11}, \, \delta_{12}, \, \delta_{13}, \, \delta_{14}, \, \delta_{15}\}. \end{cases}$$

are the equivalence classes. We put

$$p : \begin{pmatrix} 1 & 2 & 3 & 4 & 5 & 6 & 7 & 8 & 9 & 10 & 11 & 12 & 13 & 14 & 15 \\ 1 & 11 & 2 & 13 & 3 & 14 & 4 & 15 & 12 & 5 & 8 & 6 & 9 & 7 & 10 \end{pmatrix}.$$

Then we have

$$\hat{\sigma}_p : \begin{pmatrix} 1 & 2 & 3 & 4 & 5 & 6 & 7 & 8 & 9 & 10 & 11 & 12 & 13 & 14 & 15 \\ 9 & 1 & 2 & 10 & 4 & 5 & 6 & 7 & 3 & 8 & 15 & 11 & 12 & 13 & 14 \end{pmatrix}$$

and get the desired map $\hat{S}_\alpha$ (Fig. 19.5).

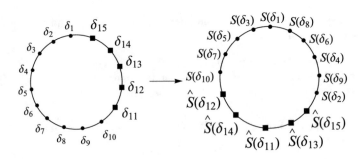

Fig. 19.5 $s = 3$, $d_1 = 3$, $d_2 = 5$, $d_3 = 4$

19.3 IPR Map

In the sequel, we assume that $\{S_\alpha^n(\delta_j) : n \geqslant 0\}$ is an infinite set for $1 \leqslant j \leqslant d-1$ and $\{S_\alpha^n(\delta_j) : n \geqslant 0\} \cap \{S_\alpha^n(\delta_k) : n \geqslant 0\} = \emptyset$ for any $1 \leqslant j < k \leqslant d-1$. This assumption corresponds to the i.d.o.c. for interval exchange maps. Hereafter, $[\overrightarrow{u, v})$ denotes the arc from u to v in natural order of $\mathbb{T}$. We consider the orbit of one fixed discontinuous point which we call the marked discontinuous point. Also we call the equivalence class Λ_j of the set of discontinuous point which includes the marked fixed point the marked equivalence class. For simplicity, we choose $j = 1$. A positive integer N_0 is said to be a critical iterate (associated with δ_1) if

1. $\{S_\alpha(\delta_1), S_\alpha^2(\delta_1), \ldots, S_\alpha^{N_0}(\delta_1)\} \cap \mathcal{I}_j \neq \emptyset$ for $1 \leqslant j \leqslant d-1$
    and either
2. (i) $S_\alpha^{N_0}(\delta_1) \in \mathcal{I}_{j_0}, \#(\{S_\alpha(\delta_1), S_\alpha^2(\delta_1), \ldots, S_\alpha^{N_0}(\delta_1)\} \cap \mathcal{I}_{j_0}) \geqslant 2$,
    and
    $$\{S_\alpha(\delta_1), S_\alpha^2(\delta_1), \ldots, S_\alpha^{N_0}(\delta_1)\} \cap [\overrightarrow{\delta_{j_0}, S_\alpha^{N_0}(0)}) = \emptyset,$$
    or,
    (ii) $S_\alpha^{N_0}(\delta_1) \in \mathcal{I}_{j_0-1}, \#(\{S_\alpha(\delta_1), S_\alpha^2(\delta_1), \ldots, S_\alpha^{N_0}(\delta_1)\} \cap \mathcal{I}_{j_0-1}) \geqslant 2$,
    and
    $$\{S_\alpha(\delta_1), S_\alpha^2(\delta_1), \ldots, S_\alpha^{N_0}(\delta_1)\} \cap [\overrightarrow{S_\alpha^{N_0}(\delta_1), \delta_{j_0}}) = \emptyset$$
    for some $1 \leqslant j_0 \leqslant d-1$ where we read $\mathcal{I}_{d-1}$ for $\mathcal{I}_{j_0-1}$ if $j_0 = 1$
    (Fig. 19.6).

Now we consider the partition V_0 generated by $\{S_\alpha(\delta_1), S_\alpha^2(\delta_1), \ldots, S_\alpha^{N_0-1}(\delta_1)\}$, which consists of $N_0 - 1$ arcs. Each element of V_0 is of the form $[\overrightarrow{S_\alpha^n(\delta_1), S_\alpha^{n'}(\delta_1)})$ for some $n, n', 1 \leqslant n, n' < N_0$, by the definition. Then for each $j, 1 \leqslant j \leqslant d-1$, there exist n_j and n_j' such that $[\overrightarrow{S_\alpha^{n_j}(\delta_1), S_\alpha^{n_j'}(\delta_1)}) \ni \delta_j$ is an element of V_0. We consider the induced transformation of S_α to $\mathbb{J} = \cup_{j=1}^{d-1} \mathbb{J}_j$ where $\mathbb{J}_j = [\overrightarrow{S_\alpha^{n_j}(\delta_1), S_\alpha^{n_j'}(\delta_1)})$. This map is called an IPR map and denoted by S_{α, N_0}. We note that the IPR map

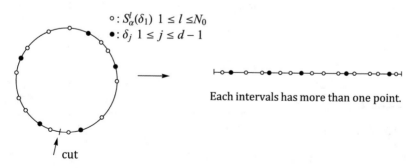

$\circ : S_\alpha^l(\delta_1)$ $1 \leq l \leq N_0$
$\bullet : \delta_j$ $1 \leq j \leq d-1$

Each intervals has more than one point.

cut

Fig. 19.6 An interval from a circle

is determined by S_α, the choice of the marked discontinuous point (here, we have chosen δ_1), and the critical iterate N_0. We will see that for any piecewise rotation map of the circle with a given $(d_1, \ldots, d_s)$ in Theorem 19.2 and a critical iterate N_0, the IPR map induces a translation surface with the vector $(d_1, \ldots, d_s)$ of the orders of singularities, where the marked singularity is determined by the choice of the marked equivalence class (equivalently the marked discontinuous point).

Since δ_j, $1 \leqslant j \leqslant d - 1$, is a discontinuous point, $S_\alpha \mathbb{J}_j$ is not an arc, however, $[\overrightarrow{S_\alpha(\delta_j), S_\alpha^{n'_j+1}(\delta_1))}$ is a sub-arc of $[\overrightarrow{S_\alpha^{n_l+1}(\delta_1), S_\alpha^{n'_j+1}(\delta_1))}$ for some l, $1 \leqslant l \leqslant d - 1$. Also $[\overrightarrow{S_\alpha^{n_j+1}(\delta_1), S_\alpha(\delta_j))}$ is a sub arc of $[\overrightarrow{S_\alpha^{n_j+1}(\delta_1), S_\alpha^{n'_{l'}+1}(\delta_1))}$ for some l', $1 \leqslant l' \leqslant d - 1$. This shows that if $[\overrightarrow{S_\alpha^{n_l+1}(\delta_1), S_\alpha^{n'_j+1}(\delta_1))}$ is an element of V_0, then

$$[\overrightarrow{S_\alpha^{n_l+1}(\delta_1), S_\alpha^{n'_j+1}(\delta_1))} = [\overrightarrow{S_\alpha^{n_l+1}(\delta_1), S_\alpha(\delta_l))} \cup [\overrightarrow{S_\alpha(\delta_j), S_\alpha^{n'_j+1}(\delta_1))} \quad (19.11)$$

holds for some j and l, $2 \leqslant j$, $l \leqslant d - 1$, see Fig. 19.7. From Lemma 1, we see $\hat{\sigma}_p(j) = l$. If $j = 1$ or $l = 1$, then we have

$$[S_\alpha(\delta_1), S_\alpha^{n'_1+1}(\delta_1)) \in V_0 \quad \text{or} \quad [S_\alpha^{n_1+1}(\delta_1), S_\alpha(\delta_1)) \in V_0$$

respectively. In these cases, we regard that one of two terms of the right hand side of (19.11) is empty. It is also possible that

$$[\overrightarrow{S_\alpha^{n_l+1}(\delta_1), S_\alpha(\delta_l))} \cup [\overrightarrow{S_\alpha(\delta_j), S_\alpha^{n'_j+1}(\delta_1))} \notin V_0.$$

holds. This happens only at j with $\mathbb{J}_j \ni S_\alpha^{N_0}(\delta_1)$. To distinguish the special case, we denote by j^* such a j i.e. $\mathbb{J}_{j^*} = [\overrightarrow{S_\alpha^{n_{j^*}}(\delta_1), S_\alpha^{n'_{j^*}}(\delta_1))} \ni S_\alpha^{N_0}(\delta_1)$.

$$S_\alpha^{n_l+1}(\delta_1) \quad S_\alpha(\delta_j) \quad S_\alpha^{n'_j+1}(\delta_1)$$

$$S_\alpha^{n_l}(\delta_1) \quad \delta_{j'} \qquad S_\alpha^{n'_l}(\delta_1) \qquad\qquad S_\alpha^{n_j}(\delta_1) \quad \delta_j \qquad S_\alpha^{n'_j}(\delta_1)$$

Fig. 19.7 $\mathbb{J}$ and V_0

Here are two cases.

- $j \neq j^*$

  We put

$$n(j) = \max\{n \geqslant 0 : [S_\alpha^{n_j-n}(\delta_1), \overrightarrow{S_\alpha^{n'_j-n}}(\delta_1)) \text{ is an element of } V_0\}.$$

Lemma 19.10 *For* $j \neq j^*$,

$$\mathbb{J}_j = \begin{cases} S_\alpha^{n(j)}[\overrightarrow{S_\alpha^{n_{\hat{o}p(1)}+1}}(\delta_1), S_\alpha(\delta_1)) & \text{if } S_\alpha^{n'_j-n(j)}(\delta_1) = S_\alpha(\delta_1) \\ S_\alpha^{n(j)}[\overrightarrow{S_\alpha(\delta_1), S_\alpha^{n'_i+1}}(\delta_1)) & \text{if } S_\alpha^{n_j-n(j)}(\delta_1) = S_\alpha(\delta_1) \\ S_\alpha^{n(j)}[\overrightarrow{S_\alpha^{n_{\hat{o}p(k)}+1}}(\delta_1), S_\alpha^{n'_k+1}(\delta_1)) & \text{if } S_\alpha^{n'_j-n(j)}(\delta_1), S_\alpha^{n_j-n(j)}(\delta_1) \neq S_\alpha(\delta_1). \end{cases}$$

Proof This follows from the fact that if $j \neq j^*$, then there is no point of the form $S_\alpha^n(\delta_1)$, $1 \leqslant n < N_0$ in $(\overrightarrow{S_\alpha^{n_j-m}}(\delta_1), S_\alpha^{n'_j-m}(\delta_1))$, $0 \leqslant m \leqslant n(j)$. Then the assertion of the lemma follows from (19.11). □

- $j = j^*$

  Since $[\overrightarrow{S_\alpha^{n_{j^*}}}(\delta_1), S_\alpha^{n'_{j^*}}(\delta_1)) \ni S_\alpha^{N_0}(\delta_1)$, we divide

$$[\overrightarrow{S_\alpha^{n_{j^*}}}(\delta_1), S_\alpha^{n'_{j^*}}(\delta_1)) = [\overrightarrow{S_\alpha^{n_{j^*}}}(\delta_1), S_\alpha^{N_0}(\delta_1)) \cup [\overrightarrow{S_\alpha^{N_0}}(\delta_1), S_\alpha^{n'_{j^*}}(\delta_1))$$

and define

$$n(j^*, l) = \max\{n \geqslant 0 : [\overrightarrow{S_\alpha^{n_{j^*}-n}}(\delta_1), S_\alpha^{N_0-n}(\delta_j)) \text{ is an element of } V_0\}$$

$$n(j^*, r) = \max\{n \geqslant 0 : [\overrightarrow{S_\alpha^{N_0-n}}(\delta_j), S_\alpha^{n'_{j^*}-n}(\delta_1)) \text{ is an element of } V_0\}.$$

Lemma 19.11 *We have*

$$[\overrightarrow{S_\alpha^{n_{j^*}-n(j^*,l)}}(\delta_1), S_\alpha^{N_0-n(j^*,l)}(\delta_1))$$

$$= [\overrightarrow{S_\alpha^{n_{\hat{o}p(k)}+1}}(\delta_1), S_\alpha^{n'_k+1}(\delta_1)) \text{ or } [\overrightarrow{S_\alpha(\delta_1), S_\alpha^{n'_k+1}}(\delta_1))$$

and

$$[S_\alpha^{N_0-n(j^*,r)}(\delta_1), \; \overrightarrow{S_\alpha^{n'_{j^*}-n(j^*,r)}(\delta_1))}$$

$$= [S_\alpha^{n_{\hat{\sigma}_p(k')}+1}(\delta_1), \; \overrightarrow{S_\alpha^{n'_{k'}+1}(\delta_1))} \; or \; [S_\alpha^{n_{\hat{\sigma}_p(k')}+1}(\delta_1), \; \overrightarrow{S_\alpha(\delta_1))}$$

for some k and k′, $1 \leqslant k, k' \leqslant d-1$.

Proof From the definition of the critical iterate, we see $S_\alpha^{N_0-n(j^*,l)}(\delta_1) \neq S_\alpha(\delta_1)$ and $S_\alpha^{N_0-n(j^*,r)}(\delta_1) \neq S_\alpha(\delta_1)$ (equivalently, $N_0 - n(j^*, l)$ and $N_0 - n(j^*, r) \geqslant 2$. This shows the assertion of this lemma. $\qquad\square$

These two lemmas give us the tower representation of S_α and show the behavior of the induced transformation S_{α,N_0} defined on $\mathbb{J}$.
We put

$$\theta_j^l = |\overrightarrow{[S_\alpha^{n_j}(\delta_1), \; \delta_j)}|$$

and

$$\theta_j^r = |\overrightarrow{[\delta_j, \; S_\alpha^{n'_j}(\delta_1))}|$$

for $1 \leqslant j \leqslant d-1$, where $|\cdot|$ denotes the length of an arc or an interval.

Proposition 19.12

(i) *If $j \neq j^*$, then we have*

$$|\mathbb{J}_j| = \begin{cases} \theta_{\hat{\sigma}_p(k)}^l + \theta_k^r & \text{if } k \neq 1, \; \hat{\sigma}_p(1), \; j^* \\ \theta_{\hat{\sigma}_p(1)}^l & \text{if } S_\alpha^{n'_j-n(j)}(\delta_1) = S_\alpha(\delta_1) \\ \theta_{\hat{\sigma}_1}^r & \text{if } S_\alpha^{n_j-n(j)}(\delta_1) = S_\alpha(\delta_1), \end{cases} \tag{19.12}$$

where k is chosen so that $S_\alpha(\delta_k) \in [\overrightarrow{S_\alpha^{n_j-n(j)}(\delta_1), \; S_\alpha^{n'_j-n(j)}(\delta_1)})$.
(ii) *If $j = j^*$, then we have*

$$|\mathbb{J}_{j^*}| = \theta_{\hat{\sigma}_p(k)}^l + \theta_k^r + \theta_{\hat{\sigma}_p(k')}^l + \theta_{k'}^r, \tag{19.13}$$

where k and k′ are chosen so that

$$S_\alpha(\delta_k) \in [\overrightarrow{S_\alpha^{n_j-n(j)}(\delta_1), {}^{N_0-n(j^*,l)}_{\;,\alpha}(\delta_1))}$$

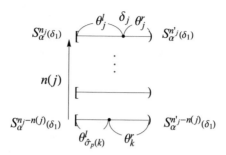

Fig. 19.8 The tower without a balcony

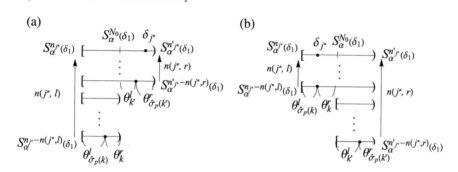

Fig. 19.9 (**a**) Balcony at the right side. (**b**) Balcony at the left side

and

$$S_\alpha(\delta_{k'}) \in [S_\alpha^{N_0-n(j^*,r)}(\delta_1), \overline{S_\alpha^{n'_{j^*}-n(j^*,r)}(\delta_1)}),$$

otherwise, $|\mathbb{J}_{j^*}| = \theta_1^l + \theta_{\hat{\sigma}_p(k')}^l + \theta_{k'}^r$ *or* $= \theta_{\hat{\sigma}_p(k)}^l + \theta_k^r + \theta_1^r$.

Proof This is a direct consequence of Lemmas 19.10 and 19.11 with the definitions of θ_j^l and θ_j^r, see Figs. 19.8 and 19.9. □

There are two natural orders ρ_r and ρ_l, which are called the right order and the left order, respectively. These are given by

$$0 < n'_{\rho_r^{-1}(1)} < n'_{\rho_r^{-1}(2)} < \cdots < n'_{\rho_r^{-1}(d-1)}$$

and

$$0 < n_{\rho_l^{-1}(1)} < n_{\rho_l^{-1}(2)} < \cdots < n_{\rho_l^{-1}(d-1)}.$$

In connection with $n(j), j \neq j^*, n(j^*, l), n(j^*, r)$, we have the following:
We see $\rho_r(j) = 1$ if

$$n'_j - n(j) = 1, \ j \neq j^*$$

or

$$n'_{j^*} - n(j^*, r) = 1,$$

equivalently,

$$S_\alpha^{n'_j - n(j)}(\delta_1) = S_\alpha(\delta_1) \qquad \text{when} \qquad j \neq j^*$$

or

$$S_\alpha^{n'_{j^*} - n(j^*, r)}(\delta_1) = S_\alpha(\delta_1).$$

Then we see ρ_r inductively by

$$\rho_r(j) = \rho_r(k) + 1 \qquad \text{if} \qquad n'_j - n(j) = n'_k + 1, \ j \neq j^*.$$

or

$$\rho_r(j^*) = \rho_r(k) + 1 \qquad \text{if} \qquad n'_{j^*} - n(j^*, r) = n'_k + 1.$$

We continue this process until we get $\rho_r(j) = d - 1$ for some $1 \leqslant j \leqslant d - 1$. On the other hand, we see $\rho_l = 1$ if

$$n_j - n(j) = 1, \ j \neq j^*$$

or

$$n_{j^*} - n(j^*, l) = 1,$$

equivalently,

$$S_\alpha^{n_j - n(j)}(\delta_1) = S_\alpha(\delta_1) \text{when} \qquad j \neq j^*$$

or

$$S_\alpha^{n_{j^*} - n(j^*, l)}(\delta_1) = S_\alpha(\delta_1).$$

Then we see ρ_l inductively by

$$\rho_r(l) = \rho_l(k) + 1 \qquad \text{if} \qquad n_j - n(j) = n_k + 1, \ j \neq j^*.$$

or

$$\rho(j^*) = \rho_l(k) + 1 \qquad \text{if} \qquad n_{j^*} - n(j^*, l) = n_k + 1.$$

Again we continue this process until we get $\rho_l(j) = d - 1$ for some $1 \leqslant j \leqslant d - 1$.

Proposition 19.13 *We have*

$$
\theta_j^l + \theta_j^r =
\begin{cases}
\theta^l_{\rho_l^{-1}(\rho_l(j)-1)} & \text{if } \rho_r(j) = 1 \\[2mm]
\theta^r_{\rho_r^{-1}(\rho_r(j)-1)} & \text{if } \rho_l(j) = 1 \\[2mm]
\theta^l_{\rho_l^{-1}(d-1)} + \rho^r_{\rho_r^{-1}(d-1)} & \\[1mm]
\quad + \theta^l_{\rho_l^{-1}(\rho_l(j)-1)} + \theta^r_{\rho_r^{-1}(\rho_r(j)-1)} & \text{if } j = j^* \\[2mm]
\theta^l_{\rho_l^{-1}(\rho_l(j)-1)} + \theta^r_{\rho_r^{-1}(\rho_r(j)-1)} & \text{otherwise.}
\end{cases}
\tag{19.14}
$$

Proof This follows from (19.12) and (19.13) with the definitions of ρ_r and ρ_l. □

If $j \neq j^*$, the tower concerning δ_j is simple. Indeed, the tower consists of $[S_\alpha^{n_j-n(j)}(\delta_1), S_\alpha^{n'_j-n(j)}(\delta_1)), [S_\alpha^{n_j-n(j)+1}(\delta_1), S_\alpha^{n'_j-n(j)+1}(\delta_1)), \ldots, [S_\alpha^{n_j}(\delta_1), S_\alpha^{n'_j}(\delta_1))$ (hight $n(j) + 1$), see Fig. 19.8. However, the tower concerning δ_{j^*} is different from other j's. For further discussion, we need an information on $n(j^*, l)$ and $n(j^*, r)$. There are three possibilities:

Case (a) $n(j^*, l) > n(j^*, r)$
Case (b) $n(j^*, l) < n(j^*, r)$
Case (c) $n(j^*, l) = n(j^*, r)$

Here Case (c) occurs only a special case which we never discuss (the minimum critical iterate under a special condition, see Lemma 19.14 below). In the sequel we only consider the Case (a) and Case (b), see Fig. 19.9.

We call the right (or the left) part of the tower the "balcony" in Case (a) (or in Case (b), respectively), i.e. from $[S_\alpha^{N_0-n(j^*,r)}(\delta_1), S_\alpha^{n'_{j^*}-n(j^*,r)}(\delta_1))$ to $[S_\alpha^{N_0}(\delta_1), S_\alpha^{n'_{j^*}}(\delta_1))$ or $[S_\alpha^{N_0-n(j^*,l)}(\delta_1), S_\alpha^{n_{j^*}-n(j^*,l)}(\delta_1))$ to $[S_\alpha^{N_0}(\delta_1), S_\alpha^{n_{j^*}}(\delta_1))$, respectively. The tower $\mathbb{J}^*$ on the top is called the tower with balcony. Then the set of $d - 1$ towers (one has the balcony) is called a castle. We denote by $\mathcal{K}_j$ each tower such that $\mathbb{J}_j$ on the top, $1 \leqslant j \leqslant d - 1$, and by $\mathcal{K}$ the castle (i.e. $\mathcal{K} = \{\mathcal{K}_j : 1 \leqslant j \leqslant d - 1\}$). We note that a castle is defined by the choice of the marked discontinuous point (here, δ_1) and the critical iterate N_0.

Lemma 19.14 *If we choose N_0 not the minimum critical iterate, then either Case (a) or Case (b) holds always.*

Proof We consider the critical iterate $N_1 > N_0$ such that there is no critical iterate in (N_0, N_1). Each Case (a) or Case (b) has two sub-cases;

Case (a-i) (a) holds and $S_\alpha^{N_0}(\delta_1) \in [S_\alpha^{n_{j^*}}(\delta_1), \delta_{j^*})$

Case (a-ii) (a) holds and $S_\alpha^{N_0}(\delta_1) \in, [\delta_{j^*}, S_\alpha^{n'_{j^*}}(\delta_1))$

Case (b-i) (b) holds and $S_\alpha^{N_0}(\delta_1) \in [S_\alpha^{n_{j^*}}(\delta_1), \delta_{j^*})$

Case(b-ii) (b) holds and $S_\alpha^{N_0}(\delta_1) \in, [\delta_{j^*}, S_\alpha^{n'_{j^*}}(\delta_1))$.

Suppose that (a-i) holds. In this case, the balcony is at the right side. If $\rho_l(j^*) \neq d - 1$, then there exists $j^{**} \neq j^*$ such that $\rho_l(j^{**}) = \rho_l(j^*) + 1$. Then we see

$$S_\alpha^{n_{j^*} + n(j^{**}) + 1}(\delta_1) = S_\alpha^{n_{j^{**}}}(\delta_1)$$

and

$$S_\alpha^{N_0 + 1 + n(j^{**})}(\delta_1) \in [S_\alpha^{n_{j^{**}}}(\delta_1), S_\alpha^{n'_{j^{**}}}(\delta_1)).$$

This implies that $N_1 = N_0 + n(j^{**}) + 1$ and the tower $\delta_{j^{**}}$ on the top is changed to the tower with balcony (at the right side) with

$$n(j^{**}, l) = n_{j^{**}} + n(j^*, l) + 1 \qquad \text{and} \qquad n(j^{**}, r) = n_{j^{**}}$$

for the new critical iterate N_1 after cutting the left (main tower) part of the tower associated to δ_{j^*} and connecting it to the bottom of the tower associated to $\delta_{j^{**}}$. As a consequence, the balcony part of the tower associated to δ_{j^*} is a new tower (without balcony) with respect to N_1. If (a-ii) holds, then we see that N_1 is the same but the balcony is the left side for the tower associated to $\delta_{j^{**}}$ with

$$n(j^{**}, l) = n_{j^{**}} \qquad \text{and} \qquad n(j^{**}, r) = n_{j^{**}} + n(j^*, r) + 1.$$

The same holds for the cases (b-i) and (b-ii), where we choose j^{**} so that $\rho_r(j^{**}) = \rho_r(j^*) + 1$. If $\rho_l(j^*) = d - 1$ in Cases (a-i) and (a-ii) or $\rho_r(j^*) = d - 1$ in Cases (b-i) and (b-ii)), we put $j^{**} = j^*$. Consequently, $n(j^{**}, l) \neq n(j^{**}, r)$ holds in any cases. $\square$

The proof of this lemma also shows the following whose idea we use in Sect. 19.5.

Proposition 19.15 *If N_1 is the next critical iterate after a critical iterate N_0, i.e. any N, $N_0 < N < N_1$ is not a critical iterate, then*

$$N_1 = N_0 + n(j^{**}) + 1$$

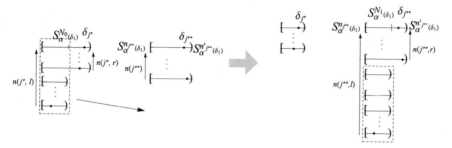

Fig. 19.10 a-i

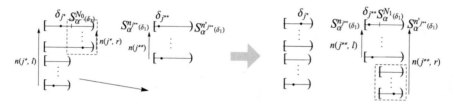

Fig. 19.11 a-ii

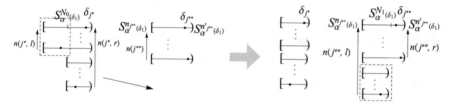

Fig. 19.12 b-i

if $j^{**} \neq j^*$ and

$$N_1 = \begin{cases} N_0 + n(j^*, r) + 1 & \text{when} & \text{Cases (a-i), (b-i)} \\ N_0 + n(j^*, l) + 1 & \text{when} & \text{Cases (a-ii), (b-ii)} \end{cases}$$

if $j^{**} = j^*$, see Figs. 19.10, 19.11, 19.12 and 19.13.

The rest of this section, we construct a translation surface from an IPR map (a discrete type construction).

Case (a) We put $\zeta_0 = \zeta_0^* = 0$,

$$\zeta_k = n'_{\rho_r^{-1}(k)} + i\theta^r_{\rho_r^{-1}(k)}$$

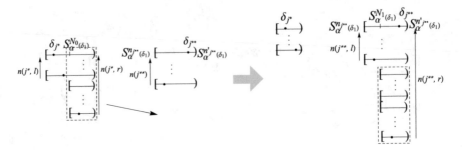

Fig. 19.13 b-ii

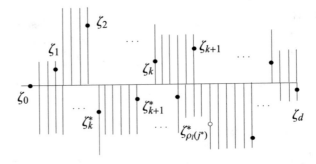

Fig. 19.14 Translation surface (discrete type) for Case (a)

for $1 \leqslant k \leqslant d - 1$, and

$$\zeta_k^* = \begin{cases} n_{\rho_l^{-1}(k)} - i\theta_{\rho_l^{-1}(k)}^l & \text{if } 1 \leqslant k < \rho_l(j^*) \\ n_{j^*} - n(j^*, r) - 1 - i(\theta_{j^*}^l + \theta_{j^*}^r - \theta_{\rho_l^{-1}(d-1)}^r) & \text{if } \rho_l^{-1}(k) = j^* \\ n_{\rho_l^{-1}(k-1)} - i\theta_{\rho_l^{-1}(k-1)}^l & \text{if } \rho_l(j^*) < k \leqslant d \end{cases}$$

for $1 \leqslant k \leqslant d$, and $\zeta_d = \zeta_d^*$ (Fig. 19.14).
Case (b) We put $\zeta_0 = \zeta_0^* = 1$,

$$\zeta_k = \begin{cases} n'_{\rho_r^{-1}(k)} + i\theta_{\rho_r^{-1}(k)}^r & \text{if } 1 \leqslant k < \rho_r(j^*) \\ n'_{j^*} - n(j^*, l) - 1 + i(\theta_{j^*}^l + \theta_{j^*}^r - \theta_{\rho_r^{-1}(d-1)}^l) & \text{if } \rho_r^{-1}(k) = j^* \\ n'_{\rho_r^{-1}(k-1)} + i\theta_{\rho_r^{-1}(k-1)}^r & \text{if } \rho_r(j^*) < k \leqslant d \end{cases}$$

for $1 \leqslant k \leqslant d$, and

$$\zeta_k^* = n_{\rho_l^{-1}(k)} - i\theta_{\rho_l^{-1}(k)}^l$$

for $1 \leqslant k \leqslant d - 1$ and $\zeta_d = \zeta_d^*$ (Fig. 19.15).

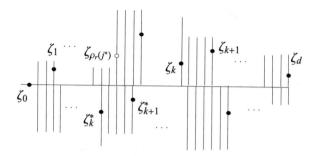

Fig. 19.15 Translation surface (discrete type) for Case (b)

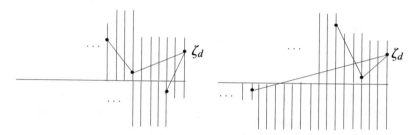

Fig. 19.16 No intersection and self intersection

Then we consider $2d$ line segments $\ell(\zeta_{j-1}, \zeta_j)$, $\ell(\zeta_{j-1}^*, \zeta_j^*)$, $1 \leqslant j \leqslant d$ where we by denote $\ell(z, w)$ the line segment from complex numbers z to w. Here we have to be careful that the line segments by (ζ_{d-1}, ζ_d) or $(\zeta_{d-1}^*, \zeta_d^*)$ may intersect to $(\zeta_{k-1}^*, \zeta_k^*)$ or (ζ_{k-1}, ζ_k), respectively, for some $1 \leqslant k \leqslant d - 1$. However, if we choose the critical iterate in proper way, then they do not intersect and these $2d$ line segments are sides of a $2d$-polygon $\mathbf{F}$, see Fig. 19.16.

Theorem 19.16 *The sets of line segments* $\{\ell(\zeta_{j-1}, \zeta_j)\}$ *and* $\{\ell(\zeta_{j-1}^*, \zeta_j^*)\}$ *are pairwise parallel by the following correspondence and have the same length. In particular, if the figure* $\mathbf{F}$ *is a $2d$-polygon, then it is a translation surface.*

Case (a)

$$\ell(\zeta_{k-1}, \zeta_k) \leftrightarrow \ell(\zeta_{\rho_l(\rho_r^{-1}(k))-1}^*, \zeta_{\rho_l(\rho_r^{-1}(k))}^*) \text{ if } 1 \leqslant k < d - 1,$$
$$\ell(\zeta_{d-1}, \zeta_d) \leftrightarrow \ell(\zeta_{\rho_r(j^*)-1}^*, \zeta_{\rho_r(j^*)}^*) \qquad \text{if } k = d.$$

Case (b)

$$\ell(\zeta_{k-1}^*, \zeta_k^*) \leftrightarrow \ell(\zeta_{\rho_l(\rho_r^{-1}(k))-1}, \zeta_{\rho_l(\rho_r^{-1}(k))}) \text{ if } 1 \leqslant k < d - 1,$$
$$\ell(\zeta_{d-1}^*, \zeta_d^*) \leftrightarrow \ell(\zeta_{\rho_r(j^*)-1}, \zeta_{\rho_r(j^*)}) \qquad \text{if } k = d.$$

Proof From the definition of n_j, n'_j, and $n(j)$, we see that

$$n_j - n(j) = n'_j - n(j)$$

always holds when $j \neq j^*$. Thus, in the case (a), we see

$$Re(\zeta_k - \zeta_{k-1}) = Re(\zeta^*_{\rho_l(\rho_r^{-1}(k))} - \zeta^*_{\rho_l(\rho_r^{-1}(k))-1})$$

if $1 \leqslant k < d - 1$. On the other hand, from (19.14), we see

$$Im(\zeta_k - \zeta_{k-1}) = \theta^r_j - \theta^r_{\rho_r^{-1}(\rho_r(j)-1)}$$

$$= \theta^l_{\rho_l^{-1}(\rho_l(j)-1)} - \theta_{\rho_l(j)}$$

$$= Im(\zeta^*_{\rho_l(j)} - \zeta^*_{\rho_l^{-1}(\rho_l(j)-1)})$$

for $j = \rho_r^{-1}(k)$ with $j \neq j^*$, $\rho_r(j) \neq 1$, and $\rho_l(j) \neq 1$. Thus two line segments $\ell(\zeta_{k-1}, \zeta_k)$ and $\ell(\zeta^*_{\rho_l^{-1}(\rho_l(j)-1)}, \zeta^*_{\rho_l(j)})$ are parallel and have the same length when $1 \leqslant k < \rho_r(j^*)$, $k \neq 1$, and $\rho_l(\rho_r^{-1}(k)) \neq 1$. When $1 \leqslant k < \rho_r(j^*)$ and $k = 1$, we have

$$Im(\zeta_1 - \zeta_0) = \theta^r_j = \theta^l_{\rho_l^{-1}(\rho_l(j)-1)} - \theta^l_j = Im(\zeta^*_{\rho_l(j)} - \zeta^*_{\rho_l^{-1}(\rho_l(j)-1)}).$$

for $j = \rho_r^{-1}(1)$. When $1 \leqslant k < \rho_r(j^*$ and $\rho_l(j) = 1$, we have

$$Im(\zeta_k - \zeta_{k-1}) = \theta^r_j - \theta_{\rho^{-1}(\rho_r(j)-1)} = -\theta^l_j = Im(\zeta^*_1 - \zeta^*_0).$$

For $k = d$ and $\rho_l(j^*) \neq 1$,

$$Re(\zeta_d - \zeta_{d-1}) = n(j^*, r) - n(j^*, l) = Re(\zeta^*_{\rho_r(j^*-1)}, \zeta^*_{\rho_r(j^*)}).$$

Moreover, since

$$Im(\zeta_d - \zeta_{d-1}) = -\theta^l_{\rho_r(d-1)} - \theta^l_{\rho_r(d-2)}$$

$$= \theta^l_{\rho_l^{-1}}(d-1) + \theta^r_{\rho_r^{-1}(\rho_r(j^*)-1)} - \theta^l_{j^*} - \theta^r_{j^*}$$

$$= Im(\zeta^*_{\rho_r(j^*-1)}, \zeta^*_{\rho_r(j^*)})$$

Thus we have the same conclusion. The other cases also follow by the same way.

□

We should note that $\zeta_0 = 0$ comes from the marked discontinuous point δ_1 (but δ_1 itself appears in $\mathbb{J}_1$) and $Re(\zeta_j)$ denotes the number of iterations of S_α to get the nearest visit along the orbit of δ_1 to δ_j. Then,

Case (a)   Each ζ_j, $1 \leqslant j \leqslant d - 1$, corresponds to the discontinuous point δ_j which is on the top of the tower $\mathcal{K}_j$ and ζ_d is a sort of virtual discontinuous point along the right order of the castle;

Case (b)   Each ζ_j^*, $1 \leqslant j \leqslant d - 1$, corresponds to the discontinuous point δ_j which is on the top of the tower $\mathcal{K}_j$ and ζ_d^* is a sort of virtual discontinuous point the left order of the castle.

In this point of view, we can induce the equivalence relation among ζ_j, $0 \leqslant j \leqslant d$, from $\hat{\sigma}_p$ with Lemma 19.1. We denote by $\Theta = \{\Theta_t, : 1 \leqslant t \leqslant s\}$ the set of the equivalence classes. It is easy to see that ζ_{j^*} is equivalent to ζ_d in this equivalence relation. Now we put

$$\tilde{\Theta}_t = \Theta_t \setminus \{\zeta_0, \zeta_d\}, 1 \leqslant t \leqslant s.$$

Then by a suitable indexing, we see

$$(|\tilde{\Theta}_1|, |\tilde{\Theta}_2|, \ldots, |\tilde{\Theta}_s|\} = (d_1, d_2, \ldots, d_s).$$

Moreover, if $\zeta_0 \in \Theta_{t_0}$ for $1 \leqslant t_0 \leqslant s$, then Θ_{t_0} gives the marked singularity. Thus we have the following.

Theorem 19.17 *Suppose that the piecewise rotation map of the circle S_α is the one constructed in Sect. 19.2 with $(d_1, d_2, \ldots, d_s)$. If the figure **F** constructed in the above is a translation surface, then its orders of singularities are given by $(d_1, d_2, \ldots, d_s)$ and the order of the marked singularity is d_{t_0} with t_0 defined in the above.*

In the above construction, only information we need are: (θ_j^l, θ_j^r), $1 \leqslant j \leqslant d - 1$ with j^*, (ρ_l, ρ_r), and "Case (a) or Case (b)", i.e. (a) $n(j^*, l) > n(j^*, r)$ or (b) $n(j^*, l) < n(j^*, r)$ holds. Thus we extend the idea in the above to the continuous parameters of the size of length. We will see it in the next section.

19.4 $\hat{\sigma}_p$ and σ_π via $\{\zeta_j, \zeta_j^*\}$ and $\{\xi_j, \xi_j^*\}$

In this section, we give $\eta_k > 0$ for $1 \leqslant k \leqslant d$. We start with $\theta = \{(\theta_j^l, \theta_j^r) : 1 \leqslant j \leqslant d - 1\}$, $\rho = \{\rho_l, \rho_r\}, j^*$, $1 \leqslant j^* \leqslant d - 1$.

Moreover, either Case (a) or Case (b) is indicated. We use $\{\eta_k\}$ instead of $\{n(j), n(j^*, l), n(j^*, r)\}$ in the definition of ζ_j and ζ_j^*.

We give $\eta_k > 0$ for all $1 \leqslant k \leqslant d$.

Case (a)   We put

$$\zeta_{\rho_r^{-1}(j)} = \sum_{k=1}^{j} \eta_{\rho_r^{-1}(k)} + i\theta_{\rho_r^{-1}(j)} \qquad \text{if } 1 \leqslant j \leqslant d-1, \qquad (19.15)$$

$$\zeta_j^* = \begin{cases} \displaystyle\sum_{k=1}^{j} \eta_{\rho_l^{-1}(k)} - i\theta_{\rho_l^{-1}(j)}^l & \text{if } 1 \leqslant j \leqslant \rho_l(j^*) - 1 \\[2mm] \displaystyle\sum_{k=1}^{\rho_l(j^*)-1} \eta_{\rho_l^{-1}(k)} + \eta_d - i(\theta_{j^*}^l + \theta_{j^*}^r + \theta_{\rho_l^{-1}(d-1)}) \\[2mm] \qquad\qquad\qquad \text{if } j = \rho_l(j^*) \\[2mm] \displaystyle\sum_{k=1}^{k'} \eta_{\rho_l^{-1}(k)} + \eta_d - i\theta_{\rho_l^{-1}(k')}^r & \text{if } \rho_l(j^*) \leqslant k' \leqslant d-1, j = k'+1. \end{cases} \qquad (19.16)$$

Case (b)   We put

$$\zeta_j = \begin{cases} \displaystyle\sum_{k=1}^{j} \eta_{\rho_r^{-1}(k)} + i\theta_{\rho_r^{-1}(j)}^r & \text{if } 1 \leqslant j \leqslant \rho_r(j^*) - 1 \\[2mm] \displaystyle\sum_{k=1}^{\rho_r(j^*)-1} \eta_{\rho_r^{-1}(k)} + \eta_d + i(\theta_{j^*}^l + \theta_{j^*}^r + \theta_{\rho_r^{-1}(d-1)}) \\[2mm] \qquad\qquad\qquad \text{if } j = \rho_r(j^*) \\[2mm] \displaystyle\sum_{k=1}^{k'} \eta_{\rho_r^{-1}(k)} + \eta_d + i\theta_{\rho_r^{-1}(k')}^r & \text{if } \rho_r(j^*) \leqslant k' \leqslant d-1, j = k'+1, \end{cases} \qquad (19.17)$$

$$\zeta_{\rho_l^{-1}(j)}^* = \sum_{k=1}^{j} \eta_{\rho_l^{-1}(k)} - i\theta_{\rho_l^{-1}(j)} \qquad \text{if } 1 \leqslant j \leqslant d-1. \qquad (19.18)$$

Both cases, we put $\zeta_0 = \zeta_0^* = 0$ and $\zeta_d^* = \zeta_d$.

Then, the same as the previous section, we construct a figure **F** by concatenating $\ell(\zeta_{j-1}, \zeta_j)$ and $\ell(\zeta_{j-1}^*, \zeta_j^*)$, $1 \leqslant j \leqslant d$.

Here we have the following.

Theorem 19.16′ *The sets of line segments* $\{\ell(\zeta_{j-1}, \zeta_j)\}$ *and* $\{\ell(\zeta_{j-1}^*, \zeta_j^*)\}$ *are pairwise parallel by the following correspondence and have the same length. In particular, if the figure* **F** *is a 2d-polygon, then it is a translation surface.*

Case (a)

$$\ell(\zeta_{k-1}, \zeta_k) \leftrightarrow \ell(\zeta^*_{\rho_l(\rho_r^{-1}(k))-1}, \zeta^*_{\rho_l(\rho_r^{-1}(k))}) \text{ if } 1 \leqslant k < d-1$$

$$\ell(\zeta_{d-1}, \zeta_d) \leftrightarrow \ell(\zeta^*_{\rho_r(j^*)-1}, \zeta^*_{\rho_r(j^*)}) \qquad \text{ if } k = d$$

Case (b)

$$\ell(\zeta^*_{k-1}, \zeta^*_k) \leftrightarrow \ell(\zeta_{\rho_l(\rho_r^{-1}(k))-1}, \zeta_{\rho_l(\rho_r^{-1}(k))}) \text{ if } 1 \leqslant k < d-1$$

$$\ell(\zeta^*_{d-1}, \zeta^*_d) \leftrightarrow \ell(\zeta_{\rho_r(j^*)-1}, \zeta_{\rho_r(j^*)}) \qquad \text{ if } k = d$$

Theorem 19.17′ *Suppose that the piecewise rotation map of the circle S_α is the one constructed in Sect. 19.2 with $(d_1, d_2, \ldots, d_s)$. If the figure **F** constructed in the above is a translation surface, then its orders of singularities are given by $(d_1, d_2, \ldots, d_s)$ and the order of the marked singularity is d_{t_0} defined in the previous section.*

The proofs of these theorems are the same as those of Theorems 19.16 and 19.17. Now we identify $\{\zeta_j, \zeta_j^* : 0 \leqslant j \leqslant d\}$ with $\{\xi_j, \xi_j^* : 0 \leqslant j \leqslant d\}$ in (19.4).

Theorem 19.18 *For any $\{\zeta_j, \zeta_j^* : 0 \leqslant j \leqslant d\}$ arising from S_α, N_0, and $\{\eta_j : 1 \leqslant j \leqslant d\}$, there exist an interval exchange map T with the length data $(\lambda_1, \ldots \lambda_d)$ and the combinatorial data π and $\{h_j, a_j : 0 \leqslant j \leqslant d\}$ which satisfies (19.2) and (19.3) such that $\xi_j = \zeta_j$ and $\xi_j^* = \zeta_j^*$ for $0 \leqslant j \leqslant d$, where $\{\xi_j, \xi_j^*\}$ is given in (19.4).*

Proof We define $(\lambda_1, \ldots, \lambda_d)$, π, and $\{h_j, a_j\}$ as follows. Then it is easy to check the assertion of the theorem.

Case (a)

$$\lambda_k = \begin{cases} \eta_{\rho_r^{-1}(k)} & 1 \leqslant k \leqslant d-1 \\ \eta_d & k = d, \end{cases}$$

$$\pi(k) = \begin{cases} \rho_l \circ \rho_r^{-1}(k) & \text{if } 1 \leqslant k < \rho_r(j^*) \\ \rho_l \circ \rho_r^{-1}(k) + 1 & \text{if } \rho_r(j^*) \leqslant k \leqslant d-1 \\ \rho_l \circ \rho_r^{-1}(j^*) & \text{if } k = d, \end{cases}$$

$$a_k = \begin{cases} 0 & \text{if } j = 0 \\ \theta^r_{\rho_r^{-1}(k)} & \text{if } 1 \leqslant k \leqslant d-1 \\ -\theta^l_{\rho_l^{-1}(\rho_l(j^*)-1)} & \text{if } j = d, \end{cases}$$

and then h_j is given by (19.2) inductively with $h_0 = h_{d+1} = 0$.

Case (b)

$$\lambda_k = \begin{cases} \eta_{\rho_r^{-1}(k)} & 1 \leqslant k \leqslant \rho_r(j^*) \\ \eta_d & k = \rho_r(j^*) \\ \eta_{\rho_r^{-1}(k-1)} & \rho_r(j^*) < k \leqslant d, \end{cases}$$

$$\pi(k) = \begin{cases} \rho_l \circ \rho_r^{-1}(k) & \text{if } 1 \leqslant k < \rho_r(j^*) \\ d & \text{if } k = \rho_r(j^* \\ \rho_l \circ \rho_r^{-1}(k-1) & \text{if } \rho_r(j^*) \leqslant k \leqslant d \end{cases}$$

$$a_k = \begin{cases} 0 & \text{if } j = 0 \\ \theta^r_{\rho_r^{-1}(k)} & \text{if } 1 \leqslant k < \rho_r(j^*) \\ \theta^l_{\rho_l^{-1}(\rho_l(j^*)-1)} + \theta^r_{\rho_r^{-1}(\rho_r(j^*)-1)} + \theta^r_{\rho_r^{-1}(d-1)} & \text{if } k = \rho_r(j^*) \\ \theta^r_{\rho_r^{-1}(k-1)} & \text{if } \rho_r(j^*) \leqslant d, \end{cases}$$

and again h_j is given by (19.2) inductively with $h_0 = h_{d+1} = 0$.   □

Now we suppose that an interval exchange map T_π with $(\lambda_j : 1 \leqslant j \leqslant d)$ is given. Moreover, suppose that $\{h_j, a_j : 0 \leqslant j \leqslant d\}$, which satisfies (19.2) and (19.3), is given. Then we consider $\{\xi_j, \xi_j^*\}$ is defined by (19.4).

Theorem 19.19 *There exist a piecewise rotation map S_α of the circle, a critical iterate N_0, and $\{\eta_j : 1 \leqslant j \leqslant d\}$ such that $\zeta_j = \xi_j$ and $\zeta_j^* = \xi_j^*$ for $0 \leqslant j \leqslant d$ where $\{\xi_j, \xi_j^* : 0 \leqslant j \leqslant d\}$ is given by (19.4) and $\{\zeta_j, \zeta_j^* : 0 \leqslant j \leqslant d\}$ is given by either (19.15) and (19.16) or (19.17) and (19.18).*

Sketch of the Proof If $a_d < 0$ (or $a_d > 0$), then we consider Case (a) (or Case (b)), respectively. In Case (a), we construct a castle with $d - 1$ towers by making correspondent:

$$\begin{cases} \beta_j \longrightarrow \delta_j \\ \theta_j^l = h_j - a_j \\ \theta_j^r = a_j \end{cases}$$

for $1 \leqslant j \leqslant d - 1$. Then $\pi^{-1}(\pi(d) + 1)$ plays the role of j^*. We define

$$\rho_r(j) = j \qquad \text{for } 1 \leqslant j \leqslant d - 1$$

and

$$\rho_l(j) = \begin{cases} \pi(j) & \text{if } 1 \leqslant \rho_l(j) \leqslant \pi^{-1}(\pi(d)+1) \\ \pi(j)-1 & \text{if } \pi^{-1}(\pi(d)+1) \leqslant \rho_l(j) \leqslant d-1. \end{cases}$$

If $a_d > 0$, then we put

$$\rho_l(j) = j \qquad \text{for } 1 \leqslant j \leqslant d-1$$

with

$$\beta_{\pi^{-1}(j)} \longrightarrow \delta_j \qquad \text{for } 1 \leqslant j \leqslant d-1$$

and

$$\rho_r(j) = \begin{cases} \pi(j) & \text{if } 1 \leqslant j \leqslant \pi^{-1}(d) \\ \pi(j+1) & \text{if } \pi^{-1}(d) < j \leqslant d-1. \end{cases}$$

Consequently we have θ, ρ and j^*. To construct S_α and find the critical iterate N_0, we need the nearest return of the orbit 0 by T_π. We refer to [I-N] the detail of the construction.   □

Thus, we have a piecewise rotation map S_α which gives the same equivalence relation. However, the permutation p associated to S is not the same as one give in Sect. 19.2, in general.

As mentioned $\Theta = \{\Theta_t : 1 \leqslant t \leqslant s\}$ arising from $\hat{\sigma}_p$ determined $(d_1, \ldots, d_s)$ which comes from $\{\Lambda_t : 1 \leqslant t \leqslant s\}$. More precisely, $j \in \Theta_t$ if and only if $\delta_j \in \Lambda_t$ for $1 \leqslant j \leqslant d-1$. On the other hand, the equivalence relation on $\{\zeta_j : 1 \leqslant j \leqslant d-1\}$ is the same as the equivalence relation on $\{\xi_j : 1 \leqslant j \leqslant d-1\}$ determined by σ_π. Indeed, we have the following

Theorem 19.20 *In the above correspondence between $\{\zeta_j, \zeta_j^*\}$ and $\{\eta_j, \eta_j^*\}$, we have*

$$\hat{\sigma}_p(j) = \begin{cases} \sigma_\pi(j) & \text{if } \sigma_\pi(j) \neq 0, d \\ \sigma_\pi^2(j) & \text{if } \sigma_\pi(j) = 0 \text{ or } d \end{cases}$$

19.5   Castles and Zippered Rectangles

In this section, we give a new definition of IPR maps (without piecewise rotation maps) and define a (continuum) castle without the critical iterate. Suppose that $\theta = \{(\theta_j^l, \theta_j^r) : 1 \leqslant j \leqslant d-1\}$, $\rho = (\rho_r, \rho_l)$ and j^*, $(1 \leqslant j^* \leqslant d-1)$ are given. We assume that (19.14) holds.

We choose $\delta_j \in \mathbb{R}$ such that

$$\delta_j - \delta_{j-1} > \max |\theta_j^l| + \max |\theta_j^r|$$

and define

$$\mathbb{J}_j = [\delta_j - \theta_j^l, \, \delta_j + \theta_j^r) \qquad \text{and} \qquad \mathbb{J} = \bigcup_{j=1}^{d-1} \mathbb{J}_j$$

for $1 \leqslant j \leqslant d - 1$. The condition on δ_j implies that $\mathbb{J}_j$ is disjoint. We define S as follows. For $x \in [\delta_j - \theta_j^l, \, \delta_j)$,

$$S(x) = \begin{cases} \delta_{\rho_l^{-1}(\rho_l(j)+1)} - \theta_{\rho_l^{-1}(\rho_l(j)+1)} + x - (\delta_j - \theta_j^l) & \text{if } \rho_l(j) \neq d - 1 \\ \delta_{j*} + \theta_{j*}^r - \theta_{\rho_r^{-1}(\rho_r(j*)-1)}^r - (\delta_j - x) & \text{if } \rho_l(j) = d - 1. \end{cases}$$

For $x \in [\delta_j, \, \delta_j + \theta_j^r)$,

$$S(x) = \begin{cases} \delta_{\rho_r^{-1}(\rho_r(j)+1)} - \theta_{\rho_r^{-1}(\rho_r(j)+1)} + x - (\delta_j - \theta_j^r) & \text{if } \rho_r(j) \neq d - 1 \\ \delta_{j*} + \theta_{j*}^l - \theta_{\rho_l^{-1}(\rho_l(j*)-1)}^l - (\delta_j - x) & \text{if } \rho_r(j) = d - 1. \end{cases}$$

For a given $\{\eta_j > 0 : 1 \leqslant j \leqslant d\}$, we define a castle associated to S as follows.

- jth tower for $j \neq j^*$,

$$\mathcal{K}_j = \{(x, y) : x \in \mathbb{J}_j, \, 0 \leqslant y \leqslant \eta_j\}.$$

For j^*, there are two choices.
- j^*th tower with balcony at the right side

$$\mathcal{K}_{j*} = \{(x, y) : x \in \mathbb{J}_{j*}, \, 0 \leqslant y \leqslant \eta_{j*} + \eta_d$$

$$\text{if } x \in [\delta_{j*} - \theta_{j*}^l, \, \delta_{j*} - \theta_{j*}^l + \theta_{\rho_l^{-1}(\rho_l(j*)-1)}^l + \theta_{\rho_r(d-1)}^r),$$

$$\eta_d \leqslant y \leqslant \eta_d + \eta_{j*} \text{ otherwise.}\}$$

- j^*th tower with balcony at the left side

$$\mathcal{K}_{j*} = \{(x, y) : x \in \mathbb{J}_{j*}, \, \eta_{j*} \leqslant y \leqslant \eta_{j*} + \eta_d$$

$$\text{if } x \in [\delta_{j*} - \theta_{j*}^l, \, \delta_{j*} - \theta_{j*}^l + \theta_{\rho_l^{-1}(\rho_l(j*)-1)}^l + \theta_{\rho_r(d-1)}^r),$$

$$0 \leqslant y \leqslant \eta_d + \eta_{j*} \text{ otherwise.}\}$$

Then $\mathbb{K} = \{\mathcal{K}_j : 1 \leqslant j \leqslant d - 1\}$ is said to be a castle. Here we note that Lemma 19.14 explains the reason that we need η_d.

Now we consider a map from an IPR map to an IPR map and also a castle to a castle (continuum case). For this reason, we explain the idea introduced by Cruz and da Rocha [2] for the discrete case.

Let $N_0 < N_1 < N_2 < \cdots$ be a sequence of critical iterates for a piecewise rotation map S_α such that any integer $N \in (N_{k-1}, N_k)$ is not a critical iterate for any $k \geqslant 1$. The correspondence $S_{\alpha,N_{k-1}} \to S_{\alpha,N_k}$ (and also a (discrete) castle $\mathbb{K}$ arising from N_{k-1} to that arising from N_k) is said to be the Cruz-da Rocha induction.

Since the critical iterate is not fixed in the discussion below, we denote by $n_{k,j}$, $n'_{k,j}$, $n(k,j)$, $j^{k,*}$, $n(k, j^{k,*}, l)$, and $n(k, j^{k,*}, r)$ instead of n_j, n'_j, $n(j)$, j^*, $n(j^*, l)$, and $n(j^*, r)$, respectively, if these are determined by the critical iterate N_k. We also denote by $\mathcal{K}_{k,j}$ a tower with δ_j on the top with respect to N_k and by $\mathbf{F}_k$ the figure $\mathbf{F}$ constructed in Sect. 19.3 with respect to the critical iterate N_k.

Proposition 19.21 *If the balcony sides of $\mathcal{K}_{k-1,j^{k-1},*}$ and $\mathcal{K}_{k,j^{k,*}}$ are different to each other (i.e. one is on the left and the other on the right), then $\mathbf{F}_k$ constructed in Sect. 19.3 is a translation surface.*

Proof We denote simply by ζ_j and ζ_j^* for vertices of $\mathbf{F}_k$. It is enough to consider the case $\mathcal{K}_{k-1,j^{k-1},*}$ has a balcony at the right side and $\mathcal{K}_{k,j^{k,*}}$ at the left side, respectively. This means

$$S_\alpha^{n_{j^{k-1},*}}(\delta_1) < \delta_{j^{k-1},*} < S_\alpha^{N_{k-1}}(\delta_1) < S_\alpha^{n'_{j^{k-1},*}}(\delta_1)$$

and

$$S_\alpha^{n_{j^{k,*}}}(\delta_1) < S_\alpha^{N_k}(\delta_1) < S_\alpha^{n'_{j^{k,*}}}(\delta_1)$$

with

$$\begin{cases} n(k, j^{k,*}, r) = n_{k-1,j^{k,*}} + n(k-1, j^{k-1,*}, r) \\ n(k, j^{k-1,*}) = n(k-1, j^{k-1,*}, l) \\ n(k, j^{k,*}, l) = n(k-1, j^{k,*}) \\ n(k-1, j^{k-1,*}, l) > n(k-1, j^{k-1,*}, r). \end{cases}$$

This implies that $\mathcal{K}_{k,j^{k-1},*}$ is the final tower in the left order at the kth level, i.e. $\rho_l(j^{k-1,*}) = d - 1$, where ρ_l is the left order with respect to N_k. Then it turns out that the size of the $Re(\zeta_d - \zeta_{d-1})$ is equal to $n(k-1, j^{k-1,*}, r)$ and the size of the $Re(\zeta_d^* - \zeta_{d-1}^*)$ is equal to $n(k-1, j^{k-1,*}, l)$. From the last inequality in the above, we see that two line segments $\ell(\zeta_d, \zeta_{d-1})$ and $\ell(\zeta_d^*, \zeta_{d-1}^*)$ never intersect to each other, see Fig. 19.16. $\square$

Proposition 19.22 *The change of the balcony side occurs infinitely often along the sequence $N_0, N_1, N_2, \ldots$.*

Proof Suppose that the balcony side with respect to N_{k-1} and N_k are on the same side. Then it is easy to see that the width of the bottom of the main part of the tower $\mathcal{K}_{k-1,j^{k-1,*}}$ is the same as the width of the bottom of the main part of the tower $\mathcal{K}_{k,j^{k,*}}$ i.e. the width of the bottoms of the main part of the towers with balcony with respect to N_{k-1} and N_k are is the same. On the other hand, the width of $\mathcal{K}_{k,j^{k-1,*}}$ (no balcony anymore) is equal to the width of the balcony part of $K_{k-1,j^{k-1,*}}$. This shows

$$\left| \left(S_\alpha^{n_{k-1,j^{k-1,*}}}(\delta_1), S_\alpha^{n'_{k-1,j^{k-1,*}}}(\delta_1) \right) \right| - \left| \left(S_\alpha^{n_{k,j^{k-1,*}}}(\delta_1), S_\alpha^{n'_{k,j^{k-1,*}}}(\delta_1) \right) \right|$$

$$= \left| \left(S_\alpha^{n_{k,j^{k,*}}}(\delta_1), S_\alpha^{n'_{k,j^{k,*}}}(\delta_1) \right) \right| - \left| \left(S_\alpha^{n_{k+1,j^{k,*}}}(\delta_1), S_\alpha^{n'_{k+1,j^{k,*}}}(\delta_1) \right) \right|$$

holds whenever the balcony side (of three towers with balcony) stays at the same side from N_{k-1} to N_k and N_k to N_{k+1}, in other words, we cut the same size when we move the balcony to the balcony of the next step unless we change the balcony side. Since there are $d-1$ towers in the castle, at some $l \geqslant 0$, the balcony side has to be changed at the change of the critical iterates N_{k-1+l} to N_{k+l}. $\qquad \square$

Now we return to our new IPR maps. For a given $\mathcal{S}$, we define a new IPR maps $\mathcal{FS}$ as follows. This is certainly a generalization of the above step from $\mathcal{S}_{\alpha,k}$ to $\mathcal{S}_{\alpha,k+1}$. We consider two cases: (Case A) $\theta^l_{\rho_l^{-1}(\rho_l(j^*)-1)} + \theta^r_{\rho_r^{-1}(d-1)} < \theta^l_{j^*}$ and (Case B) $\theta^l_{\rho_l^{-1}(\rho_l(j^*)-1)} + \theta^r_{\rho_r^{-1}(d-1)} > \theta^l_{j^*}$.

(Case A): We define

$$\hat{\theta}^l_{j^*} = \theta^l_{j^*} - (\theta^l_{\rho_l^{-1}(\rho_l(j^*)-1)} + \theta^r_{\rho_r^{-1}(d-1)}) \text{ and } \hat{\theta}^r_{j^*} = \theta^r_{j^*}$$

and the marked index j^* is changed for

$$\hat{j}^* = \begin{cases} \rho_l^{-1}(\rho_l(j^*)+1) & \text{if } \rho_l(j^*) < d-1 \\ j^* & \text{if } \rho_l(j^*) = d-1. \end{cases}$$

We also define $\hat{\rho}_l$ by

$$\hat{\rho}_l = \begin{cases} \rho_l(j) & \text{if } \rho_l(j) < \rho_l(j^*) \\ d-1 & \text{if } j = j^* \\ \rho_l(j)-1 & \text{if } \rho_l(j) > \rho_l(j^*) \end{cases}$$

and $\hat{\rho}_r = \rho_r$.

(Case B) We define

$$\hat{\theta}^r_{j^*} = \theta^r_{j^*} - (\theta^r{}_{\rho_r^{-1}(\rho_r(j^*)-1)} + \theta^l{}_{\rho_l^{-1}(d-1)}) \text{ and } \hat{\theta}^l_{j^*} = \theta^l_{j^*}$$

and the marked index j^* is changed for

$$\hat{j}^* = \begin{cases} \rho_r^{-1}(\rho_r(j^*)+1) & \text{if } \rho_r(j^*) < d-1 \\ j^* & \text{if } \rho_r(j^*) = d-1. \end{cases}$$

Either cases, we put we put $\hat{\theta}^l_j = \theta^l_j$ for $j \neq j^*$ and $\hat{j}^*$. We also define $\hat{\rho}_r$ by

$$\hat{\rho}_r = \begin{cases} \rho_r(j) & \text{if } \rho_r(j) < \rho_r(j^*) \\ d-1 & \text{if } j = j^* \\ \rho_r(j)-1 & \text{if } \rho_r(j) > \rho_r(j^*) \end{cases}$$

and $\hat{\rho}_l = \rho_l$.

Both cases, it is easy to see that the above $\{(\hat{\theta}^l_j, \hat{\theta}^r_j) : 1 \leqslant j \leqslant d-1\}$, $\hat{\rho} = (\hat{\rho}_l, \hat{\rho}_r)$, and $\hat{j}^*$ satisfy (19.14) and then we define $\mathcal{F}S$ by these $\hat{\theta}$, $\hat{\rho}$, and $\hat{j}^*$.

It is not so hard to see the following.

Proposition 19.23 *The induced map of S to $\hat{\mathbb{J}} = \cup[\delta_j - \hat{\theta}^l_j, \ \delta_j + \hat{\theta}^r_j)$ is also an IPR map and it is $\mathcal{F}S$.*

With the above $\{(\hat{\theta}^l_j, \hat{\theta}^r_j)\}$, $\hat{\rho}$, and $\hat{j}^*$, we can also define a map a castle $\mathbb{K}$ to a castle $\hat{\mathcal{F}}\mathbb{K}$ generalizing the correspondence from $\{\mathcal{K}_{k,j}\}$ to $\{\mathcal{K}_{k+1,j}\}$ i.e.

(Case A-i) $\theta^l{}_{\rho_l^{-1}(\rho_l(j^*)-1)} + \theta^r{}_{\rho_r^{-1}(d-1)} < \theta^l_{j^*}$ and $\mathcal{K}_{j^*}$ has a balcony at the right side:
We define

$$\hat{\eta}_{j^*} = \eta_{j^*} \text{ and } \hat{\eta}_d = \eta_{j^*} + \eta_d.$$

(Case A-ii) $\theta^l{}_{\rho_l^{-1}(\rho_l(j^*)-1)} + \theta^r{}_{\rho_r^{-1}(d-1)} < \theta^l_{j^*}$ and $\mathcal{K}_{j^*}$ has a balcony at the left side:
We define

$$\hat{\eta}_{j^*} = \eta_{j^*} + \eta_d \text{ and } \hat{\eta}_d = \eta_{j^*}.$$

(Case B-i) $\theta^l{}_{\rho_l^{-1}(\rho_l(j^*)-1)} + \theta^r{}_{\rho_r^{-1}(d-1)} > \theta^l_{j^*}$ and $\mathcal{K}_{j^*}$ has a balcony at the right side:
We define

$$\hat{\eta}_{j^*} = \eta_{j^*} + \eta_d \text{ and } \hat{\eta}_d = \eta_{j^*}.$$

(Case B-ii) $\theta^l_{\rho_l^{-1}(\rho_l(j^*)-1)} + \theta^r_{\rho_r^{-1}(d-1)} > \theta^l_{j^*}$ and $\mathcal{K}_{j^*}$ has a balcony at the left side:
We define

$$\hat{\eta}_{j^*} = \eta_{j^*} \text{ and } \hat{\eta}_d = \eta_{j^*} + \eta_d.$$

In all four cases, we put $\hat{\eta}_j = \eta_j$ for $j \neq j^*$.

Now we can generalize these proposition to the (continuum) castles by the similar way.

Proposition 19.22′ *If the balcony sides of $\mathcal{K}_{j^*}$ of a castle $\mathbb{K}$ and the balcony side of $K_{\hat{j}^*}$ of the castle $\mathcal{F}\mathbb{K}$ are different to each other (i.e. one is on the left and the other on the right), then $\mathbf{F}$ constructed from $\hat{\theta}$, $\hat{\eta}$, and $\hat{\rho}$ by (19.15) and (19.16) or (19.17) and (19.18) is a translation surface.*

Proposition 19.23′ *The change of the balcony side occurs infinitely often along the sequence arising from the iterations of $\mathcal{F}$, i.e. $\mathbb{K}, \mathcal{F}\mathbb{K}, \mathcal{F}^2\mathbb{K}, \ldots$.*

Now we define the notion of zippered rectangles. For a given $(\mathbf{h}, \mathbf{a})$ which satisfies (19.2) and (19.3), we define

$$R_j = \{x + iy : \beta_{j-1} \leqslant x < \beta_j, 0 \leqslant y \leqslant h_j\} \quad \text{for } 1 \leqslant j \leqslant d$$

and

$$R_+ = \bigcup_{j=1}^{d} R_j.$$

We cut and glue vertical sides of R_j in the following way. We cut (β_j, y), $a_j \leqslant y \leqslant \min(h_j, h_{j+1})$ in $R_j \cup R_{j+1}$, $1 \leqslant j \leqslant d-1$ except for the case $a_d > 0$ and $j = \pi^{-1}(d)$ where $\min(h_j, h_{j+1}) > a_j$ holds. Except for the case $a_d > 0$ and $j = \pi^{-1}(d)$, R_j and R_{j+1} are connected at (β_j, y), $0 \leqslant y \leqslant a_j$. We glue R_j and $R_{\pi^{-1}(\pi(j)+1)}$, $1 \leqslant j \leqslant d$, $j \neq \pi^{-1}(d)$ by identifying

$$\beta_j + iy \quad \text{and} \quad \beta_{\pi^{-1}(\pi(j)+1)-1} + i\left(h_{\pi^{-1}(\pi(j)+1)} - a_{\pi^{-1}(\pi(j)+1)-1} + y\right) \quad (19.19)$$

for $0 \leqslant y \leqslant h_j - a_j$. Here we recall that $\beta_0 = h_0 = a_0 = 0$. Concerning $j = d$, there are two cases. First we consider the case $a_d > 0$. Here $R_{\pi^{-1}(d)}$ and $R_{\pi^{-1}(d)+1}$ are connected at $\beta_{\pi^{-1}(d)} + iy$, $0 \leqslant y \leqslant h_{\pi^{-1}(d)}$. Then we glue R_d and $R_{\pi^{-1}(d)+1}$ by identifying

$$\beta_d + iy \quad \text{and} \quad \beta_{\pi^{-1}(d)} + iy \quad (19.20)$$

for $0 \leqslant y \leqslant a_d$. On the other hand, R_d is also glued to $R_{\pi^{-1}(\pi(d)+1)}$ by identifying

$$\beta_d + i(a_d + y) \quad \text{and} \quad \beta_{\pi^{-1}(\pi(d)+1)-1} + i\left(h_{\pi^{-1}(\pi(d)+1)} - a_{\pi^{-1}(\pi(d)+1)-1} + y\right) \quad (19.21)$$

for $0 \leqslant y \leqslant h_d - a_d$ ((19.21) is a special case of (19.19)). Next we consider the case $a_d < 0$. In this case, we glue R_d and $R_{\pi^{-1}(\pi(d)+1)}$ by identifying

$$\beta_d + iy \qquad \text{and} \qquad \beta_{\pi^{-1}(\pi(d)+1)-1} + i\left(h_{\pi^{-1}(\pi(d)+1)} - h_d + y\right) \qquad (19.22)$$

for $0 \leqslant y \leqslant h_d$. Then we glue $R_{\pi^{-1}(d)}$ and $R_{\pi^{-1}(\pi(d)+1)-1}$ by identifying

$$\beta_{\pi^{-1}(d)} + i\left(a_{\pi^{-1}(d)} + y\right) \qquad \text{and}$$
$$\beta_{\pi^{-1}(\pi(d)+1)-1} + i\left(h_{\pi^{-1}(\pi(d)+1)} - a_{\pi^{-1}(\pi(d)+1)-1} + y\right) \qquad (19.23)$$

for $0 \leqslant y \leqslant h_{\pi^{-1}(d)} - a_{\pi^{-1}(d)}$ ((19.23) is also a special case of (19.19)). For horizontal sides, we glue $x + h_j$ and $(T_\pi(x), 0)$ for $\beta_{j-1} \leqslant x < \beta_j$. Together with (19.19), (19.20), (19.21) (or with (19.19), (19.22), (19.23)) and this horizontal glueing, we have a Riemann surface. In this point of view R_j, $1 \leqslant j \leqslant d$, are called zippered rectangles originally see [9].

Now we define

$$R_- = \bigcup_{j=1}^{d} R'_j$$

where R'_j is the translation of R_j by

$$(x, y) \longrightarrow (T_\pi(x), y - h_j).$$

By the construction of R_+ and R_-, we have the following.

Proposition 19.24 $R_+ \cup R_-$ *is a double cover of the translation surface defined by (19.4)*

For a given $(\mathbf{h}, \mathbf{a})$, we introduced $(\mathbf{h}', \mathbf{a}')$ at the introduction as a correspondence between a translation surface to a new "translation surface". Here, it is clear that $\hat{R}$ which maps $(\mathbf{h}, \mathbf{a})$ to $(\mathbf{h}', \mathbf{a}')$ is a map from zippered rectangles to zippered rectangles. Recall that, for a fixed irreducible permutation π, Rauzy class is a set of permutations, which is a connected component of permutations which are the combinatorial data of all interval exchange maps appeared in $\{\mathcal{R}^n T_\pi : n \geqslant 0\}$ with all possible length data. It is known that for the set of all possible combinatorial data in the Rauzy class arising from π and all combinatorial data λ, $\mathcal{R}$ is two-to-one map and then with all possible (π, λ) with $(\mathbf{h}, \mathbf{a})$, $\hat{R}$ is one-to-one. From the discussion in this paper, it is not so hard to see that there is a natural correspondence "zippered rectangles" R_+ or R_- and a castle, where R_+ (or R_-) corresponds to a castle with a tower with balcony on the left side (or on the right side) if $a_d > 0$ (or $a_d < 0$, respectively). The former case, the tower with balcony corresponds to the concatenation of $R_{\pi^{-1}(d)}$ and $R_{\pi^{-1}(d)+1}$, on the other hand, the latter case, it corresponds to the concatenation of R_d and $R_{\pi^{-1}(\pi(d)+1)}$. The correspondence is simply made by the rotation of $\frac{\pi}{2}$. We refer [3] the detail of this correspondence.

The main point, here, is that since $\{\xi_j, \xi_j^*\}$ and $\{\zeta_j, \zeta_j^*\}$ may not produce a translation surfaces. So it is natural to define $\hat{\mathcal{R}}$ and $\hat{\mathcal{F}}$ on the set of zippered rectangles and the set of castles.

Theorem 19.25 *The extension of Rauzy induction $\hat{\mathcal{R}}$ induces $\hat{\mathcal{F}}^{-1}$ by the natural correspondence stated in the above.*

Proof The natural correspondence comes from the identification between $\{\xi_j, \xi_j^* : 0 \leqslant j \leqslant d\}$ and $\{\zeta_j, \zeta_j^* : 0 \leqslant j \leqslant d\}$. Then it is easy to see the assertion of this theorem from the definitions of $\hat{\mathcal{R}}$ and $\hat{\mathcal{F}}$. $\square$

Suppose that

$$n_Z = \min_{n>0}\{\hat{\mathcal{F}}^n \mathbb{K} : \text{the discontinuous point } \delta_{j^*} \text{ is on the top}$$
$$\text{of the main part of the tower with balcony}\}.$$

Then we define $Z\mathbb{K} = \hat{\mathcal{F}}^{n_Z+1}\mathbb{K}$.

Theorem 19.26 *If the map Z maps $\{\xi_j, \xi_j^* : 0 \leqslant j \leqslant d\}$ to $\{\hat{\xi}_j, \xi_j^* : 0 \leqslant j \leqslant d\}$, then $\{\hat{\xi}_j, \xi_j^* : 0 \leqslant j \leqslant d\}$ produces a translation surface (i.e. 2d polygon).*

Proof This follows from Proposition 19.22. $\square$

Remark 19.27 If we define

$$\mathbb{K}_0 = \{\mathbb{K} : \text{castle such that the balcony side of } \mathbb{K}$$
$$\text{is not the same side with } \hat{\mathcal{F}}^{-1}\mathbb{K}\}.$$

Then Z is a bijective map of $\mathbb{K}_0$ and it gives the inverse of the Zorich map (see [12]) on a set of zippered rectangles.

Remark 19.28 There are two natural normalizations. The first one is $\beta_d = 1$ with the sum of the area size of R_j associated zippered rectangles R_+ equals to 1. The second is $|\mathbb{J}| = |\cup_{j=1}^{d-1} \mathbb{J}_j| = 1$ with the sum of the area size of $\mathcal{K}_j$ of a castle $\mathbb{K}$ equals to 1. The first case is well discussed in many literatures. Indeed, it is known that the renormalized map of $\hat{\mathcal{R}}$ is ∞ measure preserving and ergodic. Moreover, the Zorich map is finite measure preserving and ergodic. Consequently, the second case, the same result holds since it is the inverse of map in a different projection. We denote $\bar{Z}$ the renormalized map of Z on the set of castles with $|\mathbb{J}| = 1$ and the area size of $\mathbb{K}$ equals to 1. Then this is a two-points extension of the generalized Gauss map $\mathcal{G}$ defined by Cruz-da Rocha [2]. In [2], the density of the absolutely continuous invariant measure (w.r.t. the Lebesgue measure on the parameter space of θ) for $\mathcal{G}$ is given. We will discuss the related subject in a different occasion.

Remark 19.29 For given $(d_1, d_2, \ldots, d_s)$, the vector of the orders of singularities, Boissy [1] showed that two choices of the marked singularities make the Rauzy class if the orders of singularities are the same. So far, the authors do not know there is a proof of this result only by properties of a piecewise rotation with the equivalence relation of discontinuous points associated to $(d_1, \ldots, d_s)$.

Acknowledgements The second author was partially supported by JSPS grants No. 16K13766 and JSPS Core-to-core program, "Foundation of a Global Research Cooperative Center in Mathematics focused on Number Theory and Geometry".

References

1. C. Boissy, Classification of Rauzy classes in the moduli space of quadratic differentials. Discrete Continuous Dyn. Syst. A **32**(10), 3433–3457 (2012)
2. S.D. Cruz, L.F.C. da Rocha, A generalization of the Gauss map and some classical theorems on continued fractions. Nonlinearity **18**(2), 505–525 (2005)
3. K. Inoue, H. Nakada, On the dual of Rauzy induction. Ergod. Theory Dyn. Syst. **37**(5), 1492–1536 (2017)
4. M. Keane, Interval exchange transformations. Math. Z. **141**, 25–31 (1975)
5. M. Keane, Non-ergodic interval exchange transformations. Isr. J. Math. **26**(2), 188–196 (1977)
6. M. Kontsevich, A. Zorich, Connected components of the moduli spaces of Abelian differentials with prescribed singularities. Invent. Math. **153**(3), 631–678 (2003)
7. G. Rauzy, Echanges d'intervalles et transformations induites (French). Acta Arith. **34**(4), 315–328 (1979)
8. W.A. Veech, Interval exchange transformations. J. Anal. Math. **33**, 222–278 (1978)
9. W.A. Veech, Gauss measures for transformations on the space of interval exchange maps. Ann. Math. (2) **115**(1), 201–242 (1982)
10. M. Viana, Ergodic theory of interval exchange maps. Rev. Mat. Comput. **19**(1), 7–100 (2006)
11. J-C. Yoccoz, Continued fraction algorithms for interval exchange maps: an introduction, in *Frontiers in Number Theory, Physics, and Geometry. I* (Springer, Berlin, 2006), pp. 401–435
12. A. Zorich, Finite Gauss measure on the space of interval exchange transformations. Lyapunov exponents. Ann. Inst. Fourier **46**(2), 325–370 (1996)

LECTURE NOTES IN MATHEMATICS   🐎 Springer

Editors in Chief: J.-M. Morel, B. Teissier;

Editorial Policy

1. Lecture Notes aim to report new developments in all areas of mathematics and their applications – quickly, informally and at a high level. Mathematical texts analysing new developments in modelling and numerical simulation are welcome.

   Manuscripts should be reasonably self-contained and rounded off. Thus they may, and often will, present not only results of the author but also related work by other people. They may be based on specialised lecture courses. Furthermore, the manuscripts should provide sufficient motivation, examples and applications. This clearly distinguishes Lecture Notes from journal articles or technical reports which normally are very concise. Articles intended for a journal but too long to be accepted by most journals, usually do not have this "lecture notes" character. For similar reasons it is unusual for doctoral theses to be accepted for the Lecture Notes series, though habilitation theses may be appropriate.

2. Besides monographs, multi-author manuscripts resulting from SUMMER SCHOOLS or similar INTENSIVE COURSES are welcome, provided their objective was held to present an active mathematical topic to an audience at the beginning or intermediate graduate level (a list of participants should be provided).

   The resulting manuscript should not be just a collection of course notes, but should require advance planning and coordination among the main lecturers. The subject matter should dictate the structure of the book. This structure should be motivated and explained in a scientific introduction, and the notation, references, index and formulation of results should be, if possible, unified by the editors. Each contribution should have an abstract and an introduction referring to the other contributions. In other words, more preparatory work must go into a multi-authored volume than simply assembling a disparate collection of papers, communicated at the event.

3. Manuscripts should be submitted either online at www.editorialmanager.com/lnm to Springer's mathematics editorial in Heidelberg, or electronically to one of the series editors. Authors should be aware that incomplete or insufficiently close-to-final manuscripts almost always result in longer refereeing times and nevertheless unclear referees' recommendations, making further refereeing of a final draft necessary. The strict minimum amount of material that will be considered should include a detailed outline describing the planned contents of each chapter, a bibliography and several sample chapters. Parallel submission of a manuscript to another publisher while under consideration for LNM is not acceptable and can lead to rejection.

4. In general, **monographs** will be sent out to at least 2 external referees for evaluation.

   A final decision to publish can be made only on the basis of the complete manuscript, however a refereeing process leading to a preliminary decision can be based on a pre-final or incomplete manuscript.

   Volume Editors of **multi-author works** are expected to arrange for the refereeing, to the usual scientific standards, of the individual contributions. If the resulting reports can be forwarded to the LNM Editorial Board, this is very helpful. If no reports are forwarded or if other questions remain unclear in respect of homogeneity etc, the series editors may wish to consult external referees for an overall evaluation of the volume.

5. Manuscripts should in general be submitted in English. Final manuscripts should contain at least 100 pages of mathematical text and should always include

   – a table of contents;
   – an informative introduction, with adequate motivation and perhaps some historical remarks: it should be accessible to a reader not intimately familiar with the topic treated;
   – a subject index: as a rule this is genuinely helpful for the reader.
   – For evaluation purposes, manuscripts should be submitted as pdf files.

6. Careful preparation of the manuscripts will help keep production time short besides ensuring satisfactory appearance of the finished book in print and online. After acceptance of the manuscript authors will be asked to prepare the final LaTeX source files (see LaTeX templates online: https://www.springer.com/gb/authors-editors/book-authors-editors/manuscriptpreparation/5636) plus the corresponding pdf- or zipped ps-file. The LaTeX source files are essential for producing the full-text online version of the book, see http://link.springer.com/bookseries/304 for the existing online volumes of LNM). The technical production of a Lecture Notes volume takes approximately 12 weeks. Additional instructions, if necessary, are available on request from lnm@springer.com.

7. Authors receive a total of 30 free copies of their volume and free access to their book on SpringerLink, but no royalties. They are entitled to a discount of 33.3 % on the price of Springer books purchased for their personal use, if ordering directly from Springer.

8. Commitment to publish is made by a *Publishing Agreement*; contributing authors of multiauthor books are requested to sign a *Consent to Publish* form. Springer-Verlag registers the copyright for each volume. Authors are free to reuse material contained in their LNM volumes in later publications: a brief written (or e-mail) request for formal permission is sufficient.

Addresses:
Professor Jean-Michel Morel, CMLA, École Normale Supérieure de Cachan, France
E-mail: moreljeanmichel@gmail.com

Professor Bernard Teissier, Equipe Géométrie et Dynamique,
Institut de Mathématiques de Jussieu – Paris Rive Gauche, Paris, France
E-mail: bernard.teissier@imj-prg.fr

Springer: Ute McCrory, Mathematics, Heidelberg, Germany,
E-mail: lnm@springer.com

Printed in the United States
By Bookmasters